OPTICAL CHARACTER RECOGNITION • *Shunji Mori, Hirobumi Nishida, and Hiromitsu Yamada*

ANTENNAS FOR RADAR AND COMMUNICATIONS: A POLARIMETRIC APPROACH • *Harold Mott*

INTEGRATED ACTIVE ANTENNAS AND SPATIAL POWER COMBINING • *Julio A. Navarro and Kai Chang*

FREQUENCY CONTROL OF SEMICONDUCTOR LASERS • *Motoichi Ohtsu (ed.)*

SOLAR CELLS AND THEIR APPLICATIONS • *Larry D. Partain (ed.)*

ANALYSIS OF MULTICONDUCTOR TRANSMISSION LINES • *Clayton R. Paul*

INTRODUCTION TO ELECTROMAGNETIC COMPATIBILITY • *Clayton R. Paul*

INTRODUCTION TO HIGH-SPEED ELECTRONICS AND OPTOELECTRONICS • *Leonard M. Riaziat*

NEW FRONTIERS IN MEDICAL DEVICE TECHNOLOGY • *Arye Rosen and Harel Rosen (eds.)*

ELECTROMAGNETIC PROPAGATION IN MULTI-MODE RANDOM MEDIA • *Harrison E. Rowe*

NONLINEAR OPTICS • *E. G. Sauter*

InP-BASED MATERIALS AND DEVICES: PHYSICS AND TECHNOLOGY • *Osamu Wada and Hideki Hasegawa (eds.)*

DESIGN OF NONPLANAR MICROSTRIP ANTENNAS AND TRANSMISSION LINES • *Kin-Lu Wong*

FREQUENCY SELECTIVE SURFACE AND GRID ARRAY • *T. K. Wu (ed.)*

ACTIVE AND QUASI-OPTICAL ARRAYS FOR SOLID-STATE POWER COMBINING • *Robert A. York and Zoya B. Popović (eds.)*

OPTICAL SIGNAL PROCESSING, COMPUTING AND NEURAL NETWORKS • *Francis T. S. Yu and Suganda Jutamulia*

SiGe, GaAs, AND InP HETEROJUNCTION BIPOLAR TRANSISTORS • *Jiann Yuan*

Optical Character Recognition

Optical Character Recognition

SHUNJI MORI
The University of Aizu
Fukushima City, Japan

HIROBUMI NISHIDA
Software Rearch Center, Ricoh Co., Ltd.
Tokyo, Japan

HIROMITSU YAMADA
Electrotechnical Laboratory
Tsukuba City, Japan

A WILEY-INTERSCIENCE PUBLICATION
JOHN WILEY & SONS, INC.
NEW YORK / CHICHESTER / WEINHEIM / BRISBANE / SINGAPORE / TORONTO

This book is printed on acid-free paper. ♾

Published simultaneously in Canada.

Library of Congress Cataloging-in-Publication Data:

Mori, Shunji, 1934–
Optical character recognition / Shunji Mori, Hirobumi Nishida, Hiromitsu Yamada.
p. cm. — (Wiley series in microwave and optical engineering)
"A Wiley–Interscience publication."
Includes index.
ISBN 0-471-30819-6 (cloth : alk. paper)
1. Optical character recognition devices. I. Title. II. Series.
TA1640.M67 1999
621.39'9—dc21 98-23908

Printed in the United States of America.

10 9 8 7 6 5 4 3 2 1

To my teacher Dr. Taizo Iijima
Shunji Mori

Contents

Preface

Optical character recognition is, in a broad sense, a branch of artificial intelligence and it is also a branch of computer vision. Nevertheless, it is a distinct discipline in its own right, analogous to speech recognition. If we imagine designing a robot, both reading and listening functions are indispensable if the robot is to be really intelligent. Of course, even a conventional computer must have the capacity to read input documents such as in office and library work. Furthermore, recently, prototype electronic libraries have started to come on-line. In part, the optical character recognition field has grown because of this specific application.

Not so long ago, immediately after the advent of digital computers, many researchers and engineers were misled into believing in the feasibility of intelligent computers. Language translation, game playing, reading, listening, were all actual objectives at the time and were thought to be tractable. Soon, however, it became obvious that these qualities were very difficult to attain. The difficulty had little to do with the limited performance of contemporary computers: It was rather that true intelligence is beyond the reach of any current supercomputer.

However, it was very fortunate for researchers and engineers in character recognition that the OCR problem can be decomposed into levels, and each level can have a value when partially solved. For example, in Japan on the top-right portion of a postal card three red-colored boxes with two smaller ones are printed, within which numbers are expected to be written. That is, isolated handprinted character recognition has a very practical meaning. Furthermore, the correct recognition of these numbers is not so critical compared with, for example, a banking application.

In the 1960s a major initiative in postal number recognition was started. Gradually there came success in the reading of postal numbers. At first, high performance of correct reading was not required, although further R&D was expected to improve the reading performance. This fortunate situation can be contrasted with speech recognition, in which essentially continuous speech recognition was required from the beginning. This level of difficulty corresponds to cursive script character recognition, which is a current topic of investigation. Thus OCR has grown up as an industry, aided by the rapid development of computer technology. This is rather unusual in the AI field in general. In this sense, OCR technology is still at a stage of development. However, considerable knowledge has been accumulated, so it is early, but not premature, to present a book for OCR design.

The processes occurring in OCR can be roughly divided into three parts: preprocessing, feature extraction, and discrimination. Furthermore recent real applications require analysis of a structured document. This requires image analysis, character recognition, and knowledge application. However, it is premature to write a book about all the aspects of document processing; indeed, the most essential part is character recognition. Therefore we have decided to concentrate on character recognition in this book.

In my view, discrimination has been well studied, and several books have been published on the topic. For example, Duda and Hart's *Pattern Classification and Scene Analysis* is an excellent and well-known classic book on pattern recognition. A reader can survey discrimination and pattern recognition by reading its introduction and Chapter 2 on Bayes decision theory. Other books are listed at the end of Chapter 1.

On the other hand, concerning preprocessing and feature extraction, to my knowledge there are few books that intensively and systematically describe character recognition. This is somewhat strange, since it is always said that feature extraction is very important. However, it is understandable because the systematic description of feature extraction is very difficult. Therefore it was an adventure for us to attempt to do that. In addition to feature extraction, instead of discrimination theory, we have introduced shape representation and matching. Although these are treated by books in the related field of image recognition/computer vision, we provide original descriptions on this topic as a whole.

We have benefited from the book by T. Pavlidis, *Structural Pattern Recognition,* which is close to our approach, although it is intended for a broader application area than ours. Some parts overlap in such topics as curve fitting and graph matching. Our approach is more focused on the data and is thus OCR oriented. Another important point common to the two books is the theoretical flavor of the description. This approach on our part was also influenced by my teacher Dr. T. Iijima who is an originator of OCR in Japan and a distinguished applied mathematician. He built a grand theoretical framework for character recognition based on functional analysis. However, his work is not well known abroad because he has been reluctant to write his papers in English, and his theoretical work is not easy for engineers to understand. I introduce the basic style of his work, since it is essential to the theory of image recognition.

For the theoretical work on recognition, some mathematical background is required, in general. However, we have tried to make our discussion self-contained by providing the key mathematical tools as needed. We assume, however, a familiarity with advanced calculus, linear algebra, and basic algebra, which are all usually taught in the undergraduate engineering schools.

We have made an effort for the text to be as readable as possible. However, this means that the methods employed are naturally restricted. Actually it is nearly impossible to introduce all the methods of OCR technology. In order to cover this gap between these two poles, in particular, in preprocessing and feature extraction, somewhat detailed bibliographical remarks were added at the end of each chapter. Feature extraction directly from gray scale is a good example. Where the bibliographical remarks are not exhaustive, we welcome readers' comments. Once we selected a method to be treated, we tried to make it concise and comprehensible. For this reason some questions were asked of the originators of the methods so that any ambiguous aspects could be cleared up. In this respect I wish to thank Professors T. Pavlidis and A. Rosenfeld. About half of the methods discussed are original works of researchers who were my associates in the development phases of my OCR career. By these means we have endeavored to be as precise in the descriptions as possible. Fortunately these researchers are all famous as experts in the field of OCR, and I am very proud of their accomplishments.

The authors of this book are not native English speakers. It has been an adventure for me to write a book in English. While proofing of any technical work can be a challenge even for a native reader, in support of this, Professor Pavlidis provided his associate, Dr. B. Sakoda, who was a researcher at Professor Pavlidis's lab and now works for Symbol Technologies. He did a very fine job of proofreading the English and provided helpful comments.

Since this book is on the cutting-edge technology of preprocessing, feature extraction, and their systematic description matching, there are sure to be many points to be improved or added. Already, in our attempts to keep it up-to-date, curvature detection using the quadratic B-spline was added, whose investigation by myself and my associates took almost a year. It still continues. As it became evident that pursuing such concerns further would make writing the book an endless project, I gave in. In this sense in this book we are still at the threshold of pattern recognition. Therefore it is our great pleasure to offer a book that will help students, researchers, and engineers who want to proceed over that threshold.

A general introduction to character recognition is provided in Chapter 1, where the reader can survey the essence of character recognition technology and begin to appreciate its difficulty. Chapters 2 through 6 describe preprocessing in detail. In practice, preprocessing is very important, while papers on it are comparatively few, except those dealing with mathematical morphology, which is addressed in Chapter 6. Feature extraction is a highlight of the book, and it is covered in Chapters 7, 8, and 10. In particular, in Chapter 8, feature extraction is intensively and systematically investigated in the image recognition field in general. As an intermezzo inserted between feature extraction and matching, Chapter 9 gives an algebraic description. To be sure, before matching, an object must be represented neatly as is critical in computer vision. However, it is not so easy to describe a shape systematically by strictly mathematical means. We have introduced our original work on this topic. Description or representation of a shape is covered in both Chapters 8 and 12. Finally, matching is described in Chapters 11 through 13. In particular, in Chapter 13, the discussion on elastic matching is original, with the objects selected being not only characters but also image recognition objects such as tools and maps that include characters. This is a method that bridges character recognition and image recognition.

Some parts of the material were used in the graduate school of the Computer Information Science Department at Tokyo University, in the undergraduate school of Tsukuba University, and in both the undergraduate and graduate schools of the University of Aizu.

I take great pleasure in closing these introductory remarks by acknowledging the people who have helped and assisted me. First, Professor Pavlidis encouraged me to write the book and answered my questions about his papers. Professor Rosenfeld also answered my questions about his papers. Dr. Iijima allowed me to introduce his original work and has supported me in my research life. Dr. N. Otsu of ETL also allowed me to introduce his work. Dr. N. Otsu was my associate when I worked for ETL. I used Professors Y. Noguchi's, I. Yamazoki's, K. Yamamoto's, R. Oka's, and T. Mori's works, who were my associates at ETL, and Professor K. T. Miara's work with whom I am jointly doing research. Professor J. Kanai allowed me to use a fine picture of character samples. Professor M. Yasuda, who has led in OCR development in Japan, was a colleague of mine when I worked for ETL. I used his work in which we mutually cooperated in that period. Professor C. Y. Suen, Dr. J. Tsugumo, and Dr. Y. Kurosawa provided me with precious advise. I thank Professor Suen, in particular, who is an old friend. My preliminary work on the algebraic description was fostered when I was a visiting professor in his laboratory. I further thank Professor M. Nagao who encouraged me to write the book, *Fundamentals of Image Recognition I/II,* from which I reuse some figures in this book. Permission was granted by my coauthor Professor Sakakura and Ohm Publishing Ltd., and I am very grateful to them for their generosity. This work was partly funded by the University of Aizu as a Top-down Education Project. I am thankful for the support of Dr. T. L. Kunii, formerly president of the university, who also encouraged me to write this book and M. Onoue, former vice-president of Ricoh. With his strong support I and also Dr. Nishida have spent exciting and confortable time as research managers and researchers. In this connection, I thank my associates at that time at Ricoh Research & Development Center, in particular, Mr. S. Suzuki. The text input and Mac Drawings were done by Ms. N. Fukuda, which was very tedious work indeed when I first began the writing. As I realized soon that the task was large, she assisted me and eventually did all the work for me. In this connection Mr. Y. Nakajima of my lab helped us with the computer processing. Engineer H. Hoshi and Student T. Nagai helped execute the artwork making programming tools. I wrote this preface as a representative of the three authors.

SHUNJI MORI

Aizuwakamatsu
Fukushima

Optical Character Recognition

CHAPTER ONE

Character Recognition

The objectives of this first chapter are to establish what character recognition is and to show that it is a profound problem in terms of both its science and technology. Humans have such a highly developed aptitude for pattern recognition that they tend to overlook the problems in machine pattern recognition. In this chapter the problems of character recognition will be classified first according to their level of difficulty. In this way the classes of recognition difficulties for the machine can be approached systematically.

Some of the problems that have been already solved are conceptually simple, and they can easily be demonstrated. We will be concerned with uncovering the more difficult hidden problems. For this reason the discussion in this chapter will not adhere to any traditional order involving observation, preprocessing, and recognition/discrimination. Rather, the approach taken will be just the reverse. By our way we hope to provide readers with an idea of the wide range of open problems in character recognition, and also the motivation for reading the subsequent chapters.

1.1 WHAT IS THE PROBLEM?

The objective of pattern recognition is to interpret input as a sequence of characters taken from a given set of characters. One such character set is the alphabet. An alphabetic character set has components whose shapes reflect the culture in which the character set was born. Some examples are shown in Fig. 1.1. Certain letters of these alphabets give us the impression that they are alike by their shapes. However, we know by their use that they must be different from each other. That is to say, they are different, although they have general similarities. In any case, all the characters are rich in shape. Further, just a single character from an alphabet is subject to many variations when in handwriting. Despite these variations, humans often easily recognize

FIGURE 1.1 Some examples of partial character sets. From the top, Hebrew, Arabic, Sanskrit, Russian, Greek, Roman, German (old style), Chinese (Kanji), Hiragana, Katakana, Hangul, and Chinese (Kanji) numeral. From *Language and Mathematics* by H. Noguchi, © 1997, Diamond, Tokyo.

the character. By this simple fact we can conjecture that each character in an alphabet must have an abstracted form. This abstraction must be so basic that any variations can be recognized easily. It is at least obvious that the character set memorized by the human brain is not a simple physical image.

Now, what is the abstracted shape? What is the overall impression of an alphabet? How is each character element constrained by the global characteristics of an alphabet? What is the difference between characters and shapes existing in nature?

We know that characters function as media transmitting information based on the agreement of a community. But we are interested in whether the characters are perceived by a community as rigid or mechanical in their representations of the shapes. Even though the actual shapes are not abstract, the human mind, which is very flexible, is commonly able to make the abstraction. It is not known yet exactly how this capacity for abstraction operates. An abstracted shape is like a symbol in that as shapes are matched, there are absorbed changes in size, position, rotation, and so on. In the case of printed characters, shapes vary in stroke width, ornament, style, and the like. In the case of handwritten characters, there are many more nonlinear transformations. Then there are to be considered the tremendous variations, statistically, in additional and negative noise, such as stains/smears and breaks/gaps. Nevertheless, the human mind is not so bothered by the virtually infinite variations and can identify most characters. By this logic, the representation of a character must be abstracted so that it can absorb variations. We call the abstracted representation of a character its conceptual form. Thus the essential problem of character recognition can be narrowed to understanding *the concept of a character's shape and the mechanism that identifies any instantiation of this concept.* This is the principal problem in character recognition.

At the beginning we need to state that this is an open problem for which we cannot give any solution in this book. There are yet no signs that it will be solved in the near future; in fact, it is so big a problem that it will take the next 100 years to solve it. All we can provide the reader are the knowledge and courage to challenge the problem. We do this from two perspectives, scientific and engineering. On the one hand, to solve the problem mentioned above is scientific activity; on the other, it is not necessarily true that there can be no solution in the engineering sense if the problem has not yet been understood in principle. For example, the principle of dynamics was not known when the pyramids were built. In effect it could be said that pattern recognition is like physics back in the age before Newton, though many character optical readers are commercially available. With this in mind, let us consider the problem from the engineering point of view.

The problem of character recognition as an engineering problem has changed over time. Until the mid-twentieth century the problem of character recognition concerned reading printed numerals of one font with a constant size. Nowadays the engineering problem of shape recognition involves a great many depths or levels. At first researchers were not aware of this fact because humans recognize shape very easily. As they researched shape recognition, they realized that people can recognize cursive handwritten characters with 100% recognition rate when they are written neatly, but there is yet no OCR that can read handwriting with such a high recognition rate. Thus today character recognition is still a very critical area of research.

Let us look at some levels of character recognition according to their difficulties. Table 1.1 gives a list of the problems involved. The level of difficulty is roughly divided into four sublevels, 0, 1, 2, and 3, each of which defines the problem more concretely. Levels 0 through 2 are set in terms of the shape variation and the degree of noise. Level 3 is set by a somewhat different view of the problem, segmentation. Segmentation is central in picture recognition or image understanding. It is because of this difficult problem that image understanding is not yet commercialized. Cursive handwriting character recognition is a simple instance of image understanding but is a very difficult task, as mentioned earlier. In other words, at levels 0 through 2, a character string can be easily segmented into isolated characters. This is a very important and crucial assumption from the engineering standpoint. For example, the mail numbering system in Japan uses blind color boxes, in each of which a digit is written. The OCR industry in Japan has developed using this system. Sometimes a compromise between humans and machines is necessary to develop this type of machine, as is the case of

TABLE 1.1 Level Description of Character Recognition According to Its Difficulty

Level	Sublevel	Description
0	Little variation in shape, etc.; small number of characters in the object alphabets; little noise.	
	0–0	Printed characters of a specific font with a constant size. Roman alphabets; arabic numerals.
	0–1	Constrained handprinted characters. Arabic numerals.
1	Medium variation in shape, etc.; medium noise.	
	1–0	Printed characters of multiple fonts. Number of characters in the object alphabets is less than 100.
	1–1	Loosely constrained handprinted characters. Number of characters in the object alphabets is less than 100.
	1–2	Chinese characters of few fonts.
	1–3	Loosely constrained handprinted characters. Number of characters in the object alphabets around 1000.
2	Much variation in shape; heavy noise.	
	2–0	Printed characters of multiple fonts.
	2–1	Unconstrained handprinted characters.
	2–2	Affine-translated and/or projection-translated characters.
3	Nonsegmented string of characters.	
	3–0	Touching/broken characters.
	3–1	Cursive handwriting characters.
	3–2	Characters on a textured background.

pattern recognition and artificial intelligence in general. However, the compromise must be understandable to both humans and machine. This is why setting the problem is so difficult and why setting it up well is very important.

Obviously level 0, the lowest level, is the oldest, most primitive, and completely resolved. All OCR that are commercially available now have this capability. Nevertheless, there are some qualitative variations and noise. To be honest, this is really a big unsolved problem. Therefore the definitions in Table 1.1 are intrinsically ambiguous. But we will proceed intuitively with some typical examples with which this author has had experience and which represents the current state of research. We should notice that by "variation" is meant both geometrical and statistical changes in shape in a general sense. More precisely "variation" refers to geometrical and topological changes and "noise" to statistical changes. Included under geometrical changes, for example, is size variation. However, size is excluded under the more specific condition of one font with a constant size in the case of level 0, in which the typical variation is thickness of strokes. Concerning topological variations, we assume that the connectivity of each character is preserved in level 0. In other words, there is no gap within a stroke. The typical noise is evident in a slight raggedness to a character stroke. A small number of blemishes, which must be also small, are permissible noise on a character. All the above-mentioned defects must be quantitatively significant; otherwise, we face the intrinsic difficulty of pattern recognition itself. Fortunately, recently there has been systematic research on printed text quality. In particular, OCR performance is the focus of the Information Science Research Institute (ISRI) at the University of Nevada, Las Vegas (UNLV) [1]. With current research results we can show the reader "real and difficult" data for the top-level OCR readers, which are commercially available, concerning printed texts.

What is meant by constrained handprinted characters described as level 0–1? American handprint is the typical standard. Canada and Japan have their own standards, but they are fairly close [2]. Generally, the standard shapes are printed somewhere on a form, and people are asked to fill out the form using those shapes. Sometimes these are instructions such as using a straight and vertical line for "1" and no loop for "4." In the first case the instruction is necessary to avoid confusion with "2," and in the second, an open "4" avoids confusion with a "9." Box guidelines are thus quite useful. Some typical examples are shown in Fig. 1.2 [2].

Moving to level 1, we need to illustrate what "loosely constrained" is. This is shown in Fig. 1.3. In general, the topological properties of the shapes are not preserved, but they can still be read by humans. In practice, it is difficult to read such loosely constrained handprinted characters with almost no error. At least with numerals, this is less of a problem right now. At the present time commercially available OCR machines can manage handprinted characters of level 1–1. At level 1–3 there is Kanji handprinted character recognition. Some examples of handprinted Kanji/Chinese characters are shown in Fig. 1.4. In the figure the first two characters from the left mean "character" and the next two characters mean "recognition," and so on. Therefore a group of Kanji/Chinese characters corresponds to a word in western languages, in general. In Japan there is great need for OCR machines that read Kanji, but the performance and cost of the few machines available are not low enough to be used widely. Also, although Kanji originated from Chinese characters, it is a different language.

Moving now to level 2–1, the unconstrained handprinted characters, it is at this level that we have freely handwritten that most people have difficulty reading. Some examples are shown in Fig. 1.5. Here the two digits within the small upper boxes show two results by the machine (the left side) corresponding to the handwritten numerals (the right side) depicted beneath the boxes. A serious problem is that some people write characters that are quite different from the standard forms. In the examples of Fig. 1.5 the character images of the right top row of the figure show some careless mistakes. Since for numerals no context is generally given, character recognition of unconstrained characters is a big problem in practice.

Degrees of noise are explored by Robinson [4]. Print in a magazine is a typical example of medium noise where the human reader perceives little or no irregularities. Newspaper print is another typical example of medium noise. Figure 1.6 shows

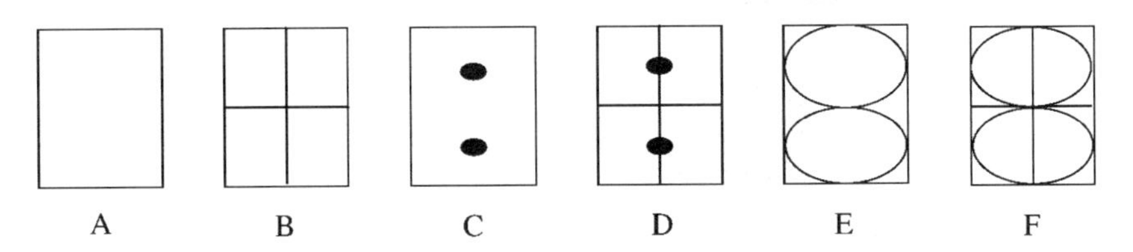

FIGURE 1.2 Six types of box guidelines. From reference [2], © 1998, CRC.

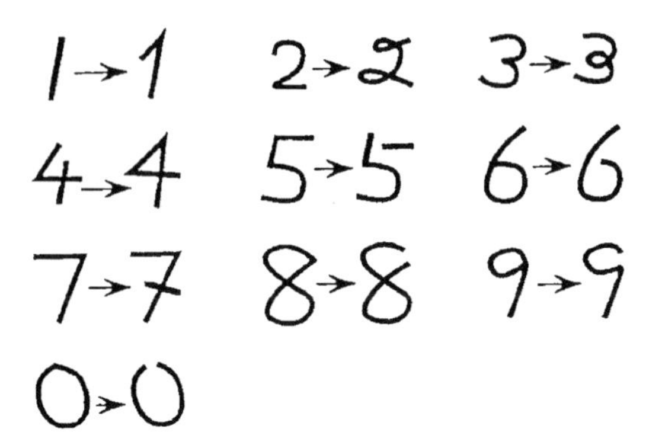

FIGURE 1.3 Some examples of loosely constrained handprinted characters.

an example of medium noise. Because the characters spread out, they look heavy. Notice that the middle character "W" is almost heavy noised. As Fig. 1.7 shows, heavy noise characters are hard to read, even by humans, without a context: "Takoma Park, Maryland." However, most Japanese university students could read these heavy noise characters without any context. We can infer, however, that they used knowledge that a Roman alphabet was being used. Figure 1.8 gives some examples of medium/heavy noise printed characters from a line printer. Note that "B" characters can be easily read as "R" characters even by humans. Also the lower left part of "J" and bottom line of "Z" are so thin that is hard for a machine to binarize them.

We have roughly sketched the spectrum of engineering problems of OCR systems. Let us return now to where we are in terms of levels. We are at level 2. Problems at level 1 have been generally solved in the engineering sense, except for level 1–3 which concerns handprinted Kanji recognition. Here we should also note the current state of the OCR technology on handprinted character recognition. The Institute for Posts and Telecommunications Policy (IPIP) held its second character recognition competition in 1996 (the first was held in 1993) [5]. The results showed the best three algorithms selected to achieve very high performance of 97.94% correct recognition rate, 0.20% substitution error rate, and 1.86% rejection rate for 5000 test data of unconstrained handprinted numeral characters obtained from handprinted three-digit postal codes in Japan. The results were analyzed in detail, and the causes of substitution/rejection errors were attributed to heavy-handed stroke, disconnected stroke, peculiar handprinting, extra segment, and unavoidable misrecognition. Samples of results that were substituted/rejected by the best three algorithms are given in Fig. 1.9. Human interpretations for the same handprinted characters as used in three-

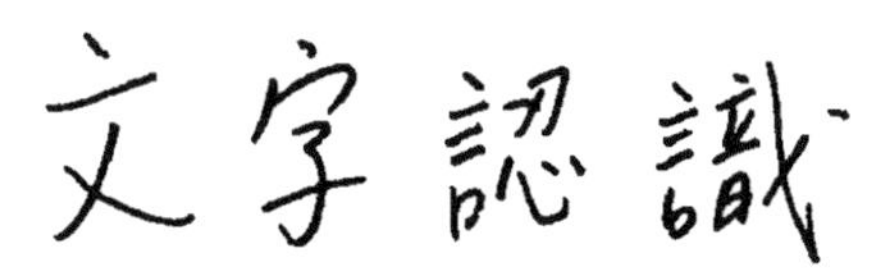

FIGURE 1.4 A string of Kanji/Chinese characters which means "character recognition" in Japanese.

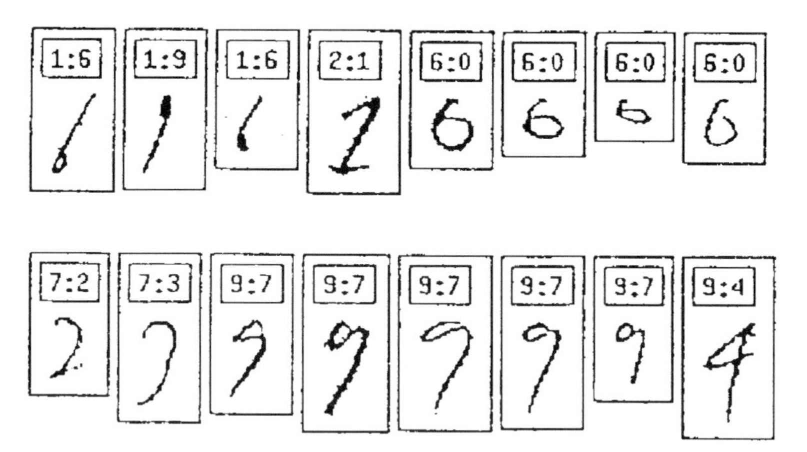

FIGURE 1.5 Some examples of unconstrained characters. From reference [3], © 1998, IEEE.

digit postal codes were estimated to have a 99.84% correct recognition rate and 0.13% substitution rates [6].

The results of systematic applications of ISRI at UNLV showed that for good-quality text data, all six top-level OCR machines gave correct readings of 99.77% to 99.13%. For middle-quality text data, correct readings ranged between 99.27% and 98.21%. For poor-quality text data, the range was extended to 97.01% and 89.34%. Some examples of these poor-quality characters are shown in Fig. 1.10. From a closer look at these character images, we can see two major reasons for the drop in recognition rates. It is due to touching and broken character images. The two tendencies were analyzed further by ISRI at UNLV [7]. They found that poor character images resulted in error in the majority of the tested OCRs. This is shown in Table 1.2. Their experiments were based on a very large and powerful database called GT1. The problem of image "segmentation" will be discussed below.

A large number of Kanji characters are based on a few simple structures which inherently cause segmentation problems. This problem will become clear in the illus-

FIGURE 1.6 Example of medium noise characters. From reference [4].

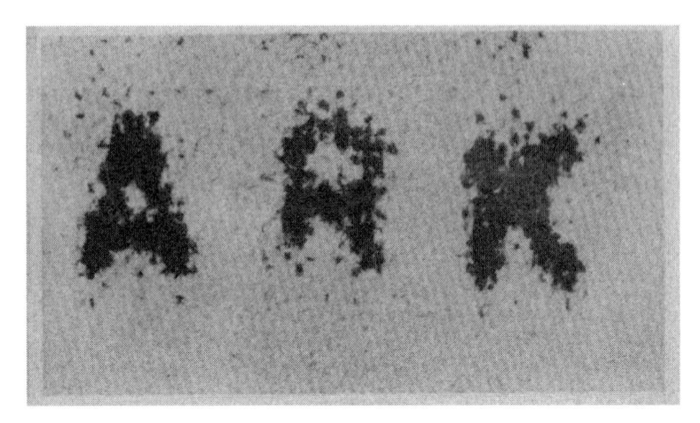

FIGURE 1.7 Example of heavy noise characters. From reference [4].

trations of Fig. 1.11. To avoid this problem a word recognition strategy must be taken, which means creating many subclasses for each category. In this sense, this problem really belongs to level 3. In this case the effect of "loosely constrained" may be taken to apply to "written components." As Fig. 1.11 shows, the Kanji character consists of three components, each of which has the same shape and meaning of tree. Two trees mean a bush, and three trees a forest, such as the typical construction of Chinese/Kanji characters. However, such separate forms are hard to keep when they are handwritten as shown at the right side of the figure. Nevertheless, good performances at level 1 are commercially available at reasonable cost despite the existing engineering problems with level 1. That is, high performance and low cost are always requested by users. Also some software packages for reading printed characters of multiple fonts (western alphabets) and of Kanji (nonmultiple fonts) are available at low cost.

So far, we have discussed levels 0 through 2. There is a common important point. That is, the characters are separated from neighboring characters and other typographical elements like lines of a table, so they are easily distinguished. At first, this seems to be a general condition, but in running text it cannot always be satisfied. In practice, engineers have trouble separating characters from a string of characters. For this purpose level 3 was established. For level 3–0, characters often touch each other or lines of a table. At level 3–1, the character contact is due to the intrinsic form of cursive script. When characters are printed or written on textured backgrounds such as is the case with maps and checks, something more than contact is involved and so classifies it as level 3–2.

0123456789ABCDEFGHIJKLMNOPQRSTUVWXYZ
0123456789ABCDEFGHIJKLMNOPQRSTUVWXYZ
0123456789ABCDEFGHIJKLMNOPQRSTUVWXYZ
0123456789ABCDEFGHIJKLMNOPQRSTUVWXYZ
0123456789ABCDEFGHIJKLMNOPQRSTUVWXYZ
0123456789ABCDEFGHIJKLMNOPQRSTUVWXYZ

FIGURE 1.8 Example of middle- and heavy-noise printed characters.

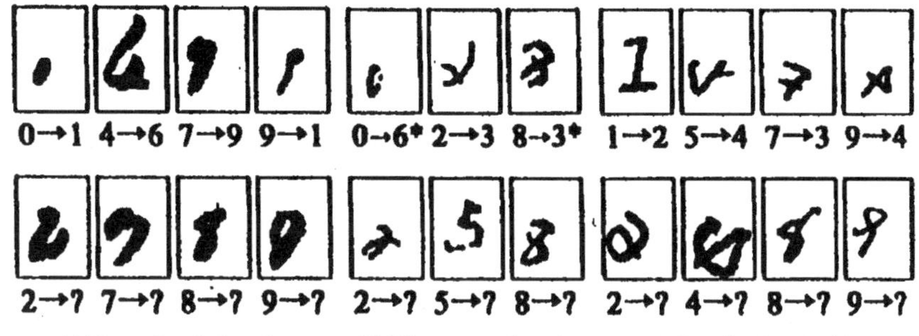

(*a*) Heavy-handed stroke (*b*) Disconnected stroke (*c*) Peculiar handwriting

FIGURE 1.9 Examples of substituted/rejected patterns that remained between algorithms P_A, P_B, and P_C in the first competition and in the second competition; only the top three samples in (*b*) are related to both competitions, where the samples marked with * were improved as the rejection in the second competition. From references [5] and [6].

At the present time, character data sets belonging to levels 2 and 3 are subjects of research, so there is no practical nor experimental system that has complete performance data on these character levels. However, for level 2, considerable research has been done, and we can expect a commercial system in the near future that has very high performance for the data of level 2, or at least an experimental system.

Since level 3 is considerably more difficult, we are limited in our *knowledge* on the language and situations where it can be effectively applied. For example, the

TABLE 1.2 Distribution of Estimated ASCII-OCR Problems

Problem Category	Number of Errors	Fraction of Total Errors
Broken characters	1872	52.1
Touching characters	734	20.4
Noise/speckle	122	3.4
Skew (or curved baseline)	49	1.4
Broken and touching	186	5.2
Broken and noise	9	0.3
Broken and skew	33	0.9
Touching and noise	2	0.1
Touching and skew	10	0.3
Total	3017	83.9
Similar symbols (1, l 0, o)	207	5.8
Wrong case	12	0.3
Stylized characters	46	1.3
Introduced spaces	79	2.2
Dropped spaces	39	1.1
Unknown cause	196	5.5
Total	579	16.1
Grand Total	3596	100

Source: From reference [7], © 1997, ISRI at UNLV.

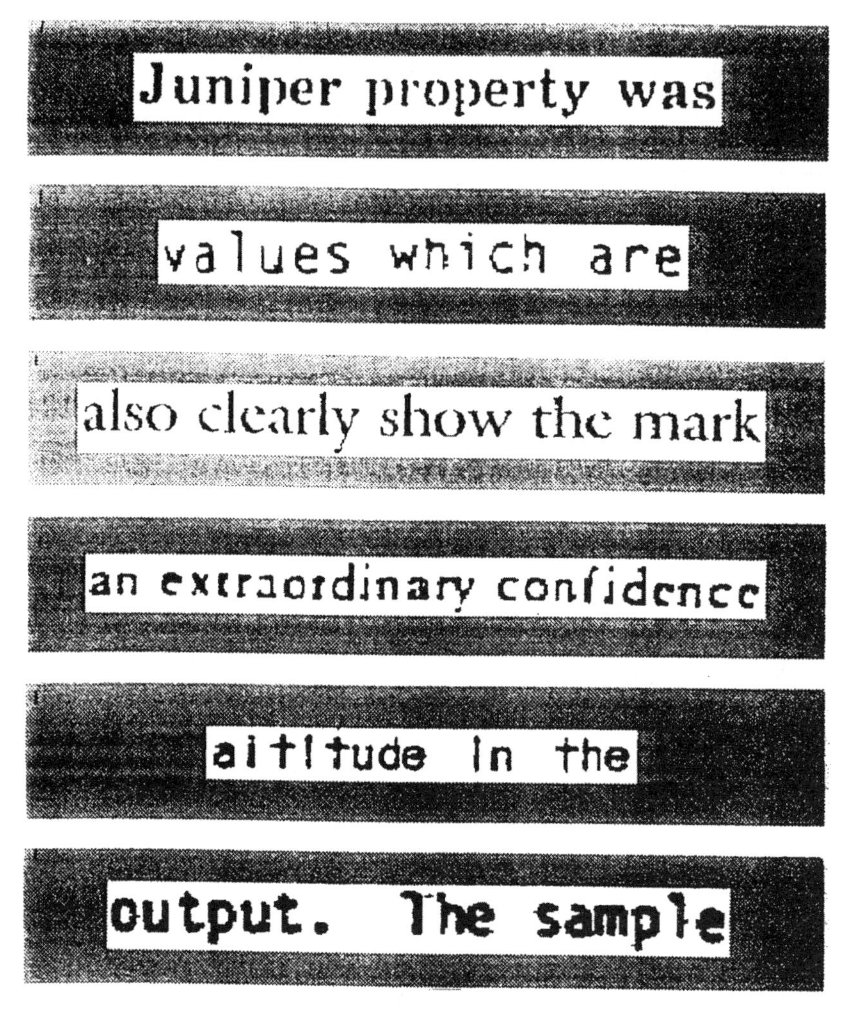

FIGURE 1.10 Examples of poor-quality characters (3 × linear magnification). From reference [7], © 1997, ISRI at UNLV.

application domains may be addresses or numerals written using the Roman alphabet, such as in bank checks. However, for level 3–1 such situations happen frequently in practice, and a partial engineering solution may be possible, although a complete solution would require a more advanced stage of technology. Therefore some methods can be considered and applied in practice. Here we note that *complete performance* means *human performance.*

This book gives concrete methods for these levels and points out remaining problems. The discussion is paired with a general presentation of the levels, since both concreteness and generality are both important. Dynamic thinking from specific considerations to generalizations, and vice versa, are of course essential to the developing research. In any case this is an area of tremendous growth with much work remaining for another generation of researchers.

FIGURE 1.11 The Chinese/Kanji characters on the left consist of three components, each conceptually connected but separated. When written out, all the three components are connected in one character, as shown on the right side.

1.2 BASIC METHOD

Here we consider some basic methods of character recognition in which the data are almost at level 0. Through the chapter we aim to provide the reader with an engineering model for character recognition.

Since at this level the data are simple, many kinds of methods will be considered; we include their historical background, state of development, and future potential.

1.2.1 Projection Method

In engineering, efficiency is the most important factor, and so it is pursued endlessly. Character recognition is no exception. However, this task is very difficult in character recognition. While some parts of the character recognition function can be constructed simply, it is very difficult to devise the eventual function simply. First let us consider the case of a projection method.

The aim of the projection method is to simplify drastically a system of character recognition by reducing two-dimensional information to one-dimension. Historically this method appeared in the early stage of OCR. Figure 1.12 illustrates its principle function. In panel (*a*) is shown a given character and the slit that scans it from left to right. The character is depicted in black, but it can be reversed so that the strength of darkness is proportional to the reflected light. By projecting the light value within the slit to the x-axis, the two-dimensional light distribution can be reduced to the one-dimensional light distribution on the x-axis. This is shown in Fig. 1.12(*b*). The resulting one-dimensional pattern is by far simpler to handle compared with the two-dimensional one. Several methods are conceivable after this information reduction, such as cross-correlation, Fourier analysis, or moment. We will later consider some experimental results using cross-correlation.

The performance of the method is shown in Fig. 1.13. As shown, "/" and "\" are not differentiated in this method, although a projection to the y-axis is also used. The projection method is weak in diagonally oriented patterns in general. In Fig. 1.13(*b*), two more complex figures are shown.

The projection method is effective for preprocessing in a character recognition system; more on this will be mentioned later. Here we are only concerned with a general illustration. When a string of characters is given, each character must be separated from the others. The projection method works well for processing in segments. It is effective for numerals and the Roman alphabet because, except for "i," their character images are connected in a topological sense. In the case of Fig. 1.13, since only a projection to x-axis is used for the segmentation, there is no problem in practice.

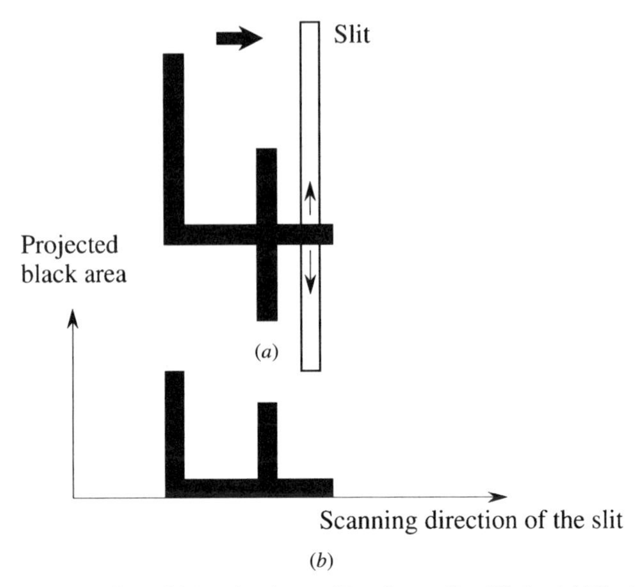

FIGURE 1.12 Illustration of 2D reduction to 1D using a slit of light. (*a*) The input numeral "4" is scanned from left to right. (*b*) The black area is projected onto x-axis in the scanning direction of the slit. From reference [8], © 1998, IEEE.

When, however, the number of characters in an alphabet is large, as with Chinese characters, the projection is used as a rough classification. The theoretical importance of projection will be fully described later.

1.2.2 Cross-Correlation Method

The projection method used to simplify a system by reducing two-dimensional information to one-dimension can fail if it is recognized that the pattern is essentially two-dimensional. There is a simple method that deals with such two-dimensional information, the cross-correlation method. This is a typical method used in matching, and it will be described in detail later.

Let the ith template, unknown input character and its domain be $g_i(x, y), f(x, y)$ and R, respectively. The similarity based on cross-correlation is defined as

$$S^i(f) = \frac{\iint_R f(x, y) g_i(x, y) dxdy}{\sqrt{\iint_R f(x, y)^2 dxdy} \sqrt{\iint_R g_i(x, y)^2 dxdy}}, \tag{1.1}$$

where $i = 1, 2, \ldots, L$; L is the number of characters of a given alphabet.

The maximum value of $S^i(f)$ is found scanning i, and if it is $S^i(f)$, then the input character f is identified as class j. The denominator of (1.1) is for the normalization of amplitude denoted as A. That is,

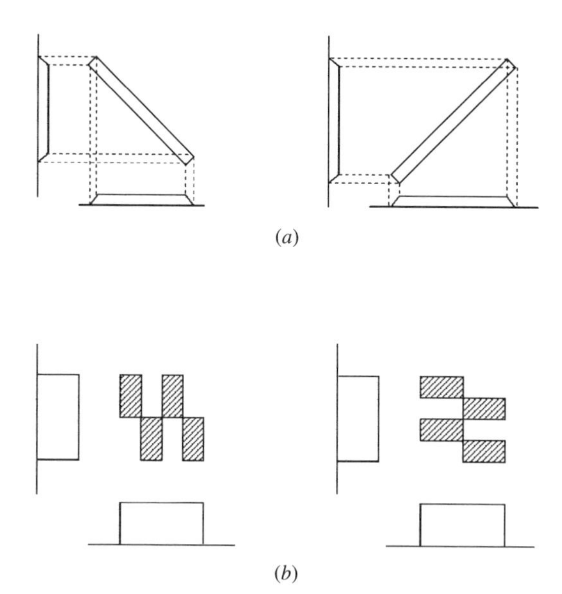

FIGURE 1.13 Some typical examples of the projection method applied to character recognition. From reference [9], © 1997, Ohm.

$$S^i(Af) = S^i(f). \tag{1.2}$$

holds for any scalar A. By convention, the integration of the numerator of (1.1) is expressed as (f, g_i), called the inner product of $f(x, y)$ and $g_i(x, y)$, and the denominator is expressed as $\|f\| \cdot \|g_i\|$, where $\|h\|$ is called norm of h and is its size or length intuitively. According to the Schwartz inequality

$$|(p, q)| \leq \|p\| \cdot \|q\|, \tag{1.3}$$

the following relations hold:

$$-1 \leq S^i(f) \leq 1. \tag{1.4}$$

Therefore we can set

$$\cos\theta = \frac{(f, g_i)}{\|f\| \cdot \|g_i\|}. \tag{1.5}$$

Thus the cross-correlation can be interpreted as a projection of $f(x, y)/\|f\|$ to $g_i(x, y)/\|g_i\|$. The functions $f(x, y)$ and $g_i(x, y)$ are interpreted as vectors in a space of infinite dimension. In practice, however, these functions are always given by finite-dimensional vectors. We use the terms "functions" and "vectors" interchangeably. The rigorous discussion will be given later.

Turning to practical considerations, we note the following: First, (1.1) is not actually used because $\|f\|$ is common to all $S_i(f)$, $i = 1, \ldots, L$. Another thing is that if the difference between the maximum value, say $S^j(f)$, and the next maximum value, say $S^k(f)$, or $\Delta = S^j(f) - S^k(f)$ is small, then the confidence that the input f belongs to the jth class is low. Therefore it is not safe to identify f as jth class with certainty. This caveat is important in particular when the input characters are numerals. In practice, some margin Δ_T is set, and if Δ is less than Δ_T, then the input character is rejected. For example, Δ_T might be chosen as 5% of the maximum output value of the cross-correlation system.

Table 1.3 compares 2D templates and their 1D projected patterns. It is clear that the two-dimensional information is much better than the one-dimensional one. The reputed advantage of the cross-correlation method is that it is robust against breaks/gaps and stains/smears. However, this is only true under the conditions illustrated in Fig. 1.14. As shown in Fig. 1.14 (*a*), if the break occurs within the stroke of a character, "L" in this case, then the value of the cross-correlation is little affected. However, as shown in Fig. 1.14 (*b*), if a boundary that opens out of the character's domain (e.g., often the end of a stroke) is broken off, the negative noise can sometimes be serious.

On the other hand, if positive noise such as a stain is added within the domain spanned by the character stroke, then the cross-correlation remains robust. But it

TABLE 1.3 The Comparison of Correlations of (*a*) 2D Patterns and (*b*) 1D Patterns

(a) 2D Patterns

	0	1	2	3	4	5	6	7	8	9
0	1.000	0.128	0.426	0.522	0.131	0.103	0.280	0.217	0.505	0.432
1	0.128	1.000	0.336	0.214	0.159	0.246	0.212	0.270	0.208	0.194
2	0.426	0.336	1.000	0.551	0.181	0.116	0.241	0.439	0.403	0.313
3	0.522	0.214	0.551	1.000	0.197	0.225	0.358	0.456	0.530	0.276
4	0.131	0.159	0.181	0.197	1.000	0.286	0.305	0.183	0.302	0.167
5	0.103	0.246	0.116	0.225	0.286	1.000	0.470	0.152	0.281	0.253
6	0.280	0.212	0.241	0.358	0.305	0.470	1.000	0.172	0.441	0.301
7	0.217	0.270	0.439	0.456	0.183	0.152	0.172	1.000	0.289	0.169
8	0.505	0.208	0.403	0.530	0.302	0.281	0.441	0.289	1.000	0.415
9	0.432	0.194	0.313	0.276	0.167	0.253	0.301	0.169	0.415	1.000

(b) 1D Patterns

	0	1	2	3	4	5	6	7	8	9
0	1.000	0.192	0.645	0.724	0.384	0.457	0.583	0.466	0.735	0.786
1	0.192	1.000	0.528	0.354	0.312	0.540	0.370	0.681	0.343	0.346
2	0.645	0.528	1.000	0.918	0.763	0.956	0.881	0.915	0.909	0.845
3	0.724	0.354	0.918	1.000	0.689	0.827	0.794	0.804	0.868	0.866
4	0.384	0.312	0.763	0.689	1.000	0.772	0.607	0.775	0.672	0.549
5	0.457	0.540	0.956	0.827	0.772	1.000	0.841	0.928	0.841	0.727
6	0.583	0.370	0.881	0.794	0.607	0.841	1.000	0.735	0.923	0.889
7	0.466	0.681	0.915	0.804	0.775	0.928	0.735	1.000	0.788	0.659
8	0.735	0.343	0.909	0.868	0.672	0.841	0.923	0.788	1.000	0.901
9	0.786	0.346	0.845	0.866	0.549	0.727	0.889	0.659	0.901	1.000

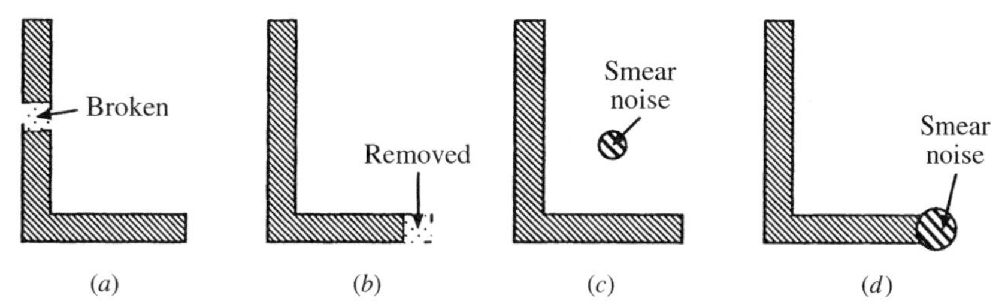

FIGURE 1.14 Illustration of the features of cross-correlation method against noises: (*a*) and (*b*) examples of negative noise and (*c*) and (*d*) are those of positive noise.

becomes negative noise if the stain occurs on a boundary that opens out of the character's domain, so cross-correlation is not reliable. In the above description, the expression "character's domain" will be disambiguated later. For the time being, an intuitive understanding is enough.

Now we turn to some quantitative aspects of the noise problem above. Let us assume that the input character f_L is exactly the same as the corresponding template g_L and that both are binary quantized, namely their pixel values are either 1 or 0. This is an important preprocessing process, and it will be fully discussed later. In terms of Fig. 1.14 (*a*), we have the value $S(g_L) - S(f_L)$ equal to δ/A_L where δ is the area of the missing part and A_L is the total area of the character image of "L." We note here that for simplicity, the superscript i in the symbol of the similarity is omitted, which is the index of the character "L" in Roman alphabet, and that the $S(g_L)$ is the value of the similarity when the input character is g_L, which is the template of the character "L." Therefore its value is 1 naturally. The $S(f_L)$ is the value of the similarity between g_L and the noisy character f_L. The difference, $S(g_L) - S(f_L)$, is small if δ is so. In the case of a break, δ is usually small enough to be insignificant.

The case shown in Fig. 1.14 (*b*) is different. First suppose that the width of the missing part is equal to a line width. After segmentation the character is normalized in its position. Several position normalization methods can be considered. The simplest one is the displacement of the character to left or right edge of the frame. In the case of the left displacement, it is the same situation as the break noise shown in Fig. 1.14 (*a*). However, in case of the right displacement, the vertical stroke of the image of "L" is shifted by the width of the stroke, which causes the loss of the overlap between the template and the input character. The loss is almost equal to a half of $S(g_L)$, namely $S(f_L) \cong 1/2S(g_L)$. This is a drastic drop in the similarity value, so there may be criticism that the normalization was too simple. Therefore let us improve it to a typical position normalization method, the so-called center normalization method. That is, the center of the input character is found and its coordinate is moved to the origin of the frame, which is usually also taken at the center of the frame. Even by this method the loss of similarity is still large, although the overlap of the vertical stroke between the template and the input character image is recovered by one-half. That is, $S(f_L) \cong 3/4S(g_L)$.

For the positive noise, there is no loss of similarity in case of Fig. 1.14 (*c*), but the loss is serious in the case of Fig. 1.14 (*d*), for the same reason as above. Therefore it is concluded that the cross-correlation method is weak against position displacement of an input character. The detailed description for an improvement will be given later. However, before that it is appropriate to mention the autocorrelation method which is widely used in the signal processing field.

Correlation Method Based on Autocorrelation For simplicity, one-dimensional waves are considered. The autocorrelation function for an input wave f(x) is defined as

$$\psi(r) = \int_{-\infty}^{\infty} f(x)f(x+r)dx. \tag{1.6}$$

For the template wave $g_i(x)$, it is defined in the same manner, as

$$\varphi(r) = \int_{-\infty}^{\infty} g_i(x)g_i(x+r)dx. \tag{1.7}$$

Accordingly, the correlation between the template autocorrelation and the input wave autocorrelation functions is defined as

$$S^i_{auto} = \frac{\int_{-\infty}^{\infty} \psi(r)\varphi(r)dr}{\|\psi\|\|\varphi\|}. \tag{1.8}$$

To see the nature of the autocorrelation function for position displacement, let us set $f(x)$ to $f(x-a)$ in (1.6). It is easily seen that

$$\psi(r; f(x)) = \psi(r; f(x-a)) \tag{1.9}$$

That is, the autocorrelation function is invariant for shift. This property of shift invariance is well-known, and it is the reason why we chose it here. Furthermore it is easily shown that the value S^i_{auto} is invariant with respect to amplitude variation of an input wave. At the beginning of OCR history, the autocorrelation function attracted the attention of researchers of OCR because of its shift invariance. For the historical discussion, see the historical remarks at the end of this chapter.

However, there is a pitfall in this approach. As is well-known, the autocorrelation function for a reflected wave, namely $f(-x)$, is the same as $f(x)$. Actually $\psi(r; f(-x)) = \int_{-\infty}^{\infty} f(-x)f(-x+r)dx = \int_{-\infty}^{\infty} f(x')f(x'+r)dx'$. That is,

$$\psi(r; f(-x)) = \psi(r; f(x)) \tag{1.10}$$

holds. This property is easily expanded to two-dimensional case so that

$$\psi(r; f(-x,-y)) = \psi(r; f(x,y)) \tag{1.11}$$

holds. Thus the reflected pattern cannot be distinguished from the original by the autocorrelation function. By this fact there can be expected to be considerable loss of pattern information in the autocorrelation function. This is shown in Table 1.4. As can be seen by the numbers, the coefficients are very close. For this reason the autocorrelation function is not used in OCR in practice as far as this author knows. However, the autocorrelation function is still interesting from the standpoint of information science.

1.2.3 Logical Matching

As mentioned before, input characters are usually binarized before processing. Binary patterns lead us, in some sense to a "logical correlation," which can be thought of as a counterpart of "analog correlation." This is quite natural considering that the process involves computer technology. The process is called logical matching or matrix matching. An example is shown in Fig. 1.15, in which an input character "A" is recognized correctly based on the logical decision that all the pixel points $W_1 \sim W_4$ lying on the background have value 0 (white) and all the pixels points $B_1 \sim B_4$ lying on the stroke have value 1 (black). This is expressed formally as

$$A = (\cap_{i=1}^{4} \overline{W}_i) \cap (\cap_{i=1}^{4} B_i) \tag{1.12}$$

TABLE 1.4 Cross-correlation Coefficients of Autocorrelation Functions Generated by Numerals

(a) Shifted to −x Direction

	0	1	2	3	4	5	6	7	8	9
0	1.000	0.903	0.851	0.900	0.860	0.874	0.906	0.882	0.934	0.932
1	0.903	1.000	0.919	0.982	0.920	0.978	0.964	0.951	0.959	0.952
2	0.851	0.919	1.000	0.975	0.983	0.968	0.929	0.977	0.937	0.918
3	0.900	0.982	0.975	1.000	0.970	0.990	0.974	0.985	0.975	0.964
4	0.860	0.920	0.983	0.970	1.000	0.959	0.956	0.994	0.949	0.933
5	0.874	0.978	0.968	0.990	0.959	1.000	0.945	0.974	0.941	0.925
6	0.906	0.964	0.929	0.974	0.956	0.945	1.000	0.976	0.988	0.985
7	0.882	0.951	0.977	0.985	0.994	0.974	0.976	1.000	0.966	0.955
8	0.934	0.959	0.937	0.975	0.949	0.941	0.988	0.966	1.000	0.997
9	0.932	0.952	0.918	0.964	0.933	0.925	0.985	0.955	0.997	1.000

(b) Shifted to +y Direction

	0	1	2	3	4	5	6	7	8	9
0	1.000	0.947	0.842	0.944	0.985	0.921	0.983	0.885	0.928	0.976
1	0.947	1.000	0.797	0.910	0.945	0.837	0.917	0.762	0.913	0.894
2	0.842	0.797	1.000	0.935	0.855	0.890	0.897	0.926	0.884	0.884
3	0.944	0.910	0.935	1.000	0.965	0.960	0.964	0.929	0.976	0.955
4	0.985	0.945	0.855	0.965	1.000	0.959	0.986	0.902	0.959	0.983
5	0.921	0.837	0.890	0.960	0.959	1.000	0.962	0.958	0.954	0.968
6	0.983	0.917	0.897	0.964	0.986	0.962	1.000	0.944	0.956	0.994
7	0.885	0.762	0.926	0.929	0.902	0.958	0.944	1.000	0.906	0.951
8	0.928	0.913	0.884	0.976	0.959	0.954	0.956	0.906	1.000	0.940
9	0.976	0.894	0.884	0.955	0.983	0.968	0.994	0.951	0.940	1.000

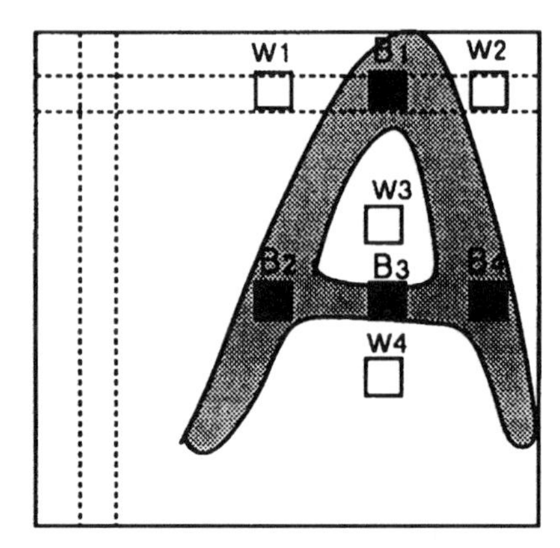

FIGURE 1.15 Logical matching. From reference [8], © 1997, IEEE.

in which, when the logical value of A is equal to 1, the input character is identified as "A"; otherwise, it does not identify any character.

A logical expression such as (1.12) is very rigid. But suppose that the part of B_3 is changed to 0. Then the logical value of A changes to 0; it does not match against the template of "A." We see that only one local change at the detecting points affects the decision decisively. To relax this point, two methods are considered. One way is to use logical OR as

$$A = W \cap B, \tag{1.13}$$

where

$$W = (\cap_{i=1}^{4} \overline{W}_i) \cup (\cap_{j=5}^{n} \overline{W}_i),$$
$$B = (\cap_{i=1}^{4} B_i) \cup (\cap_{j=5}^{m} B_i),$$

for n, $m > 5$. That is, an appropriate number of points is chosen from background and stroke regions, which can be regarded as being stable. In the new expression, the total n and m points chosen include the old ones, for background and stroke regions, respectively. The background and stroke points are divided into two parts, and they are logically connected by OR.

The other way is to introduce an analog component in order to relax the rigid logical expression. Specifically, the values on pixels are taken as analog values rather than logical values, and AND is regarded as addition. Thus expression (1.8) is improved to

$$A = \left(\sum_{i=1}^{4} W_i \leq 1 \right) \cap \left(\sum_{i=1}^{4} B_i \geq 3 \right). \tag{1.14}$$

This works well for the case where one of the black detecting points is missing, and it also applies for the white detecting points. That is, one point of the chosen white points can be black. If we proceed along this line further, it reaches the analog correlation as

$$\sum_{i=1}^{n}(-W_i) + \sum_{i=1}^{m} B_i \rightarrow \int_R f(x, y)g(x, y)dxdy, \tag{1.15}$$

where R is the range of the frame.

In the above expression, $g(x, y)$ is the template of the character, which is the black portion of the "A" in this case. In general, in the cross-correlation the black part is treated explicitly, whereas the background part is treated implicitly. To introduce a white region explicitly, we need to consider a template for the white region $g^W(x, y)$. However, the set $g^W(x, y) \cup g^B(x, y)$ is not necessarily equal to the frame R, where $g^B(x, y)$ is the template of the black part of the character, "A." Between $g^W(x, y)$ and $g^B(x, y)$, there is a gray zone in which the values are unstable. Based on this idea, (1.14) is improved as

$$A = W \cap B, \tag{1.16}$$

where

$$W = (\sum f(i, j)g^W(i, j) \leq T_W),$$
$$B = (\sum f(i, j)g^B(i, j) \geq T_B),$$

with T_W and T_B the threshold values.

This is just a hybrid of logical and analog approaches, but the white and black parts are symmetric.

It is important to understand the difference between analog and logical matchings. As the preceding discussion has shown, to understand the meaning of *analog* is essential in pattern recognition. Unlike logical matching which is local, analog matching such as cross-correlation is global. Logical matching relies on each logical value of specific pixels, namely *point information.* However, this is dangerous and premature to bring point information to a logical level. In the cross-correlation, the information is distributed across a whole area, and so it avoids the danger of a wrong decision.

Another difference between the analog and logical levels concerns the methods of synthesizing the templates. In the correlation method it is easy to make the templates, although there are many ideas and methods. A template is made by taking the average of the samples that belong to a class. In the logical matching it is very laborious to synthesize the appropriate logical expressions corresponding to a template. Automatic generation of logical expressions has been attempted, but it has failed to be used in practice. There is a famous story on that in the history of OCR research and development which appears in the historical remarks at the end of this chapter. However, the logical matching approach is still being used. See the bibliography at the end of the chapter.

Finally logical matching is weak in position displacement for reasons analogous to the cross-correlation method. However, there is a clever way to cope with this difficulty in which a shift register is used. This method will be illustrated later.

1.2.4 Crossing Method

We can view logical matching from another angle, since it extracts local features as logical values and then the extracted logical features are combined logically. Let us take the previous example shown in Fig. 1.15. In the figure B_2 and B_4 can be regarded as logical variables that indicate the existence of branching features in "A." These variables are heavily position dependent, so they are weak in preventing position displacement, in general, as we have seen. The crossing method in a broad sense is a typical and simple feature extraction method that is robust in position displacement and variations of character shape, although the technique is not perfect.

Consider now the examples in Fig. 1.16, in which the application of a sonde method is shown for characters "2," "3," and "5." Notice that the Arabic numerals "2," "3," "5," "6," "8," and "9" are regarded as being constructed because the strokes are drawn centering around two focal points. For this reason the two points are set, as shown in the figure, in a frame and marked by circles. The detecting bars (the four bars in Fig. 1.16) are radially arranged in four directions: top, bottom, right, and left. If an input character lies or crosses the bars, then "1" is counted, and otherwise, "0" is counted. These counts are listed or represented by a vector for each input character. Such real situations are shown in the panels (*a*), (*b*), and (*c*); these lists are (1, 1, 0, 1, 1, 0), (1, 1, 1, 1, 0, 0), and (1, 0, 1, 1, 0, 1), respectively. Notice that the detecting bars spanned between the two focal points are neglected because this bar cannot effectively differentiate among the numerals "2," "3," and "5," although it is effective in discriminating them from "0," for example.

This method is effective in preventing position displacements and variations in characters. One such example is shown in panel (*d*) where the "S" written must be a "5" because only numerals characters are considered here. Actually this type of variation of "5" appears frequently in writing. As a result when roman alphabet and arabic numerals are used together, some restrictions must be imposed on them. Hence,

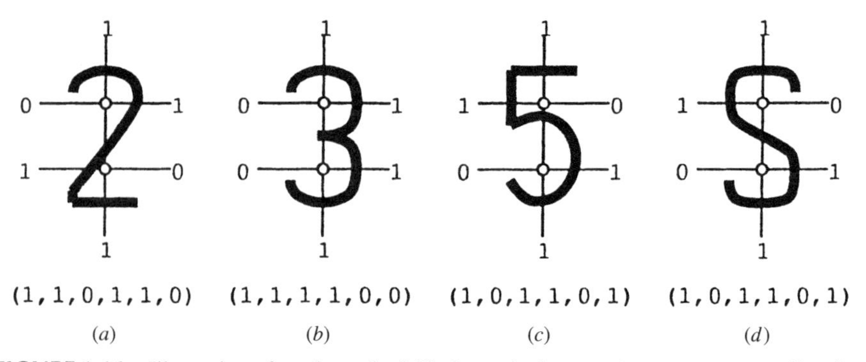

FIGURE 1.16 Illustration of sonde method: Under each character image a corresponding feature vector is drawn in which "1" and "0" denote stroke crossing and not crossing, respectively.

character standards were made for handprinted as well as printed character shapes. The sonde method can be effective, but it is unpredictable, which makes it hard to extend this method to the Roman alphabet. The problem here is that the character sets of alphabets are larger than those of numerals.

A simplified and generalized version of the sonde method is shown in Fig. 1.17. In a narrow sense, this represents the crossing method. As seen, the detecting lines are arranged parallel to the x-axis and the y-axis of the frame. The numbers of crossing times with these detecting lines for a given input character are counted. In panel (*a*), if only the detecting lines parallel to the y-axis are used, then "2" and "5" are not distinguishable, but they can be distinguished by employing the detecting lines parallel to the x-axis. This general method is not affected by the position displacement of characters, but it is as strong in resisting variations as the sonde method. In panel (*b*) some examples are shown where the input characters "2," "3," and "5" cannot be distinguished.

The crossing method is weak, however, in variation in size. The count list is contracted in such a way that runs of the same count are neglected. For example, [0, 2, 2, 2, 3, 3, 3, 0] $\rightarrow$ [0, 2, 3, 0]. A side effect of this contraction is loss of information of length, such as no distinction between "—" dash and "–" minus symbols.

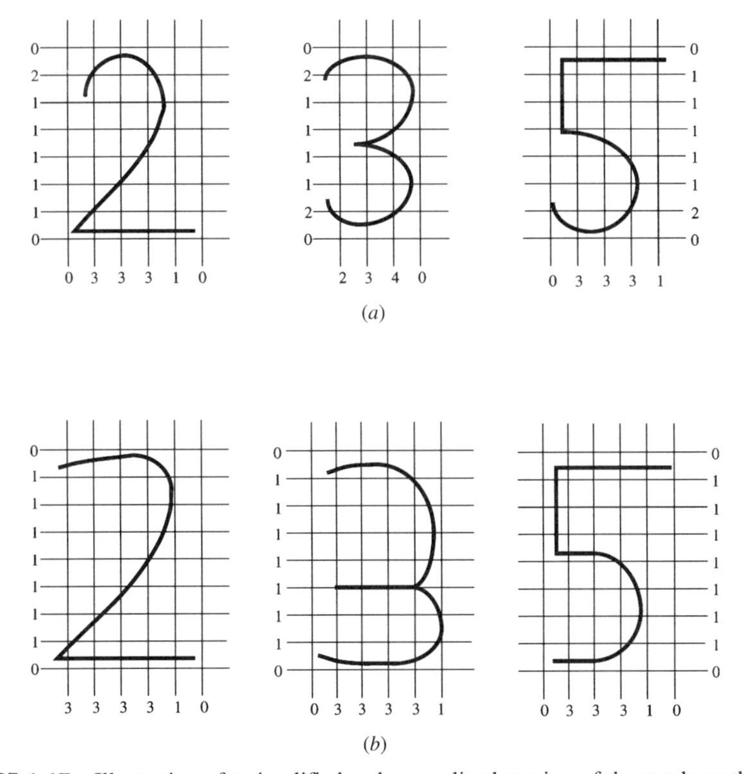

FIGURE 1.17 Illustration of a simplified and generalized version of the sonde method: (*a*) "2" and "5" images cannot be distinguished when only the y-axis parallel detecting lines are used; (*b*) "2," "3," and "5" images cannot be distinguished by the parallel detecting lines.

Both sonde and crossing methods are simple methods. But because of this simplicity they reduce the information too much, so they cannot be applied to the sets of characters in level 1. Let us begin by looking at the generalization problem of the sonde method. In the sonde method two appropriate focal points are chosen, but in the generalization process, all the points on the frame become the focal points from which the four detecting lines are spanned in the four directions, top, bottom, right, and left. Thus a list consisting of four elements is assigned at each pixel. This is shown in Fig. 1.18, in which the focal points are arranged all over the background. This method is called the characteristic loci method, or the background analysis approach. In this case the information increases drastically, and so redundant information is a problem, which will be described in detail later.

Now let us turn to the crossing method. The method extracts local features of character strokes by looking from a window that is a very thin rectangle. It differs from logical matching in that it is not completely local in the sense that it looks at the character through a line window and finds stroke segments that exist in the window. If a stroke lies perpendicular to the window, then it extracts a local feature of the stroke. However, even if the stroke shifts to the right or the left, the window can catch the stroke in the same way as before. Here we assume that the window is parallel to the x-axis. Therefore, contrary to logical matching, feature extraction is not bound to a position and so it extracts the feature invariant to the position displacement. Notice here that for a given position displacement, the crossing method is invariant if the sequences of "0" in the count list in its beginning part and ending part are contracted. That is, the shape of the distribution in the count list is invariant. The contraction of 0's at both sides of the distribution is a very simple example of normalization in feature space, and this will be described later in detail.

However, no information on the connective relationship between windows is taken, and so even if the neighboring windows have 1's, namely the sequence of counts (. . . , 1, 1, . . .), it does not mean that the strokes are connected. This loss of information is clearly illustrated in Fig. 1.19 (*a*) in which, more precisely, the detecting lines are somewhat dense, and so the count consists of a continuous spectrum. In

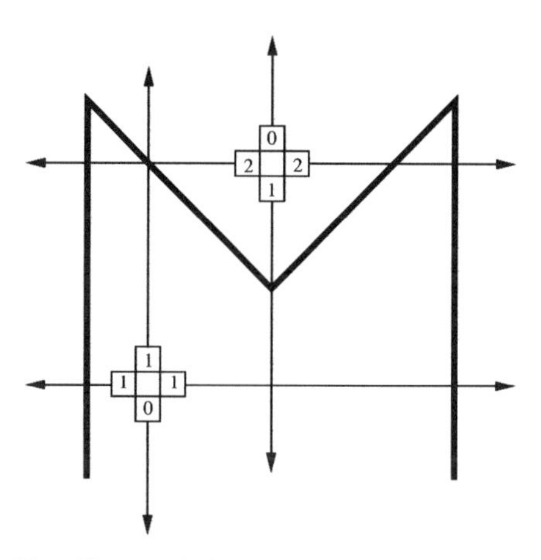

FIGURE 1.18 Characteristic loci method, or Glucksman's method.

Fig. 1.19 (*b*) another example is shown in which geometrical features (curvature) and topological features (loops) are lost. As shown, there are always pitfalls in such overly simple methods, so we must be cautious here.

On the other hand, as illustrated in Fig. 1.19 (*c*), if the diagonal detecting lines are introduced, then the discrimination power increases considerably. This point was noticed and the crossing count method including diagonal detecting lines was suc-

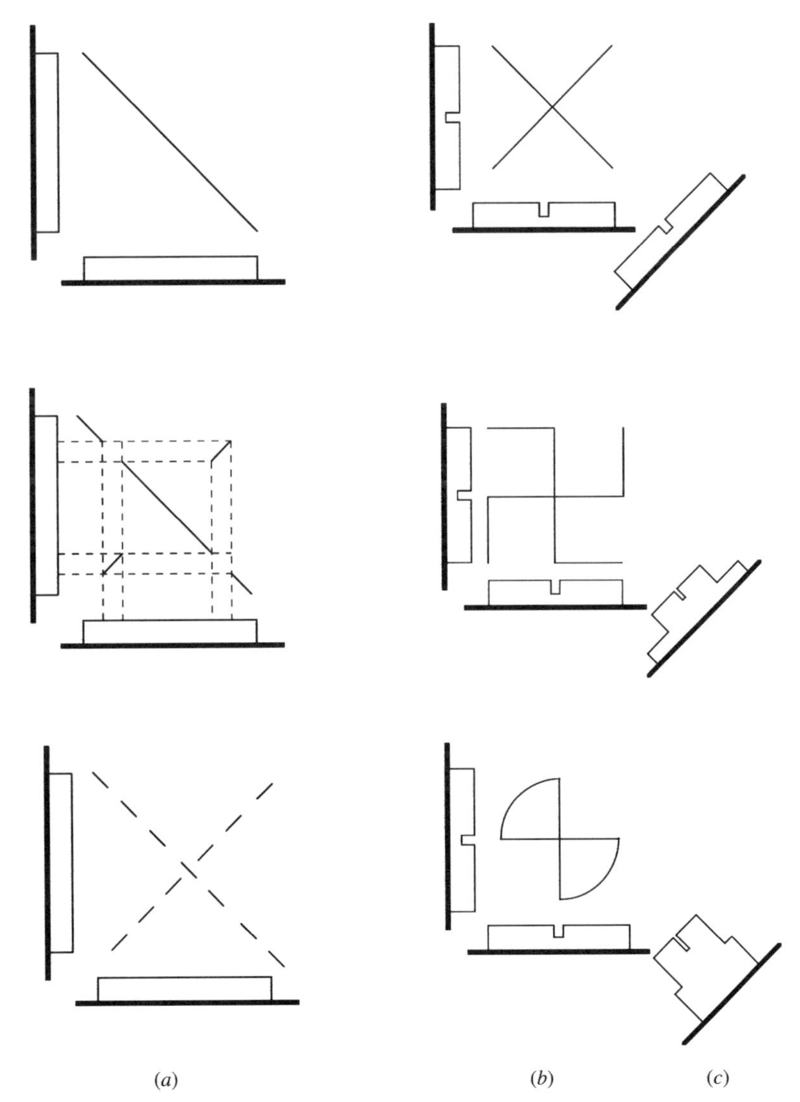

FIGURE 1.19 Examples of difficulty of crossing count feature extraction: (*a*) Three different shapes having the same crossing count projection on both *x*- and *y*-axes. (*b*) Another three totally different shapes having the same crossing count projection on both *x*- and *y*-axes. (*c*) The above three shapes have different crossing count projection to a diagonal axis. From reference [9], © 1997, Ohm.

cessfully applied to rough classification of Kanji characters [10]. This shows that we can use simple methods effectively if we understand their powers and limitations.

1.2.5 Method Based on Geometrical Features

The crossing method seems to be unpredictable, and we have seen that it does not favor important geometrical features. Such features should be searched for first in a given character. Since we know that a character image is a linear abstraction, then its geometric features consisting of end points and singular points, such as branching points, and crossing points should be easily extracted. It should be expected that such features would be preserved to guard against the variations in loosely constrained handprinted characters. A list of these features is given in Table 1.5, in which T, B, and C stand for terminal, branching, and crossing points, respectively. For each numeral in the table, the number of these features is shown. Here the shapes of the numerals are assumed to be those of the first ten standard numerals shown in Fig. 1.20.

This method of extracting features of the shapes of numerals is effective but not complete. The features T, B, and C leave the groups (1, 2, 7) and (6, 9) unresolved. Therefore we must add another geometrical feature—the direction at the end point—which can be easily extracted. The direction does not need to be precise and can be quantized into eight directions, as shown in Fig. 1.21 (*a*). In practice, this form of quantization is called a *chain code* or *Freeman code* after its inventor. Other applications of the code will be discussed later. In the direction lists for the character groups in Table 1.5, for example, [1, 3] means 1 through 3 which shows the direction range. This is because we are dealing with handprinted characters, so there are some variations in the directions. We can see how the numerals are differentiated when we take into account the directions at the end points.

Let us consider more closely the variations of these handprinted characters. First, “3” is assumed to have one branch point, which is a distinctive feature at the center of its character image. This point is very unstable and changes easily to a cusp or a corner point. Then “3” belongs to the group of (1, 2, 7). Fortunately, however, the end

TABLE 1.5 Geometric Features of Handprinted Numerals

Class	T	B	C	Direction	Positional R
0	0				
1	2			1, 5	
2	2			[1, 3], 7	[1, 3] > 7
3	3(2)	1(0)		([1, 3], [3, 5])	
4	4		1		
5	3(2)	1(0)		([3, 5], 7)	[3, 5] < 7
6	1	1		[5, 8]	
7	2			1, 1	
8	0		1		
9	1	1		[1, 4]	

0123456789

FIGURE 1.20 Handprinted standard numerals.

point direction at the under part of the character image is [3, 5], so these numerals can be distinguished. Second, "5" also has a branch point, but it is unstable, since it comes at the top part of the character image and is often angled. For this reason "5" is included in the group that has two end points and variation of "3." Fortunately, since the direction at the end point in the upper part of the character image is direction 7, "5" can be differentiated from the other numerals in the group. These variations are given in parentheses in Table 1.5. Therefore we can see that the direction at an end point is a feature that can be used effectively to distinguish handprinted numerals.

However, we cannot be confident that the discrimination is perfect at the present condition of the shape variations, since many numerals have the same feature of two end points. Therefore we have to examine variational situations. Actually there is an overlap of directional features at the end points of "2" and "5." That is, $[1, 3] \cap [3, 5] \neq \phi$. The common feature is the direction 3. Therefore another feature is needed to discriminate between "2" and "5" if we consider the variation of "5." Fortunately, it is easy to find a new feature that is simple and stable. This is positional relation among end points which differs qualitatively from the geometrical features used.

So far we did not use any relations between features in terms of their positions, that is, the positions of the features were neglected. In a sense, they were good features for shift variations, and so we wanted to avoid introducing positional attributes of features that are not flexible, as already seen in cross-correlation and logical matching. However, in character recognition we need to introduce a coordinate system in general. Actually it has been already introduced in using directional features. If we have no coordinate system, the direction cannot be defined quantitatively. A positive point of the directional features lies in their invariance to parallel displace-

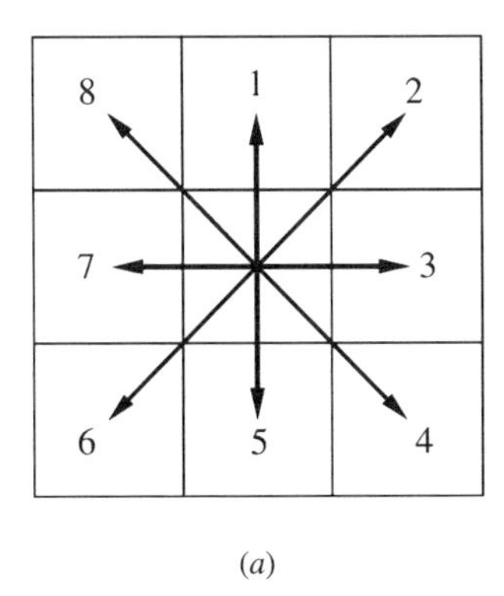

(a)

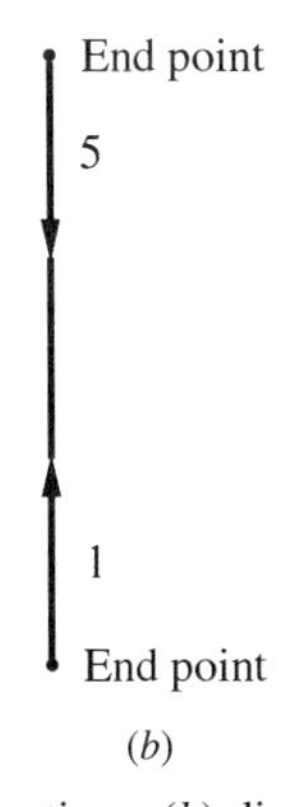

(b)

FIGURE 1.21 Quantization of direction: (*a*) Eight quantized directions; (*b*) direction at an end point taken from an end point to the interior of the line.

ment and scale variation. On the other hand, the number of end points is invariant for affine transformations and even topological transformations. Therefore this is a very stable feature. However, we still have a basic problem here as will be discussed at the end of this section. Further discussions on both theoretical and practical points will also be given later.

A typical example of the need for positional relationships is the discrimination between "6" and "9." Recall the case of "2" and "5," where the end point having the directional feature [1, 3] is located above the end point of 7 (direction) in "2," and this positional relation is just reversed in "5." Therefore they are easily discriminated. Let the relation that A is above B be denoted as $A > B$. Specifically, by introducing an appropriate coordinate system, end points are projected to y-axis, and the scalar value on the y-axis is taken as the feature.

At this point the reader might be confident of having a character recognition system for the handprinted numerals that is quite robust against variations. However, there is a pitfall, for variations often occur in loosely constrained handprinted characters. Some examples are shown in Fig. 1.22. These pairs of characters cannot be discriminated even if the positional relation introduced above is used. For this reason we need to add some features so that these examples can be differentiated. Generally, such an approach to improve a system is not a good idea, since it is a symptomatic treatment. Here we basically need to think about what the good features are.

First, we need to examine the features that have been used. Obviously they are local, except for the positional relation. The compared feature in the positional relation is local as well. As a result we have attempted to make a system using only fea-

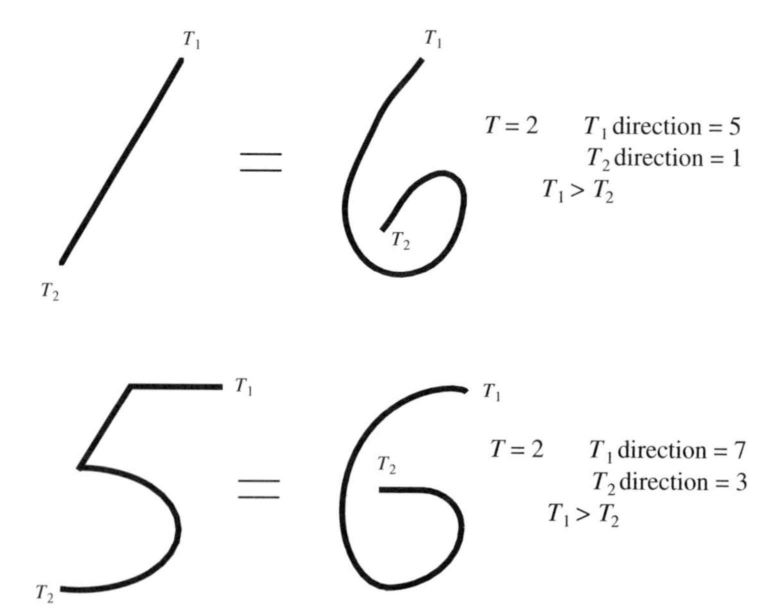

FIGURE 1.22 Examples that cannot be differentiated by end point directions nor by an end point's positional relation.

tures that can be extracted easily, but the real situation is difficult and our attempt failed for the variations shown in Fig. 1.22. Therefore eventually we need to look at global features that relate to a given character as a whole. The difficulty lies in finding an effective description of a whole character image that is as simple as possible.

Note that this method has another significant problem beside the pitfalls mentioned above. That is, we assumed that the lines are pure in a mathematical sense. This is a very general assumption. Real character images have strokes of considerable width. So the detection of end points is not as easy as may appear for idealized mathematical lines. To get ideal lines whose width is one pixel, a so-called thinning process is usually performed, and its basic idea will be discussed in this chapter.

1.2.6 Normalization

As mentioned in the discussion on cross-correlation, an input character image must coincide with a corresponding template in terms of position, size, slant, and so on. For a single font character set, position is the most important factor in the cross-correlation. In this section the normalization of an input character is considered. Since a detailed description will be given later, the objective here is to give an example of the normalization process.

For simplicity, a one-dimensional wave, $f(x)$, is assumed. Input waves are disturbed by many kinds of transformations as in the sense of statistics and geometry, in general. However, among the transformations, some simple ones are taken as follows:

1. Variation in amplitude $T_0(\alpha)f(x) = \alpha f(x)$, $\alpha > 0$.
2. Positional displacement $T_1(\beta)f(x) = f(x + \beta)$.
3. Uniform contraction and expansion $T_2(\lambda)f(x) = f(x/\lambda)$, $\lambda > 0$.

The objective of the normalization is to restore the input waves so that any such transformations have no effect on the cross-correlation matching. Invariance to transformations is achieved by the formula of cross-correlation matching defined in (1.1). Let us consider the conditions given above as they are applied to a more primitive form of cross-correlation, namely the inner product,

$$S^i(f) = \int_R f(x)g_i(x)dx, \tag{1.17}$$

and a two-dimensional image,

$$S^i(f) = \int_R f(x, y)g_i(x, y)dxdy, \tag{1.18}$$

where $f(x)$ and $f(x, y)$ are input images and $g_i(x)$ and $g_i(x, y)$ are ith templates. First we need to decide criteria for the normalization of a given variational or transformed pattern. The moments are usually taken for the criteria. The $(i+j)$th-order moment m_{ij} of an image $f(x, y)$ is given as

$$m_{ij} = \int_R x^i y^j f(x, y)dxdy. \tag{1.19}$$

For one-dimensional waves, j is set to zero.

For normalization of amplitude of a wave, it is natural to take the 0th order moment as the criterion. That is,

$$\int_R f(x)dx = \text{const.} \tag{1.20}$$

This means that the area of the $f(x)$ is taken to be constant.

A normalization of the parallel displacement, the first-order moment, is taken as the criterion as

$$\int_R xf(x)dx = 0. \tag{1.21}$$

This means that the average of the $f(x)$ is taken to be zero. For uniform contraction or expansion, the second-order moment is taken as the criterion as

$$\int_R x^2 f(x)dx = \text{const.} \tag{1.22}$$

This means that the dispersion of the $f(x)$ is taken to be constant.

All of the criteria mentioned above are intuitive and are also reasonable from a theoretical point of view, as will be explained later. However, from the practical point of view, the amplitude or gray-level normalization is usually not done explicitly. As mentioned in Section 1.2.3 on logical matching, when, as is usual, an input image is binarized, the input image can be regarded as normalized implicitly in its gray level. Therefore binarization is important in practice, as will be described later in detail.

We have already seen that for positional normalization, several kinds of criteria can be conceived. For the uniform contraction or expansion, a minimum rectangle is constructed such that the given image is included within the rectangle. It is better in practice to bring the rectangle to some constant shape and size according to a given character. We need to consider the type, since there are some special characters or symbols that will extend vertically or horizontally beyond others such as "|" and "—."

This brings us to the problem of normalization where the criteria are specified, namely to the process of normalization. We know that uniform contraction or expansion is a time-consuming process if it is done as described mathematically. Elegant formulas often do not yield fast algorithms.

Therefore let us consider some basic criteria. The purpose of normalization is to transform an input image so that it can obtain the best match from the template. The transformations are limited to those listed earlier. To be more precise, a position displacement is taken. As mentioned in Section 1.2.3 on logical matching, the character image can be assumed to be shifted directly on input. For simplicity let us take the input to be a wave, a one-dimensional pattern. The wave is represented discretely as $f(x_i)$, where x_i is an integer, $1 \leq i \leq N$. In the transformation shift, $T_1(\beta)f(x) \rightarrow f(x-\beta)$, where β is also an integer. Then $f(x_i)$ can be represented as a vector $\mathbf{f} = (f_1, f_2, \ldots, f_N)^t$, and the transformation $T_1(\beta)$ is represented as a matrix P^n, taking $\beta \rightarrow n$:

$$T_1(\beta)f(x_i) \rightarrow P^n\mathbf{f}. \tag{1.23}$$

In this equation P is a circular matrix of $N \times N$, and it has the following form:

$$\begin{pmatrix} 0 & 1 & & & \mathbf{0} \\ & 0 & 1 & & \\ & \cdots & \cdots & & \\ \mathbf{0} & & \cdots & \cdots & \\ & & & 0 & 1 \\ 1 & & & & 0 \end{pmatrix}$$

For example,

$$\begin{pmatrix} 0 & 1 & 0 & 0 \\ 0 & 0 & 1 & 0 \\ 0 & 0 & 0 & 1 \\ 1 & 0 & 0 & 0 \end{pmatrix} \begin{pmatrix} 0 \\ 1 \\ 1 \\ 0 \end{pmatrix} = \begin{pmatrix} 1 \\ 1 \\ 0 \\ 0 \end{pmatrix}.$$

Thus the equation shows that the vector $(0, 1, 1, 0)^t$ was shifted by one mesh to the top, which results in the vector $(1, 1, 0, 0)^t$. Furthermore we assume that around the two ends of the input vector, there are sufficiently many zeros to render the circulation nature of the matrix P harmless.

Our purpose is to find n such that the following inner product attains the maximum value,

$$\Phi(\mathbf{f}, \mathbf{g}) = \max_n (P^n\mathbf{f}, \mathbf{g}), \tag{1.24}$$

where Φ is a functional whose arguments are vectors $\mathbf{f}$ and $\mathbf{g}$, and $\mathbf{g}$ is a template (the index i is omitted). Equation (1.24) is generalized in the following way:

$$\Phi(\mathbf{f}, \mathbf{g}; T_i) = \max_a (T_i(\alpha)\mathbf{f}, \mathbf{g}). \tag{1.25}$$

The transformation T_i is performed as a function of α such that $(T_i(\alpha)\mathbf{f}, \mathbf{g})$ attains the maximum value. The maximization process is done to all of the templates. The normalization and matching here are integrated, and this is more effective in practice than to execute the normalization, as described mathematically, separate from the matching.

An example of integrated normalization with matching is given in Fig. 1.23. The logical matching is very weak here in its positional displacement. The integration mechanism is simple and ingenious. A binary pattern is fed to a long shift register which is folded to make a two-dimensional register array as shown in the figure. The input image is fed from the top left and down the leftmost column register; then the output of the column register is fed to the top of its neighboring column register, and the process repeats. As an input image is loaded to the array register, it is displaced from its normal position to the top and to the left as shown. If this shift in the array register continues, the input image reaches its normal position eventually, although the shift is a simple movement from left to right on the original long register. At its normal position the logical matching shown in Fig. 1.23 (*b*) obtains the best match, so the system can recognize the input character image as "4." Notice that there is no control of the shift movement in the normalization; the exact normalization is implicitly detected by logical matching. This method was used first in an OCR, IBM1418 [11].

Blurring Transformation Among the three transformations mentioned above, except for the amplitude transformation, all are geometrical transformations. The amplitude transformation is very primitive, and there is neither change in the given shape nor its position. In an abstracted figure the amplitude transformation has no meaning because it is linear, consisting of a continuous series of points in an abstracted line. In this sense an amplitude transformation is a physical phenomenon related directly to the process of observing an object. On the other hand, there is an important related transformation, called *the blurring transformation.* This results from strict consideration of the observation process. An object is observed through the system as if by a human eye, and the object's image is constructed as if by a human eye. But the constructed image is blurred in general. We assume that the blurring transformation is linear, the same as the other transformations defined. Mathematically the general form of the blurring transformation is expressed as

$$\Phi_0[g(x'); x, \sigma] = \int_{-\infty}^{\infty} \phi(g(x'); x, x', \sigma)dx', \tag{1.26}$$

where Φ_0 is a functional that takes a function as its argument and converts it to a real value and σ is a parameter of the blurring transformation. The right side of the above definition is due to the assumption of linearity. By the term "blurring" it can be inferred that ϕ is a Gauss function, which is well-known as a blurring function. This can be derived from the four conditions that the observation system must satisfy. Three conditions out of the four given below are directly related to the transformations mentioned here.

These four conditions can be described as follows:

1. When the strength of a given observed pattern becomes A times, then the observed image becomes A times,

$$\Phi_0[Ag(x'); x, \sigma] = A\Phi_0[g(x'); x, \sigma].$$

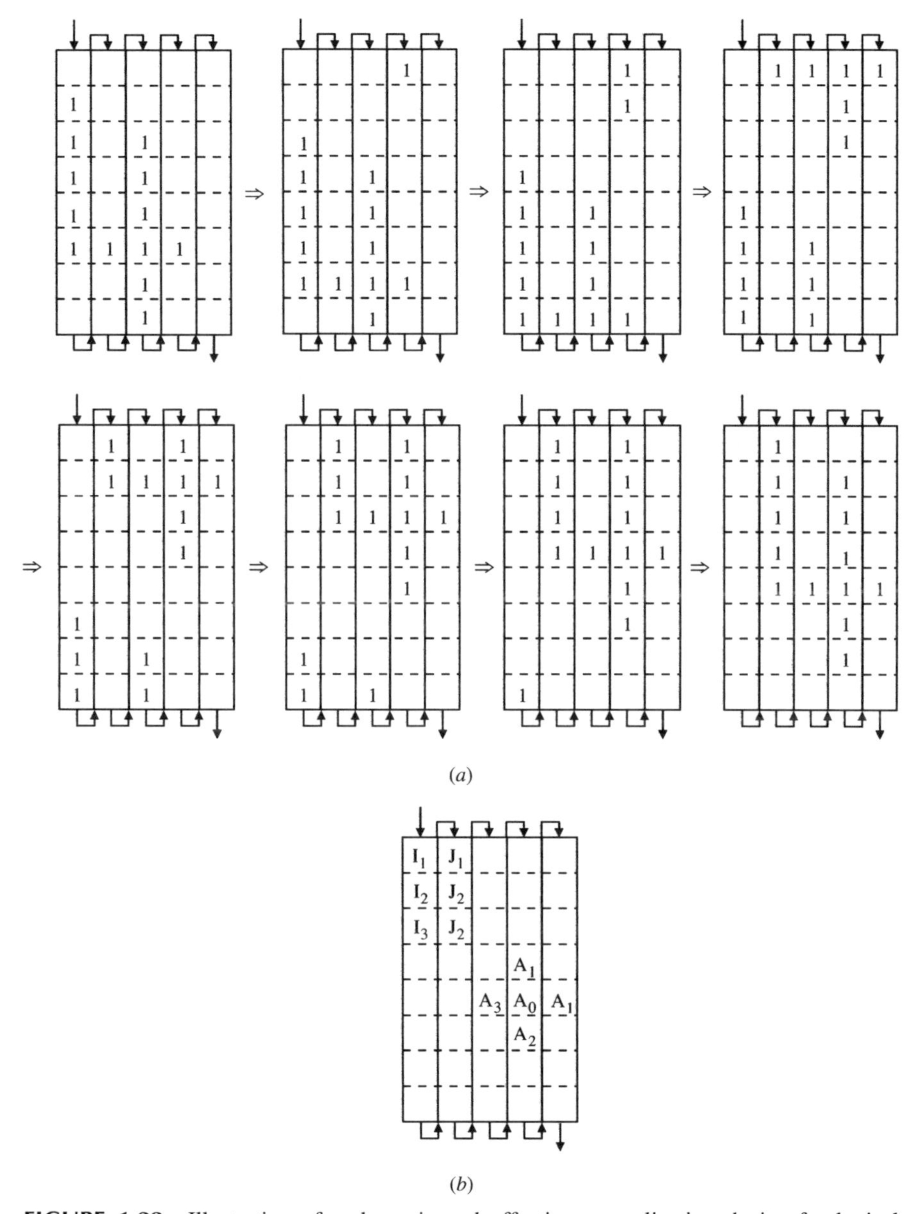

FIGURE 1.23 Illustration of a dynamic and effective normalization device for logical matching: (*a*) One cycle of pattern movement on the shift register. The movement of the pattern is performed from the top left to the bottom right. That is, the top left pattern shifts to the right by one pixel and to the bottom by one pixel. Here note that each end of the column is connected to each top of its right neighboring column: (*b*) Logical matching mark for "4."

2. When the position of the observed pattern is shifted by a, then its image is also shifted by a in the same direction,

$$\Phi_0[g(x' - a); x, \sigma] = \Phi_0[g(x'); x - a, \sigma].$$

3. When the size of the observed pattern is contracted or expanded by λ, then its image is also changed so by λ,

$$\Phi_0[g(x'/\lambda); x, \sigma] = \Phi_0[g(x'); x/\lambda, \sigma'].$$

4. When the image constructed by σ_1 is again observed by an observation system whose parameter is σ_2, then the resultant image coincides with the image of the one observation system that observed the original image with an appropriately chosen σ_3,

$$\Phi_0\{\Phi_0[g(x''); x', \sigma_1]\, x, \sigma_2\} = \Phi_0[g(x'); x, \sigma_3].$$

From these four conditions it can be proved that the form of Φ_0 must be

$$\Phi_0[g(x'); x, \sigma] = \int_{-\infty}^{\infty} g(x')\phi_m\left(\frac{x - x'}{\sigma}\right)\frac{dx'}{\sigma}, \tag{1.27}$$

where

$$\phi_m(u) = \frac{1}{2\pi}\int_{-\infty}^{\infty} e^{\zeta 2m + i\zeta u}d\zeta, \qquad m = 1, 2, \ldots. \tag{1.28}$$

The relations between σ' and (σ, λ) in condition 3 and between σ_3 and (σ_1, σ_2) in the condition 4 are given when the form of Φ_0 is determined precisely as above:

$$\sigma' = \frac{\sigma}{\lambda}, \tag{1.29}$$

$$\sigma_3^{2m} = \sigma_1^{2m} + \sigma_2^{2m}. \tag{1.30}$$

The derivation of the proof is intricate, and thus it is given in Appendix A.

Let us consider, for example, the case $m = 1$. Putting $m = 1$ in Eq. (1.30), the following relations are derived:

$$\begin{aligned}\phi_1(u) &= \frac{1}{2\pi}\int_{-\infty}^{\infty} e^{-\zeta^2 + i\zeta u}d\zeta \\ &= \frac{1}{2\pi}\int_{-\infty}^{\infty} e^{-[\zeta - i(u/2)]^2 - u^2/4}d\left(\zeta - i\frac{u}{2}\right) \\ &= \frac{1}{2\sqrt{\pi}}e^{-u^2/4}.\end{aligned}$$

Accordingly Φ_0 is given as

$$\Phi_0[g(x'); x, \sigma] = \frac{1}{2\sqrt{\pi}\sigma} \int_{-\infty}^{\infty} g(x') e^{-(x-x')^2/4\sigma^2} dx'. \tag{1.31}$$

On the other hand, when $A > 0$, $\lambda > 0$, $\sigma > 0$, for any given $f(x)$ being $f(x) > 0$, it is proved that if

$$\Phi[f(x') : x, A, a, \lambda, \sigma] > 0 \tag{1.32}$$

holds, then m must be equal to 1. Although the mathematical description is omitted here, the result is very simple and as expected. Actually a pinhole lens has the same characteristics. We can recognize the implicit meaning of the observation system in terms of linear transformations.

This theoretical consideration was given by Iijima [12]. Historically speaking, the research on the normalization of blurring was done following this basic idea. However, blurring transformation is now regarded as an essential process in the broad sense of recognition rather than normalization through much experimental evidence. Actually the normalization of blurring is not necessary, and it is known that the introduction of blurring often gives better recognition results. It should be emphasized that this pertains to recognition systems based on so-called analog pattern matching. In recognition systems based on description, the normalization of blurring is sometimes done in a very different framework. An abstracted figure is approximately obtained by a preprocess called thinning, which will be described next.

1.2.7 Thinning

Thinning is a first abstraction, and it is effective if the character image consists of abstracted lines. Further the abstractions considered are the extraction of singular points and the detection of any abrupt change in the curvature of a line. Some examples have already been discussed before. However, the problem is how to obtain the abstracted lines. The observed character images always have a thickness that is not uniform. Usually the images are mapped onto a rectangular grid so that each observed pattern is represented by a matrix, say 20×30, in which each entry has 16 or 32 gray levels. Therefore, as mentioned in Section 1.2.3 on logical matching, they are binarized, which is equivalent to the normalization of amplitude or darkness. At this stage the binarized image has no uniform thickness of width 1. Obviously the preprocess of thinning is necessary to give a binarized image a uniform width of 1 pixel.

This processing may seem simple because the input image is eroded from both sides one by one until it has the width of 1 pixel. However, we will soon see, this simple idea does not work that way. First, the original width can easily be an even number, in which case the operation brings the image to lines having 2 pixel width. Then one needs to decide which should be deleted. Either side may not be exactly the center line of the given character image. Also, as mentioned earlier, the detection of end points of a thick character image is not easy. Therefore the first stage of this simple

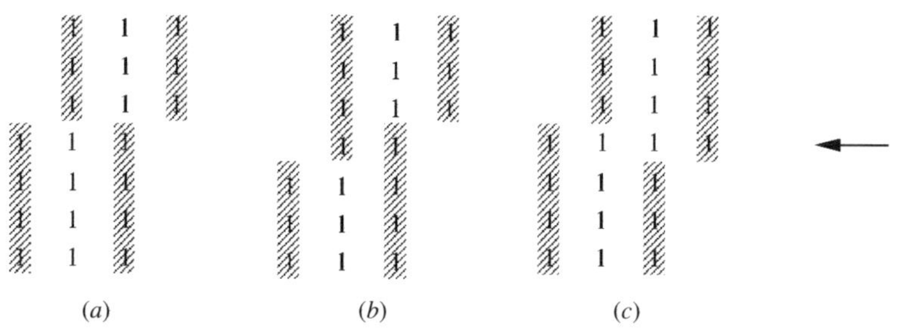

FIGURE 1.24 Illustration of difficulty of a simple eroding thinning process: (*a*) ideal case; (*b*) danger of broken line; (*c*) nonuniform thickness.

operation is to erode some neighboring parts of the end points, and this results in a shortening of the abstracted lines.

There are some other basic problems; one of which is shown in Fig. 1.24. Three different lines whose widths are equally three are shown in the figure. They are all shifted to the left by one pixel on the way down. The ideal case is shown in (*a*), which has no problem. Since there are always noises around strokes or lines, nonuniform widths occur often. In the case shown in (*b*), the simple operation of eroding results in the break of the line between the moved part and remaining part. In the case of (*c*), the width at the fault becomes two pixels. These examples suggest to us an important

(*a*) Diamond

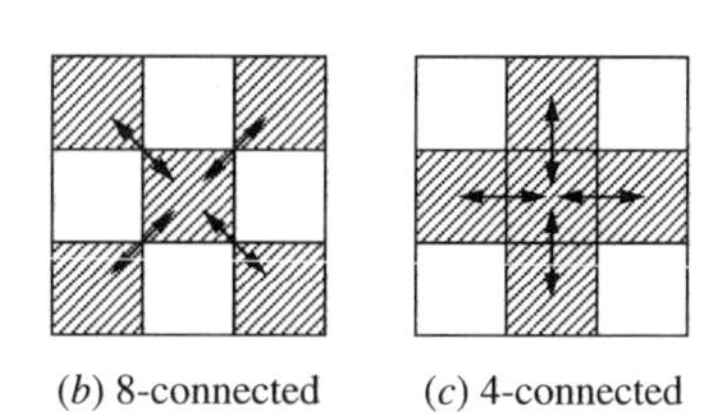

(*b*) 8-connected (*c*) 4-connected

FIGURE 1.25 Interconnections on a grid plane: (*a*) Superficial shape; (*b*) definition of 8-connected, (*c*) definition of 4-connected. From reference [13], © 1997, Ohm.

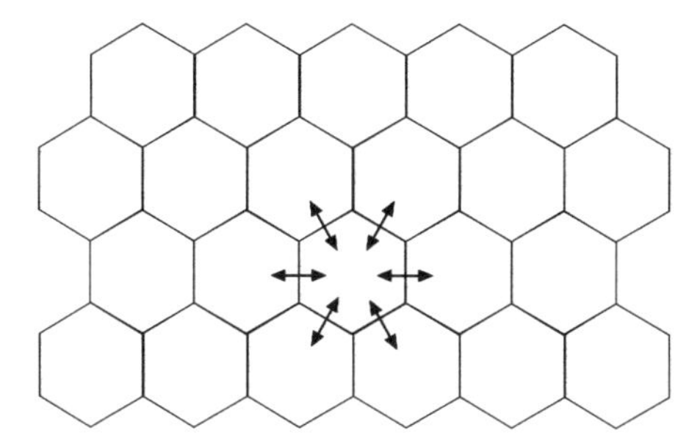

FIGURE 1.26 Apparent paradox solved on a hexagonal plane.

property of continuity in binary patterns. Before considering the erosion of a binary pattern, we have to define what is continuity of a shape on a grid plane.

In Fig. 1.25 (*a*) a diamond is drawn on a grid plane. Human beings see the boundary of the diamond in the abstracted lines. However, we can find the strange figure shown in (*a*) when looking at the diamond in terms of its black-and-white figure-ground connections. The condition for the continuity of the black ground should be equal to that of the white ground. If so, the inner white region is connected to the external white region, namely the black lines are crossed by the white ground everywhere. This is a contradiction in principle. As shown in Fig. 1.26, this apparent paradox can be solved using a hexagonal plane. In fact an observation system based on a hexagonal grid plane has long been considered, but it is not used in practice. For the square grid plane, the paradox is avoided by setting the premise that in a black line diagonal continuity is allowed, called 8-connectivity. This is the case shown in Fig. 1.24 (*b*). On the other hand, for the white ground, diagonal connectivity is not allowed, so we have what is called 4-connectivity. This is shown in Fig. 1.24 (*c*).

So far, we have discussed some intrinsic problems in a digital pattern on a grid plane. There is another basic problem in the simple eroding process. To understand it better, we will look at some examples. Figure 1.27 gives an example of the thinning

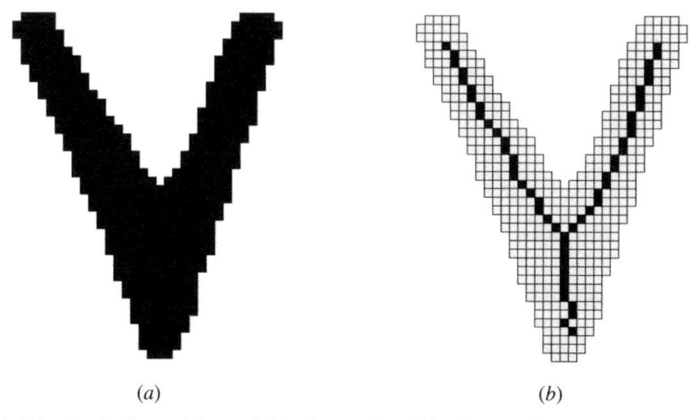

FIGURE 1.27 Intrinsic problem of simple eroding thinning method; (*a*) original image, (*b*) thinned image using the Hilditch thinning method.

process based on a simple eroding operation from both sides that is applied to an observed image "V." Clearly the bottom part of the image is changed to a short line so that "V" is deformed to "Y" by the preprocessing. This is due to the local operation of preprocessing. This is a fundamental problem in image recognition. As we have seen, at first sight the eroding process seems to be a simple problem, but now we recognize that it is a very hard one. This simple preposing method is classic, and presently some ingenious ways are being considered in which a global view may be taken into account. These developments will be discussed in later chapters.

1.3 BIBLIOGRAPHICAL REMARKS

There have been published many special issues on document analysis and character recognition, in particular in 1992 and 1993, when rapid interest in OCR systems' development was triggered by the U.S. Postal Office.

Special Issues

R. Katuri and K. O'Gorman, guest eds., "Special issue: Document image analysis techniques," *Machine Vision and Application,* vol. 5, no. 3, pp. 141–248, 1992.

R. Katuri and L. O'Gorman, guest eds., "Special issue: Document image analysis techniques," *Machine Vision and Application,* vol. 6, nos. 2–3, pp. 67–180, 1993.

L. O'Gorman and K. Katuri, guest eds., "Special issue on document image analysis systems," *Computer,* vol. 25, no. 7, July 1992.

R. Plamondon, guest ed., "Special issue: Handwriting processing and recognition," *Pattern Recognition,* vol. 26, no. 3, pp. 379–460, Mar. 1993.

T. Pavlidis and S. Mori, guest eds., "Special issue on optical character recognition." *Proc. IEEE,* vol. 80, no. 7, pp. 1,027–1,215, July 1992.

H. Tominaga, guest ed., "Special issue on postal processing and character recognition," *Pattern Recognition Letters,* vol. 14, no. 4, pp. 257–354, Apr. 1993.

K. Yamamoto, guest ed., "Special issue on document analysis and recognition," IEICE Trans. on Information and Systems, vol. E77-D, no. 7, July 1994.

S. Mori, guest ed., "Special issue on character recognition and document understanding," IEICE Trans. on Information and Systems, vol. E79-D, May 1996.

Books, Reviews and Tutorial Papers

L. D. Harmon, "Automatic recognition of print and script," *Proc. of the IEEE,* pp. 1,165–1,176, October, 1972. This describes many approaches and techniques of character recognition, which includes print, hand print and cursive script, and Roman alphanumerals, Chinese characters/Kanji and even, Hirakana, and Katakana.

J. R. Ullman, *Pattern Recognition Techniques,* London: Butterworth, 1973. This can be regarded as a classical book on character recognition.

J. R. Ullman, "Advances in Character Recognition," in K. S. Fu ed., *Application of Pattern Recognition,* CRC Press, Boca Raton, Fla., pp. 197–236, 1982. This is a further development of the book cited above, which also includes a comprehensive bibliography including many U.S. and U.K. patents.

C. Y. Suen, M. Berthod, and S. Mori, "Automatic recognition of handprinted character—The state of the art," *Proc. IEEE,* vol. 68, pp. 469–487, Apr. 1980. This is a survey of feature extraction techniques development for the recognition of isolated handprinted characters, which reflects OCR research targets at that time.

G. Nagy, "Optical character recognition: Theory and Practice," in P. R. Krishnaiah and L. N. Kanal, eds., *Handbook of Statistics,* vol. 2, pp. 621–649, 1982. This is a survey of statistical feature analysis techniques for OCR.

J. Schürmann, "Reading machines," in *Proc. 6th IJCPR,* pp. 1,031–1,044, 1982.

S. Mori, K. Yamamoto, and M. Yasuda, "Research on machine recognition of handprinted characters," *IEEE Trans. PAMI,* vol. PAMI-6, no. 4, pp. 386–405, July 1984. This is a further survey on feature extraction techniques after the survey Suen, Berthod, and Mori cited above in 1980. This also includes handprinted Chinese/Kanji Character recognition research and development, in which feature matching is emphasized.

R. H. Davis and J. Lyall, "Recognition of handwritten characters—A Review," *Image and Vision Computing,* vol. 4, pp. 208–218, 1986.

J. Mantas, "An overview of character recognition methodologies," *Pattern Recognition,* vol. 19, pp. 425–430, 1986.

V. K. Govindan and A. P. Shivaprasad, "Character recognition—A Review," *Pattern Recognition,* vol. 23, pp. 671–683, 1990.

S. N. Srihari and J. J. Hull, "Character Recognition," *Encyclopedia of Artificial Intelligence, 2nd edition.* This is a concise tutorial, which is divided into two parts; isolated character recognition and word recognition which includes character recognition in context.

S. Mori, C. Y. Suen, and K. Yamamoto, "Historical Review of OCR research and development," *Proc. of the IEEE,* vol. 80, no. 7, pp. 1,029–1,058, July 1992. This survey emphasizes the historical development of OCR in feature extraction techniques and matching techniques, in particular. This includes also the development of commercial OCR systems.

J. Schürmann, *Pattern Classification—A Unified View of Statistical and Neural Approaches,* New York: John Wiley & Sons, 1996. This takes a general and theoretical approach to pattern recognition. However, the practical examples are taken from the field of character recognition.

Collection of Papers

G. L. Fisher et al., eds., *Optical Character Recognition.* Spartan Books, 1962. This is the oldest book on optical character recognition in which several prototypes of OCR systems in the early stage are presented as well as some basic researches.

S. Impedoro and J. C. Simon, eds., *From Pixels to Features III—Frontiers in Handwriting Recognition,* Elsevier Science Publishers B.V., 1992. This is the collection of improved papers which was presented on the Second International Workshop on Frontiers in Handwriting Recognition. The book contains 42 papers covering all the most advanced topics in handwriting recognition.

S. N. Srihari, ed., "Computer Text Recognition and Error Correction," Computer Society Press, 1984. This collection of many papers (30) published in various technical journals and conferences appeared in the 1960s through 1984. Most of the papers are related to knowledge application to the error correction.

A. C. Downton and S. Impedovo, eds., "Progress in Handwriting Recognition," World Scientific Publishing. This is a Proceedings of the Fifth International Workshop on Frontiers in Handwriting Recognition held from 2nd to 5th of September 1996 at Colchester in England.

C. H. Chen, L. F. Pau, and P. S. P. Wang, eds., "Handbook of Pattern Recognition and Computer Vision," 1993, World Scientific Publishing. This handbook consists of 31 chapters using 833 pages. Except for the first and the second chapters, that is, "Image Processing Methods for Document Image Analysis" by T. M. Ha and H. Bunke, and "Pattern Approximation" by U. Kressel and J. Schürman, it covers almost all the fields of document analysis and recognition for graduate school students' and researchers' levels. The two chapters give a good and concise introduction.

R. Katsuri and K. Tombre, eds., "Graphics Recognition—Methods and Applications," 1995, Springer. This is a Proceedings of the First International Workshop on Graphics Recognition held at the Penn State University. The papers presented are very related to OCR technology, which is true in the low-level processing, vectorization, and segmentation, in particular.

OCR System

We list here some representative papers on OCR systems, and postal card reading in particular. The reader can get a good idea of the application of OCR systems.

H. Genchi, K. I. Mori, S. Watanabe, and K. Katsuragi, "Recognition of handwritten numerical characters for automatic letter sorting," *Proc. IEEE,* vol. 56, pp. 1,292–1,301, Aug. 1968.

J. Schürmann, "A multifont word recognition system for postal address reading," *IEEE Trans. on Comp.,* vol. c-27, no. 8, pp. 721–732, Aug. 1978.

R. B. Hennis, "The IBM 1975 optical page reader, Part I: System design," *The IBM J. Res. Develop.,* vol. 12, pp. 346–353, Sept. 1968.

S. Kahan, T. Pavlidis, and H. S. Baired, "On the recognition of printed characters of any font and size," *IEEE Trans. on Pattern Analysis and Machine Intelligence,* vol. PAMI-9, no. 2, pp. 294–288, 1987. This is a typical printed character recognition system in which a thinning method based on RAG is used. The thinning method is explained in detail in Chapter 5.

E. Cohen, J. J. Hull, and S. N. Srihari, "Understanding handwritten text in a structured environment: Determining ZIP codes from Address," *International Journal of Pattern Recognition and Artificial Intelligence,* vol. 5, no. 1 & 2, pp. 221–264, 1991.

F. Kimura and M. Sridhar, "Handwritten numerical recognition based on multiple algorithms," *Pattern Recognition,* vol. 24, no. 10, pp. 969–983, 1991. A complete description of the structural classifier is presented.

R. G. Gasey and D. R. Ferguson, "Intelligent Forms Processing," *IBM Systems Journal,* vol. 29, no. 3, pp. 435–450. This is a typical paper for document image analysis techniques in which OCR reading is used as one component. See also Casey's paper: "Document image analysis technique," Machine Vision and Applications, vol. 5, pp. 143–155, 1992.

C. Y. Suen, L. Lam, D. Guillavic, N. W. Strathy, M. Cheriet, J. N. Said, and R. Fan, "Bank check processing system," *International Journal of Imaging Systems and Technology,* vol.

7, pp. 392–403, 1966, Wiley. A reader can realize how the systematic image processings are necessary to segment written characters.

J. C. Simon and O. Baret, "Cursive words recognition," in *From Pixels to Features III: Frontiers in Handwriting Recognition,* S. Impedovo and J. C. Simon, eds., Elsevier Science Publishers B. V. Amsterdam, pp. 241–260, 1992. This describes how to tackle the difficult task of cursive words recognition from the engineering point of view. Simon's R&D is going well.

BIBLIOGRAPHY

[1] S. V. Rice, J. Kanai, and T. A. Nartker, "A report on the accuracy of OCR devices," Technical Repert, Information Science Research Institute of Nevada, Las Vegas, March 4, 1992.

[2] C. Y. Suen and S. Mori, "Standardization and automatic recognition of handprinted characters," C. Y. Suen and R. De Mori, eds., *Computer Analysis and Perception: Visual Signals,* vol. 1. Boca Raton, FL: CRC Press, 1981.

[3] Y. Nakajima and S. Mori, "A model-based classification in a scheme of recognition filter," *Proc. 2nd ICDAR,* Tsukuba Science City, pp. 68–71, October 20–22, 1993.

[4] J. Robinson, "Developments in character recognition machines at Rabinow engineering company," in G. L. Fisher et al., eds., *Optical Character Recognition,* Washington DC: Spartan Books, pp. 27–50, 1962.

[5] T. Tsutsumida, T. Matsui, T. Noumi, and T. Wakahara, "Results of IPTP character recognition competitions and studies on multi-expert system for handprinted numeral recognition," *IEICE Trans. Inf. Syst.,* vol. E11-D, no. 1, July 1996.

[6] T. Tsutsumida, T. Matsui, T. Noumi, and T. Wakahara, "Results of IPTP character recognition competitions and studies on multi-expert system for handprinted numeral recognition," *IEICE Trans. Inf. Syst.,* vol. E77-D, no. 7, pp. 801–809, July 1994.

[7] T. A. Nartker, J. Kanai, and S. V. Rice, "A preliminary report on OCR problems in less document conversion," Technical Report, TR-92-04, Information Science Research Institute of Nevada, Las Vegas, April 1992.

[8] S. Mori, C. Y. Suen, and K. Yamamoto, "Historical review of OCR research and development," *Proc. IEEE,* vol. 80, no. 7, pp. 1029–1058, 1992.

[9] S. Mori, "The techniques of character recognition," *Electronics,* Tokyo: Ohm, 1983.

[10] S. Naito, K. Kmori, and E. Yodogawa, "Stroke density feature for handprinted Chinese characters recognition," *IECE Trans.,* vol. J64-D, no. 8, pp. 757–764, 1981.

[11] P. H. Glauberman, "Character recognition for business machines," *Electronics,* pp. 132–136, Feb. 1956.

[12] T. Iijima, "Basic theory on normalization of pattern," *Bull. Electrotechn. Lab.,* vol. 26, no. 5, pp. 368–388, 1962.

[13] S. Mori, *Fundamentals of Image Recognition Techniques,* Tokyo: Ohm, 1986.

CHAPTER TWO

Blurring and Sampling

In this chapter we begin our consideration of the sampling problem with a simple ideal case where binary lines lie on a grid plane and obtain a reasonable result. We want to apply Shannon's sampling theorem to such a problem, but we first must deal with a difficult fact that the cross sections of these ideal binary lines are not band limited. Therefore we depend on pattern recognition and approximate the rectangular pulse by a Gaussian function; then we apply Shannon's sampling theorem. We compare the result with the simple sampling problem. Next, we consider sampling of a corner region. A corner feature is typically present in character recognition, and it is known for being difficult to detect. Turning to the rectangular pulse problem, we discuss the relation between the blurring process and the sampling process and give concrete examples of both. Experimental results are presented that show the effect of blurring. Finally we derive a formulation for sampling intervals and levels of sampled values, which is based on Shannon's information theory.

2.1 SAMPLING CONDITIONS BASED ON GEOMETRICAL CONSIDERATION

For the very simple case let us take an ideal line whose width is w and a uniform darkness d in its interior. The line's width is a parameter that changes depending on the input image, and its range is limited as $w_0 \leq w \leq w_M$. In particular, w_0 is decided by the specifications given for an OCR system. On the other hand, d is assumed to be constant regardless of the width of the line, so it is taken as 1.

These lines lie on a grid plane and whether pixel or grid square,[1] the value of the grid plane is 1 or 0 according to the area on the grid square occupied by the line on

[1] We introduce here a new term, "grid square," in place of pixel because the grid square gives a sense of area; the smallest quantized area composed by a pixel maintains the sense of a point. However, grid square, pixel, and point will be used interchangeably as appropriate.

it. In other words, when the area is greater or equal to a threshold value T, then the grid square value is 1; otherwise, it is 0. For simplicity and to emphasize the problem of interest, we don't consider analog values at each grid square.

Sampling can be regarded as the mapping of lines on R^2 (two-dimensional Euclid plane) to Z^2 (two-dimensional grid plane). By the mapping we request the preservation of the connectivity of lines. This is illustrated in the Fig. 2.1, where two connected lines are mapped onto the grid plane. The digital black lines must be connected on the grid plane at least in the sense of 8-connectedness. More specifically, we require that a line whose width is the minimum value w_0 must be mapped to the grid plane as ideally thinned, namely as width 1.

Now we will impose two conditions on the sampling.

Condition 1 Let M be the set of grid square cells representing a connected line of width w. The line may be placed anywhere on the grid plane. Then the grid square representation M is connected, and if its original width on R^2 is minimal, then it must be ideally thinned.

Condition 1 naturally preserves the continuity of a given line after sampling. Therefore we examine the most critical arrangement, which is shown in Fig. 2.2. This is where the line is arranged just vertically or horizontally in such a way that its center line lies just on the grid line. Let T be the $W \times w/2$, where W is the width of the grid square and w is the line width. If the thresholding is selected such that the area on the grid square must be greater or equal to T, then the width of the line becomes just 2 at this critical situation. However, if the line moves a little down or up, then the width of the line becomes 1. On the other hand, if the thresholding is chosen to be *area* > T instead of ≥ as above, then at the critical situation the line vanishes. However, such behavior at the critical condition can be neglected because the probability of occurrence is very small. Therefore we can say that condition 1 is satisfied if the threshold value is selected as $W \times w_0/2$, where w_0 is the smallest width of the given lines.

However, for a given w_0, we can choose T by using the given W, so the requirement for satisfying the conditions are only to set the threshold value regardless of W. In other words, no restriction is imposed on the sampling interval, W. Naturally, when w_0 is very small, such as less than 0.001 mm, there are imposed severe physical and practical limitations to detecting the brightness (irradiance). Therefore a second condition is introduced.

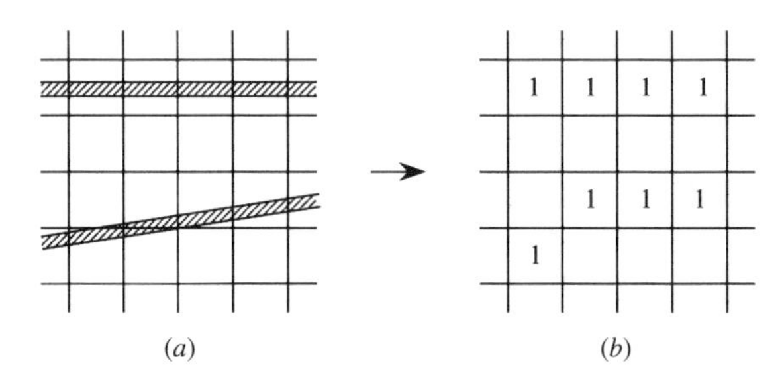

FIGURE 2.1 Mapping lines on R^2 to discrete lines on Z^2 where R^2 and Z^2 stand for a two-dimensional Euclidean plane and a two-dimensional Euclidean grid plane, respectively.

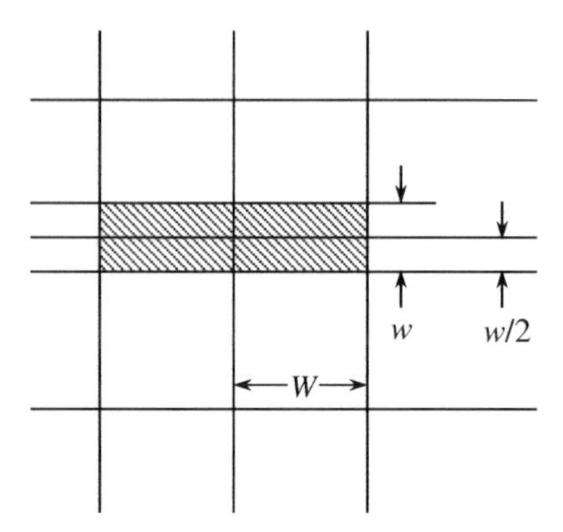

FIGURE 2.2 Critical arrangement of a line segment.

Condition 2 Let L be a connected line with n times the minimal width. Then the width of the grid square representation of L is n times the width of the grid square representation of the minimal line.

This condition is for representing the object on a sampled plane as exactly as possible. Figure 2.3 gives one such example in which this condition is not satisfied. The line l_1 is 4 times the width of line l_2, whose width is w_0, but the line l_1 has a segment whose width is 1 grid square. The other part of line l_1 has a 2 grid square width. Furthermore when l_1 and l_2 are given at the same time, the gap between l_1 and l_2 disappears. We notice that not only the black lines but also the white ground must be sampled correctly. Obviously the sampling rate is too small. If condition 2 is imposed, the sampling interval W must be taken to be equal to w_0 so that lines of width n have grid square width of n or $n + 1$ according to their arrangement on the grid plane. If a line width is close to nw_0, then it takes n width almost everywhere. On the contrary, if a line width is close to $(n + 1)w_0$, then it takes $(n + 1)$ width almost everywhere. If a line has the center value between nw_0 and $(n + 1)w_0$, it takes n or $n + 1$ width with equal probability. We notice that the threshold value T is set $w_0^2/2$ according to condition 1, that is, just half of the area of the unit grid square.

The important point to be noticed is rather that the line width cannot be uniform in its original sampling stage in general. On the other hand, if lines are separated by w_0 at least, then they are separated on the sampled plane too. Therefore conditions 1 and 2 seem to be reasonable, although they were derived from very primitive considerations. We notice that the strokes of a character image can be always approximated by a set of lines, the so-called polygonal approximation, as will be discussed fully later. Therefore the simple assumption of lines is very useful in practice. A more subtle case will be discussed later.

2.2 SAMPLING CONDITION BASED ON SHANNON'S THEOREM

Now let us compare this result with Shannon's sampling theorem [1]. For this purpose we need to consider how to apply the theorem to our case as mentioned before.

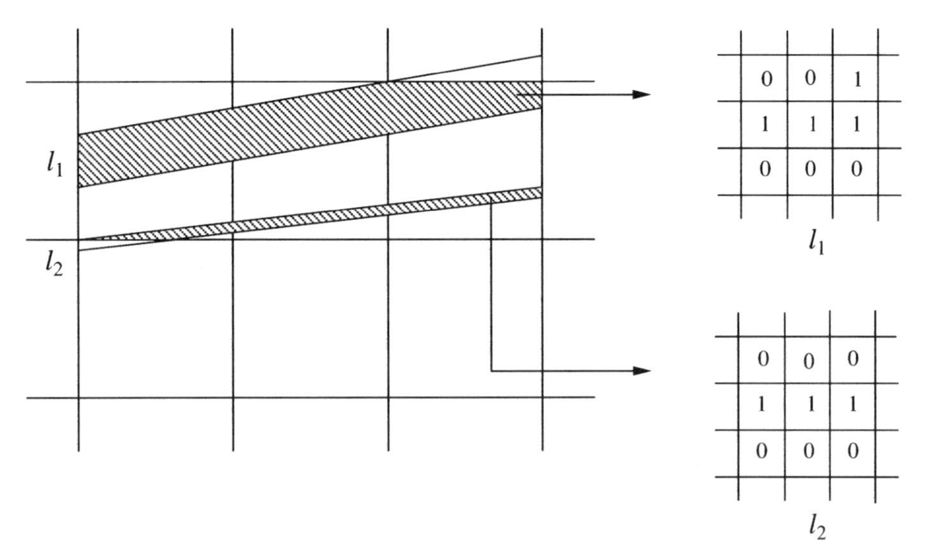

FIGURE 2.3 Case where condition 2 not satisfied.

Let us consider an image given on a two-dimensional plane. The image, $f(x, y)$, is assumed to be band limited and so the maximum frequencies are denoted as Z_u and Z_v. Then Shannon's theorem tells that the image, $f(x, y)$, can be completely reconstructed if the sampling intervals, ΔX and ΔY are taken as follows:

$$\Delta X \leq \frac{1}{2Z_u}, \tag{2.1}$$

$$\Delta Y \leq \frac{1}{2Z_v}, \tag{2.2}$$

where u and v denote the coordinates in the Fourier domain of $f(x, y)$.

In our case the image is an ideal binary pattern. For simplicity, let us consider a vertical line whose section cut parallel to the x-axis is given by a function of only x, denoted $f_p(x)$. However, the section is a rectangular pulse and its minimum width ω_0. Obviously $f_p(x)$ is not band limited. Therefore we cannot use Shannon's sampling theorem. What we can say is that we need to take ΔX and ΔY as small as possible. Of course this conclusion is counter to our goal of pattern recognition. Let us examine this difficulty from the standpoint of pattern recognition. The question is whether we need to reconstruct the rectangle pulse wave exactly. We are considering a pattern $f(x, y)$ on the two-dimensional plane. In other words, a shape on the two-dimensional plane (i.e., the shape of line section) does not matter, and it can be a triangle or a bell shape or any smooth function that has one extreme maximum point and narrow support. The rectangular pulse was introduced for convenience in our geometric and analytic application of Shannon's theorem. It does not matter what function is used so long as the above condition is satisfied. Also we use the Gaussian function for the shape of the section of a line for convenience in the analysis described below.

Therefore the following rectangle pulse whose section area is normalized to 1 is converted to Gaussian function in terms of similarity.

$$p(x) = \begin{cases} \dfrac{1}{2\alpha}, & |x| < \alpha, \\ 0, & |x| > \alpha. \end{cases} \tag{2.3}$$

The Gaussian function is assumed to take the following form:

$$G(x) = \frac{1}{\sqrt{2\pi}\sigma} \exp\left(\frac{-x^2}{2\sigma^2}\right). \tag{2.4}$$

The area of the function is normalized to 1. Both $p(x)$ and $G(x)$ are one-parameter functions, and so parameters are adjusted to maximize the similarity defined by (1.1) in Section 1.1.2. Specifically, the similarity is given as function of α/σ, and the best value of α/σ is found. According to the calculation, the best value of α/σ is given as $\alpha/\sigma = 1.4$, and the maximum similarity at the value is 0.9435. For a triangle shape, the best value of α/σ is 2.243, and the maximum similarity at the value is 0.9980. The triangle shape function is defined as

$$p(x) = \begin{cases} \dfrac{1}{\alpha}\left(1 - \dfrac{|x|}{\alpha}\right), & |x| < \alpha, \\ 0, & |x| > \alpha. \end{cases} \tag{2.5}$$

In this case we obtain a very good approximation of the triangle by a Gaussian function.

On the other hand, the Fourier transformation of the Gaussian function takes also the Gaussian form as

$$F(\zeta) = \exp\left(\frac{-\zeta^2}{2/\sigma^2}\right). \tag{2.6}$$

By this expression, Fourier components of $G(x)$ are included by 99.73% taking the angular frequency to $3/\sigma$ (rad), and so $G(x)$ is band limited at $3/\sigma$ (rad). To convert to frequency, $3/\sigma$ is divided by 2π. Therefore, in case of the Gaussian function, the sampling intervals are taken as

$$\Delta X \leq \frac{1}{2(3/2\pi\sigma)} \approx 1.047\sigma,$$

$$\Delta Y \leq \frac{1}{2(3/2\pi\sigma)} \approx 1.047\sigma. \tag{2.7}$$

There is virtually no loss of information. In our case we can set σ to $\alpha/1.4$. That is, using (2.6), we can obtain the following result:

$$\Delta X \leq (2\alpha)\frac{1}{1.4\times 2} \simeq \frac{1}{3}(2\alpha). \tag{2.8}$$

The 2α corresponds to ω_0 in the previous geometrical consideration, so the above result tells us that the sampling intervals should be taken less than $\omega_0/3$. This requirement is stricter than the simple geometrical one, which is oversimplified by taking only straight lines into account. Although a curve can be approximated by line segments, even in that approximation the neighborhood of a connection point of these line segments is problematic. This is the topic of the next section. At any rate, we can say that both the results for sampling intervals and their relations are reasonable.

2.3 SAMPLING OF A CORNER

We consider the problem of corner detection viewed from a sampling perspective. Corner detection is important in character recognition. For example, it is necessary to differentiate pairs of characters such as "*D*" and "*O*," and "*S*" and "5." This will be further considered later from the recognition standpoint rather than reconstruction point of view. We begin with the case where ΔX, $\Delta Y = \omega_0$. Our aim is to give an idea for corner detection in terms of sampling, so we restrict our discussion to the very simple case of a rectangular corner which in its arrangement is parallel to the *x*- and *y*-axes. At first the arrangement is assumed to be symmetrical about a diagonal line on the grid plane. Two images are shown in Fig. 2.4 (*a*) in which the images are binarized. The threshold value is taken as one-half of the mesh area. Obviously in the arrangement of l_1, the result of sampling is correct as shown in Fig. 2.4 (*b*), where the corner is expressed on the grid plane. However, in the arrangement of l_2, the corner is lost on the grid plane, as shown in Fig. 2.4 (*b*). More specifically, for a corner to be detected, δ must be in the range of $[\sqrt{2}/2, 1]\omega_0$. Otherwise, the corner is not detected. Such phenomena happen regardless of the setting of the threshold value. This is because, taking $\omega_0 = 1$ for simplicity, the neighboring areas of the corner point are δ, and the area of the corner point is δ^2. Therefore the equation, $\delta = \delta^2$ is only satisfied when $\delta = 1$, and due to $0 \leq \delta \leq 1$, $\delta^2 < \delta$ holds in the range of (0, 1). Hence $\delta^2 > \frac{1}{2}$ must hold for the corner to be sampled well. The above discussion tells also that this phenomenon always happens regardless of the value of ω_0. We can say, however, that the missing "1" at the corner point is neglected as $\omega_0 \rightarrow 0$.

Now we need to compare corner with noncorner images in our sampling. Assuming the simple conditions used above, we seek a condition for a noncorner image to have a sampling pattern that differs from the one-point-missing case shown in Fig. 2.4 (*b*). This is illustrated in the figure. The line image l_1 induces a false corner on the grid plane for the positive variation of δ (Fig. 2.5), and for the negative variation of

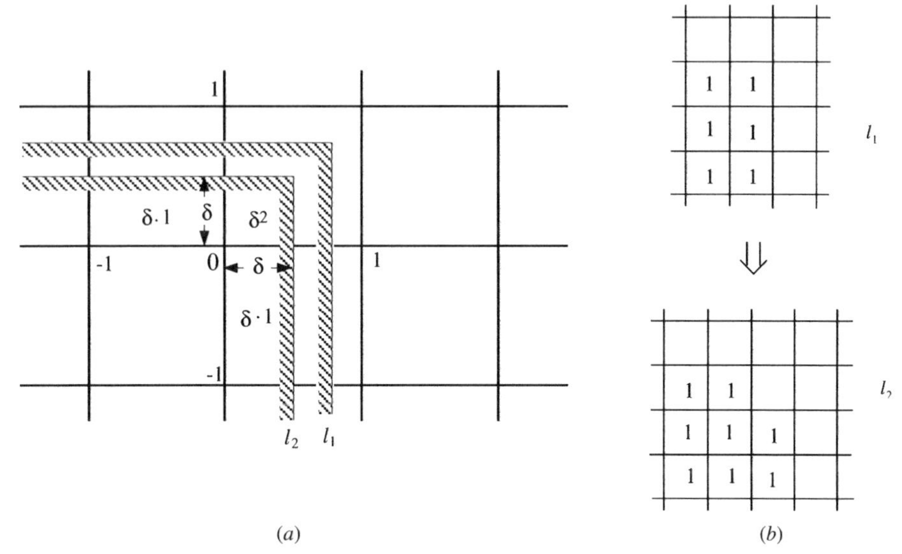

FIGURE 2.4 Sampling of a corner: (*a*) Simple images of corners designated as l_1 and l_2; (*b*) their sampling designated as l_1 and l_2, respectively. Notice that l_1 and l_2 are not lines but rectangles. However, we can regard them as lines that have sharp corners if the lines are sufficiently thick, and so we use the notation l_i, $i = 1, 2$.

δ, it exhibits the noncorner sampled pattern on the grid plane. For $\delta = 0$, the corner is unstable. Therefore we cannot distinguish the corner and the noncorner image on the sampled grid plane for such images. However, the line image l_2 exhibits the one-point-missing noncorner pattern for the positive variation of δ and the three-points-missing noncorner pattern for the negative variations of δ, as shown in Fig. 2.6. Therefore a noncorner image has to be less corner-shaped than the image l_2. For example, a noncorner image missing three points at the corner is enough to differ from the corner image on the sampled grid plane.

So far we have discussed the reconstruction of a corner on a sampled grid plane and found that it is impossible to reconstruct any given corner image from the sampled image. This is because of the very acute corners at which there is no derivative. This problem corresponds to the case of no band limiting from an analytical point of view. In this respect, if we want to recover a steeply bent line, then we need a very fine grid which is far less than minimum line width w_0. So the discrepancy between $\frac{1}{3}w_0$ and w can be understood. However, we have to reconsider this problem. The question is whether it is really necessary to detect such subtle corner features from the character recognition point of view. Actually, even if a corner is not strictly precise, we can distinguish two character samples whose chasses differ from each other, as shown Fig. 2.7. In fact the corner feature is not local but global. This point is very important and discussed fully in Section 8.7 on feature extraction. Therefore we do not necessarily have to sample given data exactly in practice. The real sampling interval is determined by the concrete objective of character recognition and the nature of the objects in practice.

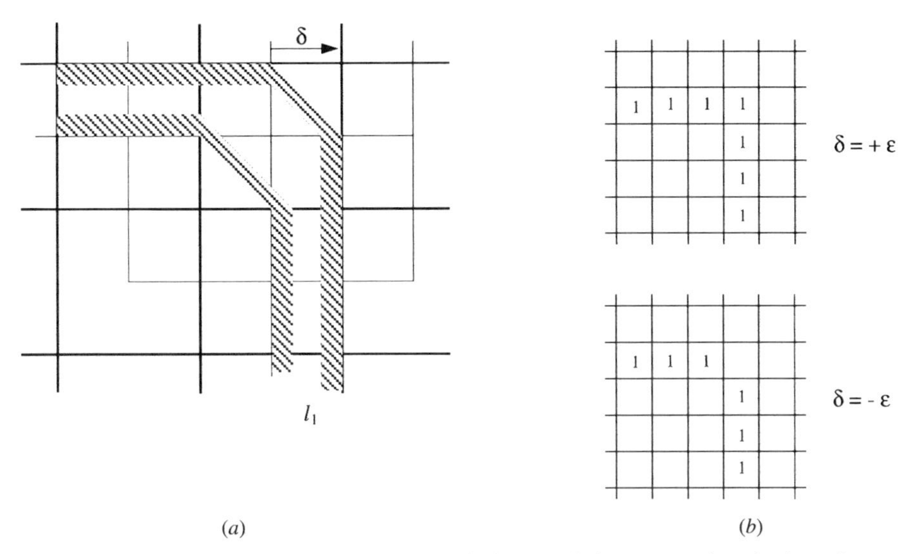

FIGURE 2.5 False corner: (*a*) Input image; (*b*) its sampled patterns when the input image moves to the right by ε (*top*) or to the left by ε (*bottom*), where ε > 0.

2.4 BLURRING AND SAMPLING

The relation between blurring effects and sampling will now be described more fully [2]. First of all an image consisting of a δ function is assumed to be an ideal pattern. This ideal thinned image is convolved by a Gaussian function with parameter σ_1 by an observation system. The result is denoted as $f_0(x)$. Furthermore the $f_0(x)$ is con-

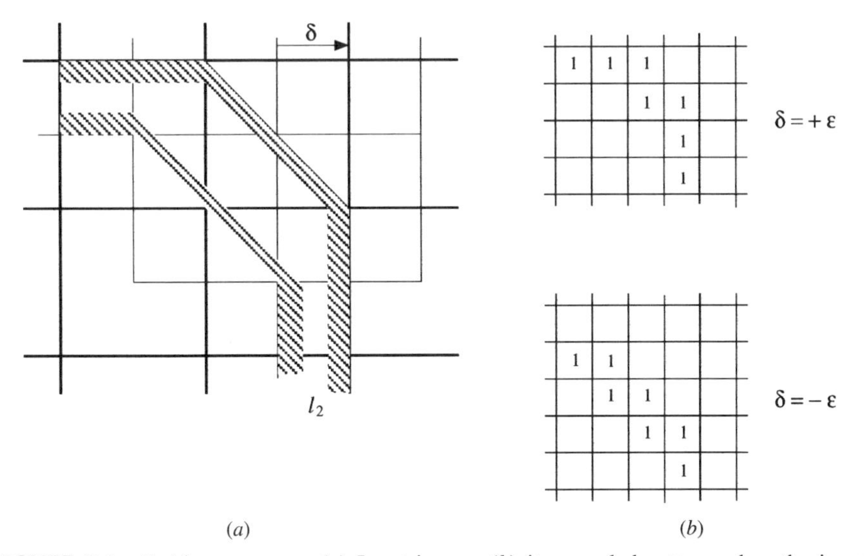

FIGURE 2.6 Stable noncorner: (*a*) Input image; (*b*) its sampled pattern when the input image moves to the right by ε (*top*) of to the left by ε (*bottom*), where ε > 0.

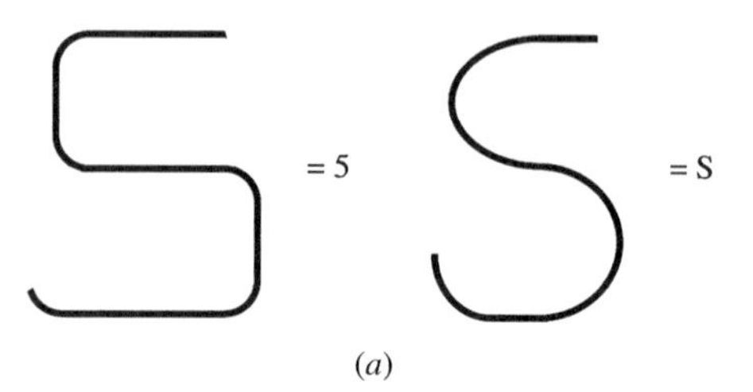

(*a*)

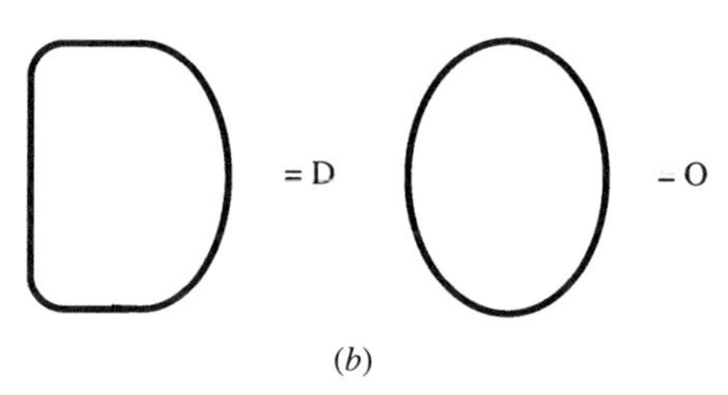

(*b*)

FIGURE 2.7 Comparison between characters having slightly rounded corners and those having completely round contours: (*a*) The "5" and "*S*" pair; (*b*) the "*D*" and "*O*" pair.

volved by a Gaussian function with a parameter σ_2, which is denoted as $f(x)$. This is described as

$$f(x;\ \sigma) = f_0(x;\ \sigma_1) * p(x;\ \sigma_2), \tag{2.9}$$

where $p(x;\ \sigma_2)$ designates a Gaussian function with parameter σ_2. Each Gaussian function form is described as follows:

$$f_0(x;\ \sigma_1) = \frac{1}{\sqrt{2\pi}\sigma_1} \exp\left(-\frac{x^2}{2\sigma_1{}^2}\right), \tag{2.10}$$

$$p(x;\ \sigma_2) = \frac{1}{\sqrt{2\pi}\sigma_2} \exp\left(-\frac{x^2}{2\sigma_2{}^2}\right), \tag{2.11}$$

$$f(x;\ \sigma) = \frac{1}{\sqrt{2\pi}\sigma} \exp\left(-\frac{x^2}{2\sigma^2}\right). \tag{2.12}$$

As is well-known, the Fourier transform of (2.9) is represented by the product of each Fourier transform; that is,

$$F(\zeta;\ \sigma) = F_0(\zeta;\ \sigma_1) \cdot P(\zeta;\ \sigma_2). \tag{2.13}$$

Furthermore the Fourier transform of a Gaussian function is again a Gaussian function, and the parameter is inverted. Therefore $F(\zeta;\ \sigma)$ is given as

$$F(\zeta;\ \sigma) = \exp\left(\frac{-\zeta^2}{2/\sigma^2}\right) \tag{2.14}$$

Using the properties of convolution in frequency domain described in (2.13) and (2.14), we can derive the following relations among the parameters σ, σ_2, and σ_2 in (2.10) through (2.12); that is,

$$\sigma^2 = \sigma_1^2 + \sigma_2^2. \tag{2.15}$$

Now let us consider the degree of blurring that can be imposed on an ideal line whose section is a rectangular pulse in terms of the sampling theorem. Specifically, we assume that the width of the line is 2ω and that the sampling interval is $2x$. To consider this problem, we take $f_0(x)$ as an equivalent Gaussian function of the rectangular pulse. According to (2.7), since σ_1 is equivalent to $\omega/1.4$ where 2ω is a line width, the following condition must be satisfied for the ω and $\Delta X \equiv 2x$:

$$\frac{2x}{1.047} \leq \frac{\omega}{1.4}.$$

On the other hand, if by the condition of the observation system, the above condition is not satisfied,

$$\frac{2x}{1.047} > \frac{\omega}{1.4},$$

then further blurring is necessary, and the necessary additive blurring parameter σ_2 is decided using the relation of (2.15) as

$$\sigma_2 \geq \sqrt{\left(\frac{2x}{1.047}\right)^2 - \left(\frac{\omega}{1.4}\right)^2}. \tag{2.16}$$

2.4.1 Effect of Blurring

Here we show a simulation on printed numerals to see the effect of the blurring using the Gaussian function [2]. Specifically, let us consider sampling printed numerals of OCR-B font at a 0.2-mm interval. Since the standard width of a character stroke is 0.35 mm, the necessary minimum blurring when sampling is decided by (2.16) as

$$\sigma_2 \geq \sqrt{\left(\frac{0.2}{1.047}\right)^2 - \left(\frac{0.35}{2} \times 1.4\right)^2} \tag{2.17}$$

Thus we obtain $\sigma_2 = 0.1444$.

Now three kinds of sampling are considered for comparison. The first is the normal one, namely sampling after blurring by $\sigma_2 = 0.1444$. The second is a simple analog way. That is, the density within a 0.2×0.2 mm pixel point taking 50% as the threshold value. Here the original data are standard patterns and so are virtually bilevel. Therefore 50% thresholding means that the grid square is occupied by a stroke for 50% or more of its area. Then it takes value 1; otherwise, it takes value 0.

To see the effect of blurring, the standard patterns were sampled changing the sampling starting point by (1/5)Δ as shown in Fig. 2.8, where Δ is sampling interval. Thus 5 × 5 = 25 samples for each category were obtained. These sampled data are correlated using the definition of similarity introduced in Section 1.2.2. Thus maximum, minimum, average values, and standard deviations were calculated for each numeral as in Table 2.1. Looking at the table, we can see the positive effect of blurring for all the numerals. However, the difference between blurring sampling and simple analog sampling is subtle. This is an experiment at the theoretical level, so this point will be investigated next. On the other hand, when compared with binarized sampling, the superiority of the first to the second is obvious at a practical level. In practice, the effect of blurring can be seen for more shifted and noisy data. This will be demonstrated in later chapters.

2.4.2 Sampling Interval and Level of Sampled Value

As seen above, analog sampling can be effective, but with some limitation. Here theoretical consideration of Shannon's information theory is given [2]. Suppose that as shown in Fig. 2.8, a region of the Euclidean plane is divided into an $m \times n$ grid square

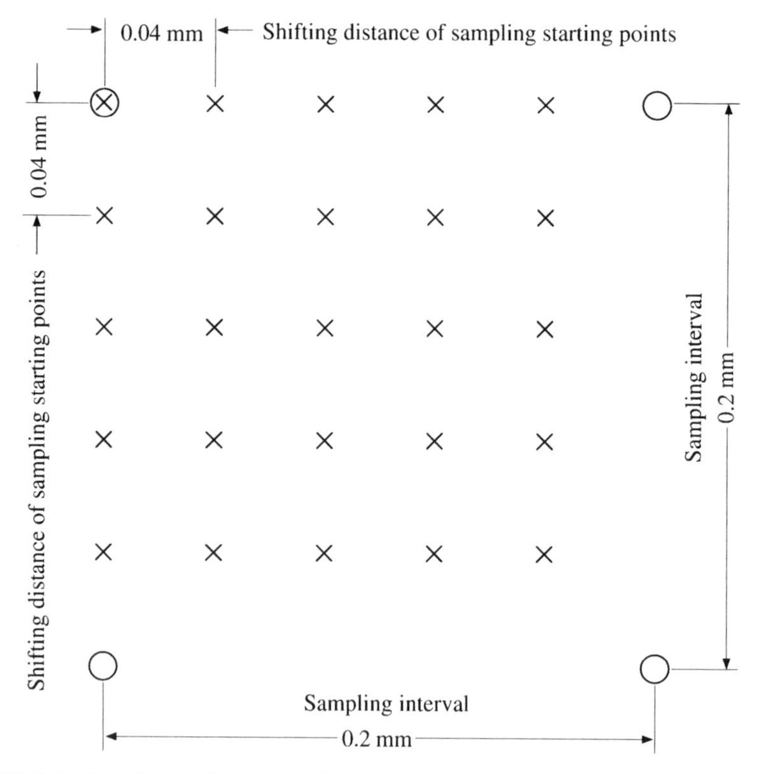

FIGURE 2.8 Sampling points marked by a circle, and sampling starting points marked by a cross. From reference [2].

TABLE 2.1 Simulation of the Blurring Effect

Numeral		Blurring	Without Blurring	Binarized
1	Maximum	0.9999	0.9993	1.0000
	Minimum	0.9908	0.9627	0.6533
	Average	0.9963	0.9855	0.9102
	Standard deviation	0.0023	0.0091	0.0821
2	Maximum	0.9998	0.9991	0.9896
	Minimum	0.9628	0.9530	0.7347
	Average	0.9900	0.9820	0.9044
	Standard deviation	0.0092	0.0118	0.0571
3	Maximum	0.9999	0.9990	0.9880
	Minimum	0.9803	0.9776	0.6890
	Average	0.9946	0.9912	0.8825
	Standard deviation	0.0048	0.0052	0.0761
4	Maximum	0.9999	0.9994	1.0000
	Minimum	0.9907	0.9810	0.8035
	Average	0.9970	0.9918	0.9254
	Standard deviation	0.0018	0.0046	0.0356
5	Maximum	0.9998	0.9987	0.9908
	Minimum	0.9835	0.9764	0.6114
	Average	0.9953	0.9898	0.8798
	Standard deviation	0.0040	0.0052	0.0625
6	Maximum	0.9998	0.9995	0.9872
	Minimum	0.9883	0.9881	0.8639
	Average	0.9966	0.9919	0.9352
	Standard deviation	0.0023	0.0045	0.0228
7	Maximum	0.9999	0.9991	0.9954
	Minimum	0.9820	0.9629	0.5138
	Average	0.9956	0.9870	0.8795
	Standard deviation	0.0035	0.0084	0.0836
8	Maximum	0.9998	0.9990	0.9849
	Minimum	0.9845	0.9731	0.8611
	Average	0.9950	0.9887	0.9255
	Standard deviation	0.0036	0.0061	0.0256
9	Maximum	0.9999	0.9994	0.9842
	Minimum	0.9841	0.9732	0.8424
	Average	0.9958	0.9898	0.9288
	Standard deviation	0.0033	0.0060	0.0247
0	Maximum	0.9997	0.9995	0.9899
	Minimum	0.9843	0.9718	0.8400
	Average	0.9947	0.9879	0.9210
	Standard deviation	0.0035	0.0073	0.0288
Mean	Maximum	0.9998	0.9992	0.9910
	Minimum	0.9831	0.9714	0.7413
	Average	0.9951	0.9886	0.9092
	Standard deviation	0.0038	0.0068	0.0502

Source: From reference [1].

and that at each grid square l levels are taken as possible differentiation levels. We note here that we use a new term "mesh" which is used frequently instead of grid square in literature. Let the probability of assigning (i, j) mesh be p_{ij}, and let the probability of taking level k at a mesh be p_k. Then according to the definitions,

$$\sum_{i=1}^{m}\sum_{j=1}^{n} p_{ij} = 1,$$

$$\sum_{k=1}^{l} p_k = 1, \tag{2.18}$$

hold.

Furthermore we assume that the event of taking a mesh and the event of taking the k level at a mesh are mutually independent. Let the probability that the ij mesh takes the k level be $\mathbf{P}_{ijk}$; then p_{ijk} can be represented as

$$p_{ijk} = p_{ij} \cdot p_k, \tag{2.19}$$

and

$$\sum_{i=1}^{m}\sum_{j=1}^{n}\sum_{k=1}^{l} p_{ijk} = 1 \tag{2.20}$$

holds.

Let us consider entropy of the region of $m \times n$, denoted as H, then H is represented as

$$H = -\sum_{i=1}^{m}\sum_{j=1}^{n}\sum_{k=1}^{l} p_{ijk} \log_2 p_{ijk}. \tag{2.21}$$

H takes its maximum value when $p_{ijk} = 1/mnl$ and its value, H_{max}, is expressed as

$$H_{max} = \log_2 mnl(\text{bits}). \tag{2.22}$$

Now suppose that the number of resolution and the levels of resolution are changed as $l \to l'$, $m \to K_1 m$, and $n \to K_2 n$, respectively. Then the maximum entropy I_{max} is expressed as

$$I_{max} = \log_2 K_1 K_2 mnl'. \tag{2.23}$$

Thus, from the condition of $I_{max} = H_{max}$, the following relation is obtained:

$$\frac{l'}{l} = \frac{1}{(K_1 K_2)}. \tag{2.24}$$

In particular, setting $K_1 = K_2 = K$, the following simple relation holds between the necessary possible differentiation levels and reduced mesh number K^2:

$$\frac{l'}{l} = \frac{1}{K^2}. \tag{2.25}$$

In fact this is the result based on Shannon's information theory, which cannot be applicable in general. However, we can gain some insight into the relation between the resolution and the number of levels. For example, as seen in Fig. 2.9, if a 6×8 bi-level pattern is analog sampled, then the question of how many levels are necessary in that analog sampling can be given in terms of Shannon's information theory. That is, setting $l = 2$ and $K = \frac{1}{2}$, the following relation is obtained:

$$\frac{l'}{2} = \frac{1}{(1/2)^2}. \tag{2.26}$$

Therefore the levels necessary are $l' = 8$. On the other hand, in this analog sampling we can obtain five levels as the maximum. Therefore the simple analog sampling is not enough in terms of Shannon's information theory.

2.4.3 Sampling under a Constant Contribution Rate

In the previous section a 2×2 region was reduced to one point with even weight to each point that belongs to each 2×2 region. This is a simple and intuitive problem, so there requires no theoretical background. In general, the problem is stated as follows: If a grid plane is sampled such that each $n \times n$ region is reduced to one point,

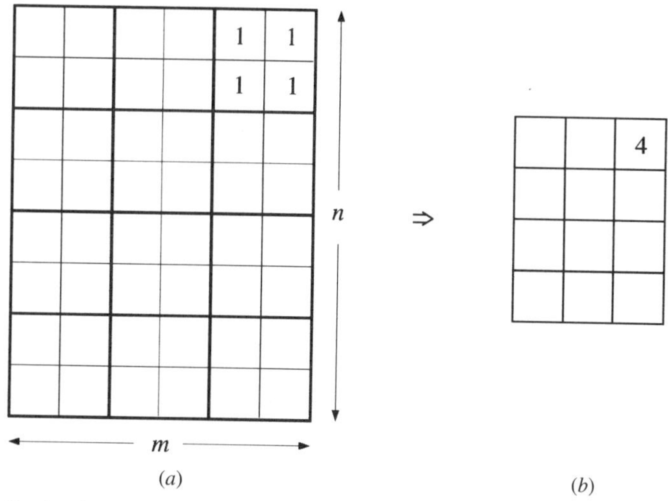

FIGURE 2.9 Analog sampling: (*a*) An original $n \times m$ grid plane; (*b*) analog sampled plane, where the resolution in space is reduced by 1/2. From reference [2].

how should we choose weights for the neighboring points. This was considered by Iijimas [9], as we now describe.

Definition of Contribution Rate For simplicity, the description is given for the one-dimensional case. We assume a weighting function $u(x)$ that satisfies the following condition:

$$\int_{-\infty}^{\infty} u(x)dx = 1, \qquad u(x) \geq 0. \tag{2.27}$$

Then a given image $f(x)$ is transformed to $g(x)$ by $u(x)$ as

$$g(x) = \int_{-\infty}^{\infty} u(x - x')f(x')dx'. \tag{2.28}$$

From the equation above, the following relation holds:

$$\int_{-\infty}^{\infty} g(x)dx = \int_{-\infty}^{\infty} f(x')dx'. \tag{2.29}$$

For any x' this is derived based on the relation

$$\int_{-\infty}^{\infty} u(x - x')dx = 1. \tag{2.30}$$

Now, if the x-axis is sampled at equal intervals, (2.27) is rewritten as

$$\sum_{m=-\infty}^{\infty} u_m = 1, \qquad u_m \geq 0. \tag{2.31}$$

Similarly (2.28) is expressed as

$$g_m = \sum_{n=-\infty}^{\infty} u_{m-n} f_n \qquad (m = 0, \pm 1, \ldots). \tag{2.32}$$

The equation above represents a transformation of $\{f_n\}$ to $\{g_m\}$. Since the following equation holds for any n,

$$\sum_{m=-\infty}^{\infty} g_m = \sum_{n=-\infty}^{\infty} f_n. \tag{2.33}$$

Here we set sampling points of $\{g_m\}$ to have k times the interval for those of $\{f_n\}$. Then the equation (2.32) is rewritten as

$$g_m^{(k)} = \sum_{n=-\infty}^{\infty} u_{km-n} f_n. \tag{2.34}$$

Now we define $\lambda_n^{(k)}$ as

$$\lambda_{\text{n}}^{(k)} = \sum_{n=-\infty}^{\infty} u_{km-n}. \tag{2.35}$$

Then the following relation holds:

$$\sum_{m=-\infty}^{\infty} g_m^{(k)} = \sum_{n=-\infty}^{\infty} \lambda_n^{(k)} f_n. \tag{2.36}$$

Comparing (2.31) with equation (2.35), we can obtain the following relation:

$$0 < \lambda_m^{(k)} \leq 1, \tag{2.37}$$

where equality on the right side holds only when $k = 1$. The coefficient $\lambda_n^{(k)}$ can be interpreted as representing the total contribution of each of f_n to the set $\{g_m^{(k)}\}$. Therefore $\lambda_n^{(k)}$ is called the *contribution rate* for the sampling point n. From (2.33),

$$\lambda_{n+k}^{(k)} = \lambda_n^{(k)} \qquad (n = 0, \pm 1, \ldots) \tag{2.38}$$

is obtained, and so $\lambda_n^{(k)}$ is a periodic series as the period k.

Sampling with a Constant Contribution Rate If a contribution rate satisfies the relation

$$\lambda_{n+1}^{(k)} = \lambda_n^{(k)} \qquad (n = 0, \pm 1, \ldots), \tag{2.39}$$

then we say that the contribution rate is constant, and the value is expressed as $\lambda^{(k)}$, called the *constant contribution rate.* By (2.38), when $k = 1$, the contribution rate is always constant.

In case of a constant contribution rate, (2.36) is written as

$$\sum_{m=-\infty}^{\infty} g_m^{(k)} = \lambda^{(k)} \sum_{n=-\infty}^{\infty} f_n. \tag{2.40}$$

On the other hand, the following relations hold:

$$\begin{aligned} \lambda^{(k)} &= \frac{1}{k} \sum_{n=1}^{k} \lambda_n^{(k)} = \frac{1}{k} \sum_{n=1}^{k} \sum_{m=-\infty}^{\infty} u_{km-n} \\ &= \frac{1}{k} \sum_{m=-\infty}^{\infty} u_m = \frac{1}{k}. \end{aligned} \tag{2.41}$$

That is, $\lambda^{(k)}$ is equal to $1/k$, so

$$\sum_{m=-\infty}^{\infty} g_m^{(k)} = \frac{1}{k} \sum_{n=-\infty}^{\infty} f_n \tag{2.42}$$

holds.

***l*-Ranked Golden Coefficient Series** Here we consider the special case where $\lambda_n^{(k)}$ is constant for k less than l. That is, $\{u_m\}$ is selected such that

$$\lambda_n^{(k)} = \lambda^{(k)} \qquad (k = 1, 2, \ldots, l) \tag{2.43}$$

hold. $\{u_m\}$ is called an *l-ranked golden coefficient series.*

In particular, the symmetrical case is expressed as

$$u_m = \begin{cases} u_{-m} = \dfrac{d_m}{D} & (m = 0, 1, \ldots, M), \\ 0 & (|m| > M). \end{cases} \tag{2.44}$$

Therefore, using (2.31),

$$D = d_0 + 2 \sum_{m=1}^{M} d_m \tag{2.45}$$

holds.

Some examples are shown below.

Example 1

$$M = 1, \quad l = 2,$$

$$u_0 = \frac{2}{4}, \quad u_1 = \frac{1}{4}.$$

Example 2

$$M = 2, \quad l = 3,$$

$$u_0 = \frac{4}{12}, \quad u_1 = \frac{3}{12}, \quad u_2 = \frac{1}{12}.$$

Some further examples for two-dimensional cases are shown in Fig. 2.10.

1	2	1
2	4	2
1	2	1

$\times \frac{1}{16}$

(*a*)

0	7	10	7	0
7	18	22	18	7
10	22	32	22	10
7	18	22	18	7
0	7	10	7	0

$\times \frac{1}{228}$

(*b*)

FIGURE 2.10 Two-dimensional and equidirectional golden coefficient series. (*a*) $l = 2$; (*b*) $l = 3$. From the Reference [9].

2.5 BIBLIOGRAPHICAL REMARKS

To the best of our knowledge, there has been little work on sampling related to OCR. I. Yamazaki's work is based on T. Iijima's basic research on normalization and I. Yamazaki and T. Iijima's research on observation mechanisms [1, 2]. Normalization, which was explained in Chapter 1, will be discussed further in Chapter 3. A concrete application of this theory was done by S. Yamamoto, A. Nakajima, and K. Nakata in their research and development of a Chinese character recognition system [3]. This is a hierarchical pattern matching system. Specifically, sampling points and gray level were reduced to $32 \times 32 \times 2(bit) \rightarrow 16 \times 16 \times 4(bit) \rightarrow 8 \times 8 \times 4(bit) \rightarrow 4 \times 4 \times 4(bit)$ in pyramidal fashion. At each reduction stage the appropriate blurring was calculated according to the theory. They achieved very successful results.

On the other hand, after the sampling on a plane, an object can be represented by its contour or boundary. In discrete contour representation, Freeman chain encoding is usually used because of its simplicity [4]. However, this encoding has an intrinsic problem that its sampling pitch is not uniform. That is, the distance between the successive points on a square grid with a spacing of d is either d or $\sqrt{2}d$. The ratio $d(\sqrt{2}/d)$ is approximated by $7/5 = 1.4 \cong \sqrt{2}$. Therefore, if we take more sampling points, using four on horizontal and vertical segments, and six along diagonal segments, we have almost a uniform sampling pitch along a contour. However, this results in a large increase in sampling points. Therefore $(5n)$th distance points are adopted, where n is an integer, and the remaining points are neglected. This is called *resampling* by B. Shahraray and D. J. Anderson [5].

In the case of blurring, this is also a standard technique in the fields of computer vision and image understanding. Another theoretical treatment showing that the smoothing filtering function is optimum for Gaussian function was given by Marr and Hildreth [6]. They based it on the fact that the smoothing function must be localized in both space and frequency domains. From a mathematical standpoint, the following formula is very important. That is, we consider a filtering of an input image $f(x)$ by convolving with Gaussian function as

$$F(x, \sigma) \equiv f(x) \otimes g(x, \sigma)$$

$$= \int_{-\infty}^{\infty} f(u) \frac{1}{\sigma\sqrt{2\pi}} \exp\left[\frac{(x-u)^2}{2\sigma^2}\right] du.$$

Then

$$\frac{\partial^n F}{\partial x^n} = f(x) \frac{\partial^n g(x, \sigma)}{\partial x^n}$$

holds.

For example, we consider the problem of edge detection, the maximum of gradient of $f(x)$. This is given by calculating the points that satisfy the following conditions:

$$F_{xx} = 0, \quad F_{xxx} \neq 0.$$

This is simply done using derivatives of the Gaussian function instead of $f(x)$ directly and is widely used in the fields mentioned above. The points that satisfy the above condition are called *zero-crossing points.* However, a problem is how to choose an appropriate parameter value of σ. If we want local edges, then smaller σ is better, but if we want global edges, then larger σ is better. This means that an appropriate single selection of parameter value is impossible in general. Therefore, to cope with this difficulty, *scale-space filtering* was conceived by Wittkin [7]. He gave a scheme of $\sigma - x$ space on which traces of zero-crossing points are delineated. This is also a hierarchical representation. A theoretical consideration was given by V. Torre and T. A. Poggio [8].

Finally we should note that some basic research has been done on sampling and quantization in image processing. Amang D. Lee, T. Pavlidis, and G. W. Wasilkowski [10] provide some important results. They used sophisticated error norms rather than using the L_2 norm. They derived the following equation to give an optimum quantization level $k*$ for the $1D$ data whose range is T and memory requirement M:

$$k* = \log \frac{M}{T} + \log h(0) + \log (2 \ln 2),$$

where $h(0)$ is the point spread function's value at zero. This is approximated as Gaussian with standard deviation σ. Then the $k*$ becomes

$$k* = \log \frac{M}{T} + \log \sigma + \frac{1}{2} \log (2\pi) + \log (2 \ln 2).$$

They gave an example substituting reasonable values for the parameters for M/T, that is, 512 samples per inch. Therefore $M/T = 512$, and so $\log M/T = 9$. If we are going to use one bit per *pixel*($k = 1$), then this is justified only if $\log \sigma = -10$, or σ equal to 1/1024 of an inch.

Their conclusion on k^* was that for less sharp σ, "gray-scale" text is preferable for digital processing. This has turned out to be true, and the relevance of the gray scale can be seen everywhere if one wants to make a more complete OCR system. This was shown more precisely by L. Wang and T. Pavlidis 10 years later [11].

BIBLIOGRAPHY

[1] I. Yamazaki and T. Iijima, "Observation mechanism for characters and figures," *Trans. IECE Japan,* vol. 55-D, no. 1, p. 15, January 1972.

[2] I. Yamazaki, "Sampling mechanism and automatic normalization of character patterns," *Res. Electrotech. Lab.,* no. 726, April 1972.

[3] S. Yamamoto, A. Nakajima, and K. Nakata, "Chinese character recognition by hierarchical pattern matching—Study of Chinese character recognition," *Trans. IECE Japan,* vol. D-117, no. 12, pp. 714–720, December 1973.

[4] H. Freeman, "On the encoding of arbitrary geometric configurations," *IRE Trans. Electronic Comp.* vol. EC-10, no. 2, pp. 260–268, June 1961.

[5] B. Shahraray and D. J. Anderson, "Uniform resampling of digitized contours," *IEEE Trans. Pattern Anal. Machine Intell.,* vol. PAMI-7, no. 6, pp. 674–681, November 1985.

[6] D. Marr and E. Hildreth, "Theory of edge detection," *Proc. Roy. Soc. London,* vol. B207, pp. 187–217, 1980.

[7] A. P. Wittkin, "Scale-space filtering," *Proc. 7th Int. Joint Conf. Artificial Intell.,* Karlsruhe, Germany, pp. 1019–1021, 1983.

[8] V. Torre and T. A. Poggio, "On edge detection," *IEEE Trans. Pattern Anal. Machine Intell.,* vol. PAMI-8, no. 2, pp. 147–163, March 1986.

[9] T. Iijima, "Sampling theory under the condition of constant contribution rate," Report of the research on fundamental theory of pattern recognition, no. 05402061, pp. 34–40, 1996.

[10] D. Lee, T. Pavlidis, and G. W. Wasilkowski, "A note on the trade off between sampling and quantization in signal processing," *J. Complexity,* vol. 3, pp. 359–371, 1987.

[11] L. Wang and T. Pavlidis, "Direct gray-scale extraction of features for character recognition," *IEEE Trans. Pattern Anal. Machine Intell.,* vol. 15, no. 10, pp. 1053–1067, October 1993.

CHAPTER THREE

Normalization

Normalization was introduced in Section 1.2.6 where, for simplicity, it was illustrated using one-dimensional patterns. In this chapter theoretical and practical methods of normalization are described in detail. First, some mathematical background is given to support the normalization methods to be presented. Next, two basic theories of normalization are introduced. One is Iijima's and the other is Amari's, which stand in contrast with each other. Then, a powerful method of linear normalization is discussed. Finally, nonlinear normalization methods are presented.

3.1 GENERAL CONSIDERATIONS AND MATHEMATICAL PREPARATION

Basically there are two kinds of transformation, linear and nonlinear. The linear transformation takes the form

$$\begin{pmatrix} x' \\ y' \end{pmatrix} = \begin{pmatrix} a & b \\ c & d \end{pmatrix} \begin{pmatrix} x \\ y \end{pmatrix} + \begin{pmatrix} a' \\ c' \end{pmatrix}, \tag{3.1}$$

$$\mathbf{r}' = T\mathbf{r} + \mathbf{a}. \tag{3.2}$$

This is called an *affine transformation.* The second term represents a *shift transformation,* which is explicitly resolved to x and y coordinates. Since the shift transformation of a two-dimensional plane can be regarded as a simple extension of the one-dimensional shift transformation, the first term is given a full discussion here. Nonlinear normalization is important when dealing with hand-printed characters, and it is described in Section 3.5.

There are likewise two approaches for normalization preprocessing. One is the normalization of a given character on an observed plane or sampled plane. The other is a normalization in feature space, where the normalization is performed after feature extraction and representation of features. These two approaches will be discussed after a mathematical review of linear transformation which is given here for the reader's convenience.

3.1.1 Representation of Linear Transformation

Affine transformations preserve only parallelism and do not preserve area and angles of a figure, in general. However, the transformation represented by a matrix T can be decomposed into dilatation/contraction, rotation, and skew. Dilatation/contraction preserves the angle but not the area. Rotation preserves both the angle and the area. From the view of invariance, rotation is a so-called orthogonal transformation that preserves the length of a vector, $|\mathbf{r}^2| = x^2 + y^2$. A reflection about the y-axis, for example, is another orthogonal transformation, and it is a 180-degree rotation in three-dimensional space. We can see such reflected characters by looking from the back at a transparent sheet on which characters are written. Any orthogonal transformation is generated by combining rotation and reflection transformations.

Now let us focus on rotation. It is well-known that the rotation is represented by the following transformation:

$$\begin{pmatrix} x' \\ y' \end{pmatrix} = \begin{pmatrix} \cos\theta & -\sin\theta \\ \sin\theta & \cos\theta \end{pmatrix} \begin{pmatrix} x \\ y \end{pmatrix}, \tag{3.3}$$
$$\mathbf{r}' = R(\theta)\mathbf{r},$$

where θ is the angle measured from the x-axis. In general, any transformation $T(|T| \neq 0)$ does not change the origin of the x, y coordinate system on the plane, so we can imagine a vector from the origin, say $\overline{op}$. θ is the angle between the vector $\mathbf{r}$ and the vector $\mathbf{r}'$, as shown in Fig. 3.1.

It is also well-known that the representation in Fig. 3.1 is a very simple transformation in terms of the Gaussian plane. That is, point p can be represented by a complex number $z = x + iy$, and the transformed point p' is represented by $z' = x' + iy'$. Then the relation between z' and z is represented as

$$z' = e^{i\theta}z. \tag{3.4}$$

Successive rotations $R(\theta_1)$ and $R(\theta_2)$ result in the rotation $R(\theta_1 + \theta_2)$:

$$z' = e^{i\theta_1}z,$$
$$z'' = e^{i\theta_2}z' = e^{i\theta_2}e^{i\theta_1} = e^{i(\theta_1 + \theta_2)}z.$$

Therefore we can consider $R(\theta)$ as a one-parameter group. Obviously it has identity $R(0)$, and every $R(\theta)$ has its inverse $R(-\theta)$. The set $R(\theta)$, $2\pi > \theta \geq 0$, is closed under its successive operation, so the rotation transformation constitutes a group. The same argument can be applied to the transformations discussed next.

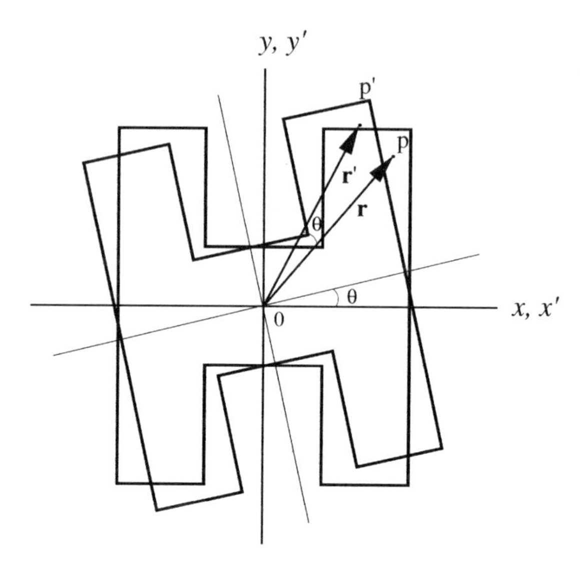

FIGURE 3.1 Rotation matrix $R(\theta)$ applied to "H."

Let us consider uniform transformations along the x- and y-axes. They are represented as

$$\begin{pmatrix} x' \\ y' \end{pmatrix} = \begin{pmatrix} \alpha & 0 \\ 0 & \beta \end{pmatrix} \begin{pmatrix} x \\ y \end{pmatrix}, \tag{3.5}$$

$$\mathbf{r}' = S\mathbf{r}. \tag{3.6}$$

S is closely related to dilatation/contraction transformation which is further generalized to the symmetric form

$$S' = \begin{pmatrix} e & f \\ f & g \end{pmatrix}. \tag{3.7}$$

This is actually a combination of rotation and the above uniform transformation. In other words, any symmetric transformation matrix can be reduced to a diagonal matrix by an appropriate rotation of the coordinate system. The proof can be found in any textbook on linear algebra.

Now let us examine the transformation S. Any vector that does not lie on an x- or y-axis changes its direction by the transformation S as

$$\tan \psi' = \frac{y'}{x'} = \frac{\beta}{\alpha} \frac{y}{x} = \frac{\beta}{\alpha} \tan \psi, \tag{3.8}$$

where ψ is the angle that the vector $\mathbf{r}$ makes with the x-axis and ψ' is a new angle of $\mathbf{r}'$ with the x-axis. This direction is called the *principal axis of S*. When $\alpha = \beta$, there is no change of direction of vectors. In this case S is simply represented as

$$S = \begin{pmatrix} \alpha & 0 \\ 0 & \alpha \end{pmatrix} = \alpha \begin{pmatrix} 1 & 0 \\ 0 & 1 \end{pmatrix} = \alpha I. \tag{3.9}$$

When $\alpha > 1$, it is dilatation; when $\alpha < 1$, it is contraction; and when $\alpha = 1$, it is nothing but an identity transformation. In this transformation S, there is no change of shape of a figure for any value of α, and only the size is changed. Let us consider the case where $\alpha\beta = 1$. Then the determinant det $(S) = 1$. Therefore no change occurs in the area of the transformed figure, although its shape has changed. A typical example is the transformation of a square, which is transformed into a parallelogram. Thus any uniform transformation is decomposed to the following form:

$$\begin{pmatrix} \alpha & 0 \\ 0 & \beta \end{pmatrix} = \begin{pmatrix} \mu & 0 \\ 0 & \dfrac{1}{\mu} \end{pmatrix} \begin{pmatrix} \lambda & 0 \\ 0 & \lambda \end{pmatrix}, \tag{3.10}$$

where det $(S) > 0$. Here we note that if det$(S) = -1$, then $\alpha\beta = -1$ and $|\alpha| = |\beta| = 1$, and the reflection is about the x- or y-axis. Any reflection can be constructed by composition of the above transformation and rotation.

In general, a linear transformation T is constructed by a symmetric transformation S' and rotation R, that is, $T = RS'$. Actually we can derive the S' by setting $S' = R^{-1}T$ giving $\tan\theta = (c - b)/(a + b)$. Solving the eigenvector and eigenvalue problem of the symmetric matrix S', the direction of the principal axis of the matrix is found as

$$\tan\psi_1 = \frac{\alpha - e}{f}, \quad \tan\psi_2 = \frac{\beta - e}{f}, \tag{3.11}$$

$$T = \begin{pmatrix} a & b \\ c & c \end{pmatrix},$$

where $S' = \begin{pmatrix} e & f \\ f & g \end{pmatrix}$ and α and β are eigenvalues of the matrix S'. The two eigenvectors are found, but they are related to each other by orthogonality. The eigenvector whose eigenvalue is larger, say α, $\alpha > \beta$, is chosen as the principal axis. We note that $\alpha \neq \beta$ holds unless $e = g$, $f = 0$, in which case it is a uniform matrix and simply dilatation or contraction where any direction can be the principal direction.

Thus we reach a structural decomposition of a general linear transformation T, which is parameterized by θ, α, β, ψ, instead of a, b, c, d, as $T \to R(\theta)S(\alpha, \beta, \psi)$, $S(\alpha, \beta, \psi) \to S_p(\psi)S_k(\mu)S_d(\lambda)$.

We note that the decomposition of a 2×2 matrix whose determinant is not zero is not unique. Another such example will be shown, which is based on Gram-Schmidt method for obtaining an orthogonal basis. Suppose that matrix A,

$$\begin{pmatrix} a_{11} & a_{12} \\ a_{21} & a_{22} \end{pmatrix} \to \begin{pmatrix} b_{11} & b_{12} \\ b_{21} & b_{22} \end{pmatrix} \equiv B,$$

is transformed so that $(b_{11}, b_{21})^t$ and $(b_{12}, b_{22})^t$ are orthogonal, where $|A| \neq 0$. Now the first column is equal to that of A and the second one is set as follows:

$$\begin{pmatrix} b_{11} \\ b_{21} \end{pmatrix} \leftarrow \begin{pmatrix} a_{11} \\ a_{21} \end{pmatrix}, \quad \begin{pmatrix} b_{12} \\ b_{22} \end{pmatrix} \leftarrow \begin{pmatrix} a_{12} \\ a_{22} \end{pmatrix} - \alpha \begin{pmatrix} b_{11} \\ b_{21} \end{pmatrix}$$

$$\mathbf{b}_1 \leftarrow \mathbf{a}_1, \qquad \mathbf{b}_2 \leftarrow \mathbf{a}_2 - \alpha \mathbf{b}_1,$$

where α is determined by the two column vectors so that the new vectors, $\mathbf{b}_1$ and $\mathbf{b}_2$ are orthogonal. This is easily determined as

$$\alpha = \frac{a_{11}a_{12} + a_{21}a_{22}}{a_{11}^2 + a_{21}^2} = \frac{\mathbf{a}_1^t \mathbf{a}_2}{\mathbf{a}_1^t \mathbf{a}_1}.$$

B is rewritten as

$$B = \begin{pmatrix} a_{11} & -\alpha a_{11} + a_{12} \\ a_{21} & -\alpha a_{21} + a_{22} \end{pmatrix} = \begin{pmatrix} a_{11} & a_{12} \\ a_{21} & a_{22} \end{pmatrix} \begin{pmatrix} 1 & -\alpha \\ 0 & 1 \end{pmatrix}.$$

Thus B is represented as $A\triangle$, where $\triangle$ is called a triangular matrix,

$$B = A\triangle.$$

On the other hand, each column vector of B is transformed to a unit vector $\mathbf{c}_j = \mathbf{b}_j/|\mathbf{b}_j|$ so that it becomes an orthogonal matrix. Thus B can be decomposed to the following form:

$$B = \begin{pmatrix} a_{11} & -\alpha a_{11} + a_{12} \\ a_{21} & -\alpha a_{21} + a_{22} \end{pmatrix} \begin{pmatrix} \frac{1}{|b_1|} & 0 \\ 0 & \frac{1}{|b_2|} \end{pmatrix} \begin{pmatrix} |b_1| & 0 \\ 0 & |b_2| \end{pmatrix},$$

$$B = OD,$$

where O is an orthogonal matrix and D is a diagonal matrix. Therefore A is decomposed as

$$A = OD\triangle^{-1},$$

where $\triangle^{-1}$ is still a triangular matrix, which represents a typical shear transformation, and O is actually the rotation transformation.

3.1.2 Infinitesimal Transformations

The transformation matrix treated has continuous parameters. This continuity gives us a very powerful tool, called *infinitesimal transformation/mapping,* that reveals the structure of a given matrix simply. Thanks to continuity, the linear transformation matrix T is represented as a function of a continuous valuable t,

$$T = T(t) = \begin{pmatrix} a(t) & b(t) \\ c(t) & d(t) \end{pmatrix}. \tag{3.12}$$

$T(t)$ can be differentiated by t and has the derivative at $t = 0$, denoted as X which is called the *infinitesimal transformation/mapping of T*,

$$\left.\frac{dT}{dt}\right|_{t=0} \equiv X \equiv \begin{pmatrix} p & q \\ r & s \end{pmatrix} = \begin{pmatrix} \dot{a}(0) & \dot{b}(0) \\ \dot{c}(0) & \dot{d}(0) \end{pmatrix}, \tag{3.13}$$

where $\dot{a}(0) \equiv da(t)/dt|_{t=0}$, according to the dot convention used in physics. It is helpful to imagine the physical analogy, and so X is the velocity of the transformation at $t = 0$.

Let us consider the successive transformation, namely the product of the two transformations S and T, $U = ST$, and its infinitesimal transformation Z. Before giving Z, we have to consider $S(0)$, and $T(0)$, or the initial values of these transformations. It is very reasonable to set these to the identity matrix I because I is a transformation of no change and all change starts from I. We use the following formula of derivative of the product of matrixes:

$$(A(t)B(t))' = A'(t)B(t) + A(t)B'(t). \tag{3.14}$$

We can set Z as $Z = X + Y$, where X and Y are infinitesimal transformations of S and T, respectively. For T^{-1}, $-X$ is obtained. Thus, as a whole, the product operation is transformed to an additive one that is easily handled, as shown below. The meaning of X can be easily seen by rewriting each matrix element in the sense of the additive operation as [1],

$$X = \begin{pmatrix} p & q \\ r & s \end{pmatrix} = \begin{pmatrix} \frac{p+s}{2} & 0 \\ 0 & \frac{p+s}{2} \end{pmatrix} + \begin{pmatrix} \frac{p-s}{2} & 0 \\ 0 & -\frac{p-s}{2} \end{pmatrix}$$
$$+ \begin{pmatrix} 0 & \frac{q+r}{2} \\ \frac{q+r}{2} & 0 \end{pmatrix} + \begin{pmatrix} 0 & \frac{q-r}{2} \\ -\frac{q-r}{2} & 0 \end{pmatrix}. \tag{3.15}$$

Setting

$$\frac{p+s}{2} = \xi_0, \quad \frac{p-s}{2} = \xi_1,$$

$$\frac{q+r}{2} = \xi_2, \quad -\frac{q-r}{2} = \xi_3,$$

X is represented as follows:

$$X = \xi_0 X_0 + \xi_1 X_1 + \xi_2 X_2 + \xi_3 X_3, \tag{3.16}$$

where

$$X_0 = \begin{pmatrix} 1 & 0 \\ 0 & 1 \end{pmatrix}, \quad X_1 = \begin{pmatrix} 1 & 0 \\ 0 & -1 \end{pmatrix},$$

$$X_2 = \begin{pmatrix} 0 & 1 \\ 1 & 0 \end{pmatrix}, \quad X_3 = \begin{pmatrix} 0 & -1 \\ 1 & 0 \end{pmatrix}. \tag{3.17}$$

Obviously X_0 represents dilatation/contraction. X_1 is a nonuniform dilatation/contraction along the x- and y-axes. We call nonuniform dilatation/contraction skew transformation. X_2 is also a skew along the directions of 45 degrees to x-axis and y-axis. The skew transformation is restricted to shape change transformation preserving the area of the shape. Note that $|\det(X_i)| = 1$, $i = 1, 2$.

We note that the reflection matrix about the x-axis has the same form as X_1. However, here we are considering a process of dynamic transformation which is represented as a function of t. Therefore these two matrices are different in their meanings. Actually when $\xi_1 \Delta t$ is small, the transformation is approximated by Taylor expansion as

$$T(\Delta t) \approx I + \xi_1 \Delta t X_1. \tag{3.18}$$

From the equation above we can imagine a skew transformation, namely extension along the x-axis and contraction along the y-axis, which makes a skew dynamically. A skew to any direction can be generated by combination of $\xi_1 X_1 + \xi_2 X_2$. To generate finite transformation, we need to extend the above equation as $T = \exp(tX)$, which is nothing but an integration of the infinitesimal transformation, $\dot{\mathbf{r}} = X\mathbf{r}$.

Infinitesimal Operators of the Transformation Group In the normalization of characters, we accept a basic assumption that there exists only one ideal pattern for each character class. Therefore a set of all the observed characters of one class consists of some deformations of the ideal pattern. The set is denoted as Ω, and the ideal pattern as $g(x)$. The deformations can be considered as consisting of two kinds. One is some transformation of an ideal pattern, where the property of the transformation is known to us. A subset of Ω, denoted as Ω_0, is constructed by such transformations on an ideal pattern. The other is some perturbation of the transformed parameters, and so it has a statistical nature. Let us consider infinitesimal transformation further, that is, the general case of r parameters and n variables [2]. Then the infinitesimals of variables dx^i ($i = 1, \ldots, n$) are given by the infinitesimals of parameters δa^σ ($\sigma = 1, \ldots, r$), where x^i is transformed by $f^i(\mathbf{x}; \mathbf{a})$ as follows in general:

$$dx^i = \sum_\sigma \left(\frac{\partial f^i(\mathbf{x}; \mathbf{a})}{\partial a^\sigma} \right)_{a=0} \delta a^\sigma \qquad (\sigma = 1, \ldots, r;\ i = 1, \ldots, n) \tag{3.19}$$

$$\equiv U^i_\sigma(\mathbf{x}) \delta a^\sigma, \tag{3.20}$$

where

$$U^i_\sigma(\mathbf{x}) = \left(\frac{\partial f^i(\mathbf{x}; \mathbf{a})}{\partial a^\sigma} \right)_{a=0}. \tag{3.21}$$

Note here that Einstein's summation notation is used.

The infinitesimal transformation $\mathbf{x} \to \mathbf{x} + d\mathbf{x}$ induces in $F(\mathbf{x})$ the transformation $F(\mathbf{x}) \to F(\mathbf{x}) + dF(\mathbf{x})$. Now

$$\begin{aligned} dF(\mathbf{x}) &= \frac{\partial F}{\partial x^i}\, dx^i \\ &= \frac{\partial F}{\partial x^i} U^i_\sigma \delta a^\sigma \\ &= \Delta a^\sigma U^i_\sigma \frac{\partial F}{\partial x^i} \\ &\equiv \Delta a^\sigma X_\sigma F. \end{aligned} \tag{3.22}$$

The operators

$$X_\sigma = U^i_\sigma \frac{\partial}{\partial x^i} \tag{3.23}$$

are called the *infinitesimal operators* of the transformation group. For example, we have already seen the following infinitesimal rotation:

$$\begin{pmatrix} x' \\ y' \end{pmatrix} = \left[I + \Delta\theta \begin{pmatrix} 0 & -1 \\ 1 & 0 \end{pmatrix} \right] \begin{pmatrix} x \\ y \end{pmatrix}$$

That is, $\delta x = -y\Delta\theta$, $\delta y = x\Delta\theta$. Therefore, from (3.19)–(3.21),

$$U(x) = \frac{\delta x}{\Delta\theta} = -y, \quad U(y) = \frac{\delta y}{\Delta\theta} = x.$$

According to (3.23), the infinitesimal operator X of the group is given as

$$\begin{aligned} X &= U(x)\frac{\partial}{\partial x} + U(y)\frac{\partial}{\partial y} = -y\frac{\partial}{\partial x} + x\frac{\partial}{\partial y} \\ &= x\frac{\partial}{\partial y} - y\frac{\partial}{\partial x}, \end{aligned} \tag{3.24}$$

which is the well-known infinitesimal rotation operator. In the same context, we can derive the infinitesimal operators for the following linear transformation, dilatation/contraction and shift:

$$x' = ax + b,$$

$$\delta x = x\Delta a + \Delta b.$$

Thus

$$X_a = x\frac{\partial}{\partial x}, \quad X_b = \frac{\partial}{\partial x}. \tag{3.25}$$

3.2 IIJIMA'S NORMALIZATION THEORY

Iijima built a voluminous and sophisticated systematic theory of normalization [3]. However, it is a basic theory rather than a practical one. Therefore we will explain its essential idea, using a one-dimensional representation that is closely related to the blurring mentioned in Section 2.6.

Three kinds of transformations were introduced as

$$\begin{aligned} &(1) \quad f(x) = Ag(x), \\ &(2) \quad f(x) = g(x-a), \\ &(3) \quad f(x) = g\left(\frac{x}{\lambda}\right). \end{aligned} \tag{3.26}$$

A set of Ω_0' was constructed by performing the above transformations on an ideal pattern, $g(x)$. However, in real situations these three kinds of transformations are mixed. That is, after the transformation (1) another transformation, say (2), is performed on the $Ag(x)$, which results in $Ag(x-a)$. Thus combined transformation gives the following form in general:

$$f(x) = Ag\left(\frac{x-a}{\lambda}\right). \tag{3.27}$$

Furthermore the above $f(x)$ itself is transformed as

$$f^*(x) = Af\left(\frac{x-a}{\lambda}\right). \tag{3.28}$$

It is desirable that Ω_0' be closed under the transformation given in (3.26), which is easily proved as follows:

$$f^*(x) \equiv A'f\left(\frac{x-a'}{\lambda'}\right),$$

$$f^*(x) = A'Ag\left\{\frac{[(x-a')/\lambda'] - a}{\lambda}\right\}.$$

Setting $A^* \equiv A'A$, $a^* \equiv \lambda' a + a'$, $\lambda^* \equiv \lambda'\lambda$,

$$f^*(x) = A^* g\left(\frac{x - a^*}{\lambda^*}\right).$$

In other words, three transformations, denoted as $T(A)$, $T(a)$, and $T(\lambda)$ are closed under the product operation. Each consists of a group, and their product also consists of a group.

Now we introduce a general transformation including blurring. It is defined as

$$F(x) \equiv \Phi(f(x'); x, A, a, \lambda, \sigma), \tag{3.29}$$

where Φ is a functional and σ is a parameter of the blurring transformation. Specifically, the blurring transformation is given as

$$\text{(4)}\quad f(x) = \Phi(g(x); x, 1, 0, 1, \sigma)$$

$$\equiv \Phi_0(g(x'); x, \sigma), \tag{3.30}$$

where the functional Φ_0 has the form

$$\Phi_0(g(x'); x, \sigma) = \int_{-\infty}^{\infty} \phi(g(x'); x, x', \sigma) dx'. \tag{3.31}$$

The actual form of the functional ϕ was given as a Gaussian functional in Section 2.2. Thus Ω_0' is expanded by introducing the blurring transformation to a set Ω_0. The following theorem is proved; that is, any pattern $f(x)$ that belongs to Ω_0 is represented as

$$f(x) = \Phi(g(x'); x, A, a, \lambda, \sigma)$$

$$= A\Phi\left(g(x'); \frac{x - a}{\lambda}, \frac{\sigma}{\lambda}\right). \tag{3.32}$$

The set Ω_0 is also closed under the introduction of the blurring transformation. Therefore all the observed character images can be regarded as being obtained by the linear transformation of (3.26) from an ideal pattern $g(x)$, which guarantees that the $g(x)$ can be inferred from any intended observed pattern $f(x)$.

For an ideal pattern, $g(\xi)$, $g(\xi, \rho)$ is defined as

$$g(\xi, \rho) \equiv \Phi_0[g(\xi'); \xi, \rho]. \tag{3.33}$$

Then any pattern $f(x)$ that belongs to Ω_0 is expressed as

$$f(x) = Ag\left(\frac{x - a}{\lambda}, \frac{\sigma}{\lambda}\right). \tag{3.34}$$

Now the problem of normalization is to decide the parameters A, a, λ, and σ, and the ideal pattern $g(\xi)$ from a given character image $f(x)$. The problem is complicated by introduction of the blurring transformation. In order to decide the parameters, some criterion is necessary as mentioned in Chapter 2. It takes the form

$$\int_{-\infty}^{\infty} f(x, \sigma)c(x)dx = \text{constant value.} \tag{3.35}$$

Usually $c(x)$ is chosen from moment functions from their intuitive meanings. For example, it is quite reasonable to take $x - a$ as $c(x)$ in order to determine the shift transformation parameter a, since it tells that its average value is set to zero, namely balanced at the origin when $f(x)$ is shifted by a, for example. However, the difficult point is that $f(x)$ is a blurred function. Therefore it is explicitly expressed as $f(x, \sigma)$ in the equation above.

It must be guaranteed that the above integration has a constant value regardless of σ. In this respect, if we take $c(x)$ as a constant value, then it is intuitively clear that the integration does not depend on σ. We have to find a series of functions that have such a property in order to solve the normalization from a theoretical point of view. To do so, considerable preparation is necessary. Therefore we follow only the scenario of the theory. The key thing is that the $g(\xi, \rho)$ satisfies the following diffusion partial differential equation.

Theorem 3.2.1

$$\left\{\frac{\partial^2}{\partial \xi^2} - \frac{\partial}{\partial \rho^2}\right\} g(\xi, \rho) = 0 \qquad (\rho > 0). \tag{3.36}$$

This is easily proved according to (1.27):

$$g(\xi, \rho) = \int_{-\infty}^{\infty} g(\xi')\phi\left(\frac{\xi - \xi'}{\rho}\right)\frac{d\xi'}{\rho},$$

where

$$\phi\left(\frac{\xi - \xi'}{\rho}\right)\frac{1}{\rho} = \frac{1}{2\pi}\int_{-\infty}^{\infty} e^{-\rho^2\xi^2 + i\xi(\xi - \xi')}d\xi.$$

On the other hand, the following equation holds,

$$\left\{\frac{\partial^2}{\partial \xi^2} - \frac{\partial}{\partial \rho^2}\right\} e^{-\rho^2\xi^2 + i\xi(\xi - \xi')} = 0.$$

Therefore

$$\left\{\frac{\partial^2}{\partial \xi^2} - \frac{\partial}{\partial \rho^2}\right\} \phi\left(\frac{\xi - \xi'}{\rho}\right)\frac{1}{\rho} = 0$$

holds. On multiplying by $g(\xi')$ both sides of the equation above and integrating it about ξ' in the range of $(-\infty, \infty)$, (3.36) results. In the same context, it can be proved that the following equation also holds:

$$\left\{\frac{\partial}{\partial x^2} - \frac{\partial}{\partial \sigma^2}\right\} f(x, A, a, \lambda, \sigma) = 0. \tag{3.37}$$

Next we introduce an invariant form and bi-orthogonal function series, $h(\xi, \rho)$ and $h^*(\xi, \rho)$ as mathematical preparation.

Theorem 3.2.2 *If the $h(\xi, \rho)$ and $h^*(\xi, \rho)$ satisfy the following equations respectively*

$$\left\{\frac{\partial^2}{\partial \xi^2} + \frac{\partial}{\partial \rho^2}\right\} h(\xi, \rho) = 0, \tag{3.38}$$

$$\left\{\frac{\partial^2}{\partial \xi^2} - \frac{\partial}{\partial \rho^2}\right\} h^*(\xi, \rho) = 0, \tag{3.39}$$

and they satisfy the boundary condition,

$$\lim_{\xi \pm \infty} \left\{ h(\xi, \rho) \frac{\partial}{\partial \xi} h^*(\xi, \rho) - h^*(\xi, \rho) \frac{\partial}{\partial \xi} h(\xi, \rho) \right\} = 0. \tag{3.40}$$

then the following integration takes a constant value regardless of ρ,

$$\int_{-\infty}^{\infty} h(\xi, \rho) h^*(\xi, \rho) d\xi. \tag{3.41}$$

Actually we can find such series of functions explicitly, that is the following theorem.

Theorem 3.2.3 *Corresponding $h(\xi, \rho)$ and $h^*(\xi, \rho)$, the following $P_n(\xi, \rho)$ and $P_n^*(\xi, \rho)$ are defined.*

$$P_n(\xi, \rho) \equiv (i)^n \left(\frac{\partial^n}{\partial \zeta^n} e^{\rho^2 \zeta^2 - i\xi\zeta} \right)_{\zeta = 0},$$

$$P_n^*(\xi, \rho) \equiv (-1)^n \frac{\partial^n}{\partial \xi^n} \left(\frac{1}{2\pi} \int_{-\infty}^{\infty} e^{-\rho^2 \zeta^2 + i\xi\zeta} d\zeta \right) \qquad (n = 0, 1, \ldots) \tag{3.42}$$

The above two series of functions satisfy the following bi-orthogonal relations:

$$\int_{-\infty}^{\infty} P_n(\xi, \rho) P_{n'}^*(\xi, \rho) d\xi = \begin{cases} n! & (n = n'), \\ 0 & (n \neq n'). \end{cases} \tag{3.43}$$

The explicit form of the series of functions, $P_n(\xi, \rho)$, is given as follows:

$$P_{2l}(\xi, \rho) = \xi^{2l} \sum_{r=0}^{l} (-1)^r \frac{(2r)!}{r!} \binom{2l}{2r} \left(\frac{\rho}{\xi}\right)^{2r}$$

$$P_{2l+1}(\xi, \rho) = \xi^{2l+1} \sum_{r=0}^{l} (-1)^r \frac{(2r)!}{r!} \binom{2l+1}{2r} \left(\frac{\rho}{\xi}\right)^{2r} \qquad (l = 0, 1, 2, \cdots) \quad (3.44)$$

The functions $P_n(\xi, \rho)$ and $P_n^*(\xi, \rho)$ are very closely related to Hermite polygonals, $H_n(x)$. This might be expected by looking at (3.42), since Hermite polygonals have the form

$$H_n(x) \equiv (-1)^n e^{x^2/2} \frac{d^n}{dx^n} e^{-(x^2/2)}. \qquad (3.45)$$

Using the above definition, the following relations are derived:

$$P_n^*(\xi, \rho) = \frac{1}{\sqrt{2\pi}(\sqrt{2}\rho)^{n+1}} H_n\left(\frac{\xi}{\sqrt{2}\rho}\right) e^{-(\xi^2/4\rho^2)}, \qquad (3.46)$$

$$P_n(\xi, \rho) = (\sqrt{2}\rho)^n H_n\left(\frac{\xi}{\sqrt{2}\rho}\right). \qquad (3.47)$$

Now the preparation is complete for normalizing a given function $f(x)$ that is an input image to a character recognition system. First of all the given function $f(x)$ is set as $f(x, \sigma)$. As stated before, any such function $f(x, \sigma)$ satisfies the partial differential equation

$$\left(\frac{\partial^2}{\partial x^2} - \frac{\partial}{\partial \sigma^2}\right) f(x, \sigma) = 0. \qquad (3.48)$$

Therefore we can regard $f(x)$ as coinciding with $f(x, \sigma)$ with some specific value of σ. In other words, if $f(x, \sigma)$ is obtained solving the equation above as a boundary condition of $f(x) = f(x, \sigma)$, then $f(x, \sigma)$ is an analytic extension of $f(x)$ to any σ.

By Theorem 3.3, the following equation holds:

$$\left(\frac{\partial^2}{\partial x^2} + \frac{\partial}{\partial \sigma^2}\right) P_n(x - a, \sigma) = 0. \qquad (3.49)$$

Furthermore, by Theorem 3.2, the following integrals take constant values regardless of σ:

$$\int_{-\infty}^{\infty} f(x, \sigma) P_n(x - a, \sigma) dx \qquad (n = 0, 1, 2, \cdots). \qquad (3.50)$$

Notice that $\lim_{x \to \pm\infty} f(x, \sigma) = 0$ and that $\lim_{x \to \pm\infty} \partial f(x, \sigma)/\partial x = 0$, so the boundary condition (3.40) holds. Therefore four $P_n(x - a, \sigma)$, $n = 0, 1, 2, 4$, are chosen as $c(x)$'s of (3.35). In other words, the following criteria are set:

$$\int_{-\infty}^{\infty} f(x, \sigma) P_0(x - a, \sigma) dx = A\lambda, \tag{3.51}$$

$$\int_{-\infty}^{\infty} f(x, \sigma) P_1(x - a, \sigma) dx = 0, \tag{3.52}$$

$$\int_{-\infty}^{\infty} f(x, \sigma) P_2(x - a, \sigma) dx = 0, \tag{3.53}$$

$$\int_{-\infty}^{\infty} f(x, \sigma) P_4(x - a, \sigma) dx = (-1) 8 A \lambda^5. \tag{3.54}$$

In these criteria $P_n(x - a, \sigma)$, $n = 0, 1, 2, 4$, are explicitly expressed as follows according to (1.47):

$$\begin{aligned} P_0(x - a, \sigma) &= 1, \\ P_1(x - a, \sigma) &= (x - a), \\ P_2(x - a, \sigma) &= (x - a)^2 - 2\sigma^2, \\ P_4(x - a, \sigma) &= (x - a)^4 - 12(x - a)^2\sigma^2 + 12\sigma^4. \end{aligned} \tag{3.55}$$

Substituting these functions into (3.51) through (3.54), the transformation parameters are derived. First, from (3.52) parameter a is derived:

$$a = \frac{\int_{-\infty}^{\infty} x f(x) dx}{\int_{-\infty}^{\infty} f(x) dx}. \tag{3.56}$$

From (3.53) parameter σ is derived,

$$\sigma^2 = \frac{1}{2} \frac{\int_{-\infty}^{\infty} (x - a)^2 f(x) dx}{\int_{-\infty}^{\infty} f(x) dx}. \tag{3.57}$$

Now, eliminating A using (3.51) and (3.54), the following equation is derived:

$$-8\lambda^4 \int_{-\infty}^{\infty} f(x) dx = \int_{-\infty}^{\infty} (x - a)^4 f(x) dx - 12\sigma^2 \int_{-\infty}^{\infty} (x - a)^2 f(x) dx + 12\sigma^4 \int_{-\infty}^{\infty} f(x) dx.$$

Substituting (1.50) into the equation above, parameter λ is derived:

$$\lambda^4 = \frac{3}{2}\left(\sigma^4 - \frac{1}{12}\frac{\int_{-\infty}^{\infty}(x-a)^4 f(x)dx}{\int_{-\infty}^{\infty} f(x)dx}\right). \tag{3.58}$$

From (3.51) parameter A is derived:

$$A = \frac{1}{\lambda}\int_{-\infty}^{\infty} f(x)dx. \tag{3.59}$$

Except for λ, these results are reasonable and intuitive. Concerning parameter a, the result is as expected; namely it is the average value of $f(x)$. The parameter σ^2 is also reasonable, since it is the dispersion of $f(x)$. The parameter A is reasonable, since A is determined by the total strength of $f(x)$ and proportional to $1/\lambda$. However, the parameter λ has a somewhat complicated meaning which causes another problem. That is, so long as $f(x) > 0$, a and σ are real values, but λ must satisfy the relation

$$12\sigma^4 \int_{-\infty}^{\infty} f(x)dx > \int_{-\infty}^{\infty}(x-a)^4 f(x)dx. \tag{3.60}$$

Setting $v^2 = 6\sigma^2$ and rewriting the relation, we reach the relation

$$\begin{aligned}&\int_0^v x^2(v^2-x^2)\{f(a+x)+f(a-x)\}dx \\ &> \int_v^{\infty} x^2(x^2-v^2)\{f(a+x)+f(a-x)\}dx.\end{aligned} \tag{3.61}$$

This relation tells us that the main part of $f(x)$ should be concentrated within the range of $(a-v, a+v)$ in order for λ to be real.

3.2.1 An Ideal Pattern

Finding an ideal pattern means arriving at the "concept" of a character. Formally an ideal pattern can be derived from

$$g(\xi, \rho) = \int_{-\infty}^{\infty} g(\xi')\phi\left(\frac{\xi-\xi'}{\rho}\right)\frac{d\xi'}{\rho}$$

as

$$\lim_{\rho\to 0} g(\xi, \rho) = \lim_{\rho\to 0}\int_{-\infty}^{\infty} g(\xi-\rho u)\phi(u)du$$

$$= g(\xi) \int_{-\infty}^{\infty} \phi(u) du$$

$$= g(\xi). \tag{3.62}$$

On the other hand,

$$\left\{ \frac{\partial^2}{\partial \xi^2} - \frac{\partial}{\partial \rho^2} \right\} g(\xi, \rho) = 0$$

and

$$\left\{ \frac{\partial^2}{\partial \xi^2} + \frac{\partial}{\partial \rho^2} \right\} P_n(\xi, \rho) = 0$$

holds. Therefore, by Theorem 3.2, we obtain the invariant form

$$C_n \equiv \frac{1}{n!} \int_{-\infty}^{\infty} g(\xi, \rho) P_n(\xi, \rho) d\xi \qquad (n = 0, 1, 2, \cdots) \tag{3.63}$$

Thus $\{C_n\}$ is a set of constant values regardless of ρ. On the other hand, the following expression is derived:

$$g(\xi, \rho) = \sum_{n=0}^{\infty} \frac{1}{n!} P_n^*(\xi, \rho) \int_{-\infty}^{\infty} g(\xi', \rho') P_n(\xi', \rho') d\xi'.$$

Substituting (3.63), we obtain

$$g(\xi, \rho) = \sum_{n=0}^{\infty} C_n P_n^*(\xi, \rho). \tag{3.64}$$

Therefore the ideal pattern $g(\xi)$ is expressed as

$$g(\xi) = \sum_{n=0}^{\infty} C_n \left\{ \lim_{\rho \to 0} P_n^*(\xi, \rho) \right\}.$$

This is a formal solution for an ideal pattern. However, the functions $\lim_{\rho \to 0} P^*(\xi, \rho)$ all converge to zero when $\xi \neq 0$ but diverge at $\xi = 0$, so they are all singular functions. In contrast, $\{C_n\}$ is equivalent to $g(\xi, \rho)$ regardless of ρ. Therefore $\{C_n\}$ is called *a characteristic constant system of patterns.* In this sense we could consider that normalization is to find $\{C_n\}$ in its abstract sense. This is an interesting point, since we have pursued an ideal pattern as a function on a two-dimensional plane that has the same basic property as the observed pattern. However, it was found that this is a very difficult process. Yet the ideal pattern can be clearly expressed by $\{C_n\}$, the abstracted space, or functional space. This gives a motivation for research of pattern recognition on a functional space. At least, it suggests that "the concept" of a character can be abstracted considerably more than the shape recognized by the naked eye. This story continues to Amari's theory of normalization on feature space.

The preceding result is stated in the next theorem.

Theorem 3.2.4 *For any given characteristic function $g(\xi, \rho)$, if*

$$C_n = \frac{1}{n!}\int_{-\infty}^{\infty} g(\xi, \rho)P_n(\xi, \rho)d\xi \qquad (n = 0, 1, 2, \cdots) \tag{3.65}$$

is constructed, $\{C_n\}$ becomes a constant value system regardless of ρ. Conversely, $g(\xi, \rho)$ is expressed by this $\{C_n\}$ as follows:

$$g(\xi, \rho) = \sum_{n=0}^{\infty} C_n P_n^*(\xi, \rho), \tag{3.66}$$

where the following four expansion coefficients take constants regardless of $g(\xi, \rho)$:

$$C_0 = 1, \quad C_1 = 0, \quad C_2 = 0, \quad C_4 = -\frac{1}{3}. \tag{3.67}$$

Note that the four parameters, A, a, λ, and σ have disappeared in the construction of $g(\xi, \rho)$. In other words, $g(\xi, \rho)$ loses four degrees of freedom when it is seen as a general function.

3.3 AMARI'S NORMALIZATION THEORY

The basic idea of Amari's theory is that normalization can be done on feature space rather than on observed space or on an image plane [4]. Amari gave the condition that a linear transformation, say τ, on the observed space is equivalent to a linear transformation on a feature space, say F.

3.3.1 Groups of Infinitesimal Transformation

Now the theory in its essential parts will be described in detail. First, the transformations on observed space are formalized in infinitesimal form. Specifically, a linear transformation operator $t : f(x, y) \rightarrow g(x, y)$ is represented by the integral

$$tg(x, y) = \int_R t(x, y; x', y')g(x', y')dx'dy', \tag{3.68}$$

where a transformation operator t is embodied by a kernel $t(x, y; x', y')$. Three kinds of operators are considered:

$$t_{00} : f(x, y) \to (1 + \epsilon)f(x, y),$$
$$t_{10} : f(x, y) \to f(x - \epsilon, y),$$
$$t_{01} : f(x, y) \to f(x, y - \epsilon).$$

The meanings of these operators are obvious, as mentioned earlier.

The corresponding infinitesimal operators, called *generators,* are deduced from the following definition:

$$\bar{t} = \frac{dt}{d\epsilon}\bigg|_{\epsilon = 0}.$$

Thus the following three infinitesimal generator are obtained according to (3.25):

$$\bar{t}_{00} = I, \tag{3.69}$$

$$\bar{t}_{10} = -\frac{\partial}{\partial x}, \tag{3.70}$$

$$\bar{t}_{01} = -\frac{\partial}{\partial y}. \tag{3.71}$$

The group generated by $\{\bar{t}_{00}, \bar{t}_{10}, \bar{t}_{01}\}$ is denoted by T_1.

Second, dilatation/contraction operators are introduced as follows:

$$t_{20} : f(x, y) \to (1 + \epsilon)f((1 + \epsilon)x, y),$$
$$t_{02} : f(x, y) \to (1 + \epsilon)f(x, (1 + \epsilon)y).$$

These operators are a little different from the usual ones seen before, but the essential parts are $(1 + \epsilon)x$ and $(1 + \epsilon)y$ terms in the arguments. The amplitude parts are for the sake of technique, in order to make the derivation of the theory easier as we will see.

Thus the infinitesimal generators obtained are

$$\bar{t}_{20} = I + x\frac{\partial}{\partial x},$$

$$\bar{t}_{02} = I + y\frac{\partial}{\partial x}. \tag{3.72}$$

The group generated by $\{t_{00}, t_{10}, t_{01}, t_{20}, t_{02}\}$ is denoted as T_2.

Third, the generator of rotation

$$t_{11} : f(x, y) \to f(x \cos \epsilon - y \sin \epsilon, x \sin \epsilon + y \cos \epsilon)$$

is introduced. In a similar manner its infinitesimal generator $\bar{t}_{11}$ is given as

$$\bar{t}_{11} = x\frac{\partial}{\partial y} - y\frac{\partial}{\partial x}. \tag{3.73}$$

The group generated by $\{t_{00}, t_{10}, t_{01}, t_{20}, t_{02}, t_{11}\}$ is denoted as T_3.

Finally, the generator of the shear

$$t'_{11} : f(x, y) \to f(x \cos \epsilon + y \sin \epsilon, x \cos \epsilon + y \cos \epsilon)$$

is introduced. The infinitesimal generator is

$$\bar{t}'_{11} = x\frac{\partial}{\partial y} + y\frac{\partial}{\partial x}. \tag{3.74}$$

As mentioned before, decomposition of the transformation is not unique, and shear is generated by rotating and dilatation/contraction operators. Therefore the shear is included in T_3. Obviously $T_1 \subset T_2 \subset T_3$ holds.

3.3.2 Admissible Feature Space

Before turning to Amari's theory, a little mathematical preparation and definitions of the notations are given here for better understanding. We are interested in features that are denoted as a vector ξ where its components are expressed by ξ^k, $k = 1, \ldots, n$. On the other hand, an image on the observed space represented as $f(x, y)$ is also represented by (x^i) $i = 1, 2, \ldots, N$ whose dimension depends on the sampling of the observed space, denoted as N. Of course n is less than N, since feature extraction has reduced this dimension. Now the linear extraction is represented as

$$\xi^k = A_i^k x^i, \tag{3.75}$$

where Einstein's summation convention is used—namely every pair of indices is repeated twice, once as a subscript and again as a superscript—which is always summed over. This convention is widely used in relativity theory and Riemannian geometry, and it makes formulations clear. As an example, the ordinary expression of the above formula is given below:

$$\begin{pmatrix} \xi^1 \\ \xi^2 \\ \vdots \\ \xi^n \end{pmatrix} = \begin{pmatrix} A_1^1 & A_2^1 & \cdots & A_N^1 \\ A_1^2 & A_2^2 & \cdots & A_N^2 \\ \vdots & & & \\ A_1^n & A_2^n & \cdots & A_N^n \end{pmatrix} \begin{pmatrix} x^1 \\ x^2 \\ \vdots \\ x^N \end{pmatrix}.$$

The n vectors $\mathbf{A}^k = (A_i^k)(k = 1, \ldots, n)$, that is, each row of the matrix A_i^k, are called *measuring vectors*. Therefore (3.75) defines the mapping from the image space or observed space to the feature space F.

The space spanned by the measuring vectors denoted as M is a subspace of the image space. The feature extraction is a projection of an image on M, called the *mea-*

suring space. (For a more detailed discussion, see Chapter 7.) The transformation defined by (3.68) is also simply represented using the preceding discrete image representation as

$$t\mathbf{x} = t^i_j x^j, \tag{3.76}$$

where t^i_j is a tensor in general. Strictly speaking, the vector $\mathbf{x} = (x^j)$ is a contravariant vector. On the other hand, $\mathbf{A} = (A_i)$ is a covariant vector where $\mathbf{A}$ represents in general such a form of vector as $\mathbf{A}^k = (A^k_i)$. The difference between covariant vector and contravariant vector corresponds to that of row vector and column vector. Since each x^i is not necessarily a component of an orthogonal axis, the difference becomes necessary. The same is true for a matrix, which is generalized to a tensor. However, more mathematically speaking, they constitute dual space. Suppose a scalar product $(\mathbf{A}, \mathbf{x})$; then the functional $(\mathbf{A}, \cdot)$ takes real value for given contravariant vector $\mathbf{x}$. Such an $(\mathbf{A}, \cdot)$ constitutes a vector space that is called a *dual space of the vector space of* $\mathbf{x}$. Therefore $(\mathbf{A}, t\mathbf{x})$ induces a conjugate transformation t^*, $(t^*\mathbf{A}, \mathbf{x})$ so that they are equal. $t^*\mathbf{A}$ is represented by its component as

$$t^*\mathbf{A} = t^i_j A_i. \tag{3.77}$$

Returning to the measuring vectors $\mathbf{A}^k$, A^k_i are transformed by t^* as

$$t^*\mathbf{A}^k = t^i_j A^k_i.$$

Now we are ready to discuss Amari's first theory of admissible transformation, which describes the relation between a linear transformation on image space and that on feature space. Specifically, concerning the relation, one question arises on whether it is possible to determine the features ξ'^k of x'^i from the features ξ^k of the original x^i. If this is possible, we have

$$\xi' = \tau\xi = (\tau^\lambda_k \xi^k), \tag{3.78}$$

where τ is a linear transformation represented by a tensor τ^λ_k in the feature space. The transformation t can be carried into effect equivalently by the induced transformation τ in the feature space. The scheme is represented by the following diagram:

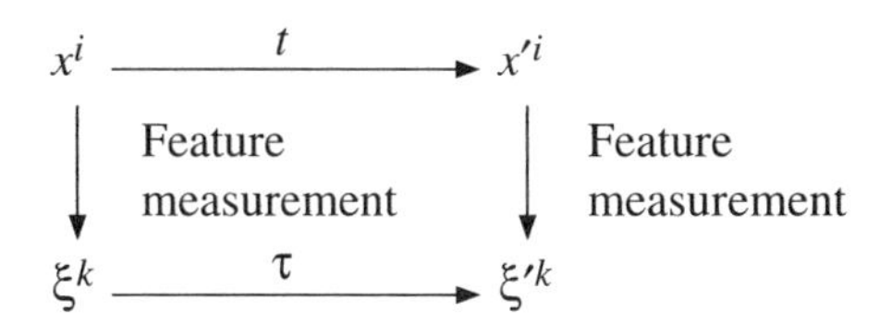

Definition 3.3.1 *When there exists a transformation τ playing the aforementioned role for a given t, we say that the feature space F admits t. When M admits all the transformations of a group T, we say that M admits T.*

Concerning admissibility, the following theorem is proved:

Theorem 3.3.1 *A necessary and sufficient condition for the measuring space M to admit a transformation t is that M be closed under the conjugate transformation t*; that is, for arbitrary* $\mathbf{A} \in M$,

$$t^*\mathbf{A} \in M$$

holds.

Proof If M admits t, there exists a transformation τ,

$$\xi'^k = \tau^k_\lambda \xi^\lambda. \tag{3.79}$$

On the other hand, ξ'^k and ξ^k are given by the feature measurements, as

$$\xi'^k = A^k_i t^i_j x^j, \tag{3.80}$$

$$\xi^\lambda = A^\lambda_j x^j. \tag{3.81}$$

Therefore

$$\xi'^k = A^k_i t^i_j x^j = \tau^k_\lambda A^\lambda_j x^j.$$

That is,

$$\tau^k_\lambda A^\lambda_j = t^i_j A^k_i \tag{3.82}$$

holds.

Since A^λ_j's are the basis vector of M, $t^*\mathbf{A}^k = (t^i_j A^k_i)$ is also included in M according to the equation above. Hence we can conclude that M is closed under t^*. The sufficiency is obvious. The τ is called *the feature transformation* corresponding to t. Let B^i_μ be the vectors satisfying

$$A^k_i B^i_\mu = \delta^k_\mu. \tag{3.83}$$

Then τ is given by

$$\tau^k_\lambda = A^k_j t^j_i B^i_\lambda. \tag{3.84}$$

Now we proceed to obtain the feature space admitting the transformation group T_i. Since an image is represented by the function $f(x, y)$, the matrix[1] A^k_i connecting the image x^i with the features ξ^k is replaced by a set of functions $A^k(x, y)$. Thus

[1] A^k_i is called a vector or a matrix. If A^k_i is considered individually, then it is called a vector. On the other hand, if A^k_i is considered as a whole, it is called a matrix: n vectors A^k_i together form an $n \times N$ matrix A^k_i.

$$\xi^k = \int_R A^k(x, y) f(x, y)dxdy. \tag{3.85}$$

The equation above tells that the measuring functions $A^k(x, y)$, $k = 1, 2, \ldots, n$ span the measuring space M_i by which T_i is admitted.

In order to obtain M_i, we need to calculate the conjugate transformations of the generators of T_i. As mentioned before, the conjugate t^* of t satisfies the equation

$$A \cdot (tx) = (t^*A) \cdot x. \tag{3.86}$$

For the generator $\bar{t}_{10}$, the following relation is obtained:

$$\begin{aligned} A \cdot (\bar{t}_{10}x) &= -\int A(x, y) \cdot \frac{\partial}{\partial x} f(x, y)dxdy \\ &= \int \left(\frac{\partial}{\partial x} A(x, y) \right) f(x, y)dxdy. \end{aligned}$$

Therefore $\bar{t}^*_{10}$ is given as

$$\bar{t}^*_{10} = \frac{\partial}{\partial x}.$$

In the same manner we have the following results:

$$\begin{aligned} &\bar{t}^*_{00} = I, \\ &\bar{t}^*_{10} = \frac{\partial}{\partial x}, \quad \bar{t}^*_{01} = \frac{\partial}{\partial x}, \\ &\bar{t}^*_{20} = -x\frac{\partial}{\partial x}, \quad \bar{t}^*_{02} = -y\frac{\partial}{\partial y}, \\ &\bar{t}^*_{11} = -x\frac{\partial}{\partial y} + y\frac{\partial}{\partial x}. \end{aligned} \tag{3.87}$$

Using the above results and Theorem 3.1, the following theorem is proved. ■

Theorem 3.3.2 *The measuring space M, which admits the group T_1, is the space spanned by the functions*

$$\{x^p y^p e^{\alpha x + \beta y}, x^p y^p e^{\bar{\alpha} x + \bar{\beta} y}\}$$

on the direct sum of such space, where p and q stand for all the integers satisfying

$$0 \le p \le k,$$

$$0 \le q \le k',$$

where k and k′ are arbitrary integers, α and β are arbitrary complex numbers, and $\overline{\alpha}$ is the complex conjugate of α.

Proof From the proof the readers will learn the reason why the infinitesimal transformations were introduced. We will find the relationship that constrains any function $g(x, y)$ belonging to the measuring space M_1. That is, $t^*g(x, y)$ must also belong to the space for all $t \in T_1$ according to Theorem 3.3.1. For $\bar{t}^*_{00}$ it is obvious, and so it suffices for M_1 to include only $\bar{t}^*_{10}g(x, y)$ and $\bar{t}^*_{01}g(x, y)$. According to (3.87),

$$\bar{t}^*_{10}g(x, y) = \frac{\partial}{\partial x} g(x, y)$$

holds. Therefore all of $(\partial/\partial x)g(x, y)$, $(\partial^2/\partial x^2)g(x, y)$, $(\partial^3/\partial x^3)g(x, y), \ldots,$ should be included in M_1, since M_1 must be closed under the operation $\partial/\partial x$. However, M_1 can be spanned by only a finite number of functions because of the finite dimension of the space. Therefore, for some r, a linear dependency relation

$$\frac{\partial^{r+1}}{\partial x^{r+1}} g = \sum_{i=0}^{r} a_i \frac{\partial^i}{\partial x^i} g \tag{3.88}$$

must hold where a_i $(i = 1, \ldots, r)$ are constant. By setting $g(x, y) \equiv h(x)l(y)$, the above equation is transformed to an ordinary differential equation with constant coefficients whose solutions are well-known, namely

$$\frac{d^{r+1}}{dx^{r+1}} h(x) = \sum_{i=0}^{r} a_i \frac{d^i}{dx^i} h(x).$$

Solving the above equation, we obtain the following independent solutions:

$$x^p e^{\alpha x}, \quad x^p e^{\overline{\alpha} x}, \qquad 0 \le p \le k,$$

where α and $\overline{\alpha}$ are the roots of the characteristic equation

$$z^{r+1} = \sum_{j=0}^{r} a_j z^j.$$

Here $r + 1$ is the multiplicity of the root α. The same analysis can be applied to $\bar{t}^*_{01}g(x, y)$, and so the theorem is proved. ■

As special cases of the theorem, we note the following results: The feature space composed of the moments of the signals,

$$\xi^{pq} = \int_R x^p y^q f(x, y) dx dy, \tag{3.89}$$

$$0 \le p \le k,$$

$$0 \le q \le k',$$

admits T_1, and the feature space composed of the Fourier component the signals

$$\xi[i, j] = \int e^{\alpha_i x + \beta_j y} f(x, y) dxdy \tag{3.90}$$

admits T_1.

Now let us move to T_2, in which t_{20} and t_{02} are included in addition to the generators of T_1. Therefore the functions belonging to M_2 are further restricted by t_{20} and t_{02}. Since

$$\bar{t}^*_{20} g = -x \frac{\partial}{\partial x} g, \quad \bar{t}^*_{10} g = \frac{\partial}{\partial x} g,$$

if $g \in M_2$, then M_2 includes all of the functions

$$x^k \frac{\partial^m}{\partial x^m} g : m \geq k \geq 0.$$

For $k = 0$ we can use the theorem above, setting g as $h(x)l(y)$, with $h(x)$ a linear combination of $x^p e^{\alpha x}$. However, for $k \neq 0$ it causes an infinite number of linear independent functions to be included in M_2, because it must be closed in the operation $x(\partial/\partial x)$. For M_2 to be closed under the operation $(x\partial/\partial x)$, the exponential part of $h(x)$ must be removed. The same argument can be applied to $\bar{t}^*_{02} g(x, y)$, namely $y(\partial/\partial y)l(y)$. Thus M_2 is spanned by

$$\{x^p, y^p\}, \qquad 0 \leq p \leq k, 0 \leq q \leq k'. \tag{3.91}$$

Moving to T_3, M_3 is closed under the rotation

$$\bar{t}^*_{11} = -x \frac{\partial}{\partial y} + y \frac{\partial}{\partial x}$$

and the shear

$$\bar{t}'^*_{11} = -x \frac{\partial}{\partial y} - y \frac{\partial}{\partial x}.$$

That is, M_3 must be closed under the operations $x(\partial/\partial y)$ and $y(\partial/\partial x)$. Taking $g(x, y) = x^p y^p$,

$$x \frac{\partial}{\partial y} (x^p y^q) = q x^{p+1} y^{q-1},$$

$$y \frac{\partial}{\partial x} (x^p y^q) = p x^{p-1} y^{q+1},$$

are obtained. Therefore, if M_3 includes $x^p y^q$ $(p + q = k)$, all of $x^{p'} y^{q'}$ $(p' + q' = k)$ should be included in it. That is, all of

$$x^p y^q, \qquad 0 \le p + q \le k, \tag{3.92}$$

together form a basis of M_3. Thus these theoretical results reveal the meaning of moment functions. They are the sole measuring functions admitting the group of the affine transformations. The Amari theory also gives the basis of the so-called invariant feature extraction, which will be described later. Amari gave a concrete procedure for normalization. We feel that the value of his theory lies in the basic results mentioned above. A more powerful normalization method in terms of both theory and practice developed by Nishida will be given next.

3.4 LINEAR NORMALIZATION

Suppose that two (binary) images O and M are given. The image O is called an "object" or "input," and the image M is called a "model" or "prototype." *Normalization* is an operation of transforming a point $\boldsymbol{x}'$ on the object to $\boldsymbol{x}$ on the model:

$$\boldsymbol{x} = T\boldsymbol{x}'. \tag{3.93}$$

To calculate T from the pair of images (O, M), we need to know the point correspondence between the model M and the object O. The problem of finding the point correspondence between two images is called *image matching*. The problem of image matching is essential in pattern recognition, and many matching methods have been explored.

One excellent approach to this problem was proposed by Henry S. Baird [5] in his Ph.D. thesis at Princeton University. He analyzed the problem of recognizing rigid shapes in a plane that had been subjected to unknown distortions. The goals of recognition are to locate the overall pattern in the image and to match each of its features with the corresponding model feature. Distortions include arbitrary translation, rotation, and scaling, and noise that is bounded within convex polygons. A pruned-tree search method is developed that makes efficient use of the Soviet ellipsoid algorithm for testing the feasibility of linear constraints. An interesting blend of theoretical analysis and practical implementation shows that the resulting algorithm has an expected runtime that is theoretically asymptotically quadratic in the number of feature points but is in practice linear for patterns with fewer than 100 points. Only knowledge of location of features is used by the algorithm, but it is easy to exploit additional information within the framework of pruned-tree search. The approach can be extended to 3D patterns and more general affine distortions.

Now let us suppose that we have found the mapping between a reference point on the model $\boldsymbol{x} = (x, y)^t$ (t is transpose of a matrix) and the corresponding point on the object $\boldsymbol{x}' = (x', y')^t$. We explicitly give T as an affine transformation:

$$T : \boldsymbol{x} \to A\boldsymbol{x} + \boldsymbol{b}, \tag{3.94}$$

where $A = (a_{ij})$ is a 2×2 regular matrix whose determinant, det A, is positive, and $\boldsymbol{b} = (b_x, b_y)^t$ is a translation vector. Therefore the relation between $\boldsymbol{x}$ and $\boldsymbol{x}'$ is expressed explicitly as

$$x = Ax' + b. \tag{3.95}$$

Let

$$\{\boldsymbol{x}_i = (x_i, y_i)^t \mid i = 1, 2, \ldots, N\}$$

be the set of reference points of the model M, and let

$$\{\boldsymbol{x}'_i = (x'_i, y'_i)^t \mid i = 1, 2, \ldots, N\}$$

be of the object O. Now, for simplicity, we assume that $\boldsymbol{x}_i$ ($i = 1, 2, \ldots, N$) corresponds to $\boldsymbol{x}'_i$. We obtain A and $\boldsymbol{b}$ that minimize

$$D_N = \sum_{i=1}^{N} \left\{ \left(\frac{x_i - a_{11}x'_i - a_{12}y'_i - b_x}{\sigma_x(i)} \right)^2 + \left(\frac{y_i - a_{21}x'_i - a_{22}y'_i - b_y}{\sigma_y(i)} \right)^2 \right\}, \tag{3.96}$$

where $\sigma_x(i)$ and $\sigma_y(i)$ are the standard deviations of distribution of (x_i, y_i). If $\sigma_x(i)$ and $\sigma_y(i)$ are unknown, they can be set to 1.

By differentiating D_N by a_{ij} ($i, j = 1, 2$), b_x, and b_y, we obtain the following system of linear equations:

$$\left\{ \begin{aligned}
&\sum_i \frac{x_i'^2}{\sigma_x(i)^2} a_{11} + \sum_i \frac{x'_i y'_i}{\sigma_x(i)^2} a_{12} + \sum_i \frac{x'_i}{\sigma_x(i)^2} b_x = \sum_i \frac{x_i x'_i}{\sigma_x(i)^2}, \\
&\sum_i \frac{x'_i y'_i}{\sigma_x(i)^2} a_{11} + \sum_i \frac{y_i'^2}{\sigma_x(i)^2} a_{12} + \sum_i \frac{y'_i}{\sigma_x(i)^2} b_x = \sum_i \frac{y'_i x_i}{\sigma_x(i)^2}, \\
&\sum_i \frac{x'_i}{\sigma_x(i)^2} a_{11} + \sum_i \frac{y'_i}{\sigma_x(i)^2} a_{12} + \sum_i \frac{1}{\sigma_x(i)^2} b_x = \sum_i \frac{x_i}{\sigma_x(i)^2}, \\
&\sum_i \frac{x_i'^2}{\sigma_y(i)^2} a_{21} + \sum_i \frac{x'_i y'_i}{\sigma_y(i)^2} a_{22} + \sum_i \frac{x'_i}{\sigma_y(i)^2} b_y = \sum_i \frac{x'_i y_i}{\sigma_y(i)^2}, \\
&\sum_i \frac{x'_i y'_i}{\sigma_y(i)^2} a_{21} + \sum_i \frac{y'_{i2}}{\sigma_y(i)^2} a_{22} + \sum_i \frac{y'_i}{\sigma_y(i)^2} b_y = \sum_i \frac{y'_i y_i}{\sigma_y(i)^2}, \\
&\sum_i \frac{x'_i}{\sigma_y(i)^2} a_{21} + \sum_i \frac{y'_i}{\sigma_y(i)^2} a_{22} + \sum_i \frac{1}{\sigma_y(i)^2} b_y = \sum_i \frac{y_i}{\sigma_y(i)^2}.
\end{aligned} \right. \tag{3.97}$$

A and $\boldsymbol{b}$ can be obtained by solving this system of equations.

We give an analysis of the matrix A and the vector $\boldsymbol{b}$ to give a quantitative description and estimation of geometric factors of pattern deformation. An affine transformation can be decomposed into the eight elementary transformations.

$$T_1(\alpha) : \boldsymbol{x} \rightarrow \boldsymbol{x} + \begin{pmatrix} \alpha \\ 0 \end{pmatrix} \qquad (x\text{-translation}), \tag{3.98}$$

$$T_2(\beta) : \boldsymbol{x} \to \boldsymbol{x} + \begin{pmatrix} 0 \\ \beta \end{pmatrix} \qquad (y\text{-translation}), \tag{3.99}$$

$$T_3(\lambda_x) : \boldsymbol{x} \to \begin{pmatrix} \lambda_x & 0 \\ 0 & 1 \end{pmatrix} \boldsymbol{x} \qquad (x\text{-dilatation}), \tag{3.100}$$

$$T_4(\lambda_y) : \boldsymbol{x} \to \begin{pmatrix} 1 & 0 \\ 0 & \lambda_y \end{pmatrix} \boldsymbol{x} \qquad (y\text{-dilatation}), \tag{3.101}$$

$$T_5(\gamma_x) : \boldsymbol{x} \to \begin{pmatrix} 1 & \gamma_x \\ 0 & 1 \end{pmatrix} \boldsymbol{x} \qquad (x\text{-shearing}), \tag{3.102}$$

$$T_6(\gamma_y) : \boldsymbol{x} \to \begin{pmatrix} 1 & 0 \\ \gamma_y & 1 \end{pmatrix} \boldsymbol{x} \qquad (y\text{-shearing}), \tag{3.103}$$

$$T_7(\mu) : \boldsymbol{x} \to \mu \boldsymbol{x} \qquad (\text{similarity}), \tag{3.104}$$

$$T_8(\theta) : \boldsymbol{x} \to \begin{pmatrix} \cos\theta & -\sin\theta \\ \sin\theta & \cos\theta \end{pmatrix} \boldsymbol{x} \qquad (\text{rotation}). \tag{3.105}$$

Obviously

$$\boldsymbol{b} = T_1(b_x) + T_2(b_y). \tag{3.106}$$

For character recognition, since the position and size of the shape are not of interest, $T_1(b_x)$, $T_2(b_y)$, and $T_7(\mu)$ have no particular meaning. Therefore, assuming that $\mu = \sqrt{\det A}$ and $\lambda_x \cdot \lambda_y = 1$, we analyze the degree of deformation in terms of $T_3(\lambda_x)$, $T_5(\gamma_x)$, $T_6(\gamma_y)$, and $T_8(\theta)$. Since the parameter μ has no meaning, there are only three independent variables in the 2×2 matrix A, whereas we are interested in four parameters λ_x, γ_x, γ_y, and θ. Therefore we give some interpretations of the affine transformation by imposing some constraints between the variables:

A: $\gamma_x = \gamma_y$.

We assume that $\gamma_x = \gamma_y$.

Theorem 3.4.1 [6] *A square matrix A can be decomposed into a product of an orthonormal matrix Q and a positive semidefinite and symmetric matrix S:*

$$A = QS. \tag{3.107}$$

Proof We assume that the matrix A is regular. The matrix A^tA can be written as

$$A^tA = \lambda_1 \boldsymbol{u}_1 \boldsymbol{u}_1^t + \lambda_2 \boldsymbol{u}_2 \boldsymbol{u}_2^t, \tag{3.108}$$

where λ_1 and λ_2 are eigenvalues of the matrix A^tA and $\boldsymbol{u}_1$ and $\boldsymbol{u}_2$ are eigenvectors corresponding to λ_1 and λ_2. Since A^tA is positive definite, its eigenvalues are positive. Therefore the square root of its eigenvalue is real, and we have a symmetric matrix

$$S = \sqrt{\lambda_1} \boldsymbol{u}_1 \boldsymbol{u}_1^t + \sqrt{\lambda_2} \boldsymbol{u}_2 \boldsymbol{u}_2^t. \tag{3.109}$$

Since eigenvectors are orthogonal to each other, we can show that

$$S^2 = \lambda_1 \boldsymbol{u}_1 \boldsymbol{u}_1^t + \lambda_2 \boldsymbol{u}_2 \boldsymbol{u}_2^t = A^t A. \tag{3.110}$$

Furthermore, for any vector $\boldsymbol{x} \neq 0$,

$$\boldsymbol{x}^t S \boldsymbol{x} = \lambda_1 (\boldsymbol{u}_1 \cdot \boldsymbol{x})^2 + \lambda_2 (\boldsymbol{u}_2 \cdot \boldsymbol{x})^2. \tag{3.111}$$

S is positive definite, since $\lambda_1 > 0$ and $\lambda_2 > 0$. Even when some eigenvalue of A^tA is 0, we can apply the method of making $S = (A^tA)^{1/2}$ and S is positive semidefinite.

If all the eigenvalues are positive, we have

$$S^{-1} = (A^t A)^{-1/2} = \frac{1}{\sqrt{\lambda_1}} \boldsymbol{u}_1 \boldsymbol{u}_1^t + \frac{1}{\sqrt{\lambda_2}} \boldsymbol{u}_2 \boldsymbol{u}_2^t. \tag{3.112}$$

Using this equation, an orthogonal matrix

$$Q = AS^{-1} = A(A^t A)^{-1/2} \tag{3.113}$$

can be calculated. The sign of det Q is the same as det A, since

$$\det Q = \det AS^{-1} = \det A \cdot \det S^{-1} \tag{3.114}$$

and det $S^{-1} > 0$. Therefore Q represents a rotation when det $A > 0$, and Q represents a reflection when det $A < 0$. (We expect always to obtain a rotation in our case.)

If the rank of A is 1, instead of the above method we use

$$Q = A \left(\frac{1}{\sqrt{\lambda_1}} \boldsymbol{u}_1 \boldsymbol{u}_1^t \right) \pm \boldsymbol{u}_2 \boldsymbol{u}_2^t, \tag{3.115}$$

where $\boldsymbol{u}_2$ is the eigenvector corresponding to the eigenvalue 0. The sign of the last term is selected so that the sign of det Q is positive. It can be easily shown that Q is an orthogonal matrix and gives the decomposition $A = QS$. ■

Furthermore, by LDU-decomposition, S can be decomposed into the following form:

$$S = \mu \begin{pmatrix} 1 & 0 \\ \gamma & 1 \end{pmatrix} \begin{pmatrix} 1 & 0 \\ 0 & \frac{1}{\lambda_x} \end{pmatrix} \begin{pmatrix} \lambda_x & 0 \\ 0 & 1 \end{pmatrix} \begin{pmatrix} 1 & \gamma \\ 0 & 1 \end{pmatrix}. \tag{3.116}$$

Therefore, by putting

$$Q = \begin{pmatrix} \cos\theta & -\sin\theta \\ \sin\theta & \cos\theta \end{pmatrix}, \tag{3.117}$$

the transformation matrix A is decomposed into the form:

D1:

$$A = \mu \begin{pmatrix} \cos\theta & -\sin\theta \\ \sin\theta & \cos\theta \end{pmatrix} \begin{pmatrix} 1 & 0 \\ \gamma & 1 \end{pmatrix} \begin{pmatrix} 1 & 0 \\ 0 & 1/\lambda_x \end{pmatrix} \begin{pmatrix} \lambda_x & 0 \\ 0 & 1 \end{pmatrix} \begin{pmatrix} 1 & \gamma \\ 0 & 1 \end{pmatrix}$$

$$= T_7(\mu) \cdot T_8(\theta) \cdot T_6(\gamma) \cdot T_4\left(\frac{1}{\lambda_x}\right) \cdot T_3(\lambda_x) \cdot T_5(\gamma). \tag{3.118}$$

Therefore the geometrical transformation can be described by a three tuple $(\theta, \gamma, \lambda_x)$.

Corollary *A square matrix A can be decomposed into a product of a positive semi-definite and symmetric matrix S and an orthonormal matrix Q,*

$$A = SQ. \tag{3.119}$$

By this corollary the matrix A can be decomposed in the following way:

D1′:

$$A = \mu \begin{pmatrix} 1 & 0 \\ \gamma & 1 \end{pmatrix} \begin{pmatrix} 1 & 0 \\ 0 & \dfrac{1}{\lambda_x} \end{pmatrix} \begin{pmatrix} \lambda_x & 0 \\ 0 & 1 \end{pmatrix} \begin{pmatrix} 1 & \gamma \\ 0 & 1 \end{pmatrix} \begin{pmatrix} \cos\theta & -\sin\theta \\ \sin\theta & \cos\theta \end{pmatrix}$$

$$= T_7(\mu) \cdot T_6(\gamma) \cdot T_4\left(\frac{1}{\lambda_x}\right) \cdot T_3(\lambda_x) \cdot T_5(\gamma) \cdot T_8(\theta). \tag{3.120}$$

B: $\gamma_y = 0.$

A square matrix A can be decomposed into a product of an orthonormal matrix Q and an upper triangular matrix R (QR-decomposition):

$$A = QR. \tag{3.121}$$

Furthermore R can be decomposed into the form

$$R = \mu \begin{pmatrix} 1 & 0 \\ 0 & \dfrac{1}{\lambda_x} \end{pmatrix} \begin{pmatrix} \lambda_x & 0 \\ 0 & 1 \end{pmatrix} \begin{pmatrix} 1 & \gamma_x \\ 0 & 1 \end{pmatrix}. \tag{3.122}$$

By putting

$$Q = \begin{pmatrix} \cos\theta & -\sin\theta \\ \sin\theta & \cos\theta \end{pmatrix}, \tag{3.123}$$

the transformation matrix $\boldsymbol{A}$ is decomposed into the following form:

D2:

$$A = \mu \begin{pmatrix} \cos\theta & -\sin\theta \\ \sin\theta & \cos\theta \end{pmatrix} \begin{pmatrix} 1 & 0 \\ 0 & \frac{1}{\lambda_x} \end{pmatrix} \begin{pmatrix} \lambda_x & 0 \\ 0 & 1 \end{pmatrix} \begin{pmatrix} 1 & \gamma_x \\ 0 & 1 \end{pmatrix}$$

$$= T_7(\mu) \cdot T_8(\theta) \cdot T_4\left(\frac{1}{\lambda_x}\right) \cdot T_3(\lambda_x) \cdot T_5(\gamma_x). \tag{3.124}$$

Therefore the geometric transformation can be described by a three tuple $(\theta, \gamma_x, \lambda_x)$.

Similarly a square matrix A can be decomposed into a product of an upper triangular matrix R and an orthonormal matrix Q:

$$A = RQ. \tag{3.125}$$

Therefore the transformation matrix A is decomposed into the following form:

D2′:

$$A = \mu \begin{pmatrix} 1 & 0 \\ 0 & \frac{1}{\lambda_x} \end{pmatrix} \begin{pmatrix} \lambda_x & 0 \\ 0 & 1 \end{pmatrix} \begin{pmatrix} 1 & \gamma_x \\ 0 & 1 \end{pmatrix} \begin{pmatrix} \cos\theta & -\sin\theta \\ \sin\theta & \cos\theta \end{pmatrix}$$

$$= T_7(\mu) \cdot T_4\left(\frac{1}{\lambda_x}\right) \cdot T_3(\lambda_x) \cdot T_5(\gamma_x) T_8(\theta). \tag{3.126}$$

The geometric transformation can be described by a three tuple $(\theta, \gamma_x, \lambda_x)$.

C: $\gamma_x = 0$.

A square matrix A can be decomposed into a product of an orthonormal matrix Q and an lower triangular matrix L:

$$A = QL. \tag{3.127}$$

Furthermore L can be decomposed into the form

$$R = \mu \begin{pmatrix} 1 & 0 \\ 0 & \frac{1}{\lambda_x} \end{pmatrix} \begin{pmatrix} \lambda_x & 0 \\ 0 & 1 \end{pmatrix} \begin{pmatrix} 1 & 0 \\ \gamma_y & 1 \end{pmatrix}. \tag{3.128}$$

By putting

$$Q = \begin{pmatrix} \cos\theta & -\sin\theta \\ \sin\theta & \cos\theta \end{pmatrix}, \tag{3.129}$$

the transformation matrix A is decomposed into the following form:

D3:
$$\begin{aligned} A &= \mu \begin{pmatrix} \cos\theta & -\sin\theta \\ \sin\theta & \cos\theta \end{pmatrix} \begin{pmatrix} 1 & 0 \\ 0 & \dfrac{1}{\lambda_x} \end{pmatrix} \begin{pmatrix} \lambda_x & 0 \\ 0 & 1 \end{pmatrix} \begin{pmatrix} 1 & 0 \\ \gamma_y & 1 \end{pmatrix} \\ &= T_7(\mu) \cdot T_8(\theta) \cdot T_4\left(\frac{1}{\lambda_x}\right) \cdot T_3(\lambda_x) \cdot T_6(\gamma_y). \end{aligned} \tag{3.130}$$

Therefore the geometrical transformation can be described by a three tuple $(\theta, \gamma_y, \lambda_x)$.

Similarly a square matrix A can be decomposed into a product of an upper triangular matrix L and an orthonormal matrix Q:

$$A = LQ. \tag{3.131}$$

The transformation matrix A is decomposed into the following form:

D3′:
$$\begin{aligned} A &= \mu \begin{pmatrix} 1 & 0 \\ 0 & \dfrac{1}{\lambda_x} \end{pmatrix} \begin{pmatrix} \lambda_x & 0 \\ 0 & 1 \end{pmatrix} \begin{pmatrix} 1 & 0 \\ \gamma_y & 1 \end{pmatrix} \begin{pmatrix} \cos\theta & -\sin\theta \\ \sin\theta & \cos\theta \end{pmatrix} \\ &= T_7(\mu) \cdot T_4\left(\frac{1}{\lambda_x}\right) \cdot T_3(\lambda_x) \cdot T_6(\gamma_y) T_8(\theta). \end{aligned} \tag{3.132}$$

Therefore the geometrical transformation can be described by a three tuple $(\theta, \gamma_y, \lambda_x)$.

D: $\theta = 0.$

A square matrix A can be decomposed into a product of a lower triangular matrix L, a diagonal matrix D, and an upper triangular matrix U (LDU-decomposition):

$$A = LDU. \tag{3.133}$$

By putting

$$L = \begin{pmatrix} 1 & 0 \\ \gamma_y & 1 \end{pmatrix}, \tag{3.134}$$

$$D = \mu \begin{pmatrix} 1 & 0 \\ 0 & \dfrac{1}{\lambda_x} \end{pmatrix} \begin{pmatrix} \lambda_x & 0 \\ 0 & 1 \end{pmatrix}, \tag{3.135}$$

$$U = \begin{pmatrix} 1 & \gamma_x \\ 0 & 1 \end{pmatrix}, \tag{3.136}$$

the transformation matrix A is decomposed into the following form:

D4:

$$\begin{aligned} A &= \mu \begin{pmatrix} 1 & 0 \\ \gamma_y & 1 \end{pmatrix} \begin{pmatrix} 1 & 0 \\ 0 & \dfrac{1}{\lambda_x} \end{pmatrix} \begin{pmatrix} \lambda_x & 0 \\ 0 & 1 \end{pmatrix} \begin{pmatrix} 1 & \gamma_x \\ 0 & 1 \end{pmatrix} \\ &= T_7(\mu) \cdot T_6(\gamma_y) \cdot T_4\left(\frac{1}{\lambda_x}\right) \cdot T_3(\lambda_x) \cdot T_5(\gamma_x). \end{aligned} \tag{3.137}$$

Therefore the geometrical transformation can be described by a three tuple $(\gamma_x, \gamma_y, \lambda_x)$

Similarly a square matrix A can be decomposed into a product of an upper triangular matrix U, a diagonal matrix D, and a lower triangular matrix L:

$$A = UDL. \tag{3.138}$$

The transformation matrix A is decomposed into the following form:

D4′:

$$\begin{aligned} A &= \mu \begin{pmatrix} 1 & \gamma_x \\ 0 & 1 \end{pmatrix} \begin{pmatrix} 1 & 0 \\ 0 & \dfrac{1}{\lambda_x} \end{pmatrix} \begin{pmatrix} \lambda_x & 0 \\ 0 & 1 \end{pmatrix} \begin{pmatrix} 1 & 0 \\ \gamma_y & 1 \end{pmatrix} \\ &= T_7(\mu) \cdot T_5(\gamma_x) \cdot T_4\left(\frac{1}{\lambda_x}\right) \cdot T_3(\lambda_x) \cdot T_6(\gamma_y). \end{aligned} \tag{3.139}$$

Therefore the geometrical transformation can be described by a three tuple $(\gamma_x, \gamma_y, \lambda_x)$.

E: $\gamma_x = \gamma_y = 0$.

A square matrix A can be decomposed into a product of two orthonormal matrices Q_1 and Q_2, and a diagonal matrix D (singular value decomposition):

$$A = Q_1 D Q_2, \qquad \det Q_1 > 0, \det Q_2 > 0. \tag{3.140}$$

By putting

$$Q_i = \begin{pmatrix} \cos\theta_i & -\sin\theta_i \\ \sin\theta_i & \cos\theta_i \end{pmatrix}, \qquad i = 1, 2, \tag{3.141}$$

$$D = \mu \begin{pmatrix} 1 & 0 \\ 0 & \dfrac{1}{\lambda_x} \end{pmatrix} \begin{pmatrix} \lambda_x & 0 \\ 0 & 1 \end{pmatrix}, \tag{3.142}$$

the transformation matrix A is decomposed into the following form:

$$\textbf{D5:} \quad A = \mu \begin{pmatrix} \cos\theta_1 & -\sin\theta_1 \\ \sin\theta_1 & \cos\theta_1 \end{pmatrix} \begin{pmatrix} 1 & 0 \\ 0 & \dfrac{1}{\lambda_x} \end{pmatrix} \begin{pmatrix} \lambda_x & 0 \\ 0 & 1 \end{pmatrix} \begin{pmatrix} \cos\theta_2 & -\sin\theta_2 \\ \sin\theta_2 & \cos\theta_2 \end{pmatrix}$$

$$= T_7(\mu) \cdot T_8(\theta_1) \cdot T_4\left(\frac{1}{\lambda_x}\right) \cdot T_3(\lambda_x) \cdot T_8(\theta_2). \tag{3.143}$$

Therefore the geometrical transformation can be described by a three tuple $(\theta_1, \theta_2, \lambda_x)$.

Since characters are oriented, the range of rotation angle θ is expected to be between –30 and 30 degrees. For the less deformed object, θ and γ are near 0 and λ is near 1. On the other hand, for the more deformed object, θ and γ are far from 0 (large absolute value) and λ is far from 1.

3.4.1 Examples

We give some examples of model shapes in Fig. 3.2 and some images along with normalized shapes in Fig. 3.3. For each object, the transformation matrix A and its decompositions are calculated for some models. In the following, we assume that $\det A = 1$.

Example 1 (Fig. 3.2 (a))

- Class *B:* $A = \begin{pmatrix} 0.929 & -0.233 \\ 0.013 & 1.074 \end{pmatrix}.$

	θ	γ_x	γ_y	λ_x
D1	7°	–0.124	–0.124	0.950
D1′	7°	–0.108	–0.108	0.923
D2	1°	–0.234	0	0.929
D2′	1°	–0.238	0	0.931
D3	12°	0	–0.167	0.910
D3′	14°	0	–0.238	0.957
D4	0°	–0.251	0.014	0.929
D4′	0°	–0.217	0.012	0.931
D5	–58°, 65°	0	0	1.140

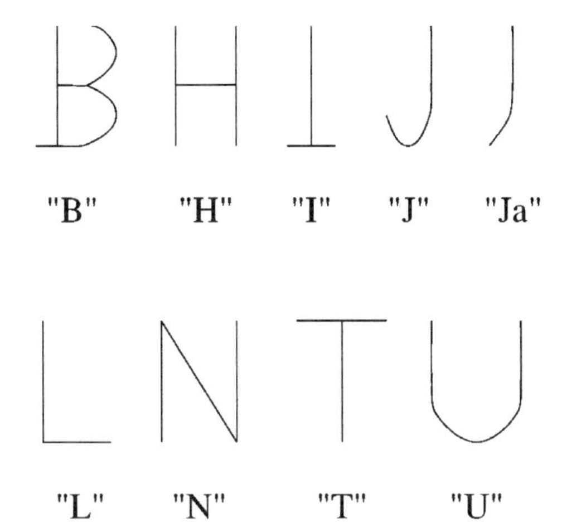

FIGURE 3.2 Some model shapes.

Example 2 (Fig. 3.3 (b))

- Class *I*: $A = \begin{pmatrix} 0.889 & -0.213 \\ -0.106 & 1.150 \end{pmatrix}$.

	θ	γ_x	γ_y	λ_x
D1	3°	−0.185	−0.185	0.899
D1′	3°	−0.173	−0.173	0.882
D2	−7°	−0.388	0	0.895
D2′	−5°	−0.339	0	0.866
D3	10°	0	−0.228	0.855
D3′	13°	0	−0.339	0.914
D4	0°	−0.240	−0.119	0.889
D4′	0°	−0.185	−0.092	0.870
D5	−63°, 66°	0	0	1.227

- Class *H*: $A = \begin{pmatrix} 0.562 & 1.647 \\ -0.463 & 0.422 \end{pmatrix}$.

	θ	γ_x	γ_y	λ_x
D1	−65°	0.108	0.108	1.73
D1′	−65°	0.477	0.477	0.657
D2	−39°	1.377	0	0.728
D2′	−48°	0.435	0	1.596
D3	−76°	0	0.253	0.588
D3′	−71°	0	0.435	1.740
D4	0°	2.931	−0.824	0.562
D4′	0°	3.903	−1.097	2.370
D5	80°, 15°	0	0	0.568

Example 3 (Fig. 3.3 (c))

- Class *N*: $A = \begin{pmatrix} 0.673 & -0.130 \\ 0.095 & 1.467 \end{pmatrix}.$

	θ	γ_x	γ_y	λ_x
D1	6°	–0.086	–0.086	0.683
D1′	6°	0.036	0.036	0.679
D2	8°	0.113	0	0.680
D2′	4°	–0.126	0	0.680
D3	5°	0	0.024	0.679
D3′	11°	0	–0.126	0.686
D4	0°	–0.193	0.141	0.673
D4′	0°	–0.088	0.065	0.682
D5	–4°, 2°	0	0	0.679

Example 4 (Fig. 3.3 (d))

- Class *Ja*: $A = \begin{pmatrix} 0.859 & 0.010 \\ -0.131 & 1.162 \end{pmatrix}.$

	θ	γ_x	γ_y	λ_x
D1	–4°	–0.058	–0.058	0.858
D1′	–4°	–0.082	–0.082	0.866
D2	–9°	–0.190	0	0.869
D2′	–6°	–0.101	0	0.855
D3	–0°	0	–0.106	0.861
D3′	–1°	0	–0.101	0.859
D4	0°	0.012	–0.153	0.859
D4′	0°	0.009	–0.113	0.861
D5	–81°, 77°	0	0	1.176

- Class *T*: $A = \begin{pmatrix} 0.865 & -0.227 \\ -0.049 & 1.168 \end{pmatrix}.$

	θ	γ_x	γ_y	λ_x
D1	5°	–0.171	–0.171	0.882
D1′	5°	–0.145	–0.145	0.858
D2	–3°	–0.338	0	0.866
D2′	–2°	–0.308	0	0.855
D3	11°	0	–0.179	0.840
D3′	15°	0	–0.308	0.894
D4	0°	–0.262	–0.057	0.865
D4′	0°	–0.194	–0.042	0.856
D5	–66°, 71°	0	0	1.226

Example 5 (Fig. 3.3 (e))

- Class *J*: $A = \begin{pmatrix} 0.460 & -0.069 \\ 0.389 & 2.121 \end{pmatrix}$.

	θ	γ_x	γ_y	λ_x
D1	10°	0.025	0.025	0.464
D1′	10°	0.578	0.578	0.520
D2	40°	2.186	0	0.602
D2′	10°	0.033	0	0.464
D3	2°	0	0.176	0.471
D3′	9°	0	0.033	0.465
D4	0°	–0.150	0.846	0.460
D4′	0°	–0.033	0.183	0.471
D5	0°, 10°	0	0	0.464

- Class *L*: $A = \begin{pmatrix} 0.633 & 0.024 \\ -0.092 & 1.577 \end{pmatrix}$.

	θ	γ_x	γ_y	λ_x
D1	–3°	–0.014	–0.014	0.633
D1′	–3°	–0.092	–0.092	0.637
D2	–8°	–0.317	0	0.640
D2′	–3°	–0.020	0	0.633
D3	–1°	0	–0.052	0.634
D3′	–2°	0	–0.020	0.633
D4	0°	0.038	–0.145	0.633
D4′	0°	0.015	–0.058	0.634
D5	–89°, 86°	0	0	1.580

- Class *U*: $A = \begin{pmatrix} 0.664 & 0.027 \\ -0.124 & 1.502 \end{pmatrix}$.

	θ	γ_x	γ_y	λ_x
D1	4°	0.029	0.029	0.664
D1′	4°	0.116	0.116	0.671
D2	11°	0.370	0	0.675
D2′	5°	0.042	0	0.664
D3	1°	0	0.075	0.666
D3′	2°	0	0.042	0.664
D4	0°	–0.041	0.187	0.664
D4′	0°	–0.018	0.083	0.666
D5	1°, 5°	0	0	0.664

From these examples we can make the following observations.

- For D1, D1′, D2, D2′, D3, D3′, D4, and D4′, the rotation angle θ always takes intuitively reasonable values. The two angles calculated in D5 do not seem intuitively right.
- For D1 and D1′, γ_x and λ_x are stable and vary within small ranges. However, γ_x (γ_y) extremely changes for D2 and D2′ (D3 and D3′).
- The decompositions D1 and D1′ take account of all the factors ($T_3(\lambda_x)$, $T_5(\gamma_x)$, $T_6(\gamma_y)$, and $T_8(\theta)$), whereas $T_6(\gamma_y)$ is not considered in the decompositions D2 and D2′, $T_5(\gamma_x)$ is not considered in D3 and D3′, $T_8(\theta)$ is not in D5, and $T_5(\gamma_x)$ nor $T_6(\gamma_y)$ is in D4.

From these observations, the decompositions D1 and D1′ are appropriate from the viewpoint of character recognition.

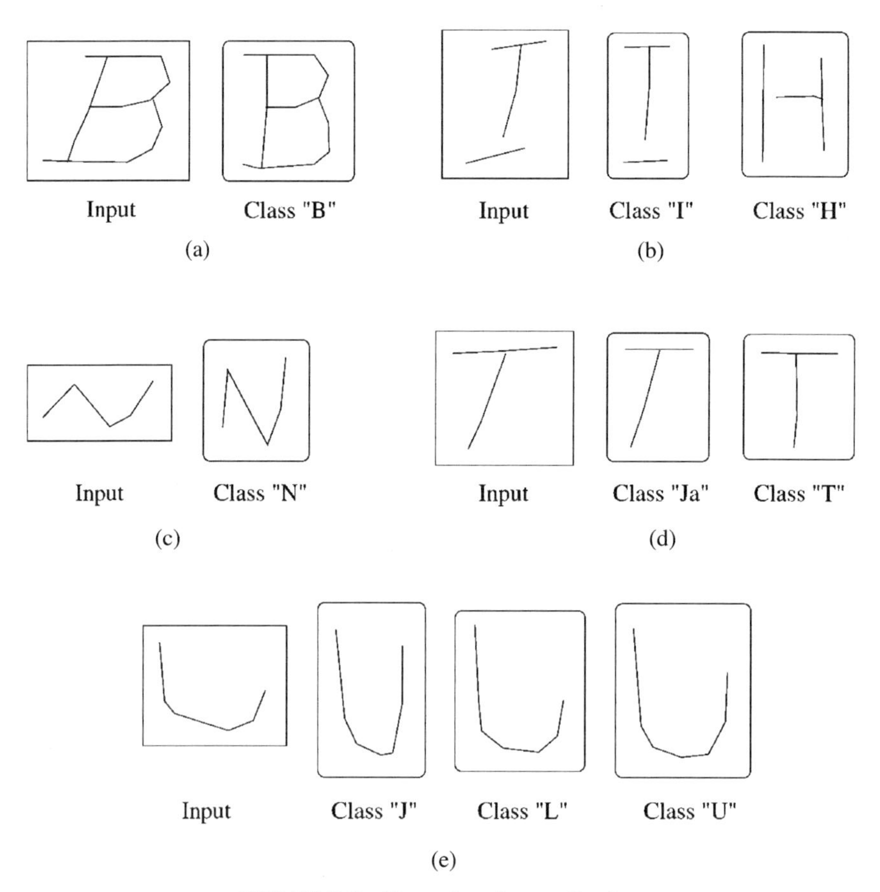

FIGURE 3.3 Examples of normalization.

3.5 NONLINEAR NORMALIZATION

So far we have mentioned linear normalization, but we sometimes need nonlinear normalization, in particular, for handprinted characters. A few examples are shown in Fig. 3.4 in which one stroke in each character is extended abnormally. In such cases linear normalization has little effect in its shape normalization. To normalize unbalanced characters, some measure for busyness/density of strokes needs to be defined. The simplest one is projection of $f(x, y)$ on the x- or y-axes. The other is crossing count. These measured values have some distribution on the x- or y-axes. The portions with the higher values are extended and the ones with the lower values are contracted so that the distribution is uniform. In this sense the basic idea of this nonlinear normalization is similar to *histogram equalization* in image processing. Hence the method described can be called *line density equalization* [7]. The idea mentioned above is formalized in a discrete sampled plane. An image there is assumed to be binary and sampled by pitch δ. That is,

$$f(x_i, y_j), \qquad i = 1, 2, \ldots, I, \quad j = 1, 2, \ldots, J, \tag{3.144}$$

where (x_i, y_j) is the ith and jth discrete sampling coordinates on the x- and y-axes, respectively. For convenience in formalization, a continuous representation is given such that

$$\begin{aligned} f(x, y) = f(x_i, y_i), \quad & x_{i-1} = x_i - \delta < x \leq x_i, \quad x_0 = 0, \\ & y_{i-1} = y_i - \delta < y \leq y_i, \quad y_0 = 0. \end{aligned} \tag{3.145}$$

That is, a step function is constructed. Using the above formalization, each crossing count measure is neatly represented. The projection of the number of lines at x_i on the x-axes, $h_X(x_i)$, is expressed as follows:

$$h_X(x_i) = \sum_{j=1}^{J} f(x_i, y_j) \cdot \bar{f}(x_i, y_{j-1}) + \alpha, f(x_i, y_0) = 0, \tag{3.146}$$

where α is a constant to control the normalization, as described later, and be assumed to zero for the time being. In the same manner the projection of the number of lines at y_i on the y-axes, $h_Y(y_j)$, is expressed as follows:

$$h_Y(y_j) = \sum_{i=1}^{I} f(x_i, y_j) \cdot \bar{f}(x_{i-1}, y_j) + \alpha, f(x_0, y_j) = 0. \tag{3.147}$$

FIGURE 3.4 Some nonlinearly transformed, distorted characters.

Examples of the distribution function are shown in Fig. 3.5 (*a*). The projected functions are also continuously represented as

$$\begin{aligned} h_X(x) &= h_X(x_i), \qquad x_i - \delta < x \leq x_i, \\ h_Y(x) &= h_Y(y_i), \qquad y_j - \delta < x \leq y_j. \end{aligned} \tag{3.148}$$

The total number of lines on each crossing count distribution are

$$\begin{aligned} N_X &= \sum_{i=1}^{I} h_X(x_i), \\ N_Y &= \sum_{j=1}^{J} h_Y(y_j). \end{aligned} \tag{3.149}$$

Now new sampling points are selected such that the new crossing count distribution is normalized. Specifically, the new sampling (x_i', y_j') is given so that the following equations hold:

$$\begin{aligned} \epsilon_X(i) \cdot h_X(x_i') &= \text{const} = \delta \cdot \frac{N_X}{I}, \\ \epsilon_Y(j) \cdot h_Y(y_j') &= \text{const} = \delta \cdot \frac{N_Y}{J}, \end{aligned} \tag{3.150}$$

where ϵ_X and ϵ_Y are variable sampling pitches. That is,

$$\begin{aligned} x_i' - \epsilon_x(i) &= x_{i-1}' \\ y_j' - \epsilon_y(j) &= y_{j-1}'. \end{aligned} \tag{3.151}$$

The meaning of (3.150) is that multiplication of each variable pitch by line density at that point takes a constant value. The values of the constants are chosen so that the number of the pixels of the normalized plane is equal to also $I \times J$. This is shown in Fig. 3.5 (*b*).

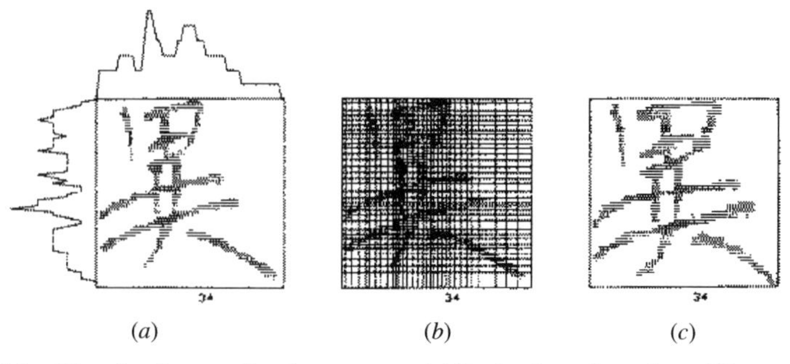

(*a*) (*b*) (*c*)

FIGURE 3.5 Line density equalization process: (*a*) Projection of number of lines to *x*- and *y*-axes; (*b*) re-sampling by the variable sampling pitches; (*c*) nonlinear normalized pattern ($\alpha = 0$).

To find new sampling points, we define the accumulation functions as

$$H_X(x) = \int_0^x \frac{h_X(x')}{\delta}\, dx',$$

$$H_Y(y) = \int_0^y \frac{h_Y(y')}{\delta}\, dy'. \tag{3.152}$$

The new sampling points are also calculated by the following equations:

$$x'_i = \left\{ x | H_X(x) = i \cdot \frac{N_X}{I} \right\},$$

$$y'_J = \left\{ y | H_Y(y) = j \cdot \frac{N_Y}{J} \right\}. \tag{3.153}$$

New uniform and local expansion/contraction are performed by treating $f'(x'_i, y'_j)$ as matrix, in which the dimension is kept the same as the original.

The discrete calculation of the normalization is performed by the equations

$$i' = \min \left\{ l | \sum_{k=1}^{l} h_X(x_k) \geq (i - 0.5) \frac{N_x}{I} \right\},$$

$$j' = \min \left\{ m | \sum_{k=1}^{m} h_Y(y_k) \geq (j - 0.5) \frac{N_x}{I} \right\}. \tag{3.154}$$

The result of the nonlinear normalization is shown in Fig. 3.5 (*c*).

Now we explain α in (3.146) and (3.147). Almost all of the characters belonging to the Roman alphabet are connected except "*i*." However, Kanji does not always have this property. In that case, when $\alpha = 0$, the white portion between black portions is crushed, so to avoid such extreme nonlinear normalization, a constant α is introduced.

In practice, when calculating the projection of crossing count, some smoothing is necessary because of ragged line boundaries. In the experiment shown later, a median filter was used such that the center value was selected among $h_X(x_{i-1})$, $h_X(x_i)$, and $h_X(x_{i+1})$ as a new value of $h_X(x_i)$.

So far we have described nonlinear normalization based on measurement of crossing count. However, simple projection also can be a measure. The projections of $f(x, y)$ on the x- and y-axes are denoted as

$$SX(i) = \sum_j f(x_i, y_j),$$

$$SY(j) = \sum_i f(x_i, y_j). \tag{3.155}$$

On these distributions of black pixels, the centroids are calculated and used to divide the image into two parts. The dividing point is a first sampling point. Then for each divided portion the centroid is calculated, by which the image is scribed into four portions, and the second and third sampling points are obtained. This procedure is continued until the number of new sampling points is equal to that of the original sampling points, which is a multiple of 2 [8]. We call this method *point density equalization.*

Some experimental results are shown in Fig. 3.6 for which the database ETL8B2 (binary, 64×63 meshes) was used. The characters shown in the first row are the original characters; linearly normalized characters are shown in the second row. The third and the fourth rows give results of the point density equalization and line density equalization, respectively. In both methods, improvements are apparent.

Further improvement can be made by taking another measurement of line busyness, namely the reciprocals of neighboring stroke distances denoted by $\alpha(i, j)$ and $\beta(i, j)$ which are the measurement of line busyness at point (i, j). More specifically, point (i, j) is a point in the ground area; functions $h(i, j)$ and $v(i, j)$, as shown in Fig. 3.7, respectively, denote horizontal and vertical distances between neighboring strokes. The measurements $\alpha(i, j)$ and $\beta(i, j)$ are given by the reciprocals of $h(i, j)$ and $v(i, j)$, respectively. If a point (i, j) is on the stroke, both values are taken very small

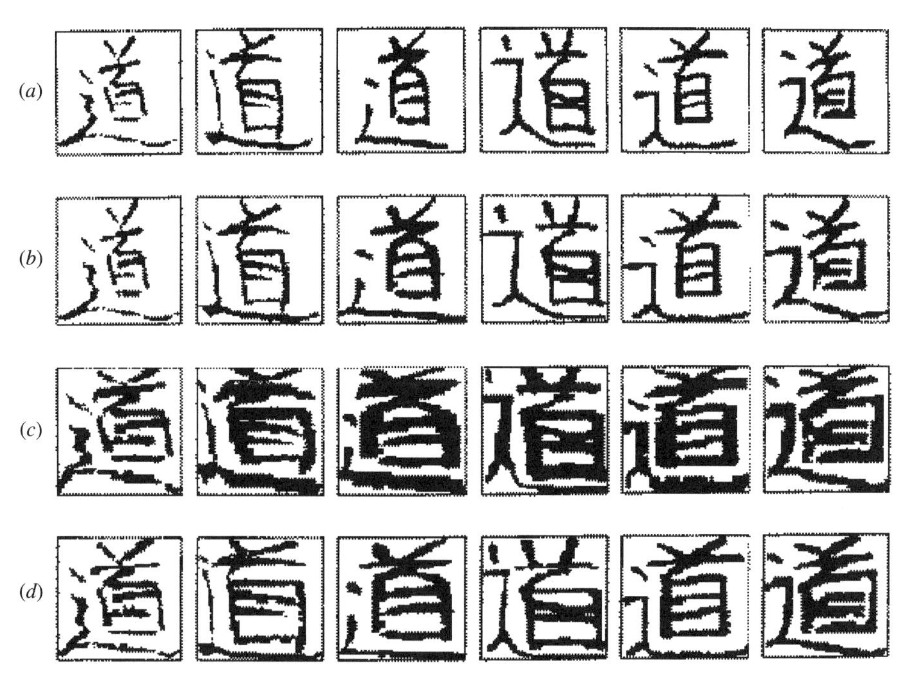

FIGURE 3.6 Sample examples of the line density equalization: (*a*) Original patterns; (*b*) linearly normalized patterns; (*c*) point density equalized patterns; (*d*) line density equalized patterns ($\alpha = 0$).

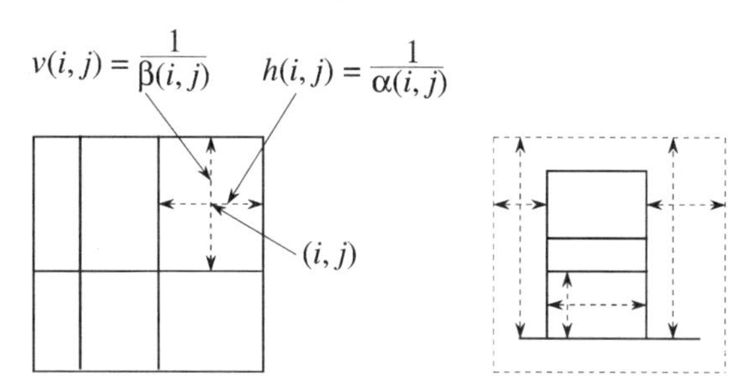

FIGURE 3.7 Features $\alpha(i, j)$ and $\beta(i, j)$ in feature projection histograms. From reference [9], © 1998, IEEE.

[9]. Experimental results are compared in Fig. 3.8. The last method seems to be the best. The effectiveness of nonlinear normalization was shown by recognition experiments conducted by Yamada, as well as Tsukumo and Tanaka, who compared it with linear normalization. The test database was ETL8 (Kanji and Hiragana), 965 characters (see Appendix B). First Yamada did two kinds of experiments. One was for studying the varying effectiveness on neatly and roughly written character sets. For the former, character set 2 and, for the latter, character set 34 were selected. The results are shown in Table 3.1. As expected, the effectiveness on the sloppy character set is higher than that on the neat one. The other experiment is for the specifically chosen character "道." The effectiveness on the set is very high, an increase in recognition rate of almost 10%. Tsukumo and Tanaka did large-scale experiments on the

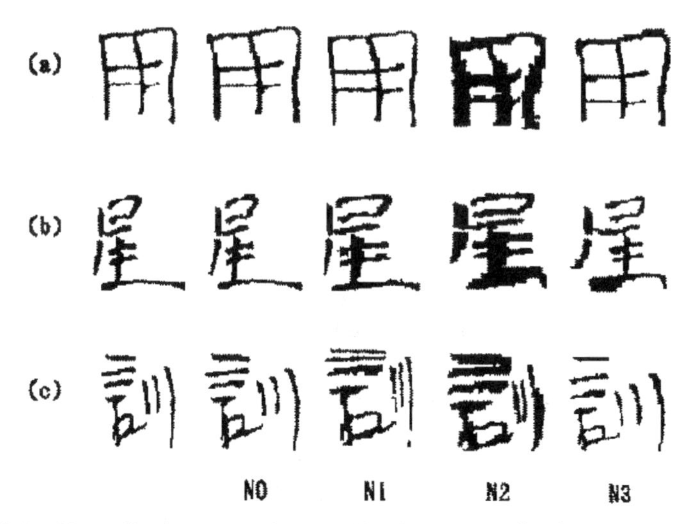

FIGURE 3.8 Normalization comparisons: (N0) Linear normalization; (N1) nonlinear normalization by line density histograms; (N2) nonlinear normalization by projection profiles; (N3) nonlinear normalization improved using the method of reference [9]. From reference [9], © 1998, IEEE.

TABLE 3.1 Comparison of Recognition Experiment Results

Data Set	Linear Normalization	Line Density Equalization ($\alpha = 1$)
2	96.6%	98.0%
34	80.9%	87.0%

same database, ETL8, using their nonlinear method, and they achieved a considerably higher recognition rate of 97.36% for 152,320 test characters. In both experiments the directional matching method was used (explained in Chapter 11 on linear matching).

Note that strictly speaking, there is a problem of defining the line density. By using the reciprocals of neighboring stroke distance, this situation was helped by Yamada who defined line density $\rho(i, j)$ as

$$\rho(i, j) = \begin{cases} \max(W/L_x, W/L_y) & \text{if } L_x + L_y < 6W, \\ 0 & \text{if } L_x + L_y \geq 6W, \end{cases}$$

where W is a width of an input image. The definitions of L_x and L_y are somewhat complex, but min (L_x, L_y) can be interpreted as a roughly approximated diameter of the inscribed circle at each point (i, j). See the bibliographical remarks below.

3.6 BIBLIOGRAPHICAL REMARKS

The earliest work on shape normalization was done by Udagawa et al. [10] using moments in a practical application. However, in 1967 it was too early to allow these authors to do a systematic experiment.

In 1970 Casey systematically researched shape normalization based on moments at the IBM Thomas J. Watson Research Center [11]. He used three moments; m_{xx}, m_{yy}, and m_{xy} and considered linear normalization such that the moments of a normalized image took values of the moments as $m_{xx} = m_{yy} = k$(const) and $m_{xy} = 0$. He showed that the normalization scheme considerably improved the recognition experiments on data sets of handprinted characters. The size of the data sets ranged from 1000 to 10,000. At about the same time and same place, G. Nagy and N. Tuong devised an interesting shape-normalized scheme based on geometrical projection and convex hull [12].

On nonlinear shape normalization (NSN) there is a systematic comparison study by Lee and Park [14]. They compared five normalization methods—linear, NSN based on dot density [8], NSN based on line density by crossing lines [7], NSN based on line density by line interval [9], and NSN based on line density by inscribed circle [15]. They used four kinds of performance criteria—recognition rate, processing speed, computational complexity, and degree of variation; and they used two kinds of directional feature matching methods in their recognition scheme. Naturally all the NSN methods gave significantly better recognition results than the linear method.

The best was NSN based on line density by inscribed criteria in terms of recognition rate. The point to be noted is that the NSN based on dot density gives a good recognition rate with high processing speed. Lee and Park used 520 Korean syllables as the data. More recently, Wakabayashi et al. [16] conducted a systematic experiment on handwritten Kanji characters, in which three NSN methods—NSN based on line density by crossing lines, NSN based on line density by line interval, and NSN based on line density by inscribed criteria together with linear—were compared. All three NSN methods gave better results than the linear method, and the best was NSN based on line density by inscribed criteria. Therefore they used it as the preprocessor for their system. They achieved a very high, record-breaking recognition rate of 99.05% for the Kanji database, ETL9B. They concluded that the contribution of the chosen NSN was very high.

BIBLIOGRAPHY

[1] T. Yamauchi and M. Sugiura, *Introduction to Continuous Group,* Tokyo: Baihukan Publishing, 1960.

[2] B. G. Wybourne, *Classical Groups for Physicists,* New York: Wiley-Interscience, 1974.

[3] T. Iijima, "Basic theory on normalization of pattern," *Bull. Electrotechn. Lab.,* vol. 26, no. 5, pp. 368–388, 1962.

[4] S. Amari, "Invariant structures of signal and feature spaces in pattern recognition problems," *RAAG Memoirs,* vol. 4, pp. 19–32, 1968.

[5] Henry S. Baird, *Model-Based Image Matching Using Location.* Cambridge: MIT Press, 1985.

[6] B. K. P. Horn, H. M. Hildin, and S. Negahdariour, "Closed-form solution of absolute orientation using orthonormal matrices," *J. Opt. Soc. Am.,* vol. 5, no. 7, pp. 1127–1135, 1988.

[7] H. Yamada, T. Saito, and K. Yamamoto, "Line density equalization of handprinted . . . A non-linear normalization for correlation method," *IECE Trans.,* vol. J670, no. 11, p. 1379, 1984.

[8] Y. Yamashita, K. Higuchi, Y. Yamada, and Y. Haga, "Classification of handprinted Kanji characters by structured segment matching method," *IECE Japan,* Technical Research Report PRL 82–12, no. 12, p. 25, 1982.

[9] J. Tsukumo and H. Tanaka, "Classification of handprinted Chinese characters using non-linear normalization and correlation methods," *Proc. 9th Int. Conf. Pattern Recogn.,* pp. 168–171, September 1988.

[10] K. Udagawa, J. Toriwaki, and K. Sugino, "Normalization and recognition of two-dimensional patterns with linear distortion by moments," *Electron. Commun. Japan,* vol. 47, pp. 34–46. June 1967.

[11] R. G. Casey, "Moment normalization of handprinted characters," IBM, *J. Res. Develop.,* vol. 14, pp. 548–557, 1970.

[12] G. Nagy and N. Tuong, "Normalization techniques for handprinted numerals," *Commun. ACM,* vol. 13, no. 8, pp. 475–481, August 1970.

[13] T. Iijima, *Theory of Pattern Recognition,* Tokyo: Morishita Publishing, 1989.

[14] S. U. Lee and J. S. Park, "Nonlinear shape normalization methods for the recognition of large-set handwritten characters," *Pattern Recogn.,* vol. 27, no. 7, pp. 895–912, 1994.

[15] H. Yamada, K. Yamamoto, and T. Saito, "A nonlinear normalization method for hand-printed Kanji character recognition—Line density equalization," *Pattern Recogn.,* vol. 23, no. 9, pp. 1023–1029, 1990.

[16] T. Wakabayashi, Y. Deng, S. Tsuruoka, F. Kimura, and Y. Miyake, "Accuracy Improvement by non-linear normalization and feature compression in handwritten Chinese character recognition," *IECE Japan,* PRU95-1, May 1995.

CHAPTER FOUR

Thresholding Selection

As we saw in Chapter 1, an analog video signal obtained from an observation device such as a CCD is typically binarized to black and white in an OCR system. This is because the processing of a binary image is very simple compared with a gray-level image. Besides this natural and practical reason, binarization models a character image as a bi-level image. This is the essential difference between a character image and a picture image in which the gray-level distribution itself carries essential information. Here we need to note that if a character image is too degraded, then it has to be treated as a pictorial image.

This chapter describes the selection of a threshold value for a given character image. In general, there are two approaches to automatic thresholding selection, statistical and model based. The statistical approach is performed in two stages: First, one extracts the statistical information from a character image; then using a given statistical distribution, one decides on a threshold value. Sometimes these two stages are mixed, so this point has to be made clear. Some comparative studies will be given.

4.1 EXAMPLE OF THRESHOLDING SELECTION

This section gives an extensive discussion on *threshold selection.* We begin by observing real data that are considerably degraded. The data are shown in Fig 4.1 as was recorded by a videcon camera. The video signal has 16 levels, which is usual in a conventional OCR system. However, the maximum gray level is 11 because of the heavy degradation. Various binarized images are shown in Fig 4.2. Notice that the pixels whose values are greater than 9 are marked by asterisks in the figure. Here we define the binarization process such that if a video signal is greater than T at each point then the binarized image takes a value 1 and otherwise 0. Looking at these images, we can see that there is considerable noise in the most degraded image so

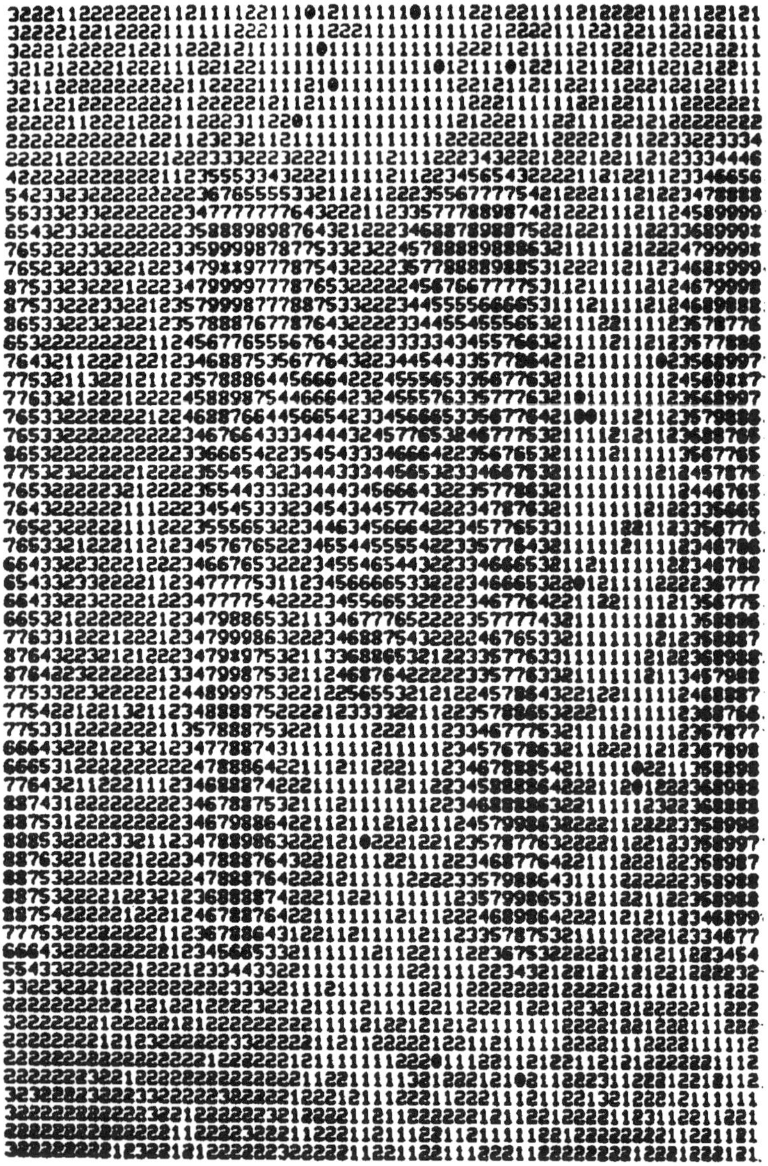

FIGURE 4.1 Original observed character "M" with 16 levels.

that it is difficult to select an appropriate threshold value. In this case only $T = 3$ is acceptable. However, we notice that $T = 2$ can also be allowed if the sampling process is performed after the binarization process. Actually the sample was taken at a 64×64 resolution. This matrix size is larger than the necessary sampling. It suffices to have approximately 20×20 resolution after the segmentation of the character image. In Fig 4.1 the domain of observation is larger than the size of the character

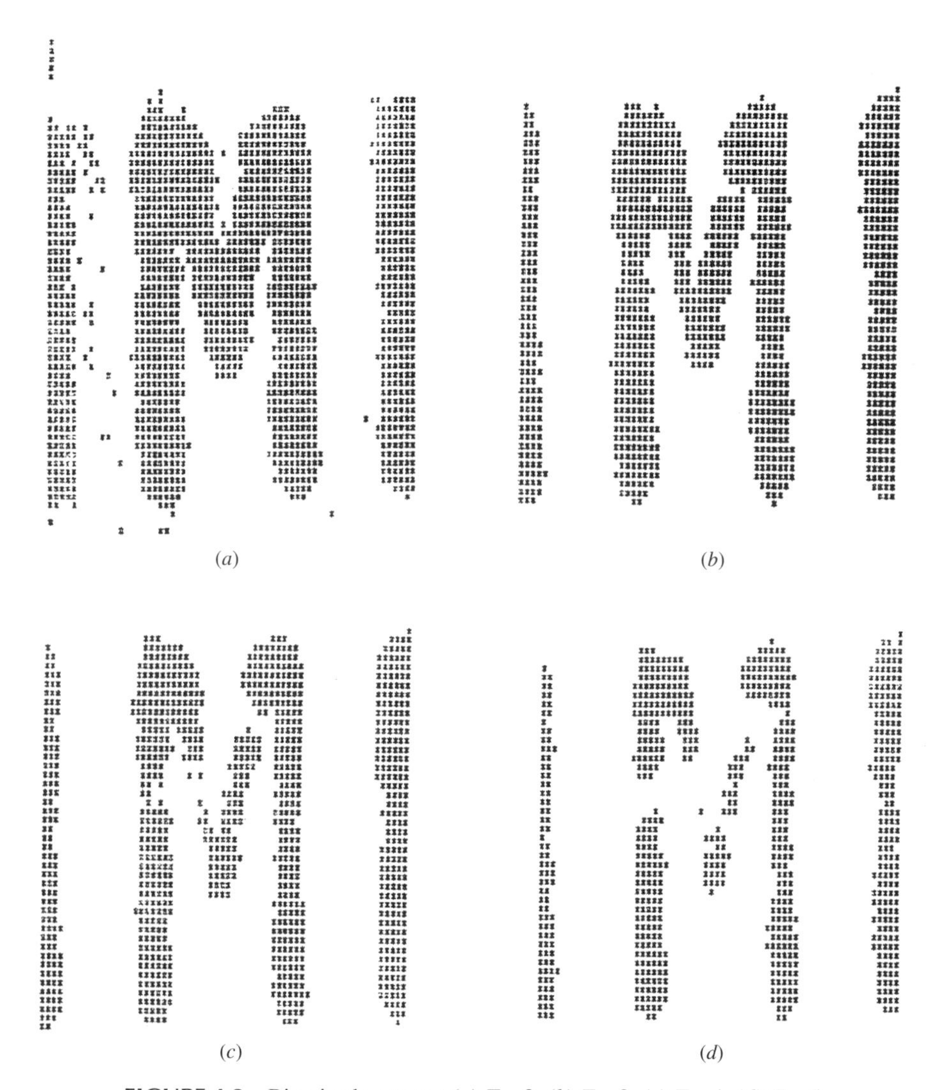

FIGURE 4.2 Binarized patterns: (*a*) $T = 2$; (*b*) $T = 3$; (*c*) $T = 4$; (*d*) $T = 5$.

image of "M" for the subsequent segmentation process after binarization. An appropriate rough sampling can sometimes be used for noise reduction. At any rate, the range of the threshold values is very narrow.

Now let us give a strategy of threshold selection for a character image. The first thing to consider is how to take statistics of the observed image. A simple method is to take an average of the gray levels of the image. Actually this is a good choice and works well as long as the image is not very degraded. In this case the decision function of the threshold value is a linear equation of the average value, say $\overline{V}$:

$$T = \alpha\overline{V} + \beta, \tag{4.1}$$

where α and β are parameters adjusted to the samples. Usually $\alpha = 1$ and $\beta = 2$ for the 16 levels of a video signal.

However, there is another problem even if character images are not so degraded. That is, a linear equation works well for a set of characters whose shapes are similar and uniform in complexity. Some examples of such sets are arabic numerals, the Latin alphabet, and Katakana. However, it does not work well for character sets with many kinds of characters such as Chinese characters and Kanji, or mixed sets of characters such as Hirakana and Kanji. The reason why the simple average does not work for such character sets is obvious. Kanji, for example, includes very simple character shapes such as "—" and also very complex characters. Therefore the simple average value can be considerably different for these character images, since a simple shape character image gives a low threshold value and a complex shape character image gives a high threshold value. The parameter α cannot be a constant.

What is wanted here is an average value of a portion of the black part of the character image. If this succeeds, we can always obtain a correct threshold value that can be adapted to the complexity of the character image. To do so, we need to know another threshold value in order to discard ground levels of an observed image, denoted $T_{\min}$. If $T_{\min}$ is set a little higher than the average ground levels, and the gray levels less than $T_{\min}$ are discarded in calculating the average value of the gray levels, then the average value of the black portion of a character image is obtained [1]. The problem then is how to chose a $T_{\min}$. The simplest case is to make $T_{\min}$ a constant, but this works only for the case where one kind of OCR sheet is used and the lighting condition is stable. In practice, an OCR system must handle many kinds of sheets whose optical properties vary greatly.

Therefore we must go back and consider how to go about taking statistical information from a character image. We recall that the average was taken for a gray-level distribution or a histogram of the gray levels. This is regarded as the simplest case of a local feature distribution on the axis of gray level. A local feature of an image can be obtained in the frame of a 3×3 mask, which is shown in Fig. 4.3. In the figure P_x is the center value, and N_i, $i = 1 \sim 8$, are the neighboring values at point x [3]. If we take only P_x at point x, then the gray-level distribution can be obtained. Using the notation of Fig. 4.3 the following local features are defined:

1. Maximum or minimum, $P_x > N_i$ or $P_x < N_i$, $i = 1 \sim 8$.
2. Gradient, $N_i > P_x > N_j$, $N_i \cong P_x > N_j$ or $N_i > P_x \cong N_i$, $i \neq j$.
3. Flat, $P \cong N_i$, $i = 1 \sim 8$.

In this way we can, for example, select the gradient feature for statistical use. A histogram of the gradient feature against the gray level provides useful statistical

N_8	N_1	N_2
N_7	P_x	N_3
N_6	N_5	N_4

FIGURE 4.3 3×3 operator and notations.

information for determining the threshold value of a given character image. Intuitively speaking, the edges of a black line segment produce a gradient feature. Therefore a decision function that detects the gray level at which the gradient distribution has a peak value can be used for selecting a reasonable thresholding value. This method has already been successfully applied to pictorial data [2].

In our case we want to detect a ground threshold value of a merged feature, choosing between max or flat and min or flat. Experimental evidence on degraded sample data suggests that the former is a better approach than the latter; that is, the condition

$$P_x \geq N_i, \qquad i = 1 \sim 8$$

should be applied. Some examples of the sample data are shown in Table 4.1, such as found for two peaks of a distribution. Obviously the lower peak corresponds to the ground and the higher peak corresponds to a black line segment.

We can try a few procedures to find $T_{\min}$. For simplicity, from this point on we will use T_W instead of $T_{\min}$ which is a more appropriate reference to the white ground. The simplest approach is to detect a peak value and set T_W to the gray level at which the peak value is obtained. However, T_W should be slightly larger than the average ground level, as mentioned before. Therefore it is reasonable to set a T_W to the peak gray level +1, which seems to work well. However, in Table 4.1 the best T_W is equal to 2 of the sample M741; this is different from the above threshold setting. So a more sophisticated procedure is necessary, one that satisfies the real sample data as shown in Table 4.1 where the desired gray levels are marked by *. On inspecting these sampled data, it becomes apparent that the threshold values can be obtained at the transition point on the gray level such that the histogram changes from convex to concave. This idea includes an important aspect of feature extraction that can be used in general. Actually the same procedure will be used in Section 4.3 on the model-based approach for threshold selection.

The first thing is to define concave and concavity. To do so, notation must be introduced for representing the histogram. Let us use $H(i)$, where i denotes a gray level and is an integer. Looking at Fig 4.4, the following relation is obtained according to the nature of convexity:

$$\overline{H}(i) = \frac{H(i-1) + H(i+1)}{2}, \tag{4.2}$$

TABLE 4.1 Histogram of the Local Feature of Either Max or Flat

	Level												
Sample ID	0	1	2	3	4	5	6	7	8	9	10	11	12
M 739	1	462	*326	24	0	4	0	9	40	**199	101	8	0
M 740	0	309	619	*76	4	0	2	0	17	67	**168	8	0
M 741	0	313	*742	28	1	1	3	13	21	**181	39	0	0

Note: The "*" indicates an optimum threshold selection position.

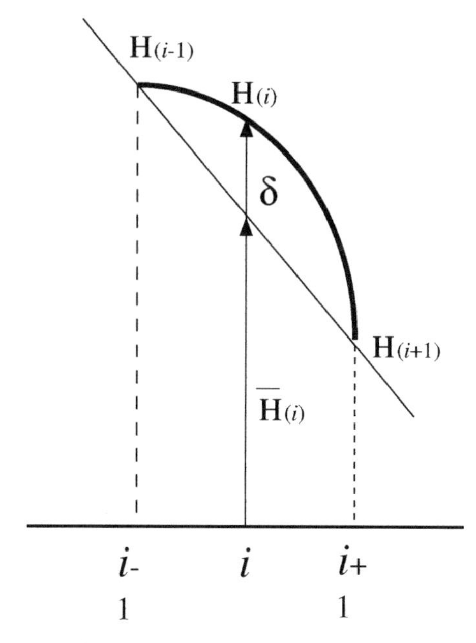

FIGURE 4.4 Definition of concavity.

$$\delta = H(i) - \overline{H}(i). \tag{4.3}$$

To be convex at *i*, δ must be positive. That is,

$$H(i) - \overline{H}(i) > 0. \tag{4.4}$$

The above inequality can be transformed as

$$H(i-1) - H(i) < H(i) - H(i+1). \tag{4.5}$$

Thus $H(i-1) + H(i+1) - 2H(i) < 0$. This says that the discrete second derivative is negative, confirming the connection with convexity. And so further notations are introduced such that

$$a = H(i-1) - H(i), \tag{4.6}$$

$$b = H(i) - H(i+1). \tag{4.7}$$

Using these notations, the following definitions for both convexity and concavity together with straight line can be made:

Definition 4.1

$a > b \cdots$ *concavity,*
$a = b \cdots$ *straightness,*
$a < b \cdots$ *convexity.*

However, to perform properly on our data, these definitions cannot be used directly. We have to expand the definition of convexity in terms of concavity. We can expand the definition of convexity by introducing a parameter:

Definition 4.1.2
$a < \gamma b \cdots$ *convexity,*
Else $\cdots$ *concavity.*

On examining the sampled data, we choose γ to be 10. For all the sampled data of 2220 items, this expanded definition of convexity works well without exception for a constant $\gamma = 10$. The experiment will be described in detail later in this chapter.

Now we introduce two kinds of decision functions.

Procedure 1 The first function we consider is a linear or nonlinear function of $\overline{V}$ (the average density value of black portions) such that

$$T = \alpha\overline{V} + \beta, \tag{4.8}$$

$$T = \alpha(\overline{V} - T_W')^4 + T_W. \tag{4.9}$$

Here $\overline{V}$ and T_W' are normalized such that their maximum values are 1 and α is taken as 40. The second function was constructed to make the widths of the binarized strokes constant. That is, the difference $\Delta = \overline{V} - T_W'$ is always less than 1. So if Δ is small, then $T \cong T_W$, and if Δ is large, $T \gg T_W$. This means that for lower contrast data, a lower threshold value is selected, and for higher contrast data, a higher threshold value is chosen. This nonlinear function works to keep the widths of the linearized strokes constant as much as possible; this has been proved clearly by the experiment.

Procedure 2 The second kind of decision function is based on the shape of a histogram of the local feature. $H(i)$ is scanned from the left, and T_W is found at the gray level of the concavity. Specifically, the expanded convexity is used, setting γ to 10, and the transition point from convexity to concavity is found as T_W. However, we cannot take a boundary point from convexity to concavity because all points are classified as either concavity or convexity according to the new definition above (Definition 4.1.2). Therefore in the convex region the closest point to the concave region is adopted as the transition point. That done, γ is set to 1, and the first convexity is found. Now, to avoid any fluctuations due to small noise, such features are only examined at the gray level whose histogram value is larger than some small value δ. Actually $H(i) \geq \delta = 10$ was chosen in which $\delta = 10$ refers to 0.25% of all the sampling points. The located histogram values for the black portion are marked by **. In these cases the convex points, say T_B, should coincide with the corresponding peak values, though not always. Thus the thresholding value T is decided as

$$T = \frac{T_B + T_W}{2} - K, \tag{4.10}$$

where K is a parameter used to adjust the threshold value and set it to 1. Such a small value of the parameter indicates that it compensates for some quantization errors in T_B and T_W.

4.2 THRESHOLDING SELECTION BASED ON DISCRIMINANT ANALYSIS

A method of threshold selection based on discriminant analysis, denoted DTSM, was proposed by Otsu [4]. So far we have emphasized statistical information rather than the decision function based on statistical information. The DTSM gives us a good method to use in deciding on an appropriate thresholding value for a statistical distribution. The method's strength is that it does not require any parameter to adjust to the sampled data. Another strong point is that it can be applied to any statistical distribution in general.

The basic idea is very simple: we can consider two classes of gray-scale levels, white and black. Therefore this becomes a classification problem, and so the discriminant analysis method can be applied.

For simplicity the simple statistical information of a gray-level histogram will be used for the formulation. Given a character image with an L gray-level scale $S = [1, 2, \ldots, L]$, let n_i be the number of pixels that have an i level. Therefore the total number of pixels N is equal to the sum $n_1 + n_2 + \cdots + n_L$. Then the following normalized histogram is considered:

$$p_i = \frac{n_i}{N} \qquad \left(i \in S,\ p_i \geq 0,\ \sum_{i=1}^{L} p_i = 1 \right). \tag{4.11}$$

Equation (4.11) can be regarded as the probability distribution of gray levels. Using the probability distribution, the total average and the total variance of a given character image, are given as follows:

$$\mu_T = \sum_{i=1}^{L} i p_i, \tag{4.12}$$

$$\sigma_T^2 = \sum_{i=1}^{L} (i - \mu_T)^2 p_i. \tag{4.13}$$

Now, let an unknown threshold value be k and a character image be divided into two classes, namely the white ground class C_1 and the black character part C_2. Then, using the assumed level k, the average and variance of gray levels in each class are calculated, each expressed as a function of the unknown level k. Since the discrimination criterion of the classification between C_1 and C_2 is expressed as a combination of the average and variance of gray levels in each class, we can obtain the best k based on the criterion. We proceed with our derivation according to the strategy mentioned above.

First, each average of gray levels is given as

$$\mu_1 = \sum_{i \in S} iP_r(i|C_1), \tag{4.14}$$

$$\mu_2 = \sum_{i \in S} iP_r(i|C_2), \tag{4.15}$$

where $P_r(i|C_j)$, $j = 1, 2$, is the probability of occurrence of level i when the i is assumed to belong to C_j. Accordingly, probabilities of class occurrences ω_i, $i = 1, 2$, are found as

$$\omega_1 = P_r(C_1) = \sum_{i \in S_1} p_i, \tag{4.16}$$

$$\omega_2 = P_r(C_2) = \sum_{i \in S_2} p_i, \tag{4.17}$$

where $S_1 = [1, 2, \cdots, k]$ and $S_2 = [k+1, k+2, \cdots, L]$. Thus $P_r(i|C_j)$ is given by normalizing p_i with ω_j as

$$P_r(i|C_j) = \frac{p_i}{\omega_j}, \qquad j = 1, 2. \tag{4.18}$$

Zero-order and first-order moments of the distribution are taken until the level k as

$$\omega(k) = \sum_{i=1}^{k} p_i, \tag{4.19}$$

$$\mu(k) = \sum_{i=1}^{k} ip_i, \tag{4.20}$$

where $\omega(L) = 1$, $\mu(L) = \mu_T$.

Using (4.14) to (4.20), we obtain the following equations:

$$\mu_1 = \sum_{i \in S_1} iP_r(i|C_1) = \sum_{i \in S_1} \frac{ip_i}{\omega_1} = \frac{\mu(k)}{\omega(k)}, \tag{4.21}$$

$$\mu_2 = \sum_{i \in S_2} iP_r(i|C_2) = \sum_{i \in S_2} \frac{ip_i}{\omega_2} = \frac{\mu_T - \mu(k)}{1 - \omega(k)}, \tag{4.22}$$

$$\sigma_1^2 = \sum_{i \in S_1} (i - \mu_1)^2 P_r(i|C_1) = \sum_{i \in S_1} \frac{(i - \mu_1)^2 p_i}{\omega_1}, \tag{4.23}$$

$$\sigma_2^2 = \sum_{i \in S_2} (i - \mu_2) P_r(i|C_2) = \sum_{i \in S_2} \frac{(i - \mu_2)^2 p_i}{\omega_2}. \tag{4.24}$$

We can confirm that the following relations hold independently of k as

$$\omega_1\mu_1 + \omega_2\mu_2 = \mu_T, \tag{4.25}$$

$$\omega_1 + \omega_2 = 1. \tag{4.26}$$

These equations state the general relations between the total and the two classes.

Now we introduce the criteria of the estimation of class separability based on discrimination analysis as

$$\lambda = \frac{\sigma_B^2}{\sigma_W^2}, \tag{4.27}$$

where

$$\sigma_W^2 = \omega_1\sigma_1^2 + \omega_2\sigma_2^2, \tag{4.28}$$

$$\sigma_B^2 = \omega_1(\mu_1 - \mu_T)^2 + \omega_2(\mu_2 - \mu_T)^2 = \omega_1\omega_2(\mu_1 - \mu_2)^2. \tag{4.29}$$

Here σ_W and σ_B are called within-class and between-class variances, respectively.

The value of λ will be large if the class separability is large, which is easily understandable according to the definitions of σ_W^2 and σ_B^2. Furthermore the following relation always holds between them:

$$\sigma_W^2 + \sigma_B^2 = \sigma_T^2. \tag{4.30}$$

Using the equation above, we can transform λ to another form as

$$\kappa = \frac{\sigma_T^2}{\sigma_W^2}, \quad \eta = \frac{\sigma_B^2}{\sigma_T^2}, \tag{4.31}$$

$$\kappa = \lambda + 1, \quad \eta = \frac{\lambda}{\lambda + 1}, \tag{4.32}$$

where η is a monotone function of λ due to $\lambda \geq 0$, and it changes 0 to 1 as λ changes from 0 to ∞. Therefore all of these criteria are equivalent. We want to choose the simplest expression among these equivalent criteria, which is η, because $\sigma_T^2 = \text{const}$. The $\eta(k)$ is expressed as

$$\eta(k) = \frac{\sigma_B^2}{\sigma_T^2}, \tag{4.33}$$

$$\sigma_B^2 = \frac{[\mu_T\omega(k) - \mu(k)]^2}{\omega(k)[1 - \omega(k)]}. \tag{4.34}$$

The optimum k^* is given as

$$\sigma_B^2(k^*) = \max_{1 \leq k < L} \sigma_B^2(k) \tag{4.35}$$

In the actual calculation the formulas

$$\omega(k) = \omega(k-1) + p_k \quad \text{and} \quad \mu(k) = \mu(k-1) + kp_k \tag{4.36}$$

can be used, and so $\sigma_B^2(k)$ is easily calculated.

Now we can proceed to apply DTSM to practical data. First of all the unimodality notion of $\sigma_B^2(k)$ must be discarded because it is not true. Fortunately, however, we have unimodality for the usual data, as shown in Fig 4.5. Therefore k^* can be determined uniquely in practice. In Fig 4.5 notice that the gray-level distributions of data (*b*) have a shallow valley, and in particular, data (*d*) do not have a valley. So we can see that even for these low-quality data, we get good results.

4.2.1 Least Squares Criterion

A very interesting property of DTSM is that the same result is obtained by a least squares criterion. This is easily seen. We generalize the problem by approximating an image by two values, α_1 and α_2, using a threshold value, here denoted k. Then we formulate the least mean square as

$$e^2(\alpha_1, \alpha_2; k) = \frac{1}{N}\left[\sum_{i \in S_1} (i-\alpha_1)^2 n_i + \sum_{i \in S_2} (i-\alpha_2)^2 n_i\right] \tag{4.37}$$

$$= \sum_{i \in S_1} (i-\alpha_1)^2 p_i + \sum_{i \in S_2} (i-\alpha_2)^2 p_i. \tag{4.38}$$

The α_j are derived as

$$\frac{\partial e^2}{\partial \alpha_j} = 2\sum_{i \in S_j} (\alpha_j - i)p_i = 0. \tag{4.39}$$

That is,

$$\alpha_j = \frac{\sum_{i \in S_j} ip_i}{\sum_{i \in S_j} p_i}. \tag{4.40}$$

This result is obvious. For a given fixed k, the minimum value is obtained when each α_j is equal to the average level of each class, μ_j. Thus, substituting these values in $e^2(\alpha_1, \alpha_2; k)$, we write $e^2(k)$ as

$$e^2(k) = \sum_{i \in S_1} (i-\mu_1)^2 p_i + \sum_{i \in S_2} (i-\mu_2)^2 p_i. \tag{4.41}$$

The optimum threshold value k^* is now obtained by calculating

$$e^2(k^*) = \min_{1 \le k < L} e^2(k) \tag{4.42}$$

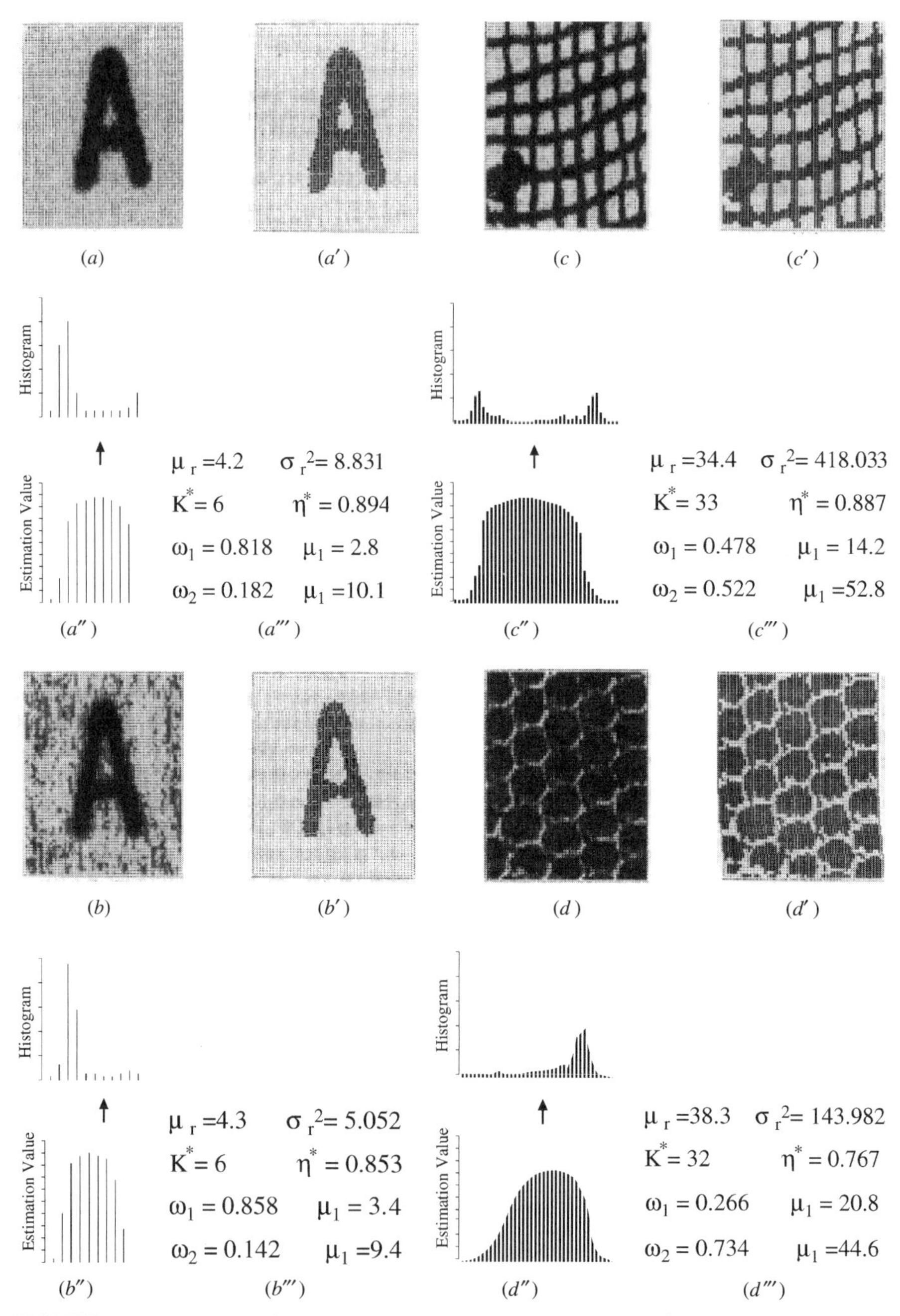

FIGURE 4.5 Examples of DTSM applications: (*x*) Input image; (*x′*) binarized image; (*x″*) histogram and estimation on value distribution; (*x‴*) parameters' values. (*x* is replaced by (*a*), (*b*), (*c*), and (*d*) according to the change of input images.) From reference [4].

According to σ_1^2, σ_2^2, and λ obtained earlier, $e^2(k)$ can be rewritten as

$$e^2(k) = \omega_1\sigma_1^2 + \omega_2\sigma_2^2 = \sigma_W^2. \tag{4.43}$$

As mentioned before, the following relation always holds:

$$\sigma_W^2 + \sigma_B^2 = \sigma_T^2 \tag{4.44}$$

when σ_T^2 is constant. Therefore the maximization of σ_B^2 is equivalent to the minimization of σ_W^2. Our claim has been proved. For use in a second experiment later, another interesting property of the DTSM is introduced. For the continuous distribution, the derivation of the solution of the least mean square is straightforward. The conditions for the minimum value are expressed as

$$\frac{\partial e^2}{\partial \alpha_j} = 0 \qquad (j = 1, 2), \tag{4.45}$$

$$\frac{\partial e^2}{\partial k} = 0. \tag{4.46}$$

For the continuous model, $e^2(k)$ is expressed as

$$e^2(\alpha_1, \alpha_2; k) = \int_0^k (g - \alpha_1)^2 p(g)dg + \int_k^L (g - \alpha_2)p(g)dg. \tag{4.47}$$

For α_j, the same results as before are obtained, namely $\alpha_j = \mu_j$. For k,

$$\frac{\partial e^2}{\partial k} = (k - \alpha_1)^2 p(k) - (k - \alpha_2)p(k) = 0 \tag{4.48}$$

holds. Since in the optimum solution $\alpha_j = \mu_i$, the following results can be obtained:

$$p(k) = 0 \tag{4.49}$$

and

$$k = \frac{\mu_1 + \mu_2}{2}. \tag{4.50}$$

The latter result is intuitive.

4.3 ADAPTIVE THRESHOLDING SELECTION BASED ON TOPOGRAPHICAL IMAGE ANALYSIS

So far the thresholding selection methods have been based on the statistics of local features, and the threshold values selected were constant over the domain in which the statistics were taken. Usually the domain is the same as the range in which a character image is included. However, the range is not always uniform in reflectance or gray scale. In other words, the range can be shaded. Therefore threshold values must be selected that adapt to such situations. We introduce for such adaptive thresholding selection a method based on topographical image analysis developed by Pavlidis and Wolberg [5].

We chose this method among the several methods of *adaptive thresholding selection* cited at the end of this chapter because it is sound both mathematically and physically. That is, it is based on a model of image distortion where the convolution of a step function $h(t)$ with a blurred function (a bell-shaped function, in general) is convex where $h(t)$ is high and concave where $h(t)$ is low. This meets our intuition and is easily proved (see the paper cited above). Therefore ideally an original step function (i.e., binary image) can be recovered by analyzing the image at each point to determine whether its local is convex or concave. However, real images are not so simple. In general, we can consider this problem in the following mathematical framework. According to our model, the given image was originally bi-level, and so the task is to decide whether a point was white or black. This can be formulated by computing the conditional probabilities for given neighbors:

$$P_b(x|N(x)) \equiv \text{probability (point } x \text{ is black | values of neighbors of } x).$$

Naturally

$$P_w(x|N(x)) \equiv \text{probability (point } x \text{ is white | values of neighbors of } x)$$

is given by

$$P_w(x|N(x)) = 1 - P_b(x|N(x)).$$

Of course these probabilities are not known, but there must be some points where P_b or P_w is very close to one or zero. Otherwise, we cannot do anything in principle. These points can be used as initial points whether black or white, and they influence their neighboring points. Therefore ambiguous values of P_b and P_w at other points change somewhat cooperatively, and we can expect that at each step of the iteration the number of points having the correct value increases so that the above computation eventually converges.

Now let us attempt the computation more exactly. Note that the intensity value of each point is reflectance and so a white point has greater value than a black point, as shown in Fig 4.6. This convention is typically used in the field of computer vision. So far we have depended on a reverse convention used in the field of character recognition. Here we comply with Pavlidis and Wolberg's [5] convention and notation.

As Fig 4.6 shows, by considering the vertical axis of the plane, we can analyze the topographical features of a given three-dimensional image. For any point the maximum change of value (reflectance) must be obtained to find convexity or concavity.

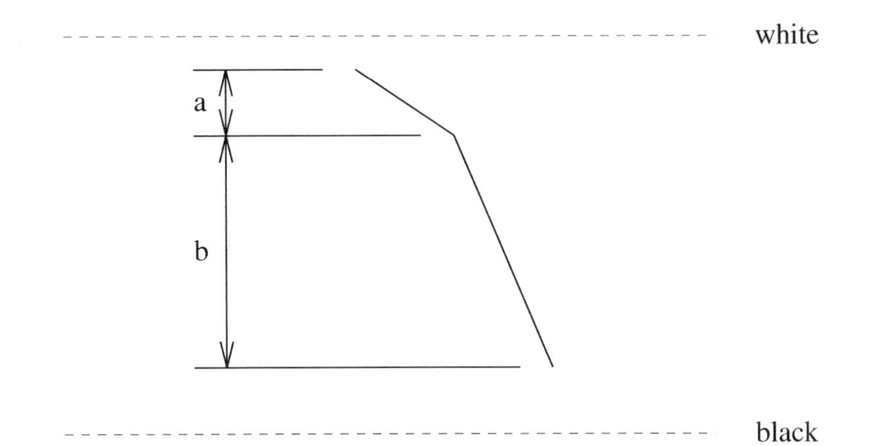

FIGURE 4.6 Gray-level convention of topographical image analysis: Both a and b are negative so $a > b$. From reference [5], © 1998, IEEE.

Concerning this point, we have already introduced the detection of these features in the Section 4.1, although there our interest was in two dimensions (value, distance). In this case one direction denoted D is chosen, and the subsequent values are defined as

$$a = \nu(P_c) - \nu(P_a(D)), \tag{4.51}$$

$$b = \nu(P_b(D)) - \nu(P_c), \tag{4.52}$$

where $\nu(P_c)$ denotes the center value of a point, and $\nu(P_a(D))$ and $\nu(P_b(D))$ are two extreme points along direction D. These a and b correspond to b and a, respectively, in Section 4.1. As before, we can determine the property of convexity or concavity comparing these values. This process is illustrated in Fig 4.7. Since in this case the situation is three-dimensional, the appropriate D must be decided by calculating the variation which is defined as

$$f(d) = || a(d) | - | b(d) ||, \tag{4.53}$$

where d is a given direction.

Actually four directions through a center point are chosen: horizontal, vertical, and two diagonal directions. The maximum value of $f(d)$ is found because the d that gives the maximum value is chosen as D. The interpretations of $f(d)$, a, b are obvious. That is, if both a and b are small, then it is likely that the center point is located at a plateau. Otherwise, if a exceeds b, then the intensity function is convex, or if a is less than b, it is concave. We note here that this convex and concave expression is just opposite to what was described in Section 4.1. This is because reflection is taken as the positive direction.

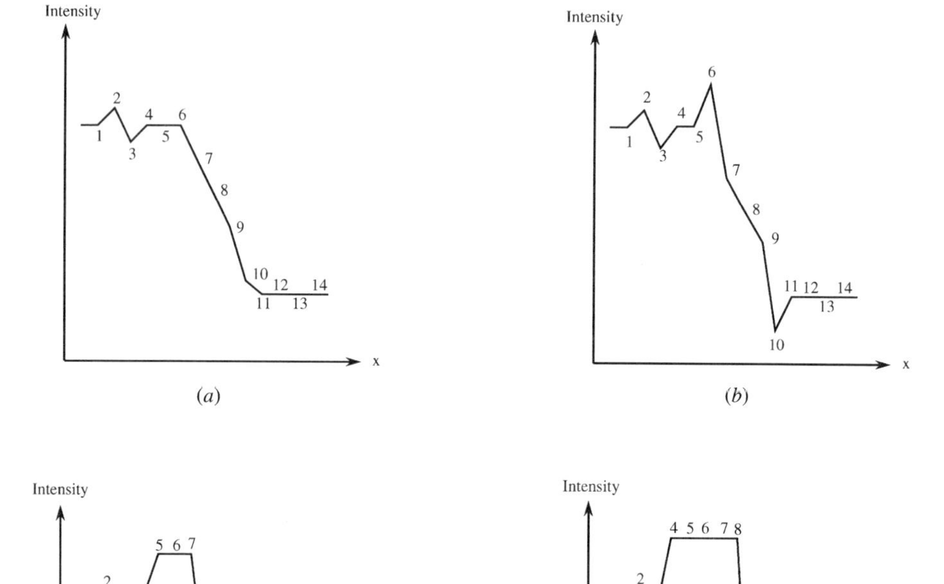

FIGURE 4.7 Application of the iterative algorithm: (a) Original profile; (b) results after procedure A; (c) profile after one application of procedure B; (d) profile after two applications of procedure B. From the reference [5], © 1998, IEEE.

The iterative process is easily implemented in parallel fashion. Therefore for a sequential implementation a copy of the image is necessary. We use $u(p)$ to denote the values of the copy and $v(p)$ the original values. The process is divided into two procedures in essence.

Procedure A For each point of the image do the following:

1. Find that D the direction where $f(d)$ as given by (4.53) is maximum.
2. Compute a and b according to (4.51) and (4.52).
3. If both a and b are small in absolute value or equal to each other, set $u(p_c) = v(p_c)$.

Else do:

4. If a is greater than b, then set $u(p_c) = \text{white}$.
5. Else if b is greater than a, then set $u(p_c) = \text{black}$.
6. For all points set $v(p_c) = u(p_c)$ (or use, instead, the copy as the original for the next iteration).

Figure 4.7 (*b*) shows the result of the application of procedure A to the input image shown in Fig 4.7 (*a*). Points where the intensity function shows a large variation and a high curvature are pushed to extreme values/goal values, black or white. This operation distorts the form of the intensity function, so re-application of the same procedure is impossible. For example, point 7 in Fig 4.7 (*b*) is in a concave area. Therefore subsequent passes are made with a modification of procedure *A* so that points with values other than white or black are attracted only by neighbors having one of these values. Since procedure A results in an increase of a or b only by making points take one of the goal values, after the application of procedure A, any points for which a subsequent comparison finds significant values of a or b must be a good value or at least one of the neighbors along D must have a good value. Actually points with the significant values excluding those points that have good values are the results of procedure A and were not significant when procedure A was applied. The next problem is how to deal with these points. Looking at the Fig 4.7 (*b*), we can understand that a geometrical feature of concavity or convexity has no meaning. Instead the absolute value $|a|$ or $|b|$ compared with $|b|$ or $|a|$ is important. For example, when $|a| \gg |b|$, P_c should take the value $P_a(D)$, which is a goal value according to the discussion above. Now the second procedure can be described.

Procedure B For each point of the image do the following:

1. If a point is already white or black, set $u(p_c) = v(p_c)$.
2. Find that D the direction where $f(d)$ as given by (4.53) is maximum.
3. Compute a and b according to (4.51) and (4.52).
4. If both a and b are small in absolute value or equal to each other, set $u(p_c) = v(p_c)$.
5. If $a + b$ equals white − black, then
   a. If $|a| > |b|$, then set $u(p_c) = \text{black}$.
   b. Else set $u(p_c) = \text{white}$.
6. Else
   a. If $|a|$ is greater than $|b|$, then set $u(p_c) = v(p_a(D))$.
   b. Else if $|a|$ is less than $|b|$, then set $u(p_c) = v(p_b(D))$.
7. For all points set $v(p_c) = u(p_c)$ (or use, instead, the copy as the original for the next iteration).

Step 5 in procedure B means that a point found to lie between white and black is pushed toward the nearest neighbor. Figure 4.7 (*c*) and (*d*) show the results of one and two applications of procedure B.

(*a*)

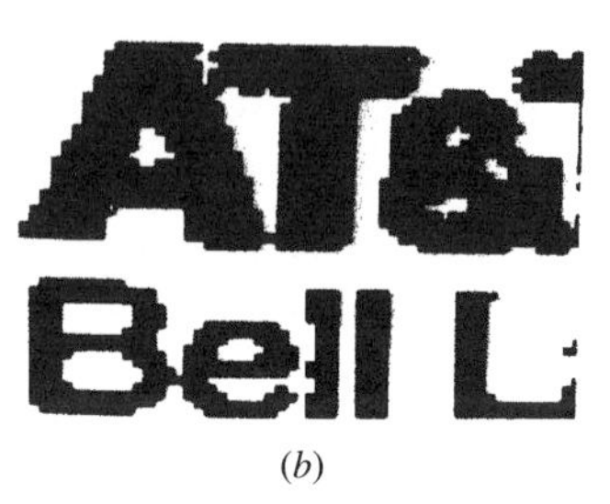

(*b*)

FIGURE 4.8 (*Left*) Input. (*Right*) Output after two iterations with Procedure B and two more with Procedure C. From reference [5], © 1998, IEEE.

FIGURE 4.9 Thresholding the example of Fig 4.8 (a). (*Left*) Threshold value set at 100. (*Right*) Threshold value set at 90. (Maximum range of the digitizer is 0–255.) From reference [5], © 1998, IEEE.

According to the results, two iterations of procedure B are enough to complete the binarization. Thus the method can be handled by a very fast parallel processor. However, to accelerate the domino-effect segmentation Procedure C was introduced. That is, it is the same as steps 1 to 7 of Procedure B but with $u(p_c)$ replaced everywhere by $v(p_c)$. The photographs of the original documents and the results are shown in Fig 4.8. However, the documents are very low quality, so it is difficult to select a uniform threshold value, such as is shown in Fig 4.9. Note that in these figures higher threshold values increase the black portion and lower values the white portion, which is just reverse of convention. The good results shown can be obtained only by adaptively chosen threshold values.

4.4 EXPERIMENTS

In this section two experiments on thresholding value selection are described. Both experiments are based on nonadaptive, namely the uniform thresholding value selection method from a practical point of view. The first includes a comparative study, but the size of samples used is relatively small. The second one is a systematic statistical analysis that uses a very large number of samples.

4.4.1 Experiment I

Two kinds of data were used.

Data Set 1 Printed arabic numeral characters. To obtain degraded quality print data, a typewriter ribbon was used. The print quality was classified into six levels. The first level is the print quality obtained by using a new ribbon. The second, the third, the fourth, the fifth, and sixth ones were obtained by using a ribbon 100,000, 200,000, 300,000, 400,000, and 500,000 times, respectively. At each level 37 sample characters were chosen. Therefore the total number of samples was $37 \times 6 \times 10 = 2220$. Naturally samples in the sixth level were very faint. However, the samples belonging to the second and third levels are not so bad. A videcon camera was used as a scanner, with effective levels and sampling size of 11 and 64×64, respectively. There were three levels of variation on the ground (white) level.

Data Set 2 Handprinted arabic numeral characters were written using a ballpoint pen that skips. The total number of samples is 200, with effective level and sampled size 7 and 64×64, respectively. The print quality of these samples is poor. Actually the poor print quality of ballpoint pens is well known in the field of OCR application.

In both experiments it was difficult to estimate the binarized data quantitatively. However, we calculated the dispersion of total area of black in a binarized sample for each category. Then an average value of the dispersions of ten classes was chosen as an evaluation value of each method for data set 1. The evaluation values are listed in Table 4.2. For data set 2 it was more difficult to find a reasonable evaluation value, so we relied on human judgment. We could not use the dispersion method because the sample sizes of each category were not constant. Specifically, we asked some people to select a range of threshold values by looking at each binarized result. Then the number of samples that are out of the threshold value range was counted for each method. The evaluation values are also listed in Table 4.2. As noted before, the threshold selection method can be resolved into two steps. One is the feature distribution step, and the other is the decision step. The gray-level feature distribution is the most typical one, and it is abbreviated as *g_l_dist* in the table. A local maximum and local flat distribution is abbreviated as *m_fl_dist,* a gradient distribution as *grad_dist,* and a Laplacian distribution as *Lap_dist.*

TABLE 4.2 Evaluations of Threshold Selection Methods Based on the Experiment I

Method	Data Set 1: Average of Dispersion	Data Set 2: Count of Out of Range
Complexity constrained	75	*1
medium/g_l_dist	290	55
l_f_aver/g_l_dist	120	*2
l_f_aver/m_fl_dist	236	0
nl_f_aver/m_fl_dist	153	0
con_cav_trans/m_fl_dist	240	10
DTSM/g_l_dist	296	67
DTSM/Lap_dist	*3	32
bottom_valley/g_l_dist	346	108
bottom_valley/Lap_dist	*4	58
max/grad_dist	387	45

* Denotes that data are not given.

In the decision function, the maximum value of a feature distribution is the most simple to compute, and it is abbreviated as max. Linear and nonlinear functions of average of a black portion are abbreviated as *l_f_aver* and as *nl_f_aver,* respectively. The *con_car_trans* is the abbreviation of convexity/concavity transition, which was explained in detail under procedure B. A typical decision method is finding the bottom point of a valley of some feature distribution, and it is abbreviated as *bottom_valley.* In the experiment for data set 2, the parameters adjusted for data set 1 were used. The complexity-constrained method measures the complexity of a character, and in this way the complexity of the black area is controlled to remain constant as much as possible. This method is only applied to the printed latin alphabet. However, the dispersion of black area was used to evaluate the threshold selection method. The original dispersion of black area of the data character set used was naturally the same in each method if it was calculated directly on the gray-level image data. The next is the *l_f_aver/g_l_dist* method, where the T_W was manually selected and adapted to data set 1. Therefore the entries *1 and *2 are omitted in Table 4.2. The *median/g_l_dist* method is a simple method in which the median was used to select a threshold value from the gray-level distribution. The third method is *nl_f_aver/m_fl_dist,* which is good because of nonlinearity as mentioned before. The reasons why the entries *3 and *4 are missing were not mentioned in [3]. The same decision method, DTSM, was applied to two kinds of feature distributions: *g_l_dist* and *Lap_dist.* The *l_f_aver/m_fl_dist* and *nl_f_aver/m_fl_dist* gave relatively good results for both data sets 1 and 2. However, these methods' weak point was that two parameters had to be fit in order to obtain T_w, although learning was effective. In contrast, *DTSM/g_l_dist* and *max/grad_dist* did not need to fit any parameter, although the results were not better than those of the parameterized methods. For this reason the DTSM is preferred in general.

4.4.2 Experiment II

The data set used is ETL-8 included Kanji characters (see Appendix B). As mentioned before, a simple statistical approach is to take the average values of all the gray levels for a given character image. In this case a parameter α was used to modify the average value, that is,

$$T = (\text{total average of gray levels}) + \alpha,$$

where $\alpha > 0$.

For a character set that has a uniform complex shape, this simple method works well but not if the images are degraded. For the Kanji character set, which is not uniform in terms of complexity, it does not work well even for not-so-degraded images. This was explained before. Therefore we need to change α adaptively to complexity. To do so, a combination of DTSM and the simple average value method was considered, which will be explained [6].

The problem with the DTSM method when applied to a gray-level distribution is illustrated in Fig 4.10. As the figure shows, normalized gray-level distributions for

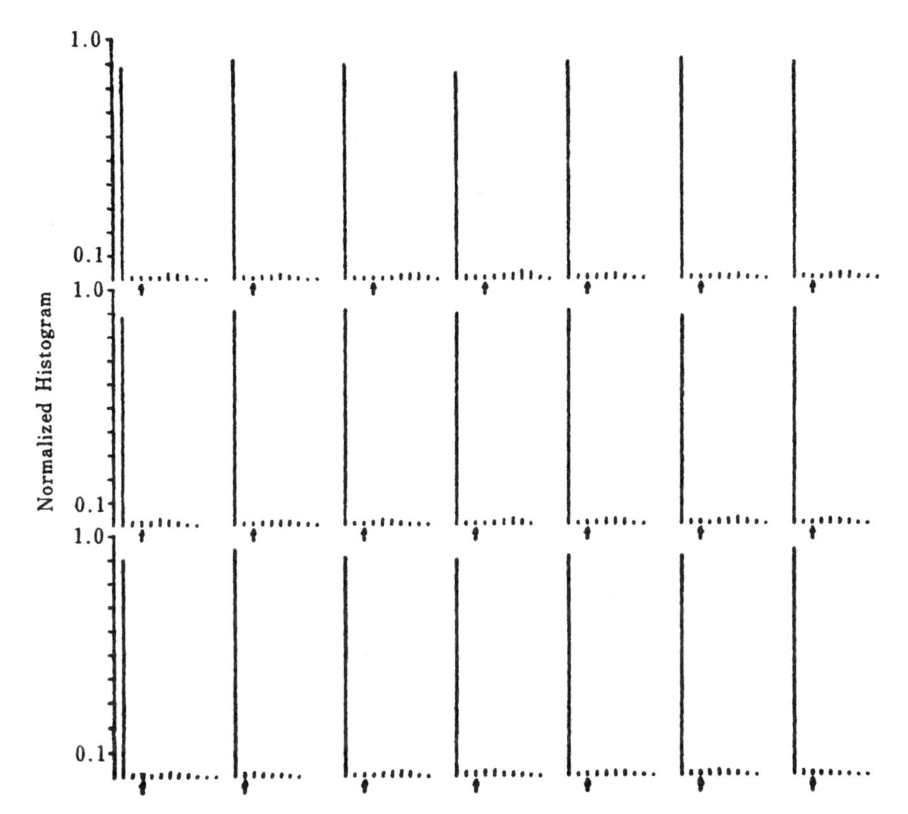

FIGURE 4.10 Normalization histograms of samples "愛" s (21 samples; arrows indicate k^*, values). From reference [6].

the data of 21 characters of "愛", in which the arrow marks indicate the optimum k^*, were found by DTSM. Looking at the data, we can see the following properties:

1. The number of points indicating white levels is very large.
2. The transition of the distribution changing from white to black is not smooth.
3. The distribution indicating black levels is almost flat compared with that of white levels.

From this experimental observation, a model of gray-level distribution was constructed as shown in Fig 4.11 (*a*). The model has $L = 16$ levels and the normalized histogram such that $p_0 = p_1 = p_w$, and $p_2 = p_3 = \cdots = p_{L-1} = p_b$, where $2p_w + (L-2)p_B = 1$ according to $\sum_{i=0}^{L-1} p_i = 1$. Then the value of $(L-2)p_B$, namely $14p_B$ is taken continuously smaller from 1 to 0. For this parameterized model the optimum k_α is calculated using DTSM based on its continuous model. That is, $k_\alpha = (\mu_0 + \mu_1)/2$ as explained in Section 4.2 on DTSM; it is plotted in Fig 4.11 (*b*). This tells us that when $14p_B \to 0$, the k_α approaches the level value 5.5 and then drops discontinuously to level value 0.5 at $14p_B = 0.003$. However, the value 0.003 is not realistic from a

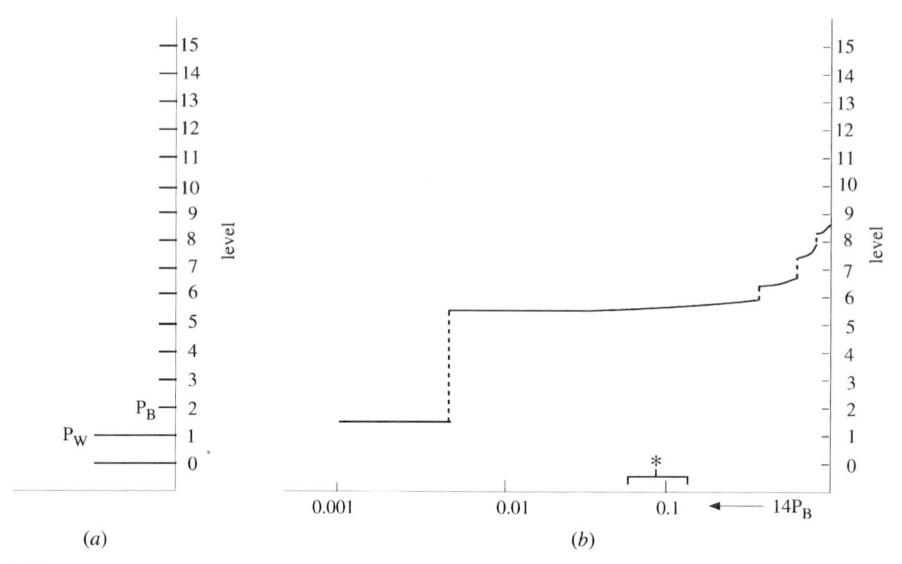

FIGURE 4.11 Peculiar behavior of DTSM: (*a*) A model of a density distribution; (*b*) transition of the optimal threshold. In the region marked by * lies the stroke rate against the ground in actual data. From reference [6].

practical standpoint. This discontinuity occurs because, in deciding the optimum k_α, the distribution changes from unimodal to bimodal. The optimum approaching the level value 5.5 is retained if the model is changed only a little as $P_0 = P_w$ and $P_1 = P_2 = \cdots = P_{L-1} = P_B$. The discontinuity does not appear in this model because it is a very peculiar phenomenon. We have found that the DTSM cannot take an optimum value less than 5.5, about $L/3$.

Therefore, in order to move a threshold value that is less than $L/3$ limited by DTSM to a lower threshold value, a total average level mentioned before, denoted as μ_T is introduced to determine a new threshold value T^* as

$$T^* = \mu_T(1 - \lambda) + k_\alpha\lambda \qquad (0 \leq \lambda \leq 1).$$

This means that T^* takes some value between μ_T and k_α. The T^* is rewritten as

$$T^* = \mu_T + (k_a - \mu_T)\lambda.$$

Thus T^* has the form $\mu_T + \alpha$, and α is changed adaptively depending on the data while λ is chosen based on the experiment. Therefore our aim is achieved. Of course T^* has no side effect when the gray-level distribution is normal, that is symmetric.

Some examples of the result are shown in Fig 4.12 (*b*) where $\lambda = 0.25$. The results of the test samples using DTSM are shown in Fig 4.12 (*a*) in which a deficiency of DTSM, namely its high threshold values, can be seen.

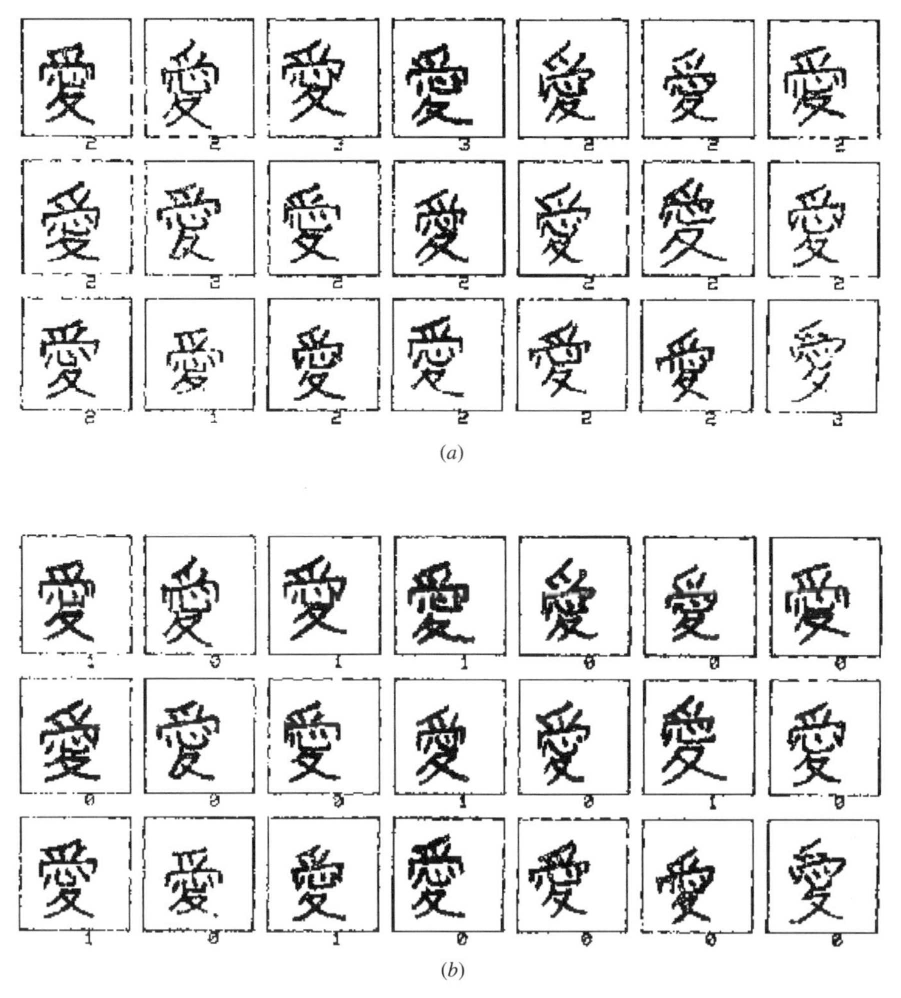

FIGURE 4.12 Samples binalized by the DTS method and its improved one: (*a*) DTS method, the numbers under the boxes denote the optimum k^*s; (*b*) the improved method, $\lambda = 0.25$, the numbers denote threshold values selected. The range of threshold value T is 0, ~15. Note here that if gray level $f(x, y) \leq T$ then $f(x, y) \rightarrow 0$, and if $f(x, y) > T$ then $f(x, y) \rightarrow 1$. From reference [6].

4.5 BIBLIOGRAPHICAL REMARKS

Concerning threshold selection techniques, there is a simple survey written by Weszka [7]. Since it was published in 1977, it is limited to an early stage in the development of these techniques. However, it gives a good basic idea of threshold selection. As mentioned in the text, a gray-level histogram does not always show a valley but there may be a "shoulder," which A. Rosenfeld and P. D. Torre note to be useful in finding the threshold value for a gray-level histogram without a valley [8]. Specifically, they showed that both valleys and shoulders correspond to concavities on the histogram, so concavities can be used to find appropriate threshold values too. In the text we described Otsu's threshold selection method in detail. However, we need to add some caveats. The unimodality of the criterion function is kept for standard images. Otsu tried to prove this but failed. In doing the experiments, he discovered counterevidence. This point was analyzed in detail by J. Kittler and J. Illingworth [9]. On the other hand, J. Weszka and A. Rosenfeld did research on "threshold evaluation" [10]. They proposed two kinds of evaluation of goodness of thresholded image. One is based on a business criterion, and the other is on an error criterion. The business criterion is intuitive, and a gray-level co-occurrence matrix [11] was primarily used, though a Laplacian operator was also considered. They showed that the two criteria gave almost the same and good results for both synthetic and real data.

There are two comparative studies of binarization methods performed by Trier and Taxt [12] and by Trier and Jain [13]. These studies focus on document images and so mainly dynamic thresholding methods are compared. In their first study the five criteria used were as follows:

1. Broken line structures.
2. Broken symbols, text, and so on.
3. Blurring of lines, symbols and text.
4. Loss of complete objects.
5. Noise in homogeneous areas.

However, they were not pleased with the evaluation criteria, and proposed instead a goal-directed evaluation, where an image understanding module is used for quantitative evaluation of results of a low-level image processing routine. Their experimental OCR module in this second study consisted of elliptic Fourier descriptors, which is explained in Chapter 7. The data used were part of a hydrographic map on which many number of numerals were written so that there was an obvious bias in the gray level.

They tested four global binarization methods using dynamic thresholding methods for the data:

1. Abutaleb's method [14].
2. Kapur et al.'s method [15].
3. Kittler and Illingworth's method [16].
4. Otsu's method.

The method of Otsu performed the best, though most of the numerals were smeared out. This problem further demonstrated the necessity of employing dynamic threshold methods. They tested the following 11 locally adoptive binarization methods:

1. Bernsen's method [17].
2. Chow and Kaneko's method [18, 19].

3. Eikvil et al.'s method [20].
4. Mardia and Hainsworth's method [21].
5. Niblack's method [22].
6. Taxt et el.'s method [12].
7. Yanowitz and Bruckstein's method [24].
8. White and Rohrer's dynamic threshold algorithm [25].
9. Parker's method [26].
10. White and Rohrer's integrated function algorithm [25].
11. Trier and Taxt's method [27].

The error rates in the results were compared with the reject rates. Accordingly Niblack's method was ranked the best, but only slightly better than Yanowitz and Bruckstein's method. Concerning execution time, the former was shortest, 3 seconds, and the latter was longest, 98 seconds, using Silicon Graphics Indy workstation for a 512×512 pixel image. In both methods a postprocessing was done, such as was proposed by Yanowitz and Bruckstein. Niblack's method and others without the postprocessing were not better than Yanowitz and Bruckstein's method (naturally the preprocessing was used only for this method). Therefore the postprocessing is very important. The idea of Niblack's method is simple, which is that the threshold value is varied over the image based on the local mean and local standard deviation.

BIBLIOGRAPHY

[1] M. R. Bartz, "The IBM 1975 optical page reader, Part III: Video thresholding system," *IBM J. Res. Develop.*, pp. 354–363, September 1968.

[2] S. Watanabe, "An automated apparatus for cancer processing: CYBEST," *CGIP,* vol. 3, pp. 350–358, 1974.

[3] S. Mori, M. Koseki, and M. Doh, "Some techniques of threshold selection," *Bull. Electrotechn. Lab.,* vol. 41, no. 10, pp. 773–782, 1977.

[4] N. Otsu, "An automatic threshold selection method based on discriminant and least squares criteria," *Trans. IECE Japan,* vol. J63-D, no. 4, pp. 349–356, April 1980.

[5] T. Pavlidis and G. Wolberg, "An algorithm for the segmentation of bilevel images," *Proc. IEEE Comp. Soc. Conf. Comp. Vision and Pattern Recogn.,* pp. 570–575, June 22–26, 1986.

[6] T. Saito and H. Yamada, "An improvement of the discriminant threshold selection method," *J. Info. Soc. Japan,* vol. 22, no. 6, 1981.

[7] J. S. Weszka, "A survey of threshold selection techniques," *Comp. Graphics Image Proc.,* vol. 7, pp. 259–265, 1978.

[8] A. Rosenfeld and F. D. Torre, "Histogram concavity analysis and aid in threshold selection," *IEEE Trans. Syst., Man, Cybern.,* vol. SMC-3, no. 3, pp. 231–235, March/April 1983.

[9] J. Kittler and J. Illingworth, "On threshold selection using clustering criteria," *IEEE Trans. Syst., Man, Cybern.,* vol. SMC-15, no. 5, pp. 652–655, September/October 1985.

[10] J. S. Weszka and A. Rosenfeld, "Threshold evaluation techniques," *IEEE Trans. Syst., Man, Cybern.*, vol. SMC-8, no. 8, pp. 622–629, August 1978.

[11] R. M. Haralick, K. Shanmugan, and J. H. Dinstein, "Textual feature for image classification," *IEEE Trans. Syst., Man, Cybern.*, vol. SMC-3, pp. 610–621, November 1973.

[12] Ø. D. Trier and T. Taxt, "Evaluation of binarization methods for document images," *IEEE Trans. Pattern Anal. Machine Intell.*, vol. 17, no. 3, pp. 312–315, March 1995.

[13] Ø. D. Trier and A. K. Jain, "Goal-directed evaluation of binarization methods," *IEEE Trans. Pattern Anal. Machine Intell.*, vol. 17, no. 12, pp. 1191–1201, December 1995.

[14] A. S. Abutaleb, "Automatic thresholding of gray-level pictures using two-dimensional entropy," *Comp. Vision, Graphics Image Proc.*, vol. 47, pp. 22–32, 1989.

[15] J. N. Kapur, P. K. Sahoo, and A. K. C. Wong, "A new method for gray-level picture thresholding using the entropy of the histogram," *Comp. Vision, Graphics Image Proc.*, vol. 29, pp. 273–285, 1985.

[16] J. Kittler and J. Illingworth, "Minimum error thresholding," *Pattern Recogn.*, vol. 19, no. 1, pp. 41–47, 1986.

[17] J. Bernsen, "Dynamic thresholding of grey-level images," *Proc. 8th Int. Conf. Pattern Recogn.*, Paris, pp. 1251–1255, 1986.

[18] C. K. Chow and T. Kaneko, "Automatic detection of the left ventricle from cineangiograms," *Comp. Biomed. Res.*, vol. 5, pp. 388–410, 1972.

[19] Y. Nakagawa and A. Rosenfeld, "Some experiments on variable thresholding," *Pattern Recogn.*, vol. 11, no. 3, pp. 191–204, 1979.

[20] L. Eikvil, T. Taxt, and K. Moen, "A fast adaptive method for binarization of document images," *Proc. 1st Int. Conf. Document Anal. Recogn.*, pp. 435–443, Saint-Malo, France, 1991.

[21] K. V. Mardia and T. J. Hainsworth, "A special thresholding method for image segmentation," *IEEE Trans. Pattern Anal. Machine Intell.*, vol. 10, no. 6, pp. 919–927, 1988.

[22] W. Nilblack, *An Introduction to Digital Image Processing*, pp. 115–116. Englewood Cliffs, NJ: Prentice Hall, 1986.

[23] T. Taxt, P. J. Flynn, and A. K. Jain, "Segmentation of document images," *IEEE Trans. Pattern Anal. Machine Intell.*, vol. 11, no. 12, pp. 1322–1329, July 1983.

[24] S. D. Yanowitz and A. M. Bruckstein, "A new method for image segmentation," *Comp. Vision, Graphics Image Proc.*, vol. 46, no. 1, pp. 82–95, April 1989.

[25] J. M. White and G. D. Rohrer, "Image thresholding for optical character recognition and other applications requiring character image extraction," *IBM J. Res. Dev.*, vol. 27, no. 4, pp. 400–411, July 1983.

[26] J. R. Parker, "Gray level thresholding in badly illuminated images," *IEEE Trans. Pattern Anal. Machine Intell.*, vol. 13, no. 8, pp. 813–819, 1991.

[27] Ø. D. Trier and T. Taxt, "Improvement of 'integrated function algorithm' for binarization of document images," *Pattern Recogn. Lett.*, vol. 16, no. 3, pp. 277–283, March 1995.

CHAPTER FIVE

Thinning

Thinning was introduced briefly in Chapter 1, where its intrinsic difficulty was discussed. In this chapter a detailed description will be developed beginning with the basic concept of connectivity. The connectivity number is introduced as a convenient indicator of the connectivity status of a given point on a line on a grid plane. Using the connectivity number, the so-called classical thinning method will be illustrated. To cope with the difficulty of thinning, new approaches will be illustrated, which are based on a global point of view. Comparison between the classical and modern thinning methods will be shown experimentally.

5.1 BASIC CONCEPT OF CONNECTIVITY

Connectivity on a grid plane was already discussed in the first chapter. There we noticed two kinds of connectivities, 4-connectedness and 8-connectedness. Here we review a more rigorous concept of connectivity on the two-dimensional Euclidean plane, R^2. Some important mathematical terms that will be used throughout the book are introduced as simply as possible.

First, consider Fig 5.1, in which some black regions (gray-colored) are drawn on a finite subset of R^2, the square X. The black regions X_i, $i = 1 \sim 4$, have a common feature, *connectivity*. Furthermore X_1 and X_2 have another common feature, *simple connectivity*. The concept of connectedness is visually obvious. That is, the regions become one blob. However, we need a rigorous definition of connectivity. Intuitively one might put it as follows: Let M be a metric space, and X a subset of M. X is path connected iff, for every pair of points a, b in X, there exists a continuous $f: I \rightarrow X$ with $f(0) = a$ and $f(1) = b$, where I denotes a unit interval [0, 1]. This definition is called *path connected*. It considers a well-known metric in Euclidean space. We can imagine a path between any two points on X_i, $i = 1 \sim 4$, that has the property of black color at every point.

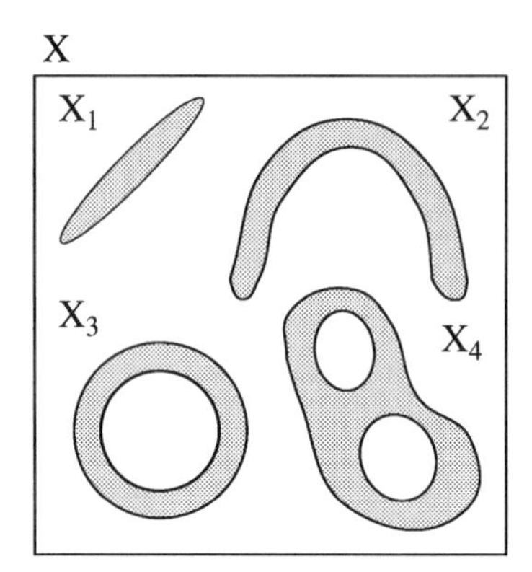

FIGURE 5.1 Some connected components and simply connected ones.

Next one might question what is the difference between $\{X_1, X_2\}$ and $\{X_3, X_4\}$. This is visually clear. Each component of the set $\{X_3, X_4\}$ has at least one *hole.* In contrast, both X_1 and X_2 have no holes at all. So they are called *simply connected.* However, a rigorous definition of simple connectivity is not so easy. To do so, we need to introduce another important concept of *homotopy equivalence,* which is illustrated in Fig 5.2. Within the space Y of a square, there is a hole. Space Y is connected, so we can draw many paths between points P and Q. However, let us consider a path, say l, moving to another path l' continuously such that l is superposed on l'. This is obviously impossible. Therefore we can say that there exists at least one hole in the space Y.

Specifically, let two paths l and l' be such that they connect points P and Q. If there is a continuous mapping $F(x, y)$ from a square $I \times I = \{(x, y)|0 \leq x \leq 1, 0 \leq y \leq 1\}$ to Y such that $F: I \times I \rightarrow Y$, $F(x, 0) = l(x)$, $F(x, 1) = l'(x)$, $F(0, 0) = P$, and $F(1, 1) = Q$, then path l and l' are called homotopes and denoted as $l \sim l'$, and $F(x, y)$ is called a homotopy of l and l' and denoted as $l \overset{F}{\sim} l'$. The meaning of $l \sim l'$ is the intuitive description of "moving and superposing l on l'." Furthermore the relation $l \sim l'$ is an equivalence relation. Therefore we can classify a set of all paths connecting points P and Q accord-

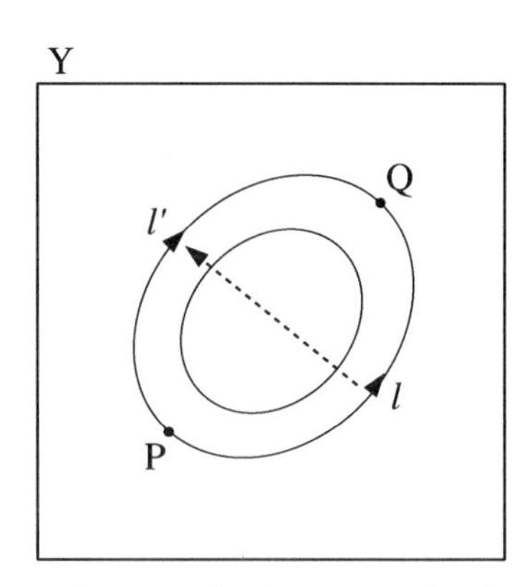

FIGURE 5.2 The difference between simply connected and not simply connected paths.

ing to this equivalence relation, which is denoted as $[l] = \{l' | l' \sim l\}$, called *homotopy class.* However, this homotopy class depends on the starting point P and the end point Q. We consider whether or not this point dependency can be removed. First, we can replace l' as l'^{-1}. That is, l'^{-1} is a reversed path of l'. Thus a loop is constructed that starts from P and ends at P, so point Q is virtually eliminated. Now all of the loop starting from P can be considered, which is classified by the homotope equivalence relation, denoted as $\pi_1(X; P)$. It is known that $\pi_1(X; P)$ constitutes a group where the multiple operation is that of homotopy class. The group is called a *fundamental group* of X having P as a base point. The group depends still on a point P, though it can be eliminated. It is known that the fundamental group is decided regardless of point P and invariant values in topology. Thus we can rewrite $\pi_1(X; P)$ as $\pi_1(X)$. For the fundamental group then, the simply connected space consists only of an identity element e of the fundamental group. Speaking intuitively, a loop can be contracted to one point if there is no hole in the loop. It is intuitively obvious that $l \sim e \cdot l$, and $l \sim l \cdot e$. To prove the statement above, considerable mathematical description is necessary, which can be found in any standard book on topology; for example, see Hocking and Young's book [28]. We give only a flavor of the fundamental group. This is because it represents a very good example of obtaining continuous/analog values by a discrete/symbol technique. This important point will be discussed later.

Returning to Fig 5.1, let us define some terms. First each connected region is called a *connected component.* The number of connected components in a given space X is an important feature that is invariant to continuous transformation, that is, topologically invariant. Topology is often said to be "rubber-sheet geometry," since the properties studied are invariant under continuous transformation. Naturally simple connectedness is a topological invariant. Furthermore the number of holes in a connected component is also topologically invariant. By the way, the region $X - \{X_1, X_2, X_3, X_4\}$ is called *background* of $\{X_i\}$, $i = 1 \sim 4$. Therefore the three holes are included in the background.

5.2 CONNECTIVITY NUMBER

In the previous section we described connectivity on a continuous plane, the Euclidean plane R^2. Now we turn to a grid plane and introduce a very useful "value" of connectivity, called the *connectivity number* [1]. First we assume that the thinning is done locally by an eroding process from the boundary of a line, as mentioned before. For this purpose the status of connectivity at each point must be defined locally, which is done considering a 3×3 frame. The general representation of a 3×3 frame and notations are shown in Fig 5.3 (*a*). We consider the connectivity at the center point whose value is denoted as x_0. Intuitively speaking, if a center point belongs to a simply connected set, then the point can be removed. For example, if the center point of the 3×3 pattern shown in panel (*b*) is removed, there is no problem in the connectivity of the line as a whole. It is assumed that the right side of the line is shown there. However, in the case of panel (*c*), the line is broken if the point is removed. That is, the point cannot be removed. The connectivity number gives a quantitative measure to such occurrences. By examining the role of the connectivity number we will be able to give its exact definition. That is, we count the number of connected components among $x_1 \sim x_8$ neighbors when the center of the 3×3 frame is removed. For example, it is 1 and 2 in case of panels (*b*) and (*c*), respectively. In the Fig 5.3 the connectivity number can be regarded as the number of connected components when the center of the 3×3 frame is removed. However, this simple rule does not work well for more complicated cases, as shown in Fig. 5.5.

In Fig. 5.3 (*a*)–(*c*), the numbers were irrespective of 4-connectedness and of 8-connectedness. However, in the case of panel (*d*) the problem is that the connectivity number is 2 and 1 for 4-connectedness and 8-connectedness, respectively. In panel (*e*), a complicated case, it is obvious in the case of 4-connectedness. However, in the case of 8-connectedness, the situation is somewhat more complicated; then each black pixel is represented by a black dot and the connections between the pixels are represented by line segments. Notice that in the 8-connectedness case, a hole is constructed at the center. Therefore it is not simply connected at all. Therefore we need to construct two numbers of connectivity denoted as N_c^4 and N_c^8 for 4-connectedness and 8-connectedness, respectively.

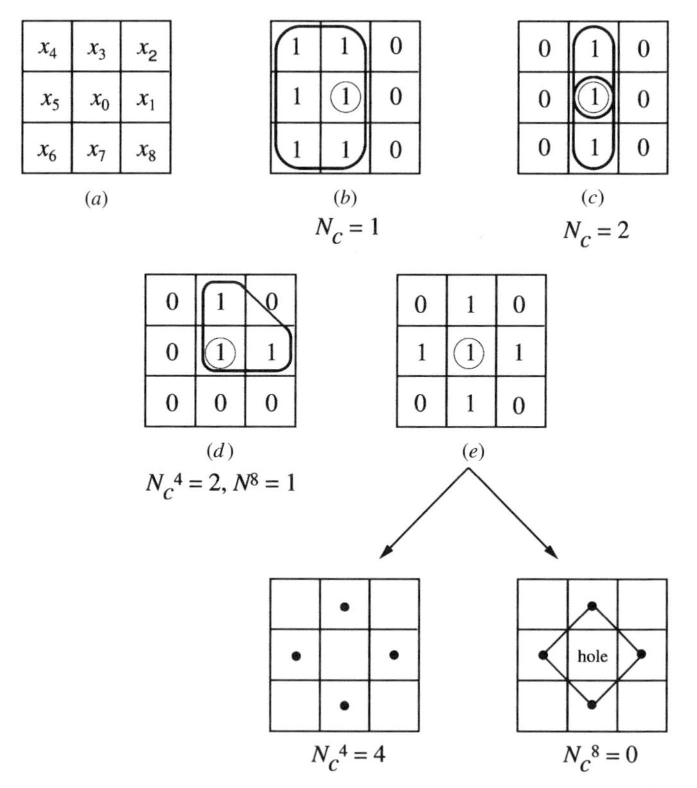

FIGURE 5.3 Yokoi's connectivity number: (*a*) Notation; (*b*)–(*e*) some typical examples. In particular, (*d*) illustrates the case of $N^8 = 1$, the case of $N^4 = 2$ being obvious and (*e*) illustrates the difference between N_c^4 and N_c^8. From reference [29], © 1997, Ohm.

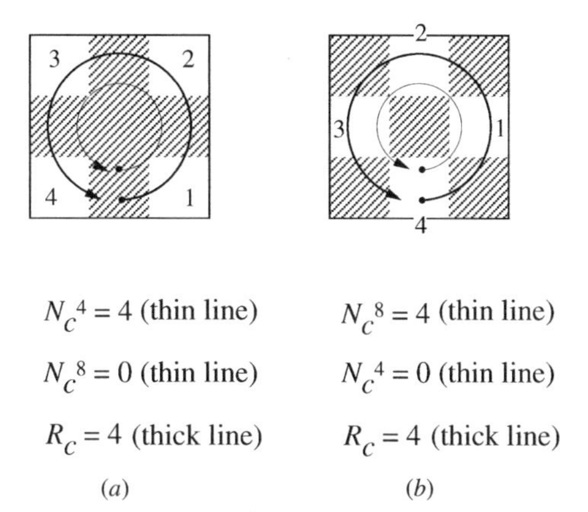

$N_C^4 = 4$ (thin line) $N_C^8 = 4$ (thin line)

$N_C^8 = 0$ (thin line) $N_C^4 = 0$ (thin line)

$R_C = 4$ (thick line) $R_C = 4$ (thick line)

(*a*) (*b*)

FIGURE 5.4 Two intuitive ways for giving connectivity numbers: Thick lines are for R_c. Thin lines are for Yokoi's crossing numbers. From reference [29], © 1997, Ohm.

The formalization of the two kinds of numbers were neatly given by Yokoi et al. [1] as follows:

$$N_c^4 = \sum_{k \in S_1} (x_k - x_k x_{k+1} x_{k+2}), \tag{5.1}$$

$$N_c^8 = \sum_{k \in S_1} (\bar{x}_k - \bar{x}_k \bar{x}_{k+1} \bar{x}_{k+2}), \tag{5.2}$$

where S_1 is the set of integers 1, 3, 5, 7 that has the 4-connected neighbors of the center point. We name the number defined above *Yokoi's connectivity number.* On the other hand, historically speaking, another definition of the number of connectivity was given by Rutovitz as follows [2]:

$$R_c = \frac{1}{2} \sum_{k=1}^{8} |x_k - x_{k+1}|. \tag{5.3}$$

Usually $2R_c$ is called *Rutovitz's connectivity number,* but since the summation in the formula above always gives an even number, we define the Rutovitz's number as shown in Fig. 5.4 in order to compare with Yokoi's number. The obvious intuitive meaning here is that it counts the number of crossings from black to white, such as when walking around the center as in Fig 5.4. In the figure the meaning of Yokoi's number is shown by the thin line. Notice that the thin line just passes the connecting points in the 8-connectedness region. Therefore in the panel (*a*), for example, the thin line lying on the black region does not pass any white region in the meaning of 8-connectedness, but it crosses from black to white four times in the meaning of 4-connectedness along the thick line. Some different examples are shown in Fig 5.5 using both Yokoi's and Rutovitz's numbers. Notice that there is no difference between 4- and 8-connectedness in using Rutovitz's number. Therefore Yokoi's number has more general applicability than Rutovitz's number. The numbers of connec-

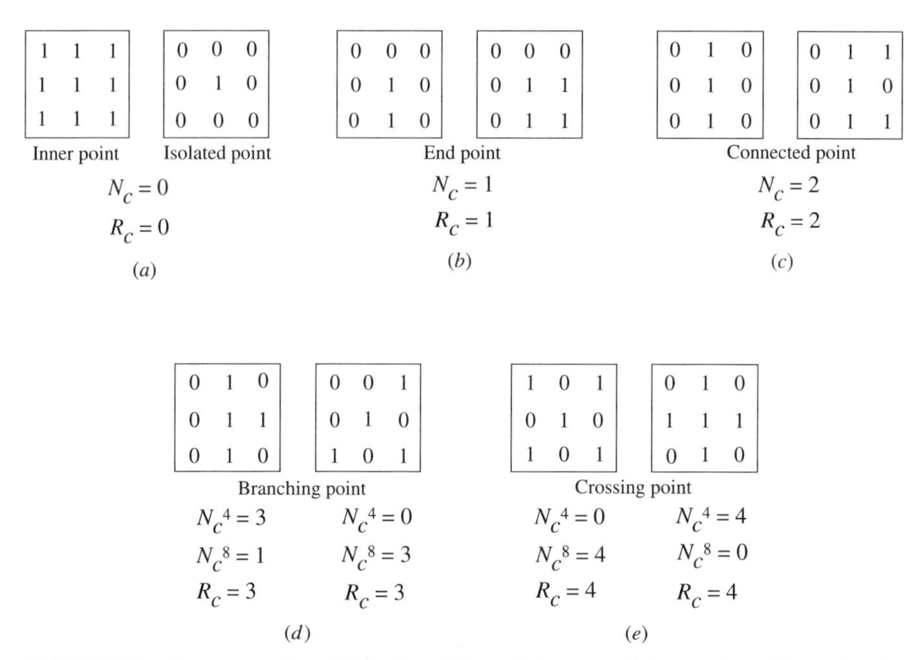

FIGURE 5.5 Same examples of Yokoi's and Rutovitz's connectivity numbers: The right side of the branching point case (*d*) is a simple interpretation of "the number of connected components when the center of 3 × 3 frame is removed," so it does not work well. If this rule is applied, it gives $N_c^4 = N_c^8 = 3$. Rather one should apply the intuitive rule of Fig. 5.4. From reference [29], © 1997, Ohm.

tivity are most useful in the thinning process. That is, when value of the numbers is equal to one, this means that the center point belongs to a simply connected region. Therefore we can delete the center point. Otherwise, we cannot delete the center point because, if the point is deleted, it tells us that the connected region is really broken into pieces. We give only a simple description here. A rigorous one is found in reference [1]. The main idea is that if we can say that the connectivity number of a center black point of a given 3 × 3 frame in a given line is equal to one, then the point can be removed. Naturally this point's deletability depends on whether 4- or 8-connectedness is considered.

5.3 HILDITCH'S THINNING METHOD

The classical thinning method is Hilditch's method. Hilditch was one of the originators of "thinning" [3]. She determined four conditions necessary for thinning to occur:

1. *Thinness.* The thinned line's width must be one pixel.
2. *Position.* The thinned line must lie along the center of the original line.
3. *Connectivity.* The thinned line must keep the connectivity of the original line.

4. *Stability.* At each step of the thinning process the thinned line cannot be eroded away from its end points.

So far conditions 1 through 3 have entered our discussion. However, we will analyze these conditions more quantitatively using a figure which we introduced in Chapter 1. For convenience, it is shown again as Fig 5.6, in which in the panel (a) the middle of the figure represents a 2-pixel width. This illustration contradicts condition 1 of thinness. Now let us consider the exact meaning of a 1-pixel width.

An ideal thinned line allows us to follow it from its one end point to another end point along a single path. In Fig 5.6, panel (*a*), there are two passes. Now recall that a pass depends on its definition of connectivity. Accordingly, if 8-connectedness is adopted, certainly there can be two paths, but if 4-connected is adopted, only one pass can exist. Therefore for the former, one more point can be removed, and for the latter, no point can be removed. If all the 4 (8)-connected points are removed, then such condition is called *complete 4 (8)-connected.* Otherwise, the situation is called *incomplete 4 (8)-connected.* In this sense Yokoi's number of connectivity can be fully used. For condition 3 and its relation to condition 4 on stability, a little illustration may be necessary. Namely at the two ends of a thick line, the local situations must be the same as at the sides. Therefore, as the end sections are eroded step by step, the result is a shortening of the line. Condition 4 is imposed to prevent such a line-shortening process.

5.3.1 Thinning Algorithm

Algorithms on thinning are roughly classified as two kinds. One is sequential, and the other is parallel. In general, sequential algorithms are good at preserving continuity, and parallel algorithms are good at bringing the thinned line nearly to the center of the line. In the sequential algorithms further there are two variants: one is the boundary-following type and the other is the raster-scanning type. The former peels a given line layer by layer in tracing its boundary. It is based on a boundary-tracing algorithm and so is intuitively appealing. Hilditch's algorithm is sequential and raster-scanning.

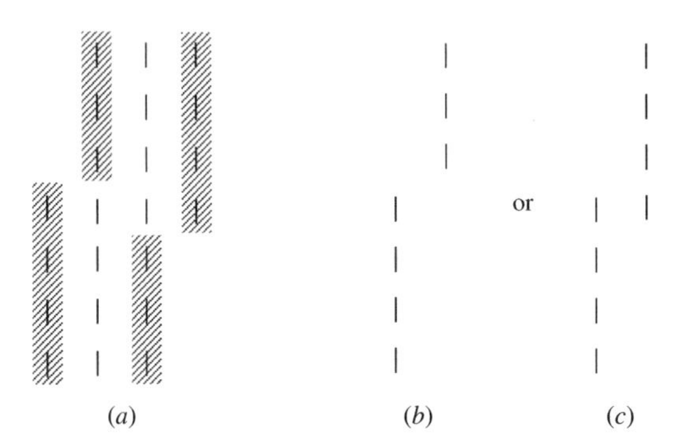

FIGURE 5.6 Illustration of incomplete 4(8)-connectedness: (*a*) mechanically eroded line, where the eroded boundaries are shaded, (*b*) complete 8-connectedness, and (*c*) complete 4-connectedness. From reference [29], © 1997, Ohm.

The essence of the algorithm will be described. First, we assume that a black figure is given by "1" and the ground is given by "0." Let the black image be denoted Q. The potentially deletable points are checked by several conditions and the resulting potentially deletable points are marked by "−1." The marked region is called R. After one frame is scanned, the region R is set to "0." Then $Q - R$ is newly set to Q, on which the same operation is applied until a stop condition holds. A strength of the sequential method is that R can be used effectively, as will be soon seen.

Now let us describe the algorithm in detail. A key point of the algorithm is to set up the conditions for the removal of a pixel. By changing the conditions, dozens of algorithms can be conceived. First of all we require that the conditions for the removal of a pixel be determined only by its neighboring points. The conditions can be divided into two parts. One is that the condition is determined by the initial state of an image. Since the image is not yet scanned, each point within the frame in which the image lies has the value "1" or "0." The conditions 1 through 4 discussed below belong to this first class of conditions. In the second class, the neighboring points of a noticed point are changed to "−1" by the scanning. The second class of conditions reflects this dynamic setting. The conditions 5 and 6 belong to the second class.

Conditions for Removal of a Point A point p will be removed if it satisfies all the following conditions.

First Class

Condition 1 Point p belongs to Q.

Condition 2 Point p lies on the edge of Q; that is,

$$\sum_{k \in S_1} \bar{x}_k \geq 1$$

holds, where S_1 is $\{1, 3, 5, 7\}$ as before.

For the notation used above refer to Fig. 5.3. Figure 5.7 illustrates this removal process. The scanning in the figure is from top left to bottom right. An example of conditions 1 and 2 is shown in panel (*a*). The point marked by a circle denotes the point being scanned at that moment. Only the boundary points are removed. Recall here the remark for the first class conditions, that the conditions are calculated using the initial state of a given image. Therefore value x_i for any i does not take −1 at any point. Further we note that if this point were set to 0, then at the first scanning the black points would be eroded deeply into the interior. Thus R's function would be set into effect, peeling layer by layer a thick line.

Condition 3 Point p is not the tip of a thin line; that is,

$$\sum_{k=1}^{8} x_k \geq 2$$

holds.

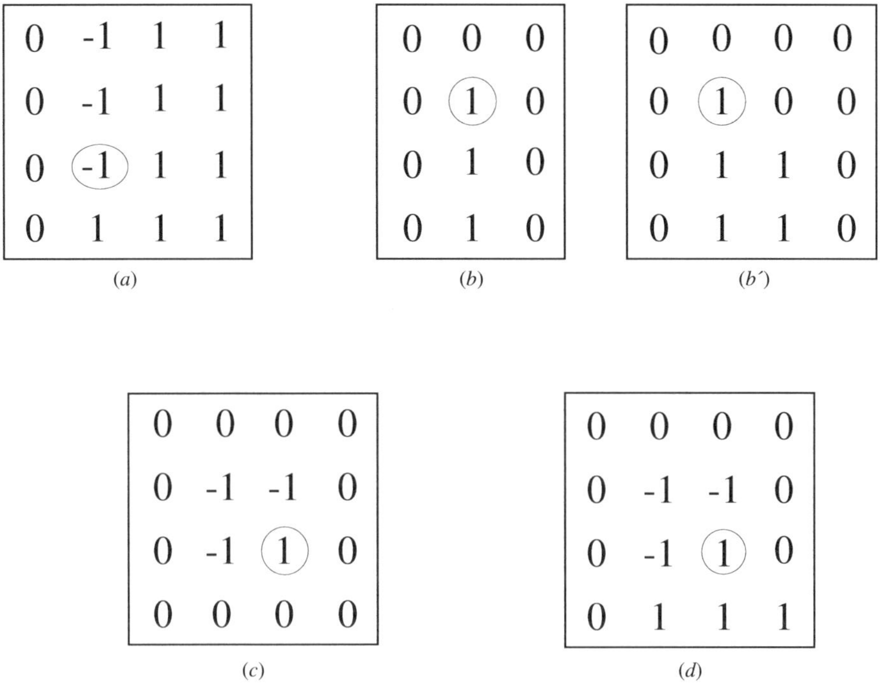

0	-1	1	1
0	-1	1	1
0	(-1)	1	1
0	1	1	1

(*a*)

0	0	0
0	(1)	0
0	1	0
0	1	0

(*b*)

0	0	0	0
0	(1)	0	0
0	1	1	0
0	1	1	0

(*b*′)

0	0	0	0
0	-1	-1	0
0	-1	(1)	0
0	0	0	0

(*c*)

0	0	0	0
0	-1	-1	0
0	-1	(1)	0
0	1	1	1

(*d*)

FIGURE 5.7 Removal conditions of Hilditch's thinning algorithm. From reference [29], © 1997, Ohm.

Examples of condition 3 are shown in panels (*b*) and (*b*′), where (*b*) is a normal tip and (*b*′) is not. Globally the point marked by a circle in (*b*′) is an end point, but it is removed. Therefore the thinned line will be shortened by two pixels except for the ideal case shown in panel (*b*). Notice that we assume that a line has two end points. This is a typical example of the limitation of the local operation for thinning.

Condition 4 Point p's removal does not alter the connectivity; that is,

$$N_c^8 = 1$$

holds.

Here Yokoi's number of connectivity is used. In the Hilditch's original paper, a complex formula was used. Naturally all the points marked by a circle satisfy this condition.

Second Class

Condition 5 The removal of point p does not remove an isolated small blob; that is,

$$\sum_{k=1}^{8} x_k' \geq 1$$

holds for a given center point to be removed.

The algorithm is constructed not to change the number of connected components. Therefore condition 5 is necessary. In practice, any very small blob is usually removed, but the danger is that this can lead to important information such as a "dot" or a "period" being lost. In the above formula, the prime at x_k' indicates that if that point is included in R, then it is set to "0." Such an example is shown in panel (c). Obviously the above condition does not hold, $\sum_{k=1}^{8} x_k' = 0$. Therefore the point is kept.

Condition 6 Point p's removal in conjunction with any one of its neighbors that has been removed does not alter the connectivity of Q; that is,

$$x_i \neq -1 \quad \text{or} \quad N_c^8(i) = 1 \quad \text{for } i = 3, 5.$$

Condition 6 deals with a two-line width. An example is shown in panel (d). Looking at the neighboring circled point, we can see that $x_3 = -1$. Therefore the next condition is checked. The meaning of $N_c^8(3)$ is that if x_3 is included in R, then N_c^8 is calculated setting $x_3 = 0$. The result is that $N_c(3) = 1$, where the -1's are regarded as 1 except at point x_3, and so $N_c(5)$ is checked in the same manner. Since x_5 is included in R, then N_c^8 is calculated by setting $x_5 = 0$. The result is that $N_c(5)$ is 2. The final result is that the point cannot be removed. If condition 6 does not exist, then the point will be removed because the point is obviously a boundary point, not an end point or an isolated point.

5.3.2 Experimental Results

In Fig. 5.8 two examples are shown. One is Hirakana and representative of characters rich in curved lines. The other is Kanji and a representative of characters straight lined but somewhat complex. In terms of thickness and boundary noise, for these types of lines the Hilditch algorithm works well. When a boundary is very noisy, a smooth operation is employed and is effective. However, there is an intrinsic problem, namely "V" is sometimes changed to "Y," as described before. This is not easily improved in the thinning framework mentioned above. This is because in a local operation the framework must be changed and not the Hilditch algorithm. This point will be described next.

5.4 PAVLIDIS'S VECTORIZER BASED ON LAG

Now we introduce a new thinning method that overcomes the intrinsic difficulties of Hilditch's method and its variations based on local operations. This class of thinning algorithms is called *pixel-based thinning.*

The input data to the new method is first encoded using *run length encoding* (RLE). This coding is widely used as a basis of image compression techniques, and

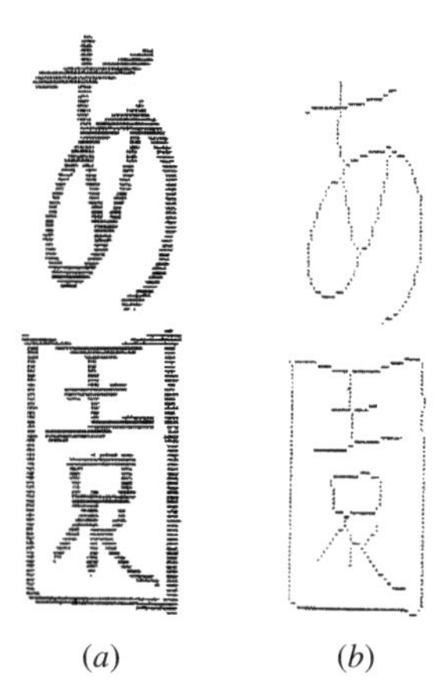

FIGURE 5.8 Examples of Hilditch's thinning algorithm: (*a*) Original character images; (*b*) results. From reference [29], © 1997, Ohm.

it is especially effective for binary images. For convenience, RLE is illustrated in Fig. 5.9 (*a*). We assume here that a frame in which a figure is embedded is raster-scanned from top left to bottom right. At each row of a frame, white or black intervals are converted to the representation of (color, length). For example, the first row and the second row of the image shown in panel (*a*) are encoded as (0, 13), and [(0, 1), (1, 3), (0, 9)], respectively. If the leftmost column is white, then the encoding becomes more compact. The third row of the frame is encoded as (1, 3, 4, 3, 2). In this case the code consists of only a sequence of lengths, which suffices for a bi-level image. Of course our assumption of starting with white does not lose its generality.

In this RLE the connectivity along each row in the horizontal direction is explicitly represented. However, the connectivity in the vertical direction is not represented explicitly. In this sense RLE is further encoded to the *line adjacency graph* (LAG). The nodes of the graph represent intervals and edges/branches nodes are connected if the corresponding intervals are on adjacent lines, namely their projections in the scan direction overlap and their pixels have the same color. One example is shown in Fig. 5.9 (*b*) where each node is located at the midpoint of the corresponding interval. Because of discreteness of a grid plane, the midpoint of an interval whose length is an even integer is taken by a pixel biased to the right. Looking at panel (*b*), it seems like it may be easy to obtain a thinned pattern once an original pattern is represented to LAG. However, the panel (*b*) is the ideal case. The real case is shown in panel (*c*), which is the case of LAG. Since the obtained LAG is considerably distorted, we need to modify the result so that it meets human intuition. This is shown by the broken lines superposed on the figure. How to do it is the problem. Pavlidis [4] solved the problem using the first method of a new generation of thinning techniques based on a global view and processing rather than a local view and pixel operations.

Now let us study Pavlidis's algorithm. The basic idea is to integrate the LAG to a more compact representation, which consists of only two kinds of nodes, path and junction. Intuitively path and junction correspond to stroke and crossing or branch-

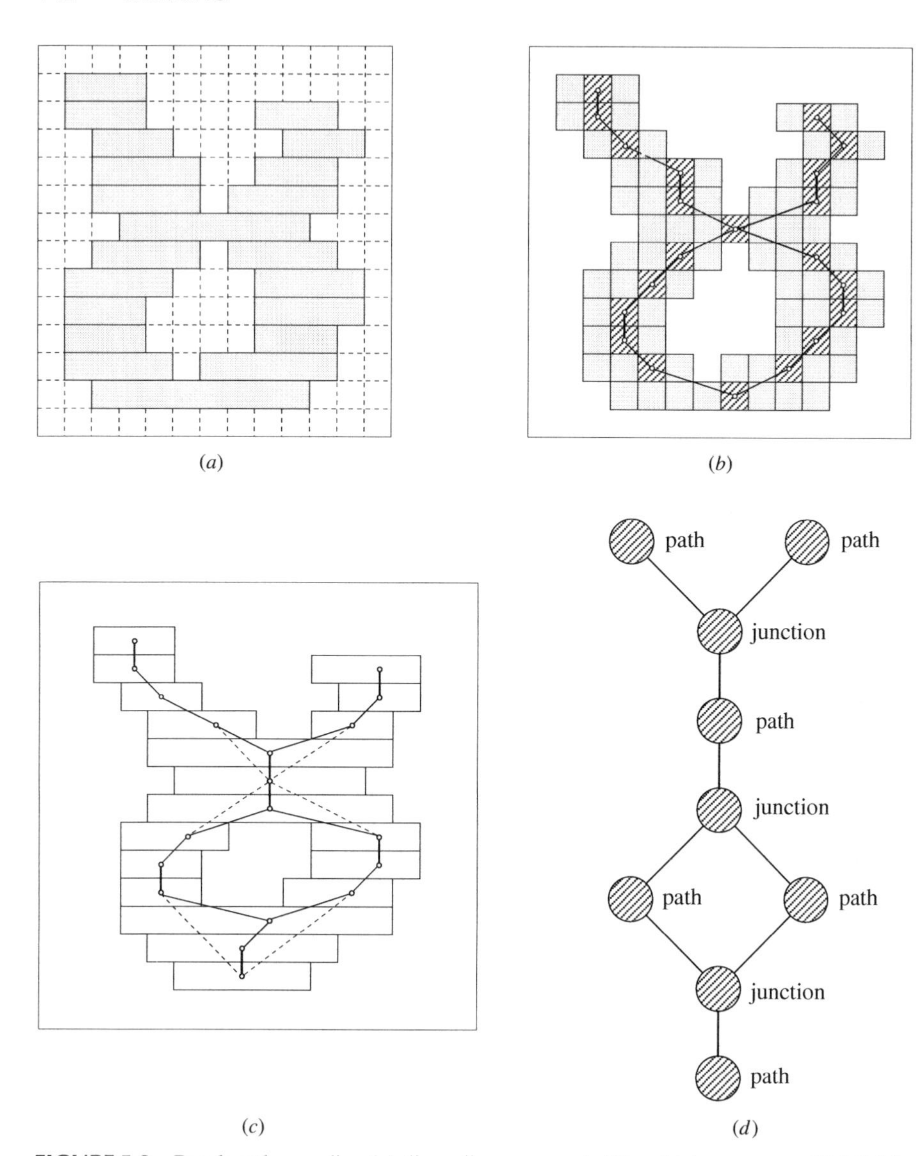

FIGURE 5.9 Run length encoding (*a*); line adjacency graph (*b*) and (*c*); compressed LAG (*d*).

ing part. After this rough classification, paths and junctions are examined to construct vector components using width and collinearity information. Then some vector components are merged into one vector. In this sense the algorithm is more than preprocessing and closer to feature extraction, so it gives compact structural information. However, still ambiguous parts remain, each of which is inferred from its neighboring vector components and contour information.

In a detailed discussion some definitions relating LAG are necessary. For a given interval-node, the *above degree* is defined as the number of intervals touching it on the line above and the *below degree* is defined as the number of intervals touching it on the line below. A pair of numbers is denoted (a, b), where a stands for above and b for below. A path is a set of connected nodes whose degrees are all (1, 1) except, possibly, for the first and last nodes which may have degrees (0, 1) and (1, 0), respectively. Otherwise, a node is labeled as a junction. The paths and junctions are connected as they are in the LAG. Thus a new graph which is homeomorphic to the LAG is constructed, which is referred to as the *compressed* LAG or c-LAG for short. Figure 5.9 (*d*) shows the c-LAG in panel (*c*).

5.4.1 Path Node Analysis

We begin with the simplest case which is shown in Fig. 5.10 (*a*). The figure shown consists of only one path, which can be divided by looking for a sequence of segments having almost uniform widths. In this case there are obviously two such sequences. Vertical sequence vectorizing is easily performed constructing a vector passing through the midpoints of the segments. The other sequence, however, cannot be applied this way. We need, naturally, the ratio S of height over width to be defined. If the ratio S exceeds 2, then the vector of the group of segments is defined as joining the centers of the first and last segments. The vector obtained is obviously vertical. On the other hand, if S is under 0.65, then the vector of the group of segments is defined as a horizontal line through the middle of the group. These examples are shown in Fig. 5.10 (*c*). However, we cannot decide definitely whether a given group of segments is vertical or horizontal from only knowing the ratio S. There are other subtle cases that must be carefully treated, so we refer the reader to the paper [4].

Looking at panel (*d*) in detail, we might, naturally, consider merging the upper and lower vectors, both of which are indicated by dotted lines. The following rule is used for the merging vectors: Let the end points of the first vector be P_1 and P_2, and those of the second vector P_3 and P_4 be as shown in Fig. 5.11. A new vector is defined by joining the first point of the first vector P_1 to the last point of the second vector P_4. Then the distance of the other end points from the new line is checked so that some absurd configurations such as in the right part of Fig. 5.11 can be excluded. The horizontal distance of the other two points is also checked. On the other hand, even if a given group of segments has almost the same width, a problem arises if the centers of the first and the last are connected by a straight line. Such a case is illustrated in Fig. 5.10 (*e*). If the maximum distance of the other centers from that line exceeds a given tolerance, then the line is centers subdivided into two parts at the segment where the maximum error occurred.

5.4.2 Compound Vectorization

It is now time to describe the advantages of Pavlidis's method over Hilditch's method. The typical problems are shown in Fig 5.12 (*a*) and (*b*). For example, the bottom path of panel (*b*) will be eroded to a single line segment in the case of Hilditch's algorithm, as noted before. Such an unclear path is dealt with by the following algorithm:

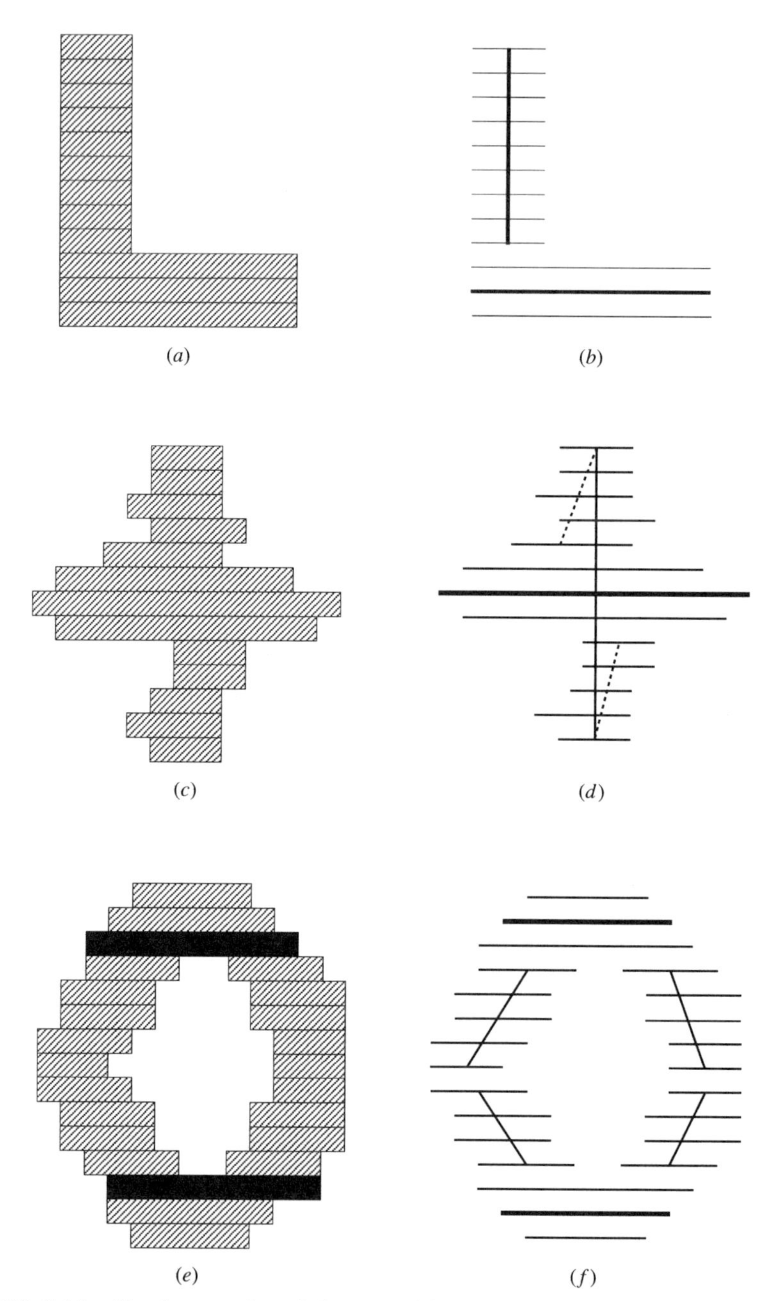

FIGURE 5.10 Simple examples of the vectorizing process: (*a*) Analysis by widths; (*b*) merging vectors; (*c*) collinearity of centers. From reference [4], © 1998, Academic Press.

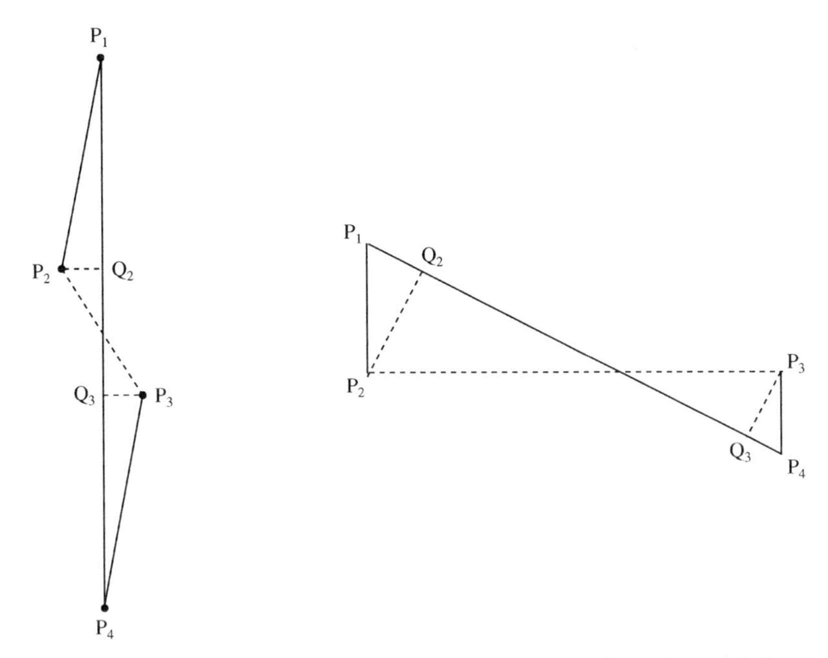

FIGURE 5.11 Rules for marging vectors. All three segments P_2Q_2, P_3Q_3, and P_2P_3 must be small to decide in favor of merging. From reference [4] © 1998, Academic Press.

STEP 1: Compute:

Define:

na degree above of top interval.
nb degree below bottom interval.
jna degree above of junction on top of top interval.
jnb degree below of junction below bottom interval.

STEP 1A: Compute **na** and **nb**.

STEP 1B: If **na** > 0 compute **jna** from the LAG, otherwise set **jna** = 0.

STEP 1C: If **nb** > 0 compute **jnb** from the LAG, otherwise set **jnb** = 0, /* from now on we forget about **na** and **nb***/

STEP 2: If both **jna** and **jnb** are less than two **return**.

STEP 3: If **jna** exceeds one, **then** select two points on the top segment lined up with the midpoints of the intervals of the Path nodes above [Fig. 5.13 (*a*)], **else** select two identical points on the top segment at its middle [Fig. 5.13 (*b*)]. Let the two points be denoted $\mathbf{P_1}$ and $\mathbf{P_2}$.

STEP 4: Select points on the bottom segment in a similar way as in step 3 using **jnb**. Denote the points by $\mathbf{P_3}$ and $\mathbf{P_4}$.

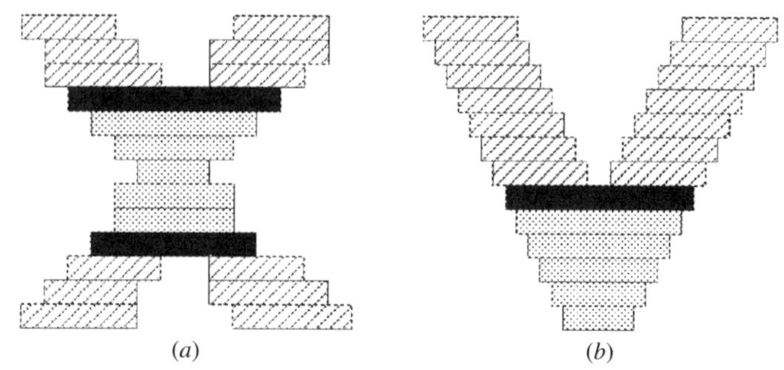

FIGURE 5.12 Typical problems of ambiguous regions indicated as grayed and darkly drawn. From reference [4], © 1998, Academic Press.

STEP 5: If the lines $\overline{\mathbf{P_1P_4}}$ and $\overline{\mathbf{P_2P_3}}$ are both within 10 degrees from the vertical direction, **then** construct a single vertical vector for the **PATH** and **return**.

STEP 6: **If** either of the two lines is within 30 degrees of the horizontal **return**.

STEP 7: Add the lines $\overline{\mathbf{P_1P_4}}$ and $\overline{\mathbf{P_2P_3}}$ to the vector list.

The algorithm can be better understood by referring to Fig. 5.13. If the c-LAG has only one node and this node has an unclear vectorization, then we cannot use the neighboring key features such as a hole and branching. Therefore a best guess is

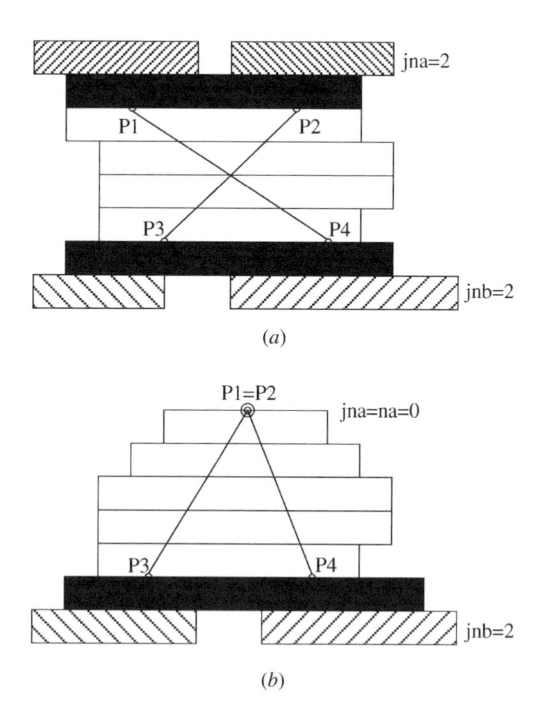

FIGURE 5.13 Vectorization of a node of the c-LAG by examining its neighbors using the method described in the text. From reference [4], © 1998, Academic Press.

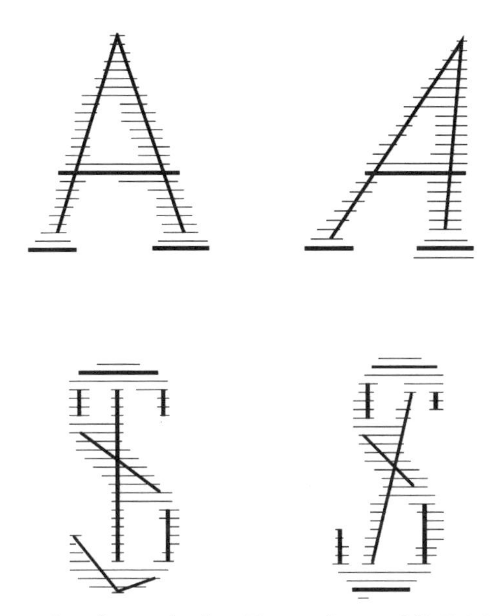

FIGURE 5.14 Examples of vectorization. From reference [4], © 1998, Academic Press.

made and the blob is reexamined later by contour analysis, for example. Such a case can actually happen. Suppose that a character "0" is heavily printed and the hole is filled up. Some successful examples given in Fig. 5.14 show the advantages of Pavlidis's algorithm over the pixel-based thinning algorithms, and the processing speed is also very fast. However, the method is quite sensitive to the widths of strokes which are measured by only one-directional raster scanning. Therefore more subtle rules are needed for detecting width change, but these are not described. In other words, there seems to be a problem of stability from an engineering point of view.

5.5 CROSS SECTION SEQUENCE GRAPH

The basic idea of Pavlidis was further advanced by Suzuki and Mori [5]. Their method is based on a so-called cross section sequence graph rather than on LAG. Their strategy for thinning is clearly described by the two terms *regular region* and *singular region.* A regular region means a straight or curved line segment, and a singular region means an end point region, a corner, a junction, or crossing region. Thus a character region can be divided into two kinds of regions, regular and/or singular regions. Naturally regular regions are stable; on the contrary, the singular regions are not usually, and they can be ambiguous, as mentioned in Pavlidis's algorithm. However, a singular region is usually surrounded by some regular regions, so a singular region can be analyzed by its adjacent regular regions.

The names *regular* and *singular* regions in the character recognition field were coined by Simon and Baret [6] in their nonpixel thinning method. They use multi-oriented runs consisting of two types, slanted as well as both horizontal and vertical runs. The cross section sequence graph (denoted CSSG) can be obtained by generalized runs with arbitrary orientation. In this sense CSSG can be regarded as an extension of Simon and Baret's method. A generalized run can extract the smallest possible singular region and thus limit the distortion of the thinning/skeleton, whereas a regular region is extracted to be as large as possible.

The thinning process based on CSSG is illustrated conceptually in Fig. 5.15. The image character described by the CSSG is composed of regular regions called a *cross section sequence* and singular regions. In order to extract a cross section sequence, a cross section is introduced across the line almost perpendicular to its direction. The cross section can be easily obtained because the edge direction is almost perpendicular to the direction of the line. The sequences of cross sections are extracted as regular regions, and remaining regions are extracted as singular regions. They are connected bichromatically and constructed as a graph called the *cross section sequence graph.* Since a singular region can be analyzed from adjacent regular regions, we can extract a singular point correctly from the point of view of stroke connectivity. Now the details of the algorithms will be illustrated systematically.

5.5.1 Smoothing

Usually smoothing is done in terms of gray levels taking an average value of neighboring gray values, for example, 3×3. However, smoothing is very direct here; the angles of sections are smoothed, which is very effective. Moreover the smoothing is very fast because it is done only along boundaries. Since such smoothing is not common, it will be described in some detail [7].

First, the boundary points are identified; these are also black points. Next, at each boundary point a 3×3 Sobel operator is applied, and from the proportion of two edge values obtained from the orthogonally paired operators, an angle of the edge is cal-

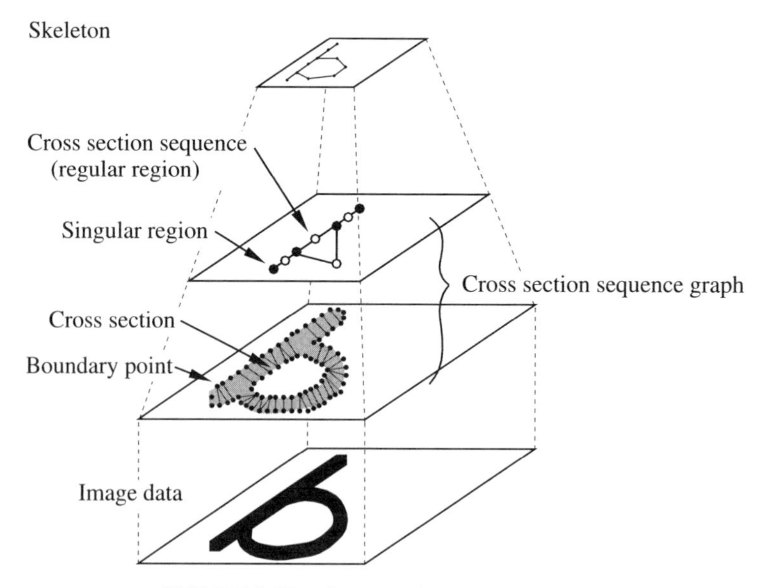

FIGURE 5.15 Cross section sequence graph.

culated, where the direction of the edge is taken inner normal from the boundary. Here we are not concerned with the edge value itself. Finally, we try to smooth the fluctuation of the edge directions directly along the boundary, which are represented by angles. Some definitions are given as follows:

p_i: ith boundary point counting from the starting point.
ang(p_i): Boundary direction at p_i represented by an angle measured clockwise from the x-axis.
dang(p_i): Difference between **ang**(p_{i-1}) and **ang**(p_i) measured from the front edge direction at p_i, clockwise for positive.

Because of difference between angles, there are two cases, one of which is restricted to the following range,

$$-\pi < \mathbf{dang}(p_i) < \pi.$$

The definition of **dang**(p_i) is illustrated in Fig. 5.16.

Related to the continuity of boundary directions, the relations among the **ang**(p_i), **ang**(p_{i-1}), and **ang**(p_{i+1}) are examined. In the ideal case they are on an ideal straight line or an ideal curved line. Then the following relation holds:

$$T = (\mathbf{dang}(p_i) \leq 0 \cap \mathbf{dang}(p_{i+1}) \leq 0) \\ \cup (\mathbf{dang}(p_i) \geq 0 \cap \mathbf{dang}(p_{i+1}) \geq 0), \quad (5.4)$$

where T means logically true. These cases are shown in Fig. 5.17. In the ideal case, no smoothing is needed. However, if the logical expression of (5.4) does not hold, then the smoothing is performed.

The point p_i is called a *smooth target point.* Two such situations are illustrated in Fig. 5.18. A smoothing target point series is defined as a point series at each end point on which the smooth target point lies, denoted as PS_j, where j means jth series. However, we will omit the suffix j to simplify the notation. Formally it can be expressed as

$$PS = \{p_s, p_{k+1}, \ldots, p_e | p_k, (k = s, \sim e) \text{ are smooth target points}\}.$$

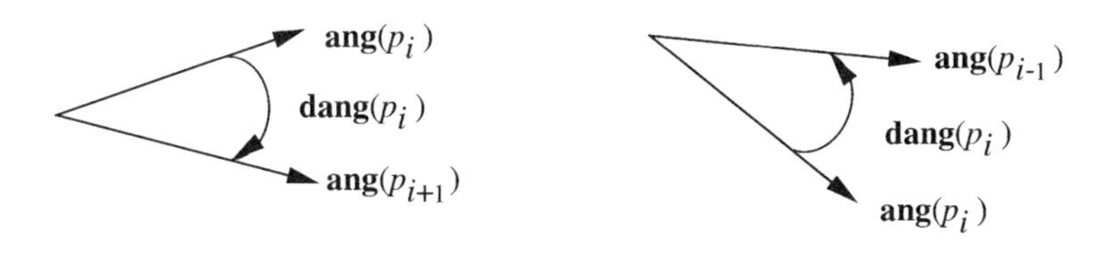

FIGURE 5.16 Definition of **dang**(p_i).

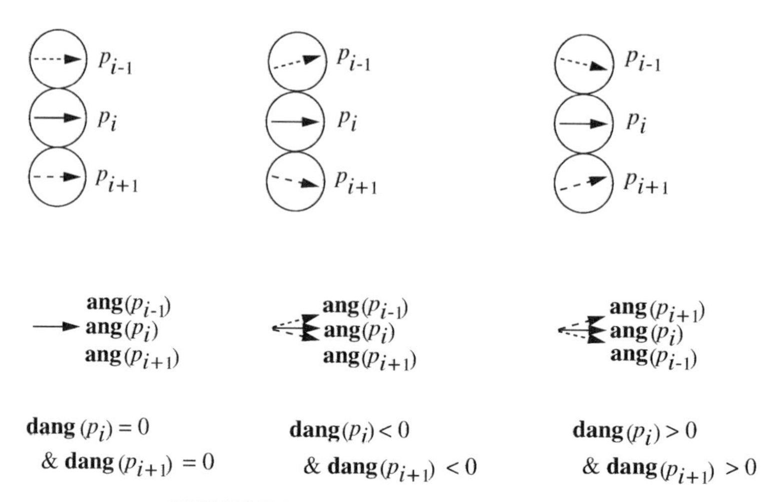

FIGURE 5.17 Some no-smoothing cases.

Here *PS* is smoothed according to the geometric property of its neighbor, concavity or convexity. However, there are two cases in the geometric property.

CASE 1 The sign of **dang**(p_{s-1}) is not different from that of **dang**(p_{e+2})

In this case *PS* has either concavity or convexity. Therefore smoothing is taken so that its concavity or its convexity is kept. The angles of the points included in *PS* should be in the range, denoted as *R(PS)*, between **ang**(p_{s-1}) and **ang**(p_{e-1}). The **ang**(p_k)($k = s \ldots e$) are smoothed so that *R(PS)* is equally divided by the number of points included in *PS*. If *R(PS)* is small enough, then the geometric property will be

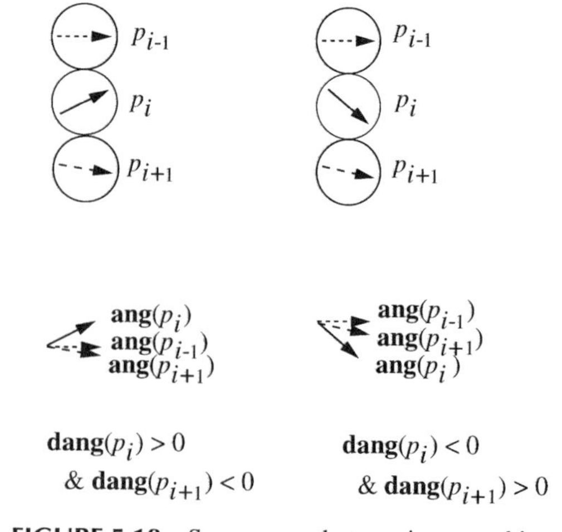

FIGURE 5.18 Some cases that require smoothing.

a straight line. Thus the smoothing is performed such that all the **angle**$(p_k)(k = s, \ldots, e)$ have the direction that bisects $R(PS)$.

CASE 2 The sign of **dang**(p_{s-1}) is different from that of **dang**(p_{e+2}).

In this case the PS must represent geometric properties of concavity or convexity, although this happens rarely. Therefore complete smoothing is impossible in general, and the following simple method is introduced:

$$\mathbf{ang}(p_k) \Leftarrow \mathbf{ang}(p_{k-1}) + \frac{(\mathbf{dang}(p_k) + \mathbf{dang}(p_{k+1}))}{2}. \tag{5.5}$$

The above setting means that the new **ang**(p_k) takes the angle bisect between **ang**(p_{k-1}) and **ang**(p_{k+1}). In the simple case the setting of (5.5) can be applied to all the boundary points uniformly. If more accurate smoothing is requested, an iteration of the above smoothing process is performed.

5.5.2 Thinning

The flow of the thinning algorithm is illustrated in Fig. 5.19. The thinning is divided into five steps: boundary extraction, cross section sequence construction, cross section sequence extension, singular region extraction, and modification of the cross section sequence graph. In panel (*b*) the initial construction of CSS after boundary extraction is shown in which some absurd cross sections and some thin density parts of cross sections can be found. In panel (*c*) the thin density parts of the cross sections are improved to be more dense. Finally, in panel (*d*), the absurd cross sections are deleted. Now let us illustrate each step of the thinning process. Before that, however, some definitions will be given here.

Definitions Given line image data can be either bilevel or multilevel. We assume that a line image has N contours, each of which is traced in such a way that the interior of the image lies on the left side of the trace. On the jth contour, boundary points are indexed cyclically from a specified starting point as $p_i(i = 0, 1, \ldots, n-1)$, and $p_i = p_j(i \bmod n_j)$, where n is the number of points on the jth contour. Here we omitted the suffix j as above. The set of boundary points in a line image is denoted by P. We assume that P is 8-connected.

The following terminologies will be defined:

Opposite Boundary Point Let a be a boundary point. We can traverse the black region from a along its direction, as if a continuous metric of a Euclidean plane, until it meets a boundary point b lying on the opposite side as shown in Fig. 5.20 (*a*). If it meets a background pixel b' that does not belong to P, without meeting any boundary point as shown in the figure, the nearest boundary point $b \in P$, which is a 4-connected neighbor of b', is selected instead of b'. Thus b is defined as an opposite boundary point for a, denoted by **opp**(a).

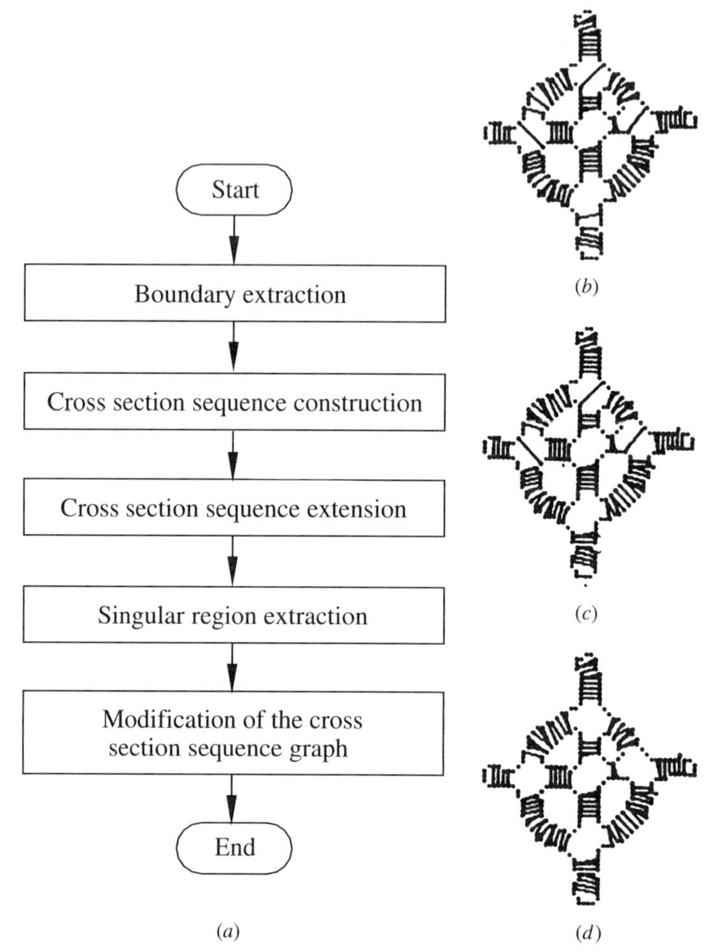

FIGURE 5.19 The flow of the algorithm: (*a*) Flowchart of the algorithm; (*b*) initial construction of cross section sequences after boundary extraction; (*c*) extended cross section sequences; (*d*) modified cross section sequence graph.

Successor (Predecessor) Point of a Boundary Point The points p_{i+1} and p_{i-1} are called the *successor* point of p_i and *predecessor* point of p_i, respectively. They are denoted by $p_{i+1} = \mathbf{succ}(p_i)$ and $p_{i-1} = \mathbf{pred}(p_i)$, respectively.

Cross Section Let $(p, q)(p, q \in P, p \neq q)$ be an ordered pair of boundary points. $C(p, q)$ is defined as a cross section if there exists no white ground pixel on the path between p and q. Note that $C(q, p)$ is different from $C(p, q)$ because their orientations differ by 180 degrees. A boundary point can be a member of one or more cross sections, so cross sections can be constructed on high curvature regions. Let $C(= C(p, q))$ and $C'(= C(p', q'))$ be the cross sections. C' is defined as the next cross section of C if $p' \in \{p, \mathbf{succ}(p)\}$ and $q' \in \{q, \mathbf{pred}(q)\}$, and $C' \neq C$. It is denoted by $C' = \mathbf{next}(C)$.

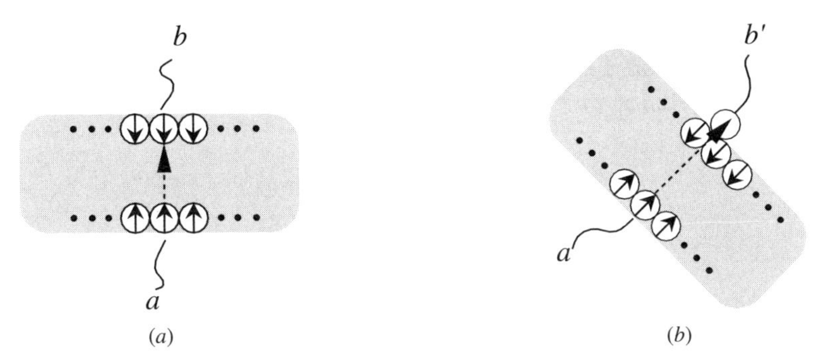

FIGURE 5.20 Finding an opposite boundary point: (*a*) Boundary point found is the traversal; (*b*) white ground pixel found in the traversal.

Similarly C is defined as the previous cross section of C' if $p \in \{p', \mathbf{pred}(p')\}$ and $q \in \{q', \mathbf{succ}(q')\}$, and $C \neq C'$. It is denoted by $C = \mathbf{prev}(C')$. This is illustrated in Fig. 5.21. However, we note that one point p is shared by the two cross sections, for example, $C = C(p, q)$ and $C' = \mathbf{next}(C) = C'(p, q)$, where $q' = \mathbf{pred}(q)$.

Cross Section Sequence Let L be a list of cross sections, $L = (C_0, C_1, \ldots, C_{k-1})$ $(k > 0)$. L is called a cross section sequence if for any pair of C_i and $C_{i+1} (i = 0, 1, \ldots, k-2)$ the relationships $C_{i+1} = \mathbf{next}(C_i)(C_i = \mathbf{prev}(C_{i+1}))$ and the $\mathbf{next}(C_{k-1}) \in \{\phi, C_0\}$ and the $\mathbf{prev}(C_0) \in \{\phi, C_{k-1}\}$ are satisfied. C_0 is defined as the tail of L, and C_{k-1} is defined as the head of L. They are denoted by $C_0 = \mathbf{tail}(L)$ and $C_{k-1} = \mathbf{head}(L)$, respectively.

Singular Region A singular region is defined as a black connected component which is surrounded by heads, tails, and boundary points that are not included in any cross section sequence. Now we are prepared to describe each step of the thinning algorithm.

Boundary Extraction The boundary extraction is an important technique, but it is not central here. We only say that the so-called nonsequential boundary processing technique is used; see reference [8].

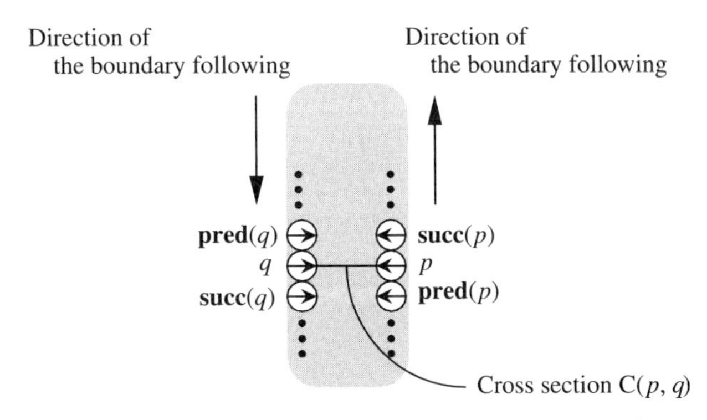

FIGURE 5.21 Relationships of boundary points and cross section $C(p, q)$.

Construction of the Cross Section Sequence After the boundary extraction, perpendicular cross sections are found, each taken almost perpendicular to the direction of the line. Let $\mathbf{cs}(a, b)(a, b \in P)$ be the logical function to find a cross section,

$$\mathbf{cs}(a, b) = \begin{cases} 1 & \text{if } [(a = \mathbf{opp}(b)) \cap (D(\mathbf{opp}(a), b) < Th1)] \\ & \cup[(b = \mathbf{opp}(a)) \cap (D(\mathbf{opp}(b), a) < Th1)], \\ 0 & \text{otherwise,} \end{cases}$$

where $D(p, q)$ is the number of the boundary points between p and q, excluding p and q, and $Th1$ is a threshold parameter, which is usually set to 2. These situations are illustrated in Fig. 5.22 for the ideal case. If the pair of boundary points (a, b) is found such that $\mathbf{cs}(a, b)$ is 1, then the cross section $C(a, b)$ is constructed. For a low curvature line, we can connect $C(a, b)$ one by one along the two line directions, as shown in the figure. However, after the simple procedure, there remain cross sections to be constructed as cross section sequences. Next we describe how to organize the rest.

Cross section sequences are constructed by repeating the procedure that picks out a cross section that has not been examined and concatenates it with the adjacent cross section. A label of a boundary point is introduced to indicate the cross section sequence it belongs to. Let $LBL(p)$ be the label of p such that p belongs to the $LBL(p)$th cross section sequence if $LBL(p)$ is nonnegative, and belongs to no cross section sequence otherwise. For all the boundary points the label is set to -1 initially.

Let $C(p, q)$ be the current cross section that $LBL(p) < 0$ and $LBL(q) < 0$. The concatenation is performed as shown in Fig. 5.23, which illustrates how to select the adjacent cross section on the next side of $C(p, q)$. Let $p'(= \mathbf{succ}(p))$ and $q'(= \mathbf{pred}(q))$ be the boundary points, which are the neighbors of p and q, respectively.

Then there can be three types of adjacent cross sections: $C(p', q')$, $C(p, q')$, and $C(p', q)$. Let $Tp'q'$, Tpq', and $Tp'q$ be the logical variables to show their existence,

$$Tp'q' = \begin{cases} 1 & \text{if } LBL(p') < 0 \cap LBL(q') < 0 \cap \mathbf{cs}(p', q') = 1, \\ 0 & \text{otherwise;} \end{cases}$$

$$Tpq' = \begin{cases} 1 & \text{if } LBL(q') < 0 \cap \mathbf{cs}(p, q') = 1, \\ 0 & \text{otherwise;} \end{cases}$$

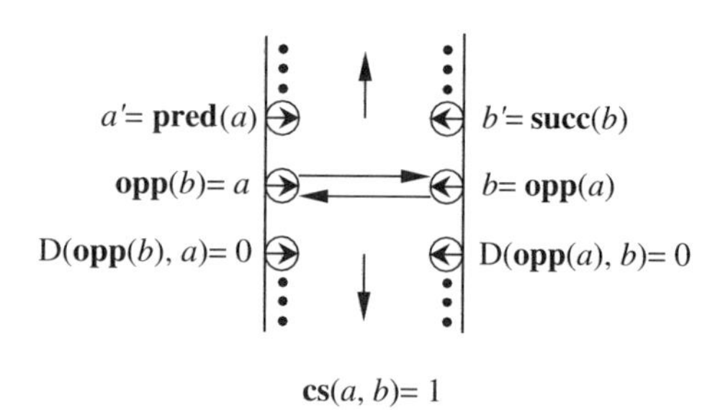

FIGURE 5.22 Logical function $\mathbf{cs}(a, b)$ for the ideal case.

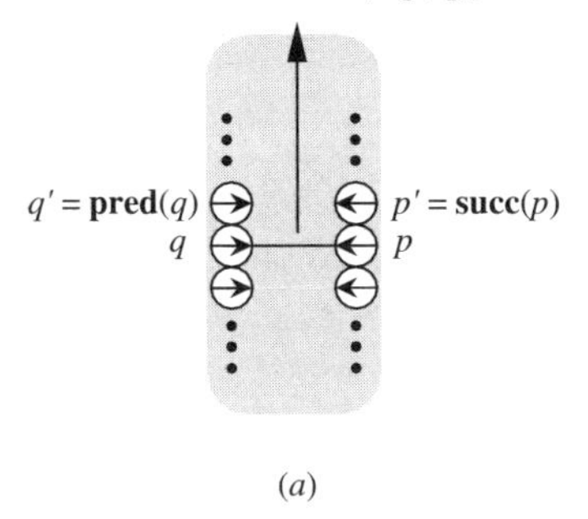

(a)

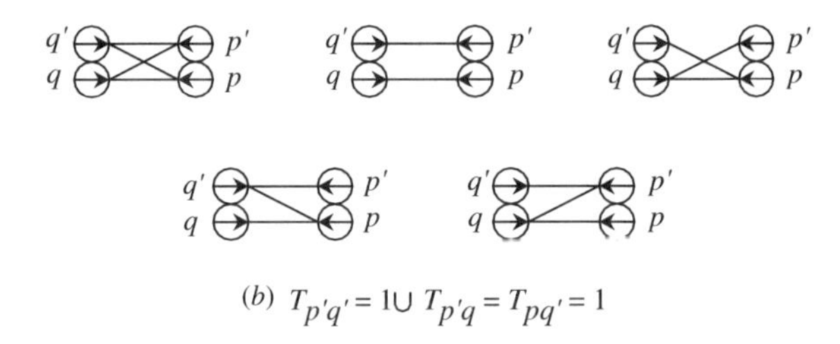

(b) $T_{p'q'} = 1 \cup T_{p'q} = T_{pq'} = 1$

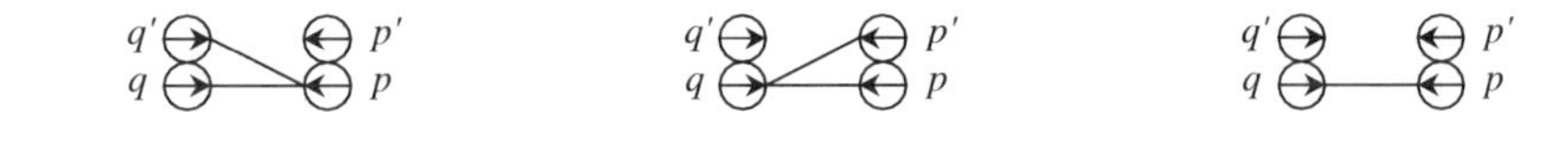

(c) $T_{pq'} = 1 \cap T_{p'q'} = T_{p'q} = 0$ (d) $T_{p'q} = 1 \cap T_{p'q'} = T_{pq'} = 0$ (e) $T_{p'q'} = T_{p'q} = T_{pq'} = 0$

FIGURE 5.23 Selection of adjacent cross section in the cross section sequence extraction process. (*a*) Boundary points and cross sections; (*b*)–(*e*) rules of the selection. In (*b*) $C(p', q')$ is selected as the next $C(p, q)$. In (*c*) $C(p, q')$ is selected as the next $C(p, q)$. In (*e*) concatenation is terminated.

$$Tp'q = \begin{cases} 1 & \text{if } LBL(p') < 0 \cap \mathbf{cs}(p', q) = 1, \\ 0 & \text{otherwise.} \end{cases}$$

Then the rules to select the adjacent cross section are described as follows:

1. Select $C(p', q')$ if $Tp'q' = 1 \cup Tpq' = Tp'q = 1$.
2. Select $C(p, q')$ if $Tpq' = 1 \cap Tp'q' = Tp'q = 0$.
3. Select $C(p', q)$ if $Tp'q = 1 \cap Tp'q' = Tpq' = 0$.
4. Stop the concatenation otherwise.

Since the line thickness is not at all restricted in these procedures to adapt *any* given line image and since these procedures are blind from the global point of view, there can be two pairs of cross section sequences that cross each other, as shown in Fig. 5.24. Then the thicker one is removed, and the labels of its boundary points are set to −1 again.

Extension of the Cross Section Sequence Since noise may be contained in the actual image, the region to be extracted as one cross section sequence is often divided into more than one cross section sequence, as shown in Fig. 5.25. Therefore it is necessary to find the gaps and to extend such cross section sequences. Although this process is important, we omit it to save space. Reference [5] provides a good discussion.

Extraction of Singular Regions and Singular Points There are several kinds of singular regions. The typical case is a branching region which will be illustrated in detail. We begin with a singular point and construct a CSSG which is directly converted into a skeleton. Figure 5.26 illustrates how to obtain a singular point using a scalar potential in which the closer a point is to the vectors that show the directions of the adjacent regular regions, the higher its evaluated value. All the black pixels in the singular region are examined, and the pixel that maximizes the evaluated value is selected as the singular point.

Let $L_k (k = k_1 \ldots k_n, n > 0)$ be the cross section sequences that connect to the singular region S, where n is the number of connections of S. Let $\mathbf{V}_k$ be the vector tangent to the boundary of the cross sections near S. It represents the direction of L_k. Let $\mathbf{W}_k$ be the vector $(x - x_k, y - y_k)$, where (x, y) are the coordinates of a black pixel inside S and (x_k, y_k) are the coordinates of the center of the head (L_k), or tail (L_k). Then the evaluated value F is calculated by

$$F = \frac{1}{n} \sum_{k=k_1}^{k_n} \frac{(\mathbf{V}_k, \mathbf{W}_k)}{|\mathbf{V}_k||\mathbf{W}_k|} = \frac{1}{n} \sum_{k=k_1}^{k_m} \cos \theta_k \tag{5.6}$$

where $(\mathbf{V}_k, \mathbf{W}_k)$ is the inner product of $\mathbf{V}_k$ and $\mathbf{W}_k$. Using the expression $\cos \theta_k$ is intuitively appealing. It means that the intersection point of the three lines can be approximated by finding a point that has the minimum distance to the extrapolated lines $\hat{L}_k (k = 1, 2, 3)$ measured by a approximate sum of θ_k^2. Note that $\cos \theta_k \cong 1 - \theta_k^2$ if θ_k is

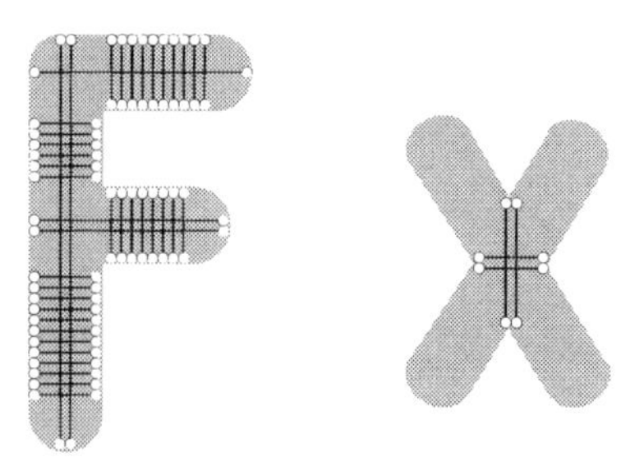

FIGURE 5.24 Some crossing pairs of cross section sequences.

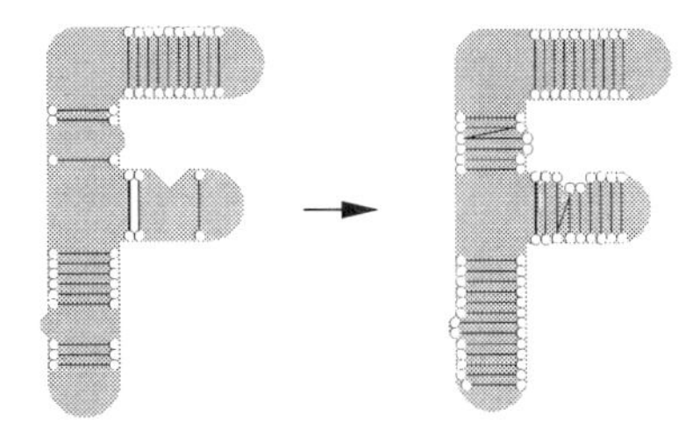

FIGURE 5.25 Cross section sequence extension.

small. In practice, the black pixels inside the boundary box of S are examined. Recall the example in Chapter 1 on detecting a singular point which concerned an intrinsic problem of pixelwise thinning whereby a thick "V" was deformed to "Y" by the Hilditch's thinning method. Now the same image data are used, and the result is shown in Fig. 5.27.

Skeletonization Procedure The procedure of constructing a skeleton from the CSSG in the form of a thin graph is quite simple because the singular points have already been described in the CSSG. Let L be the cross section sequence. Sampling several cross sections from L including the head (L) and tail (L), the procedure creates vertices at their center points and connects them appropriately.

For a singular region, the procedure proceeds depending on the degree of connections. Let S be a singular region and **deg**(S) the degree of connections of S. If **deg**(S) is greater than 1, the singular point of S is used as a vertex and is connected with vertices of adjacent heads or tails. There is no ambiguity in selecting adjacent vertices because they are determined by the connections in the CSSG. If **deg**(S) is equal to 0, the region S is considered a blob. In practice, it is important to have the capability of finding regions as small as a point and a dot.

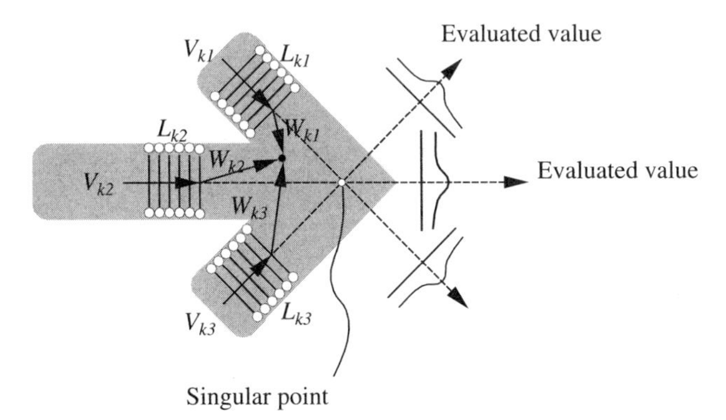

FIGURE 5.26 Inference of a singular point. The closer the point is to the direction of the tangent vectors, the higher is its evaluated value. The point that is selected as the singular point is the one that maximizes the evaluated value.

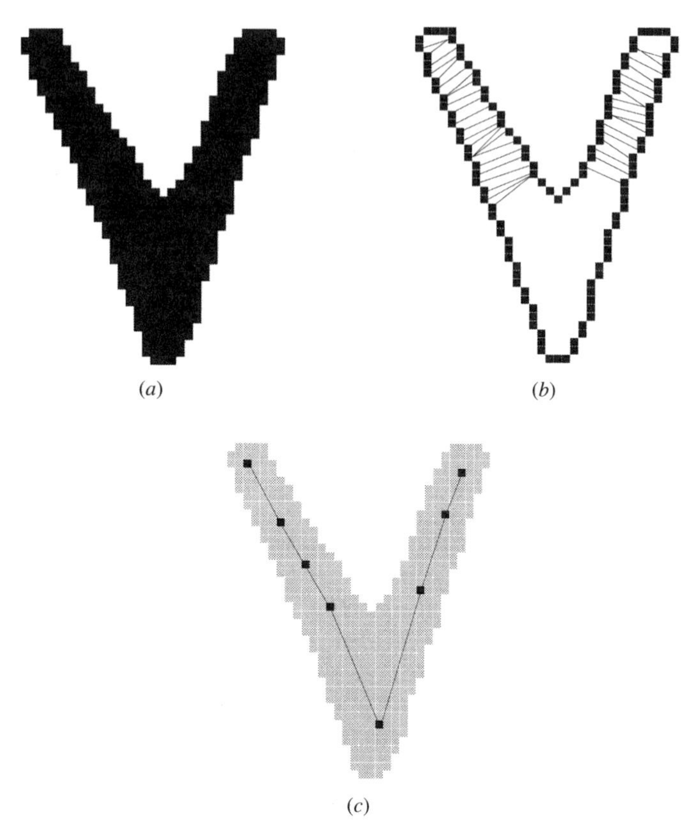

FIGURE 5.27 Experimental result of singular point detection on a singular region: (*a*) Original data; (*b*) cross sections and the singular region; (*c*) singular point and the skeleton.

5.5.3 Experiments

The thinning algorithm including the smoothing has been implemented in the *C* programming language on a SUN Sparc-station IPC (15.8 MIPS) computer. First of all, some comparison experiments are shown in Fig. 5.28 using typical test images. The compared algorithms are Suzuki and Mori's method, Hilditch's method, and Guo and Hall's method [9]. Hilditch's method is well-known as a basic algorithm as stated before, and Guo and Hall's method is a method that has been developed recently. In the image data, "Q," "X," and "I" are images used in reference [10], and first two "H's" are images used in references [11] and [12], respectively. The effect of rough sampling in constructing the final skeleton can be seen in "Q," since a fixed sampling interval was used. This does not affect its recognition, since the singular point is extracted correctly for stroke extraction. We can see that Suzuki and Mori's method works well on "X" and "I" which had been known as critical test images [12], especially for parallel pixelwise thinning. Furthermore we can see the advantages of the method over the other methods. That is, it is robust against noise such as spurs and small holes, and it does not cause excessive erosion of an open-ended stroke even when the line thickness is large.

Another set of experimental results on line images with more complicated junctions is shown in Fig. 5.29. The strength of this global approach to thinning is clearly demonstrated.

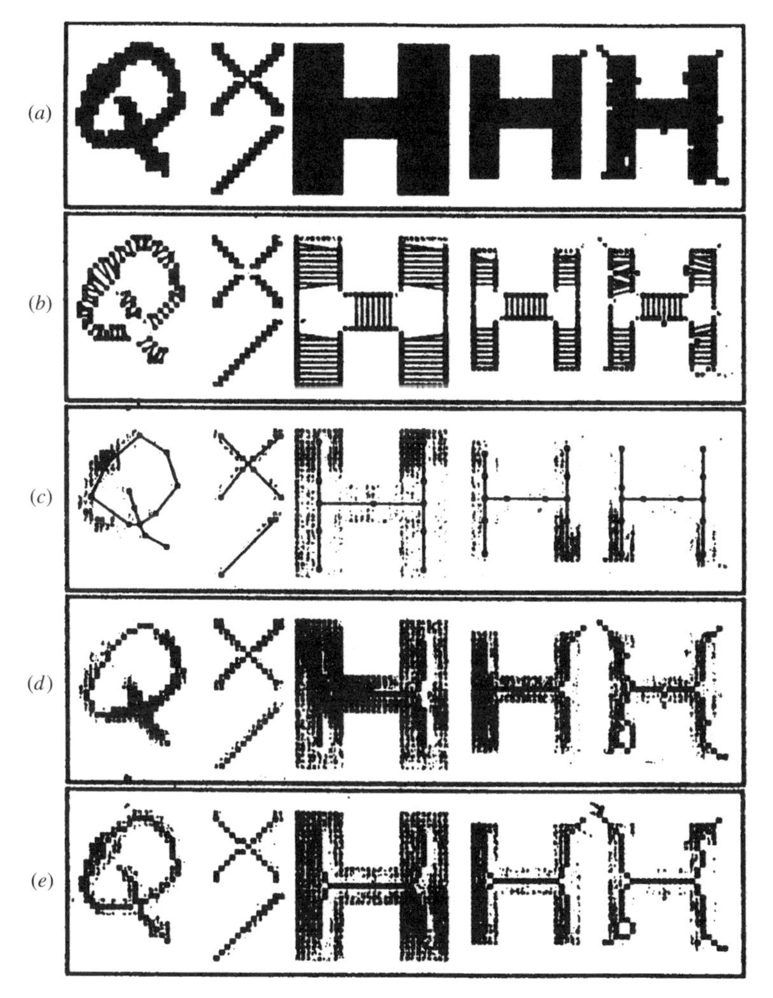

FIGURE 5.28 Experimental results of typical images. "Q," "X," and "I" are the images used from reference [2]. "X" and "I" are the critical images for parallel pixelwise thinning. The first two "H's" are images from references [7] and [8], respectively. The last "H" is used to examine the robustness against noise such as spurs and small holes. The boxes show (*a*) image data, (*b*) cross section sequence graph; (*c*) thin line graph of skeleton constructed from the cross section sequence graph, (*d*) skeleton based on Hilditch's method, and (*e*) skeleton based on Guo and Hall's method. From reference [5], © 1998, World Scientific.

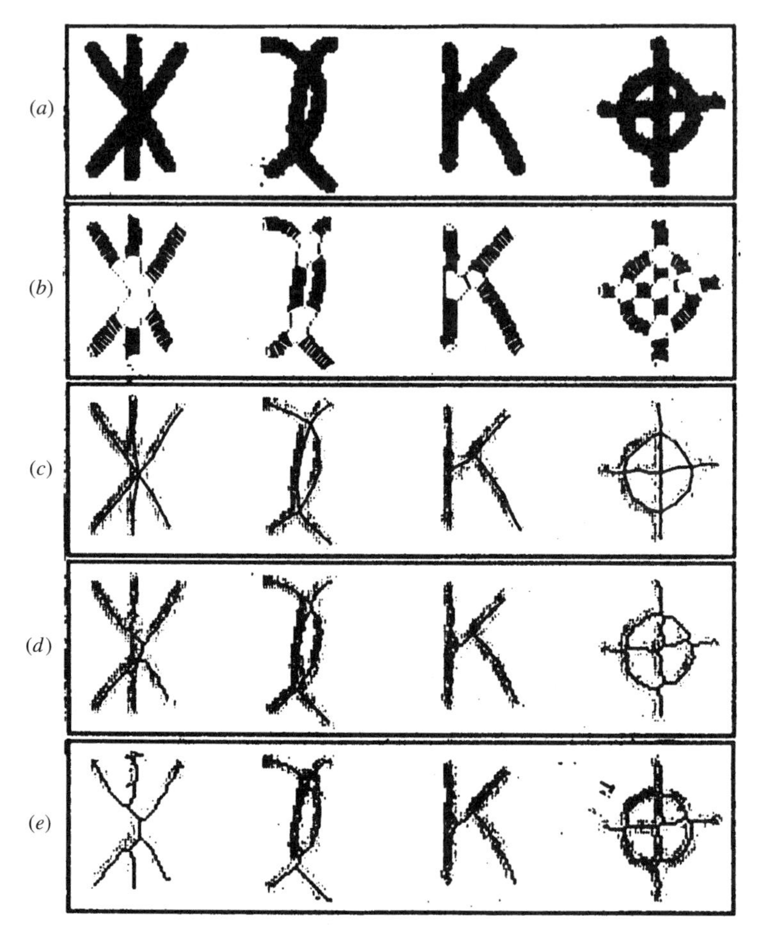

FIGURE 5.29 Experimental results of typical images. The boxes show (*a*) image data, (*b*) cross section sequence graph, (*c*) thin line graph of skeleton constructed from the cross section sequence graph, (*d*) skeleton based on Hilditch's method, and (*e*) skeleton based on Guo and Hall's method. From reference [5], © 1998, World Scientific.

Recognition Experiment To examine the effectiveness of the method, recognition experiments on handprinted numerals were conducted. Various kinds of distorted characters were written by 250 people. The examples are shown in Fig. 5.30.

The total number of characters amounted to about 13,500. We compared it with Hilditch's method. In fairness, some postprocessing was used, such as elimination of small sticks and merging two adjacent branch points into one crossing point such as is usually done. For a recognition method the quasi-topological curve analysis method was used, which will be fully discussed in Chapter 9.

The experimental results are shown in Table 5.1. The numbers in parentheses are the result of another experiment where a morphological filtering was used for a character once rejected. Morphological filtering is fully described in Chapter 6. A

FIGURE 5.30 Examples of characters used in the recognition experiments.

dilation operation with the structuring element of a 3 × 3 rhombus was used twice for a character with the estimated line thickness of 4. Similarly an erosion operation with the same structuring element was used twice for a character with the estimated line thickness of 6 or more, and once for a character with the estimated line thickness of 5.

TABLE 5.1 Experimental Results

	Pixelwise Method	Proposed Method
Recognition	94.2 (96.5)%	98.2 (99.4)%
Rejection	5.2 (2.8)%	1.6 (0.3)%
Substitution	0.6 (0.7)%	0.2 (0.3)%
Throughput	8.3 (—) cps	22.8 (—) cps

Note: Numbers in parentheses give the results of morphological filtering used for a rejected character.

We can see that the method is better than a pixelwise method in both cases. Some typical examples of the differences between the two methods are shown in Fig. 5.31. We see that singular points are extracted correctly by the method used in Fig. 5.31 (*c*) and (*g*), while it is one of the intrinsic problems of pixelwise methods as mentioned before. This is due to the fact that a singular region can be analyzed from the adjacent regular regions. Furthermore the proposed method is robust against noise such as small

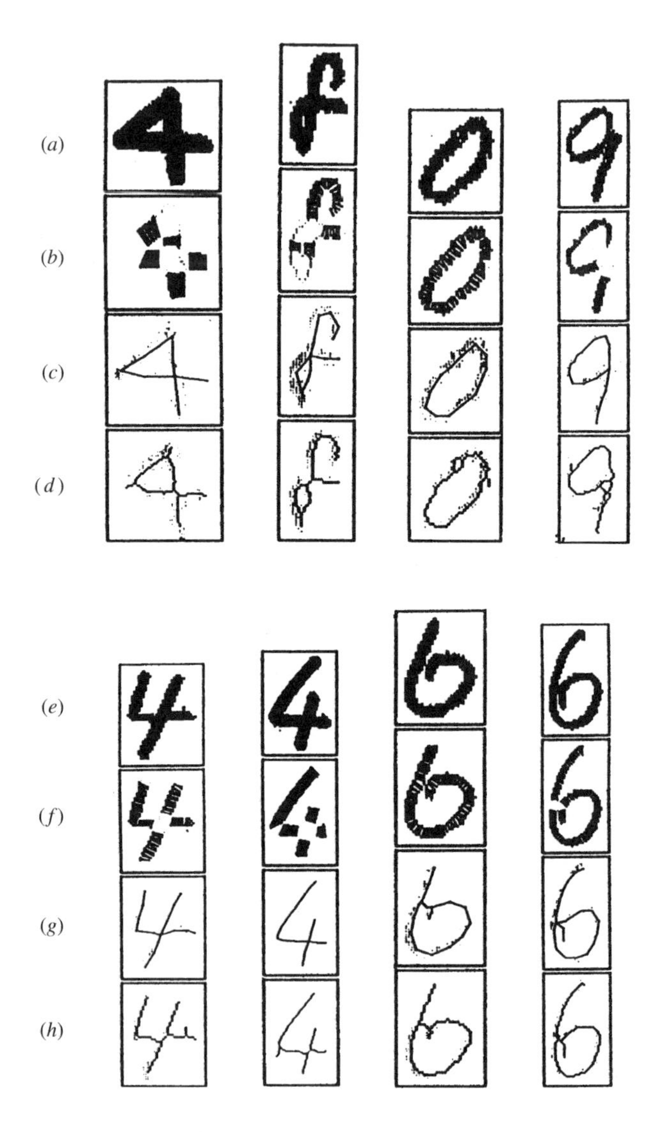

FIGURE 5.31 Typical differences between the two methods. The boxes show (*a, e*) image data, (*b, f*) cross section sequence graph, (*c, g*) thin line graph of skeleton constructed from the cross section sequence graph, and (*d, h*) skeleton based on Hilditch's pixelwise skeletonization. From reference [5], © 1998, World Scientific.

holes caused by fading and spurs, since line image data are processed globally, Fig. 5.31 (*c*) and (*g*). However, some problem with lines that touch can be seen in panels (*d*) and (*h*). Some kind of label should be introduced in the skeleton graph to represent touching lines as an exception to the usual connections. It may be possible to detect touching lines by examining the changes of line thickness on the CSSG. For pixelwise methods it is not a simple task, since the line image is not described in a structured form.

As for the throughput, we can see that the CSSG method is much faster than Hilditch's method. The time consumed in the recognition process is not included for the throughput in Table 5.1, though it takes only about 10% of the skeleton processing. This is usual. Preprocessing consumes much time, so special hardware implementations have been developed. The software is available commercially.

5.6 BIBLIOGRAPHICAL REMARKS

The terms "thinning" and "skeltonization" have been used almost synonymously in the literature. However, we chose the term "thinning" because from the recognition point of view, reconstructibility is not a necessary condition of the desired preprocessing. On the contrary, skeltonization requires reconstructibility as its essential property so that it can be used in image compression. This point will be discussed later in this section.

Now we return to work on thinning. Even if we restricted ourselves to thinning, there are over 300 papers to consider! In the 1970s the number was about 30, all on the so-called peeling method on which there were many variations. In that connection, Tamura conducted a systematic comparative study of thinning methods in which he selected six methods of Rutovitz, Hilditch, and Deutch-1 [20], Deutch-2 [21], and Yokoi and Tamura [22] as representative ones. He evaluated these methods based on eight criteria: line width, center position, stability, isotropy, strength in terms of noise, distortion at intersection, T type distortion, and L type distortion. Now we are in the 1990s, with the number of the papers on thinning increased by one order since 1970s. This reflects the wider range of applications besides character recognition such as drawing and printed circuit boards. Responding to this changed status, L. Lam, S.-W. Lee, and C. Y. Suen wrote a very concise review [23]. The highlight of their paper is a systematic description of the parallel thinning algorithm which is treated comparatively with the sequential one. We omitted the parallel thinning algorithm because it cannot be as easily explained as the sequential one, nor as compactly and systematically as in their paper. In this sense we recommend reader to read [23].

Shortly after L. Lam and C. Y. Suen [24] conducted comparative evaluation experiments on thinning methods limited to only parallel algorithms, which were chosen because of the importance of parallel algorithms considering highly developed VLSI technology. Ten algorithms were selected as representative ones and recognition experiments on preprocessed data to which these algorithms applied. The recognition algorithm was based primarily on structural classification and relaxation matching secondarily. The results of recognition rates of the experiments showed that the recognition rates are almost the same, namely that the standard deviation are small in both two kinds of database. They used two databases of isolated numerals extracted from

U.S. zip codes written on real-life mail pieces. Database-1 consists of 6000 samples, and Database-2 consists of 12,000 numerals from the BR set of the CEDAR CDROM 1 [25]. The average correct recognition rates (standard deviations) for Database-1 and Database-2 are 96.87 (0.537) and 97.70 (0.215), respectively. They concluded from their comparative study that the fact that recognition rates are comparable indicates, for most characters, that their correct classification depends more on the OCR system than on the thinning algorithm used. They based their interpretation on their observation that to some extent the thinning methodology presently employed is saturated, although minor improvement can be made. Therefore we introduced a different methodology of thinning which may be called a second-generation method.

As mentioned at the beginning of this chapter, related to thinning preprocessing is a transformation called *medial axis transformation* (usually abbreviated MAT) which was originated by H. Blum [13]. For a general bilevel shape (silhouette), this transformation enables a simple description and some features to be extracted accordingly. Originally the scheme was conceived as on a continuous plane, and later it was implemented on a grid plane by A. Rosenfeld and J. L. Pfaltz [27]. The results are very similar to those of a thinning preprocessing for similar shapes. The difference lies in its reconstruction ability, since an original image can be reconstructed from its MAT. In this sense MAT can be considered an image compression. In Rosenfeld and Pfaltz's formalization on a grid plane, they introduced an important transformation called *distance transformation,* which is the basis on which MAT is constructed. Each point, $P(x, y)$, of a bilevel image S is labeled by its distance. Specifically each distance is measured from its nearest ground point to point $P(x, y)$. Formally it can be expressed as

$$T(x, y) = \min(d[p(x, y), \overline{S}]),$$

where $T(x, y)$, $d[\cdot]$, and $\overline{S}$ denote the distance transformation label at $P(x, y)$, a distance measure, and the ground of S, respectively. Simple distance measures like *city-block distance, chessboard distance,* and *octagonal distance* were used. The distance measures discussed by A. Rosenfeld and J. L. Pfaltz [26] were still far from Euclidean distance, so U. Montanari [14] obtained a new distance considerably close to Euclidean distance. Ten years later Per-Erik Danielsson [15] succeed in obtaining a completely Euclidean distance, which was sequentially given. Soon afterward H. Yamada [16] presented a parallel method for Euclidean distance.

The many methods of MAT developed are usually called "skeltons." The terms "skeleton" and "thinning" can be confusing. For this reason we introduce only the method devised by Arcelli, Cordella, and Levialdi [17]. Their method gives nice results that are reconstructible and continuous. We note that the resulting thin lines sometimes are not continuous in usual skeleton (MAT) methods. Efforts to expand the above scheme to a multi-level image were done by Rutowitz [2], and Levi and Montani [18]. Finally a general scheme was given by Yokoi, Toriwaki, and Fukumura [19].

Last we want to note that we introduce a thinning method focused on singular points and based on MAT. This recalls G. Boccignone, A. C. Hianese, L. P. Cordella and A. Marcelli's work on recovering dynamic information from handwriting [27]. Usual pixel-based thinning methods have trouble in detecting singular points correctly. To overcome this difficulty, they map on-line writing to an off-line written image; that is the movements made by humans in drawing characters are described as piecewise continuous trajectories, which implies continuity conditions on each curve piece forming a character. As a first approximation to the thinned line representation, MAT is used, which is approximated by a polygon, called PMAT. This is further corrected heuristically at break points. Thus they succeeded in reconstructing the stroke sequence according to the most expected sequence followed by a person writing a character. Recent representative work is collected in *Thinning Methodologies for Pattern Recognition,* edited by Suen and Wang [30]. Hu and Li [31] present a nonpixelwise method based on RAG, which was found on solving a typical problem in the conventional thinning, namely, so-called X-crossing preservation.

BIBLIOGRAPHY

[1] S. Yokoi, J. Toriwaki, and T. Fukumura, "On the topological nature of sampled binary image," *Trans. IECE Japan,* vol. 56-D, no. 11, pp. 33–61, 1973.

[2] D. Rutovitz, "Pattern recognition," *J. Roy. Stat. Soc.,* vol. 129, ser. A, pp. 504–530, 1966.

[3] C. J. Hilditch, "Linear skeletons from square cupboards," in B. Meltzer and D. Michie, eds., *Machine Intelligence IV,* Edinburgh: University Press, pp. 403–420, 1969.

[4] T. Pavlidis, "A vectorizer and feature extractor for document recognition," *Comp. Vision, Graphics, Image Proc.,* vol. 35, pp. 111–127, 1986.

[5] T. Suzuki and S. Mori, "Structural description of line images by the cross section sequence graph," *Int. J. Pattern Recogn. Artificial Intell.,* vol. 7, pp. 1055–1076, 1993.

[6] J. C. Simon and O. Baret, "Handwriting recognition as an application of regularities and singularities in line pictures," in C. Y. Suen, ed., *Frontiers in Handwriting Recognition,* Montreal, Quebec: CEN-PARML, Concordia University, pp. 23–36, 1990.

[7] T. Suzuki and S. Mori, "A thinning method based on all structure," in C. Y. Suen, ed., *Frontiers in Handwriting Recognition,* Montreal, Quebec: CEN-PARMI, Concordia University, pp. 39–52, 1990.

[8] W. H. H. J. Lunscher and M. P. Beddoes, "Fast binary-image boundary extraction," *Comp. Vision Graphics Image Proc.,* vol. 38, pp. 229–257, 1987.

[9] Z. Guo and R. W. Hall, "Parallel thinning with two subiteration algorithms," *Common. ACM,* vol. 32, no. 3, pp. 359–373, 1989.

[10] N. J. Naccache and R. Shinghal, "An investigation into the skeletonization approach of Hilditch," *Pattern Recogn.,* vol. 17, no. 3, pp. 279–284, 1984.

[11] C. Arcelli and G. Sann iti di Baja, "A width-independent fast thinning algorithm," *IEEE Trans. Pattern Anal. Math. Intell.,* vol. 7, no. 4, pp. 463–474, 1985.

[12] H. E. Lu and P. S. P. Wang, "A comment on a fast parallel algorithm for digital patterns," *Common. ACM,* vol. 29, no. 3, pp. 236–239, 1986.

[13] H. Blum, "A transformation for extracting new descriptors of shape," in W. Dunn, ed., *Model for the Perception of Speech and Visual Form,* Cambridge: MIT Press, pp. 364–380, 1964.

[14] U. Montanari, "A method for obtaining skeletons using a quasi-Euclidean distance," *J. ACM,* vol. 15, no. 4, pp. 600–624, 1968.

[15] P.-E. Danielsson, "Euclidian distance mapping," *Comp. Graphics Digital Proc.*, vol. 14, pp. 227–248, 1980.

[16] H. Yamada, "Complete Euclidian distance transformation by parallel operation," *Proc. Int. Joint. Conf. Pattern Recogn.*, pp. 69–71, 1984.

[17] C. Arcelli, L. P. Cordella, and S. Levialdi, "From local maximum to connected skeletons," *IEEE Trans. Pattern Anal. Machine Intell.*, vol. PAMI-3, no. 2, pp. 134–143, March 1981.

[18] G. Levi and U. Montanari, "A gray-weighted skeleton," *Info. Control*, vol. 17, pp. 62–91, 1969.

[19] S. Yokoi, J. Toriwaki, and T. Fukumura, "On generalized distance transformation of digitized pictures," *IEEE Trans. Pattern Anal. Machine Intell.*, vol. PAMI-3, no. 4, pp. 424–443, 1981.

[20] D. S. Deutch, "Comments on a line thinning scheme," *Comput. J.*, vol. 12, p. 412, 1969.

[21] D. S. Deutch, "Thinning algorithms on rectangular, hexagonal, and triangular arrays," *Common ACM*, vol. 15, no. 9, pp. 827–837, 1972.

[22] H. Tamura, "A comparison of line thinning algorithms from a digital geometry viewpoint," *Proc. 4th Int. Joint Conf. Pattern Recogn.*, Kyoto, pp. 715–719, 1978.

[23] L. Lam, S-W. Lee, and C. Y. Suen, "Thinning methodologies—A comprehensive survey," *IEEE Trans. Pattern Anal. Machine Intell.*, vol. 14, no. 9, pp. 869–885, September 1992.

[24] L. Lam and C. Y. Suen, "An evaluation of parallel thinning algorithms for character recognition," *IEEE Trans. Pattern Anal. Machine Intell.*, vol. 17, no. 9, pp. 914–919, September 1995.

[25] R. Fench and J. J. Hull, "Concerns in creation of image databases," *Proc. 3rd Int. Workshop Frontiers in Handwriting Recogn.*, pp. 112–121, Buffalo, NY, May 1993.

[26] A. Rosenfeld and J. L. Pfaltz, "Sequential operations in digital picture processing," *J. ACM*, vol. 13, no. 4, pp. 471–494, October 1966.

[27] G. Boccignone, A. Chianese, L. P. Cordella, and M. Marcell, "Recovering dynamic information from static handwriting," *Pattern Recogn.*, vol. 26, no. 3, pp. 409–418, 1993.

[28] J. G. Hocking and G. S. Young, *Topology*, New York: Dover, 1961.

[29] S. Mori and T. Sakakura, *Fundamentals of Image Recognition I*, Tokyo: Ohm, 1986.

[30] C. Y. Suen and P. S. P. Wang, editors, *Thinning Methodologies for Pattern Recognition*, World Scientific, 1994.

[31] G. Hu and Z. N. Li, "An *X*-crossing preserving skeltonization algorithm," in C. Y. Suen and P. S. P. Wang, ed., *Thinning Methodologies for Pattern Recognition*, World Scientific, 1994, pp. 66–89.

CHAPTER SIX

Theory of Preprocessing

Mathematical Morphology is a nonlinear algebraic system of image operations based on two basic operators *dilation* and *erosion*. Mathematical morphology operations tend to simplify image data, preserving their essential image characteristics and eliminating irrelevancies. As the identification of objects, object features, and assembly defects correlate directly with shape, mathematical morphology is a natural processing approach to deal with the machine vision recognition process and the visually guided robot problem.

6.1 BINARY MORPHOLOGY

The language of mathematical morphology is that of set theory. Sets in mathematical morphology represent the shapes that are manifested on binary or gray tone images. The set of all the black pixels in a black-and-white image constitutes a complete description of the binary image. Sets in Euclidean 2-space denote foreground regions in binary images. Sets in Euclidean 3-space may denote time-varying binary imagery or static gray-scale imagery as well as binary solids. Sets in higher-dimensional spaces may incorporate additional image information, such as color, or multiple perspective imagery. Mathematical-morphological transformations apply to sets of any dimensions, such as Euclidean N-space, or those like its discrete or digitized equivalent, the set of integers Z^N. For simplicity we will refer to either of these sets as E^N.

Those points in a set to be morphologically transformed are considered as the selected set of points, and those in the complement set are considered as not selected. Therefore morphology from this point of view is binary morphology. We begin our discussion with the binary morphological operations of dilations and erosion.

Let $A \subseteq E^N$ be an Euclidean N-space and $B \subseteq E^N$ be the structuring element. Translation of A by the point $t \in E^N$ is denoted by A_t and defined by

$$A_t = \{c \in E^N \mid c = a + t \text{ for some } a \in A\}. \tag{6.1}$$

6.1.1 Binary Dilation

Dilation is the morphological transformation that combines two sets using vector addition of set elements. If A and B are sets in N-space E^N with elements $a = (a_1, \ldots, a_N)$ and $b = (b_1, \ldots, b_N)$, respectively, then the dilation of A by B is the set of all possible vector sums of pairs of elements, one coming from A and one coming from B.

The *dilation* of A by B is denoted by $A \oplus B$ and defined by

$$A \oplus B = \{c \in E^N \mid c = a + b \text{ for some } a \in A \text{ and } b \in B\} \tag{6.2}$$

Let $A \subseteq E^N$ be an Euclidean N-space and $B \subseteq E^N$ be the structuring element. *Translation* of A by the point $t \in E^N$ is denoted by A_t and defined by

$$A_t = \{c \in E^N \mid c = a + t \text{ for some } a \in A\}. \tag{6.3}$$

Then

$$\begin{aligned} A \oplus B &= \{c \in E^N \mid c = a + b \text{ for some } a \in A \text{ and } b \in B\} \\ &= \bigcup_{b \in B} A_b. \end{aligned} \tag{6.4}$$

Thus the dilation of A by B can be computed as the union of translation of A by the element of B. Figure 6.1 illustrates the translation operation, and Fig. 6.2 illustrates an example of dilation operation. The coordinate system we use for all the examples is (row, column).

Properties of Binary Dilation Let A, B, and C be subsets of E^N.

1. The dilation operation is commutative:

$$A \oplus B = B \oplus A. \tag{6.5}$$

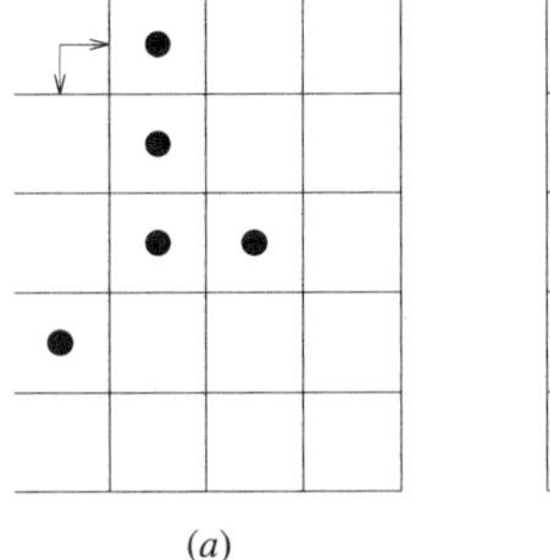

(*a*)

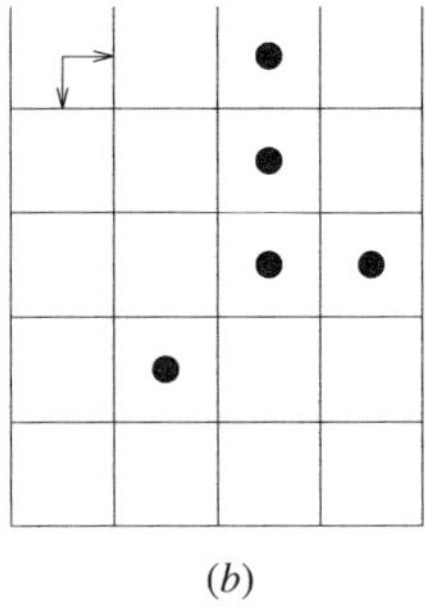

(*b*)

FIGURE 6.1 An example of translation: (*a*) image $A = \{(0, 1), (1, 1), (2, 1), (2, 2), (3, 0)\}$; (*b*) $x = (0, 1)$ and $(A)_x = \{(0, 2), (1, 2), (2, 2), (2, 3), (3, 1)\}$.

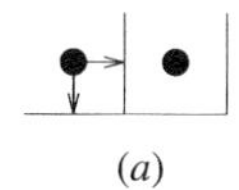

(*a*)

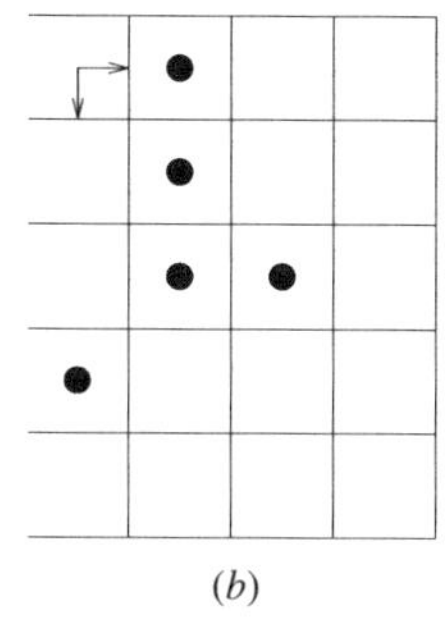

(*b*)

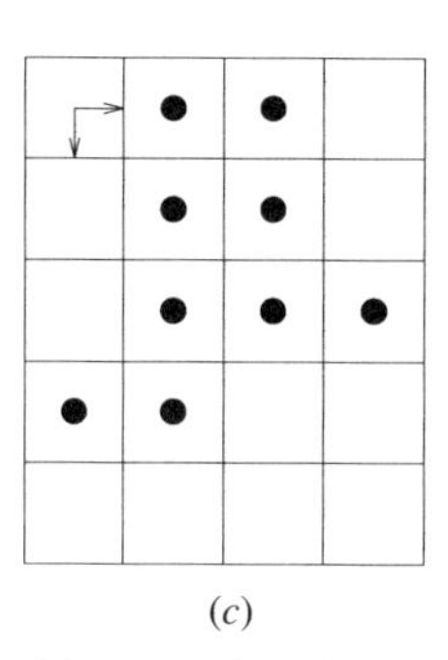

(*c*)

FIGURE 6.2 An example of dilation operation: (*a*) structuring element $B = \{(0, 0), (0, 1)\}$; (*b*) image $A = \{(0, 1), (1, 1), (2, 1), (2, 2), (3, 0)\}$; (*c*) $A \oplus B = \{(0, 1), (1, 1), (2, 1), (2, 2), (3, 0), (0, 2), (1, 2), (2, 2), (2, 3), (3, 1)\}$.

The dilation operation is commutative, but A and B are handled quite differently in practice. The first operand A is considered to be the image undergoing analysis, whereas the second operand B is referred to as the *structuring element,* to be considered as constituting a single shape parameter of the dilation transformation. We will refer to A as the image and B as the structuring element.

2. The *chain rule* for dilations is

$$(A \oplus B) \oplus C = A \oplus (B \oplus C). \tag{6.6}$$

The form $(A \oplus B) \oplus C$ represents a considerable saving in number of operations to be performed when A is the image and $B \oplus C$ is the structuring element. A brute force dilation $B \oplus C$ may take as many as N^2 operations, whereas first dilating A by B and then dilating the result by C could take as few as $2N$ operations, where N is the number of elements in B and in C.

3. Translation invariance of dilation takes the form

$$A \oplus B_t = (A \oplus B)_t, \tag{6.7}$$

$$\begin{aligned} &A \oplus B_1 \oplus \cdots \oplus (B_n)_x \oplus \cdots \oplus B_N \\ &= (A \oplus B_1 \oplus \cdots \oplus B_n \oplus \cdots \oplus B_N)_x. \end{aligned} \tag{6.8}$$

4. Suppose that the structuring element B compensates for a shift in the image A by taking B to be shifted in the opposite direction. Then the shift in B compensates for the shift in A.

$$(A_x) \oplus (B)_{-x} = A \oplus B, \tag{6.9}$$

$$(A)_x \oplus B_1 \oplus \cdots \oplus (B_n)_{-x} \oplus \cdots \oplus B_N$$
$$= A \oplus B_1 \oplus \cdots \oplus B_n \oplus \cdots \oplus B_N. \tag{6.10}$$

5. The dilation operation is necessarily extensive when the origin belongs to the structuring element. Extensivity means that the dilated result contains the original.

$$\text{If } 0 \in B, \text{ then } A \oplus B \supseteq A.$$

Figure 6.3 illustrates this property.

6. Dilation is increasing; that is, containment relationships are maintained through dilation:

$$\text{If } A \subseteq B, \text{ then } A \oplus C \subseteq B \oplus C.$$

7. The order of an image intersection and a dilation operation cannot be interchanged. The result of intersecting two images followed by a dilation of the intersection result is contained in the intersection of the dilation of the two images.

$$(A \cap B) \oplus C \subseteq (A \oplus C) \cap (B \oplus C), \tag{6.11}$$

$$A \oplus (B \cap C) \subseteq (A \oplus B) \cap (A \oplus C). \tag{6.12}$$

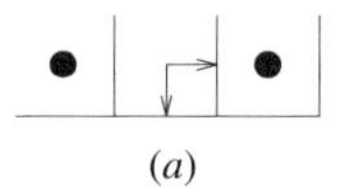

(*a*)

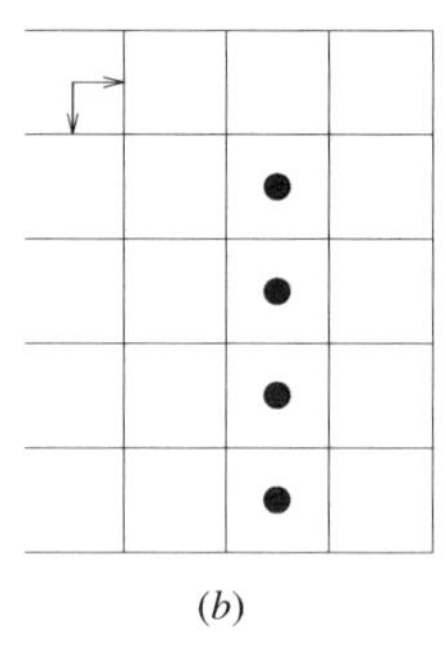

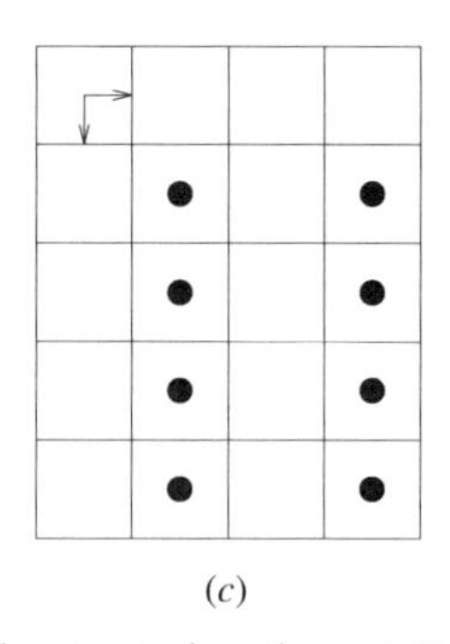

(*b*) (*c*)

FIGURE 6.3 When the origin (0, 0) is not in the structuring element B, it may happen that the dilation of A by B has nothing in common with A: (*a*) structuring element $B = \{(0, -1), (0, 1)\}$; (*b*) image $A = \{(1, 2), (2, 2), (3, 2), (4, 2)\}$; (*c*) $A \oplus B = \{(1, 1), (2, 1), (3, 1), (4, 1), (1, 3), (2, 3), (3, 3), (4, 3)\}$.

8. The order of image union and dilation can be interchanged. The dilation of the union of two images is equal to the union of the dilations of these images:

$$A \oplus (B \cup C) = (A \oplus B) \cup (A \oplus C), \tag{6.13}$$

$$(A \cup B) \oplus C = (A \oplus C) \cup (B \oplus C). \tag{6.14}$$

Those properties can be proved with some basic techniques of set theory. For illustrative purposes we give a proof for the second property.

Proof of $(A \oplus B) \oplus C = A \oplus (B \oplus C)$ We have $x \in A \oplus (B \oplus C)$ if and only if there exists $a \in A$, $b \in B$, and $C \in C$ such that $x = a + (b + c)$. Now $x \in (A \oplus B) \oplus C$ if and only if there exist $a \in A$, $b \in B$, and $C \in C$ such that $x = (a + b) + c$. But $a + (b + c) = (a + b) + c$, since addition is associative. Therefore $(A \oplus B) \oplus C = A \oplus (B \oplus C)$. ■

Dilation by disk-structuring elements correspond to isotropic swelling or expansion algorithm common to binary image processing. Dilation by small squares (3×3) is a neighborhood operation known by the name "fill," "expand," or "grow."

6.1.2 Binary Erosion

Erosion is the morphological dual to dilation. It is the morphological transformation that combines two sets using the vector subtraction of set elements. If A and B are sets in Euclidean N-space, the erosion of A by B is the set of all elements x for which $x + b \in A$ for every $b \in B$. Some image processing people use the term "shrink" or "reduce" for erosion.

The *erosion* of A by B is denoted by $A \ominus B$ and is defined by

$$\begin{aligned} A \ominus B &= \{x \in E^N \mid x + B \in A \text{ for every } b \in B\} \\ &= \{x \in E^N \mid B_x \subseteq A\} \\ &= \{x \in E^N \mid \text{for every } b \in B, \text{ there exists an } a \in A \text{ such that } x = a - b\} \\ &= \bigcap_{b \in B} A_{-b}. \end{aligned} \tag{6.15}$$

The structuring element B may be visualized as a probe that slides across the image A, testing the spatial nature of A at every point. If B translated to x can be contained in A (by placing the origin of B at x), then x belongs to the erosion $A \ominus B$. Also, erosion of an image A by a structuring element B is the intersection of all translations of A by the point $-b$, where $b \in B$. Figures 6.4 and 6.5 illustrate examples of erosion operation.

Properties of Binary Erosion Let A, B, and C be subsets of E^N.

1. The erosion transformation is generally conceived of as a shrinking of the original images. In set terms, the eroded set is often considered to be con-

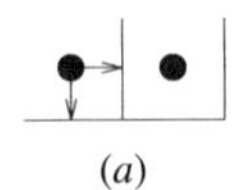

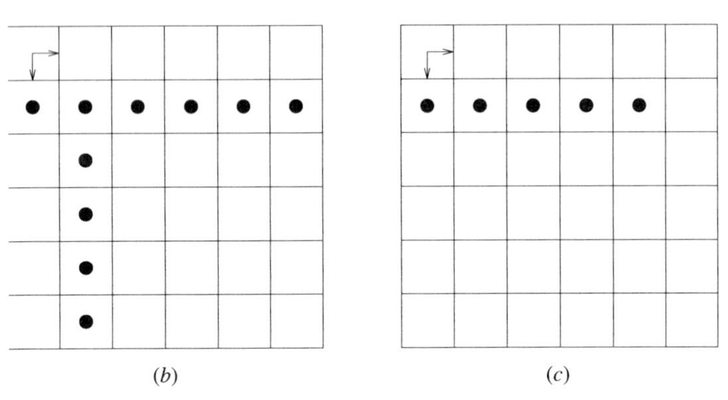

FIGURE 6.4 Example of erosion operation: (*a*) Structuring element $B = \{(0, 0), (0, 1)\}$; (*b*) image $A = \{(1, 0), (1, 1), (1, 2), (1, 3), (1, 4), (1, 5), (2, 1), (3, 1), (4, 1), (5, 1)\}$; (*c*) $A \ominus B = \{(1, 0), (1, 1), (1, 2), (1, 3), (1, 4)\}$.

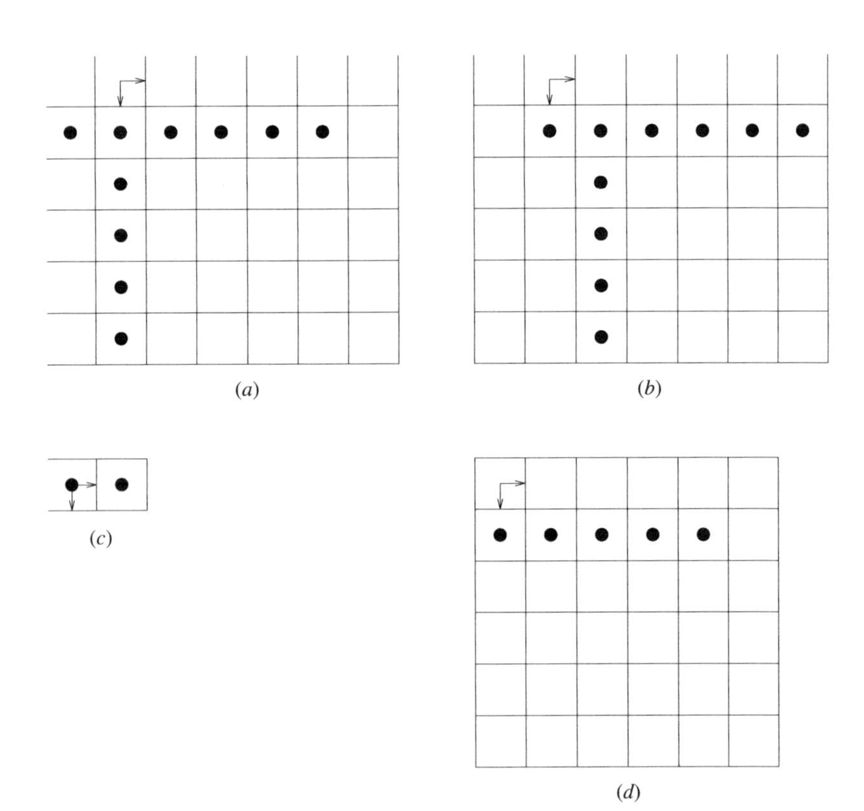

FIGURE 6.5 Erosion computed as an intersection of translations of A: (*a*) $A_{-(0, 1)}$; (*b*) $A_{(0, 0)}$; (*c*) structuring element $B = \{(0, 0), (0, 1)\}$; (*d*) $A \ominus B = A_{(0, 0)} \cap A_{-(0, 1)}$.

tained in the original set. A transformation having this property is called anti-extensive. However, the erosion transformation is necessarily anti-extensive only if the origin belongs to the structuring element:

$$\text{If } 0 \in B, \text{ then } A \ominus B \subseteq A.$$

Figure 6.6 illustrates this property.

2. Translation invariance of erosion:

$$A_x \ominus B = (A \ominus B)_x, \tag{6.16}$$

$$A \ominus B_x = (A \ominus B)_{-x}. \tag{6.17}$$

3. Erosion is increasing; that is, if image A is contained in image B, then the erosion of A is contained in the erosion of B:

$$\text{If } A \subseteq B, \text{ then } A \ominus C \subseteq B \ominus C.$$

Figure 6.7 illustrates an instance showing the increasing property of erosion.

4. If A and B are structuring elements and B is contained in A, then the erosion of an image D by A is necessarily more severe than the erosion by B; that is, D eroded by A will necessarily be contained in D eroded by B:

$$\text{If } A \supseteq B, \text{ then } D \ominus A \subseteq D \ominus B.$$

Figure 6.8 illustrates an instance showing that larger structuring elements have an effect more severe than smaller ones on the erosion process.

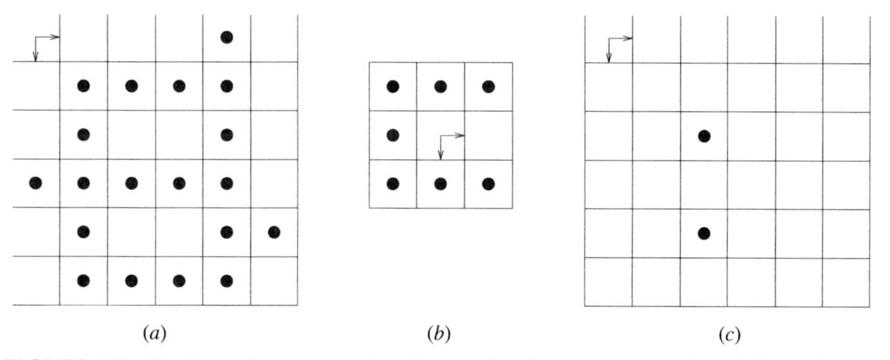

FIGURE 6.6 Eroding with a structuring element that does not contain the origin can lead to a result that has nothing in common with the set to be eroded: (*a*) Image A; (*b*) structuring element B; (*c*) $A \ominus B$.

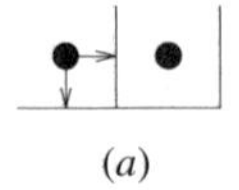

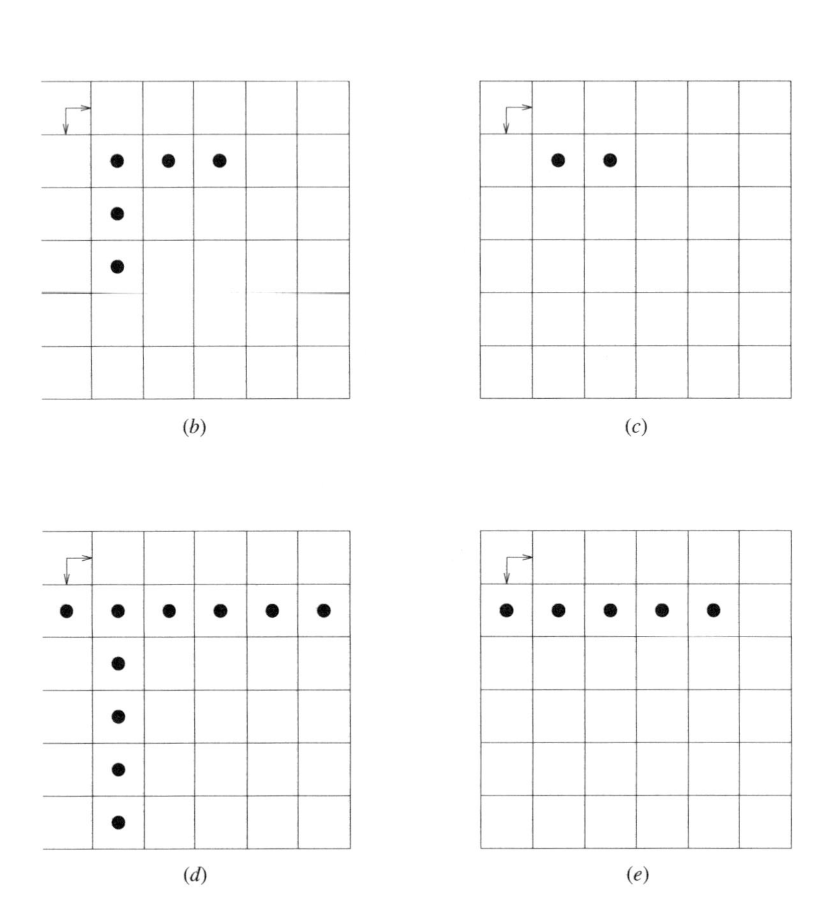

FIGURE 6.7 Increasing property of erosion: (*a*) Structuring element K; (*b*) image A; (*c*) $A \ominus K$; (*d*) image B; (*e*) $B \ominus K$.

5. The erosion of the intersection of two images is equal to the intersection of their erosions:

$$(A \cap B) \ominus C = (A \ominus C) \cap (B \ominus C). \tag{6.18}$$

Figure 6.9 illustrates the relationship $(A \cap B) \ominus C = (A \ominus C) \cap (B \ominus C)$.

6. Although the dilation of the unions of two images is equal to the union of their dilations, the relationship for the erosion transformation is one of containment:

$$(A \cup B) \ominus C \supseteq (A \ominus C) \cup (B \ominus C). \tag{6.19}$$

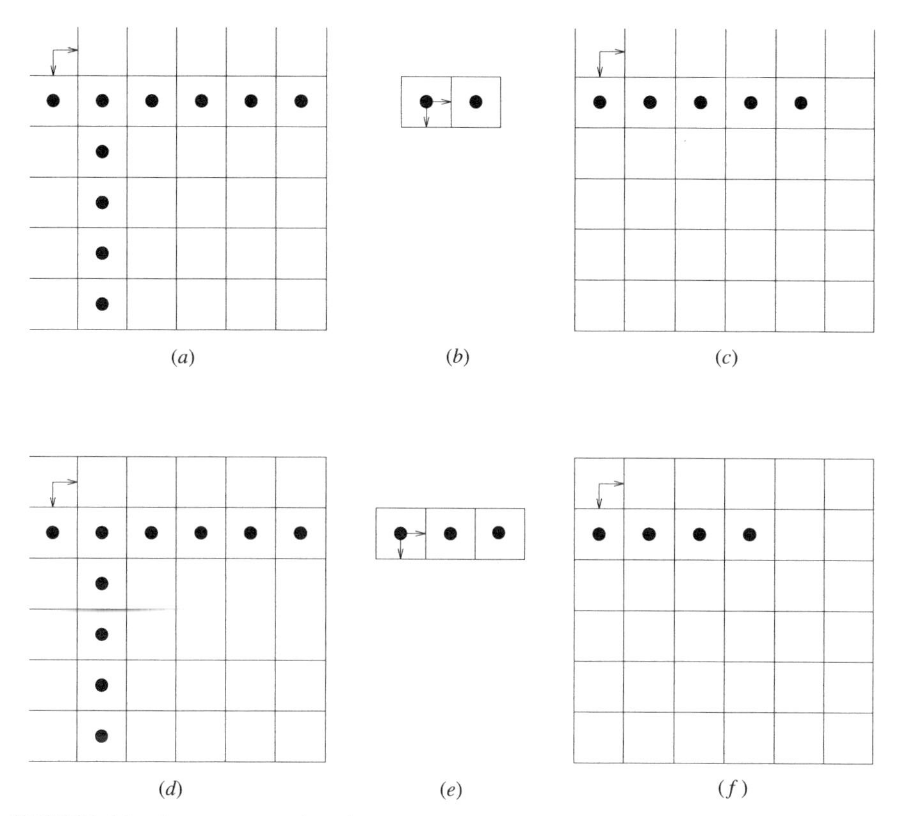

FIGURE 6.8 Larger structuring elements have an effect more severe than smaller ones on the erosion process: (*a*) Image D; (*b*) Structuring element B; (*c*) $D \ominus B$; (*d*) image D; (*e*) structuring element A; (*f*) $D \ominus A$.

Figure 6.10 illustrates an instance where $(A \cup B) \ominus C$ strictly contains $(A \ominus C) \cup (B \ominus C)$.

7. Erosion is not commutative:

$$A \ominus B \neq B \ominus A. \tag{6.20}$$

8. Structuring elements can be decomposed through union into simpler structuring elements to simplify the erosion transformation:

$$A \ominus (B \cup C) = (A \ominus B) \cap (A \ominus C). \tag{6.21}$$

Figure 6.11 illustrates an instance of the relationship $A \ominus (B \cup C) = (A \ominus B) \cap (A \ominus C)$.

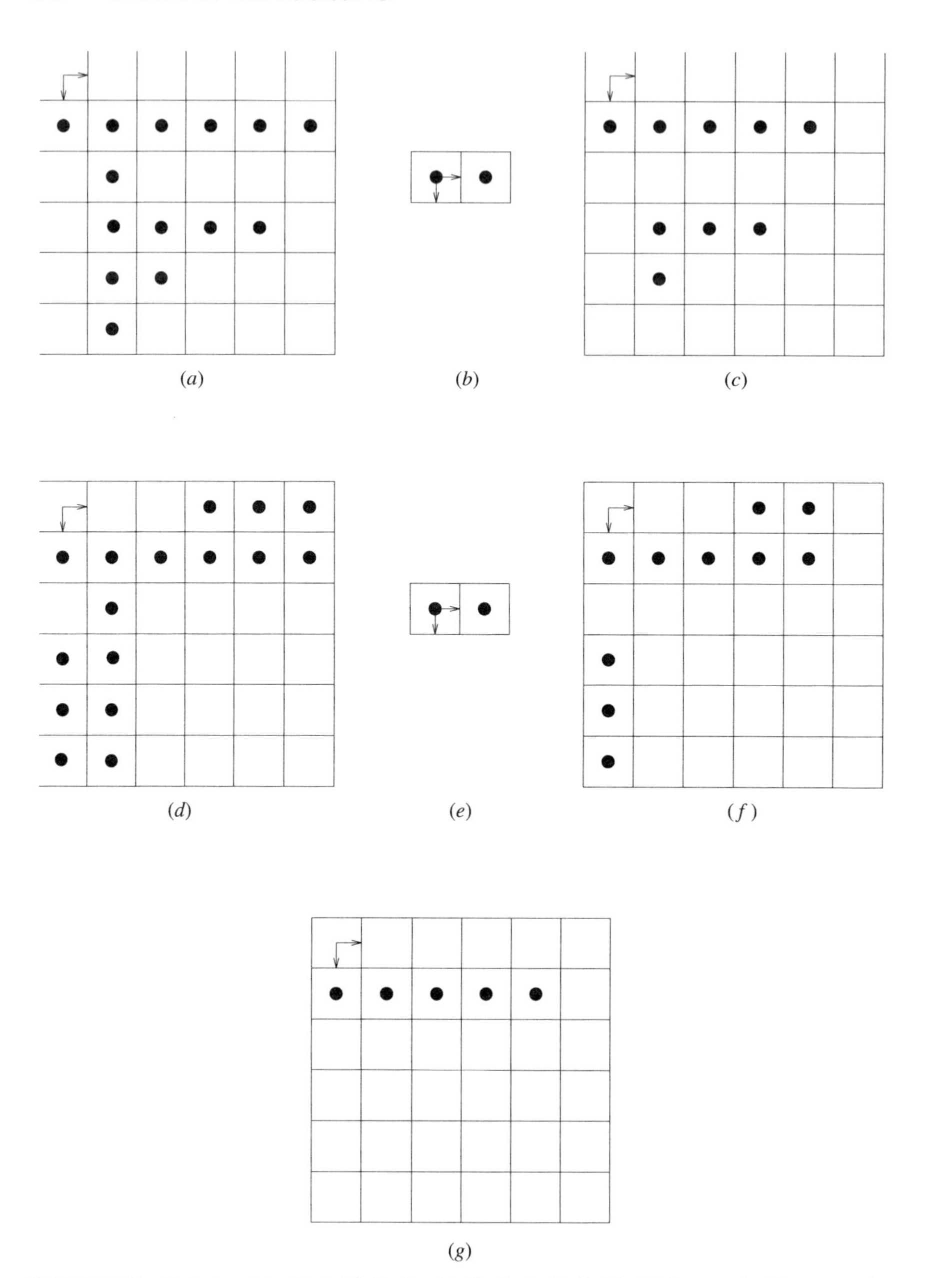

FIGURE 6.9 Relationship $(A \cap B) \ominus C = (A \ominus C) \cap (B \ominus C)$: (*a*) Image A; (*b*) structuring element C; (*c*) $A \ominus C$; (*d*) image B; (*e*) structuring element C; (*f*) $B \ominus C$; (*g*) $(A \cap B) \ominus C = (A \ominus C) \cap (B \ominus C)$.

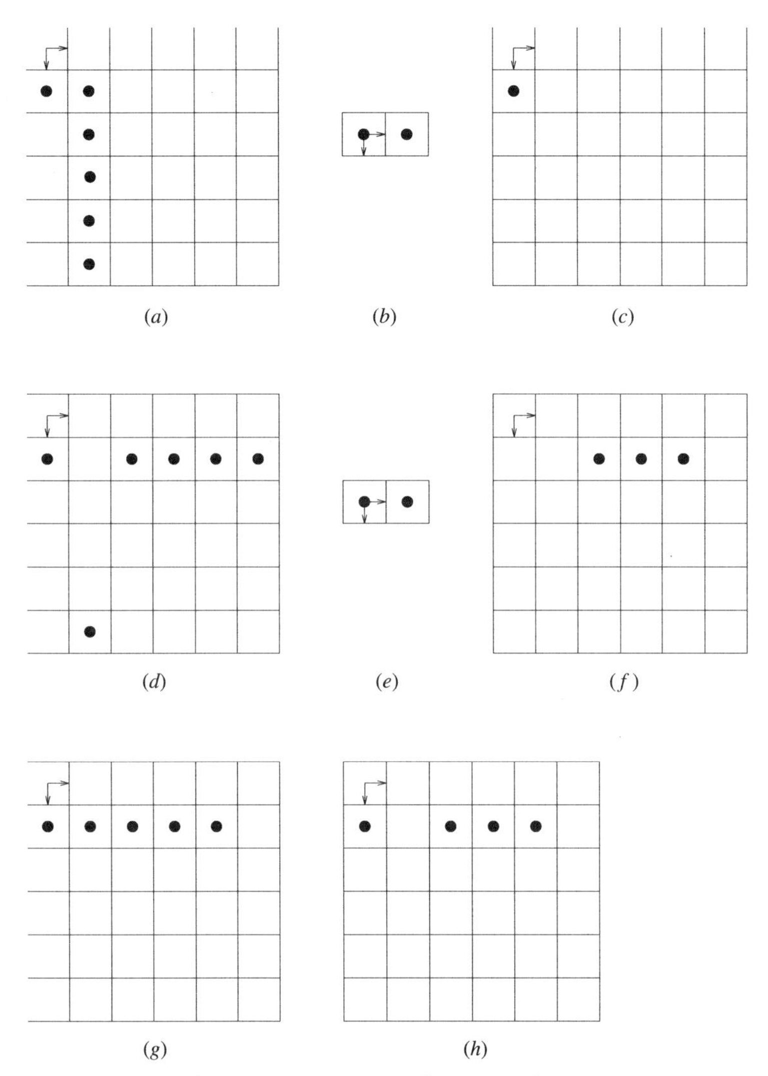

FIGURE 6.10 $(A \cup B) \ominus C$ strictly contains $(A \ominus C) \cup (B \ominus C)$: (*a*) Image A; (*b*) structuring element C; (*c*) $A \ominus C$; (*d*) image B; (*e*) structuring element C; (*f*) $B \ominus C$; (*g*) $(A \cup B) \ominus C$; (*h*) $(A \ominus C) \cup (B \ominus C)$.

9. The intersection decomposition leads to a containment relationship:

$$A \ominus (B \cap C) \supseteq (A \ominus B) \cup (A \ominus C). \tag{6.22}$$

Figure 6.12 illustrates an instance where $A \ominus (B \cap C)$ strictly contains $(A \ominus B) \cup (A \ominus C)$.

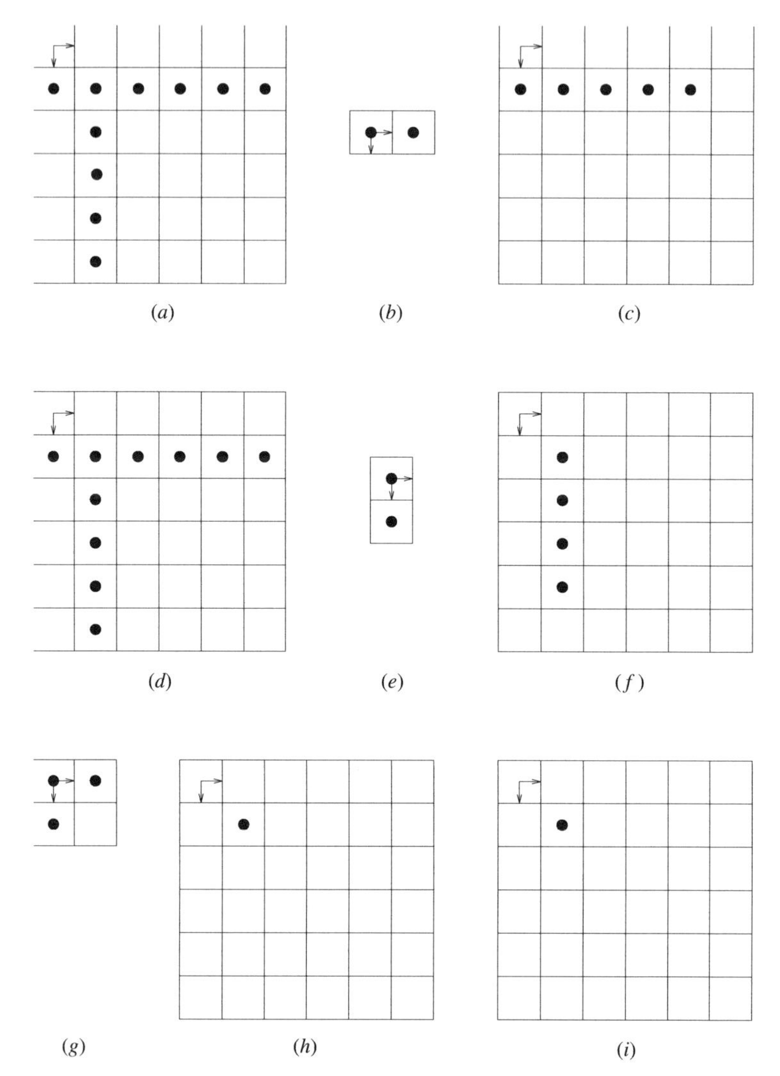

FIGURE 6.11 Illustration of the relationship $A \ominus (B \cup C) = (A \ominus B) \cap (A \ominus C)$: (*a*) Image A; (*b*) structuring element B; (*c*) $A \ominus B$; (*d*) image A; (*e*) structuring element C; (*f*) $A \ominus C$; (*g*) $B \cup C$; (*h*) $A \ominus (B \cup C)$; (*i*) $(A \ominus B) \cap (A \ominus C)$.

10. A chain rule holds when the structuring element is decomposable through dilation:

$$A \ominus (B \oplus C) = (A \ominus B) \ominus C, \tag{6.23}$$

$$A \ominus (B_1 \oplus \cdots B_K) = (\cdots (A \ominus B_1) \ominus \cdots \ominus B_K). \tag{6.24}$$

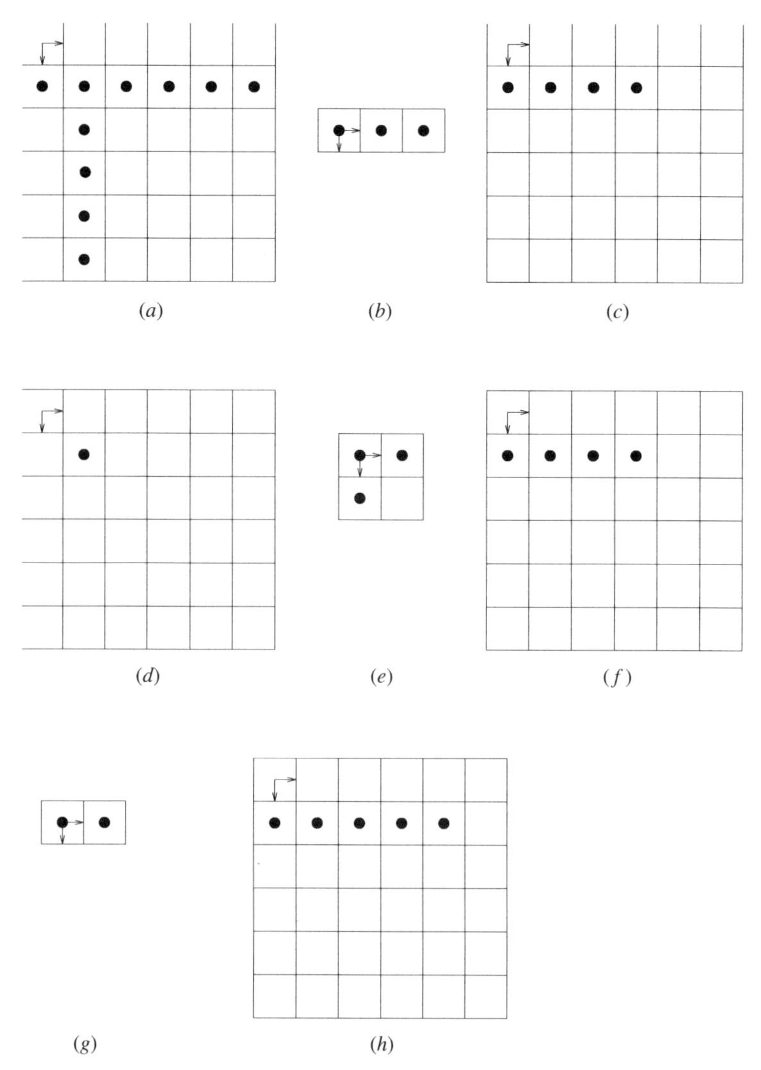

FIGURE 6.12 $A \ominus (B \cap C)$ strictly contains $(A \ominus B) \cup (A \ominus C)$: (a) Image A; (b) structuring element B; (c) $A \ominus B$; (d) image $A \ominus C$; (e) structuring element C; (f) $(A \ominus B) \cup (A \ominus C)$; (g) $B \cap C$; (h) $A \ominus (B \cap C)$.

11. When performing erosion and dilation, performing erosion first is more severe than performing dilation first:

$$A \oplus (B \ominus C) \subseteq (A \oplus B) \ominus C. \tag{6.25}$$

Figure 6.13 illustrates an instance where $A \oplus (B \ominus C)$ strictly contains $(A \oplus B) \ominus C$.

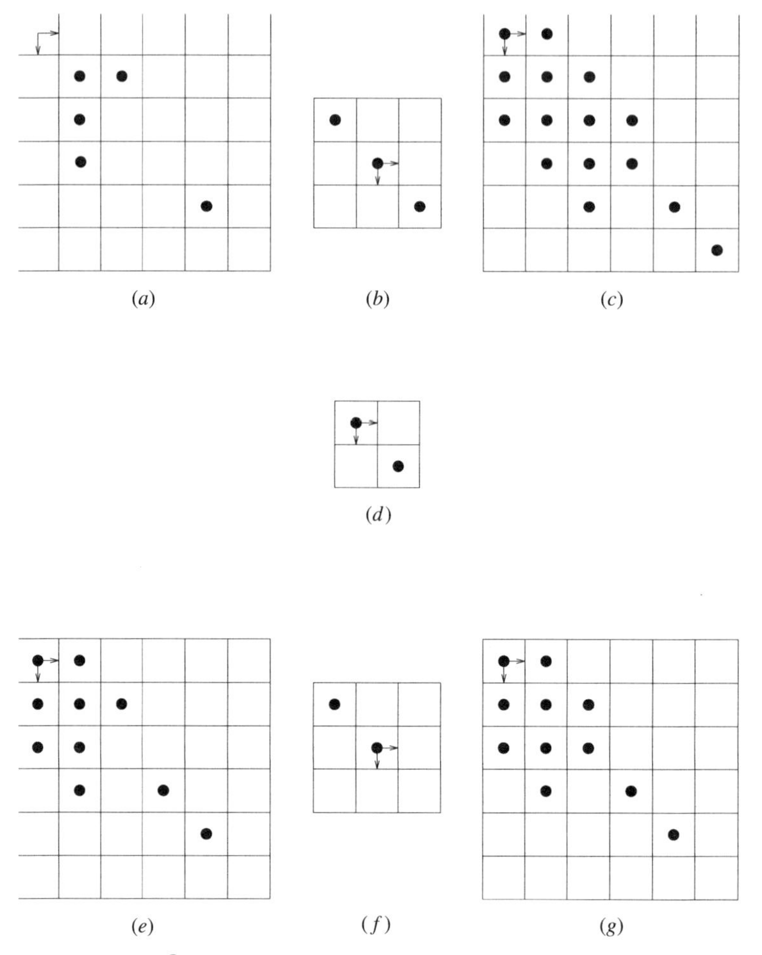

FIGURE 6.13 $A \oplus (B \ominus C)$ is strictly contained in $(A \oplus B) \ominus C$: (*a*) Image A; (*b*) structuring element B; (*c*) $A \oplus B$; (*d*) structuring element C; (*e*) $A \oplus (B \ominus C)$; (*f*) $B \ominus C$; (*g*) $(A \oplus B) \ominus C$.

6.1.3 Duality between Dilation and Erosion

Dilation and erosion transformations are markedly similar in that what one does to the image the other does to the ground. Recall that two operators are called *dual* when the negation of a formulation employing the first operator is equal to that formulation employing the second operator on the negated variables. In morphology, negation of a set is considered in a geometric sense: that of reversing the orientation of the set with respect to its coordinate axes. Such reversing is called *reflection*. The duality of dilation and erosion employs both logical and geometric negation because of the different roles of the image and structuring element in an expression employing these morphological operators.

Theorem 6.1.1

$$(A \ominus B)^c = A^c \oplus \check{B}, \tag{6.26}$$

where

$$\check{B} = \{x \mid for\ some\ b \in B,\ x = -b\}, \tag{6.27}$$

$$A^c = \{x \in E^N \mid a \notin A\}. \tag{6.28}$$

Figure 6.14 illustrates the relationship $(A \ominus B)^c = A^c \oplus \check{B}$.

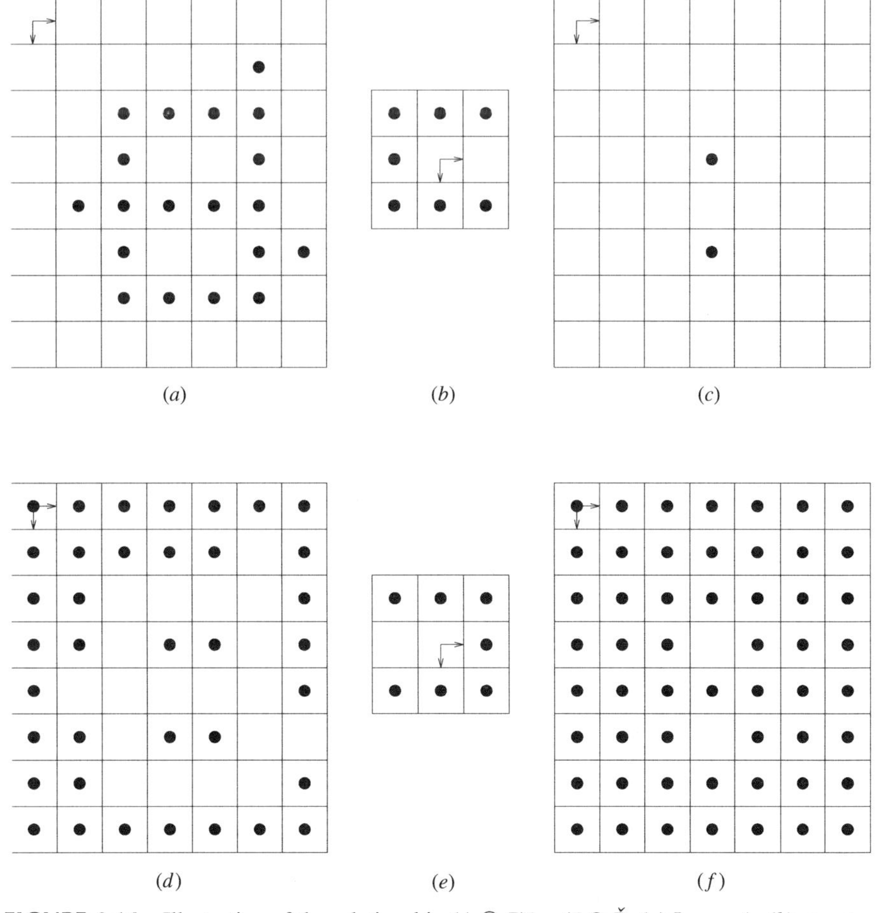

FIGURE 6.14 Illustration of the relationship $(A \ominus B)^c = A^c \oplus \check{B}$: (*a*) Image *A*; (*b*) structuring element *B*; (*c*) $A \ominus B$; (*d*) image A^c; (*e*) structuring element *B*; (*f*) $A^c \oplus \check{B}$.

Corollary

$$(A \oplus B)^c = A^c \ominus \check{B}. \tag{6.29}$$

Although dilation and erosion are dual, this does not imply that we can freely perform cancellation on morphological equalities. For example, if $A = B \ominus C$, then dilating both sides of the expression by C results in $A \oplus C = (B \ominus C) \oplus C \neq B$. However, a containment relationship is maintained:

$$A \subseteq B \ominus C \Leftrightarrow B \supseteq A \oplus C, \tag{6.30}$$

$$A \subseteq (\cdots (B \ominus C_1) \ominus \cdots) \ominus C_N,$$

$$\text{if and only if} \quad (\cdots (A \oplus C_1) \oplus \cdots) \oplus C_N \subseteq B. \tag{6.31}$$

6.2 OPENING AND CLOSING

In practice, dilations and erosions are usually employed in pairs, either dilation of an image followed by the erosion of the dilated result, or image erosion followed by dilation. In both cases the result of iteratively applied dilations or erosions is an elimination of specific image detail smaller than the structuring element without the global geometric distortion of unsuppressed features. For example, opening an image with a disk structuring element smoothes the contour, breaks narrow isthmuses, and eliminates small islands and sharp peaks and capes. Closing an image with a disk structuring element smoothes the contours, fuses narrow breaks and long thin gulfs, eliminates small holes, and fills gaps on the contours.

Of particular significance is the fact that image transformations employing iteratively applied dilations and erosions are idempotent; that is, their reapplication effects no further changes to the previously transformed result. The practical importance of idempotent transformation is that they comprise complete and closed stage of image analysis algorithms because shapes can be naturally described in terms of what structuring elements can be opened or closed and yet remain the same. Their functionality corresponds closely to the specification of a signal by its bandwidth. Morphologically filtering an image by an opening or closing operation corresponds to the ideal nonrealizable bandpass filters of conventional linear filtering. Once an image is ideal bandpass filtered, further ideal bandpass filtering does not alter the result.

The *opening* of image B by structuring element K is denoted by $B \circ K$ and is defined by

$$B \circ K = (B \ominus K) \oplus K. \tag{6.32}$$

The opening of A by K selects out precisely those points of A that match K in the sense that the point can be covered by some translation of the structuring element K which itself is entirely contained in A.

$$A \circ K = \{x \in A \mid \text{for some } t \in A \ominus K, x \in K_t \text{ and } K_t \subseteq A\}. \tag{6.33}$$

Opening an image with a disk-structuring element smoothes the contours, breaks narrow isthmuses, and eliminates small islands and sharp peaks or capes.

Figure 6.15 is an instance of opening transformation of a noisy character image. The structuring element is a 3 square as shown in Fig. 6.15 (*a*). Figure 6.15 (*b*) is the image, and (*c*) is the result of the opening transformation. The linear noise with two or less pixel width has been removed by the opening transformation with this 3×3 square structuring element.

The *closing* of image B by structuring element K is denoted by $B \bullet K$ and is defined by

$$B \bullet K = (B \oplus K) \ominus K. \tag{6.34}$$

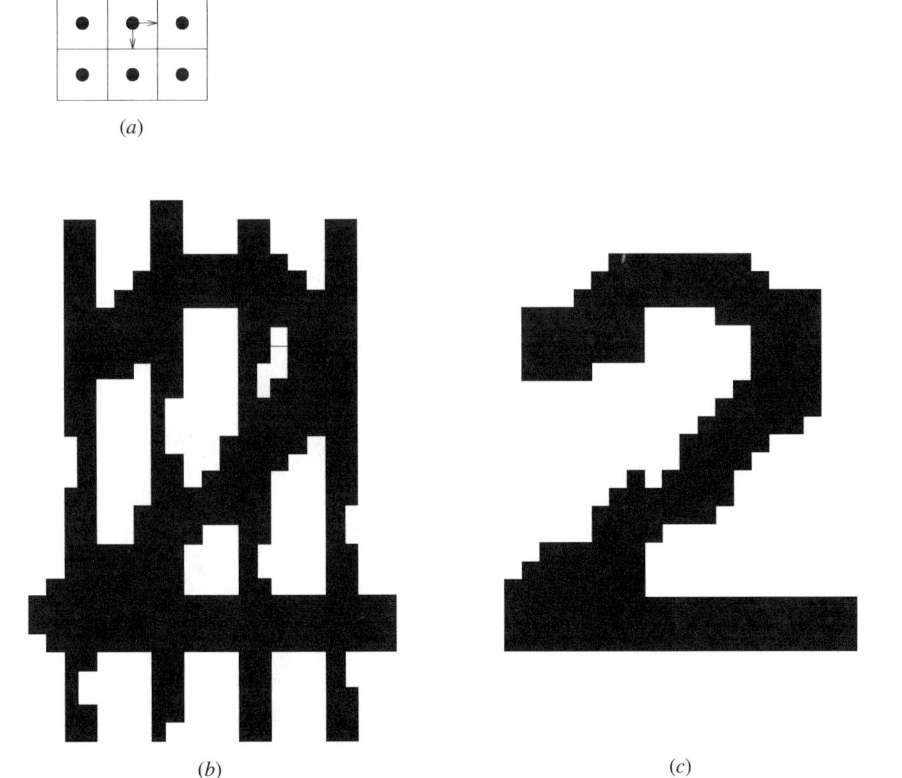

FIGURE 6.15 Opening transformation of a noisy character image: (*a*) Structuring element B; (*b*) image A; (*c*) $A \circ B$.

Closing an image with a disk-structuring element smoothes the contours, fuses narrow breaks and long thin gulfs, and eliminates small holes, and fills gaps on the contours.

Figure 6.16 is an instance of closing transformation of a noisy character image with the 3×3 square structuring element. Figure 6.16 (*a*) is the image, and (*b*) is the result of the closing transformation.

If B is unchanged by opening with K, we say that B is *open* with respect to K, whereas if B is unchanged by closing with K, then B is *closed* with respect to K.

Properties

1. The class of sets that are unaltered by erosion followed by dilation with a given structuring element K consists of all sets that can be expressed as some set dilated by K:

$$A \oplus K = (A \oplus K) \circ K = (A \bullet K) \oplus K \qquad (6.35)$$

2. Idempotency of closing is

$$(A \bullet K) \bullet K = A \bullet K. \qquad (6.36)$$

3. The class of sets that are unaltered by dilation followed by erosion with a given structuring element K consists of all sets that can be expressed as some set eroded by K:

$$A \ominus K = (A \circ K) \ominus K = (A \ominus K) \bullet K. \qquad (6.37)$$

4. Idempotency of opening is

$$(A \circ K) \circ K = A \circ K. \qquad (6.38)$$

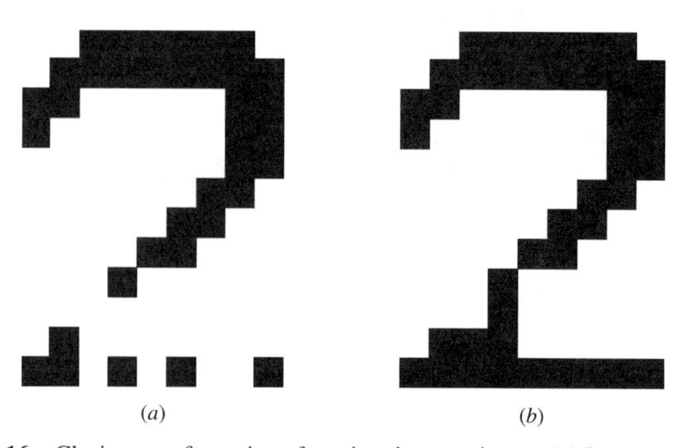

FIGURE 6.16 Closing transformation of a noisy character image. (*a*) Image A; (*b*) $A \bullet B$.

5. Opening and closing are invariant to translations of the structuring elements.

$$A \circ B = A \circ (B)_x, \tag{6.39}$$

$$A \bullet B = A \bullet (B)_x. \tag{6.40}$$

6. The opening transformation is anti-extensive:

$$A \circ B \subseteq A. \tag{6.41}$$

   Figure 6.17 illustrates an instance of the anti-extensivity of the opening transformation.

7. The closing operation is extensive:

$$A \subseteq A \bullet B \tag{6.42}$$

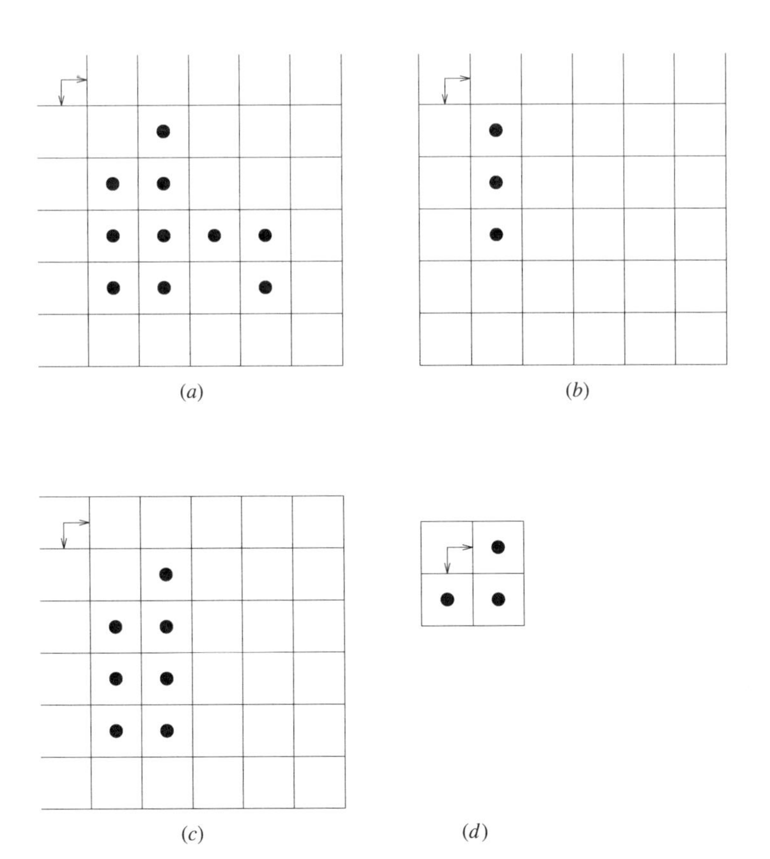

FIGURE 6.17 Anti-extensivity of the opening transformation: (*a*) Image *A*; (*b*) $A \ominus B$; (*c*) $A \circ B = (A \ominus B) \oplus B$; (*d*) structuring element *B*.

Figure 6.18 illustrates an instance of the extensivity of the closing transformation.

8. Duality of opening and closing is

$$(A \circ B)^c = A^c \bullet \check{B}, \tag{6.43}$$

$$(A \bullet B)^c = A^c \circ \check{B}. \tag{6.44}$$

9. The opening of A by B is the union of all translations of B that are contained in A:

$$\begin{aligned} A \circ B &= \{x \in A \mid \text{for some } y,\ x \in B_y \subseteq A\} \\ &= \cup_{\{y \mid B_y \subseteq A\}} B_y. \end{aligned} \tag{6.45}$$

10. The closing of A by B is the complement of the union of all translations of $\check{B}$ that are contained in A^c:

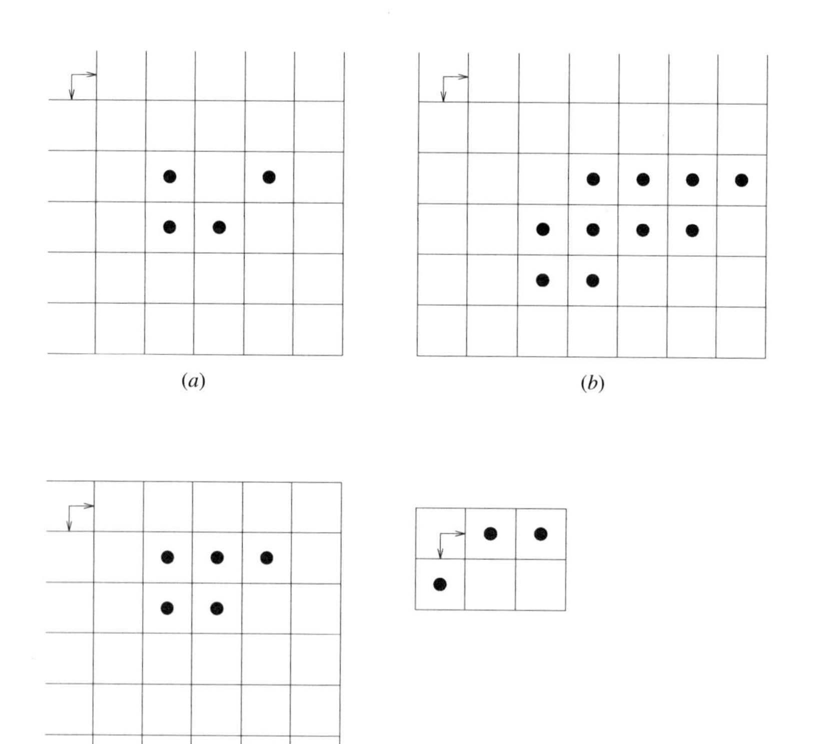

FIGURE 6.18 Extensivity of the closing transformation: (*a*) Image A; (*b*) $A \oplus B$; (*c*) $A \bullet B = (A \oplus B) \ominus B$; (*d*) structuring element B.

$$A \bullet B = (A^c \circ \check{B})^c = [\cup_{\{y | \check{B}_y \subseteq A^c\}} \check{B}_y]^c. \tag{6.46}$$

11. A geometric characterization of the closing operation is

$$\begin{aligned} A \bullet B &= \{x \in E^N \mid x \in \check{B}_y \text{ implies } \check{B}_y \cap A \neq \phi\} \\ &= \cap_{\{y \mid \check{B}_y \cap A \neq \phi\}} \check{B}^c_y. \end{aligned} \tag{6.47}$$

6.3 GRAY-SCALE MORPHOLOGY

The binary morphological operations of dilation, erosion, opening, and closing are all naturally extended to gray-scale imagery in the following way: First, we introduce the concept of the *top surface* of a set and the related concept of the *umbra* of a surface. Then gray-scale dilation is defined as the surface of the dilations of the umbras. From this definition we can proceed to the representation that indicates that gray-scale dilation can be computed in terms of a maximum operation and a set of addition operations. A similar plan is followed for erosion which can be evaluated in terms of a minimum operation and a set of subtraction operations.

Of course, having a definition and a means of evaluating the defined operations does not imply that the properties of gray-scale dilation and erosion are the same as binary dilation and erosion. To establish that the relationships are identical, we explore some of the relationships between the umbra and surface operation. Our explanation shows that umbra and surface operations are essentially inverses of each other. Then we illustrate how the umbra operation is a homomorphism from the gray-scale morphology to the binary morphology. Having the homomorphism in hand, all the interesting relationships follow by appropriately unwrapping and wrapping the involved sets or functions.

6.3.1 Gray-Scale Dilation and Erosion

We begin with the concepts of surface of a set and the umbra of a surface. Suppose that a set A in Euclidean N-space is given. We adopt the convention that the first $N-1$ coordinates of the N-tuples of A constitute the spatial domain of A and the Nth coordinate is for the surface. For ordinary gray-scale imagery, $N = 3$. The top or top surface of A is a function defined on the projection of A onto its first $(N-1)$ coordinates. For each $(N-1)$-tuple x, the top surface of A at x is the highest value y such that the N-tuple $(x,y) \in A$.

Let $A \subseteq E^N$ and

$$F = \{x \in E^{N-1} \mid \text{for some } y \in E, (x, y) \in A\}.$$

The *top* or *top surface* of A, denoted by $T[A] : F \rightarrow E$, is defined by

$$T[A](x) = \max \{y \mid (x, y) \in A\}. \tag{6.48}$$

For any function f defined on some subset F of Euclidean $(N-1)$-space, the umbra of f is the set consisting of the surface f and everything below the surface.

Let $F \subseteq E^{N-1}$ and $f : F \to E$. The *umbra* of f, denoted by $U[f]$, $U[f] \subseteq F \times E$, is defined by

$$U[f] = \{(x, y) \in F \times E \mid y \leq f(x)\}. \tag{6.49}$$

Figure 6.19 illustrates a discretized one-dimensional function f defined as a domain consisting of seven successive column positions and a finite portion of its umbra which lies on or below the function f. The actual umbra has infinite extent below f.

Having defined the operations of taking a top surface of a set and the umbra of a surface, we can define gray-scale dilation. The gray-scale dilation of two functions is defined as the surface of the dilation of their umbras.

Let $F, K \subseteq E^{N-1}$ and $f : F \to E$ and $k : K \to E$. The *dilation* of f by k is denoted by $f \oplus k$, $f \oplus k : F \oplus K \to E$, and is defined by

$$f \oplus k = T[U[f] \oplus U[k]], \tag{6.50}$$

$$(f \oplus k)(x) = \max\, \{f(x-z) + k(z) \mid z \in K,\ x - z \in F\}. \tag{6.51}$$

Figure 6.20 illustrates a second discretized one-dimensional function k defined as a domain consisting of three successive column positions and a finite portion of its umbra which lies on or below the function k: The dilation of the umbra of f and k and the surface of the dilation of the umbras of f and k are shown.

The definition for gray-scale erosion proceeds in a similar way to the definition of gray-scale dilation. The gray-scale erosion of one function by another is the surface of the binary erosion of the umbra of one with the umbra of the other.

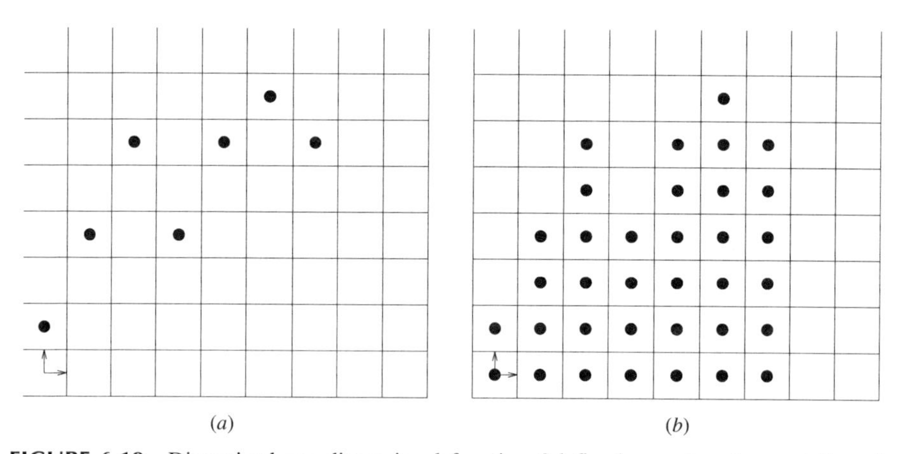

FIGURE 6.19 Discretized one-dimensional function f defined as a domain consisting of seven successive column positions and a finite portion of its umbra which lies on or below the function f: (*a*) Function f; (*b*) umbra $U[f]$.

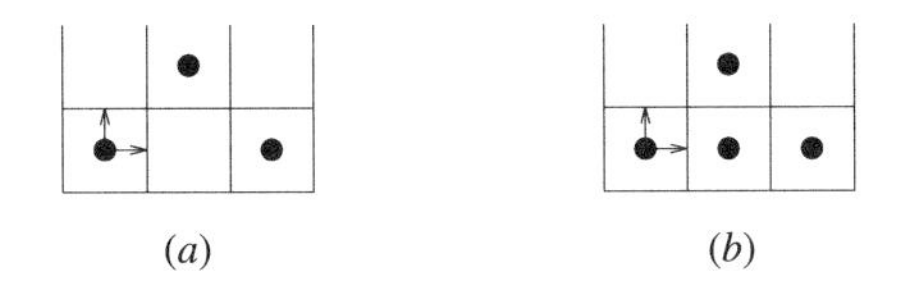

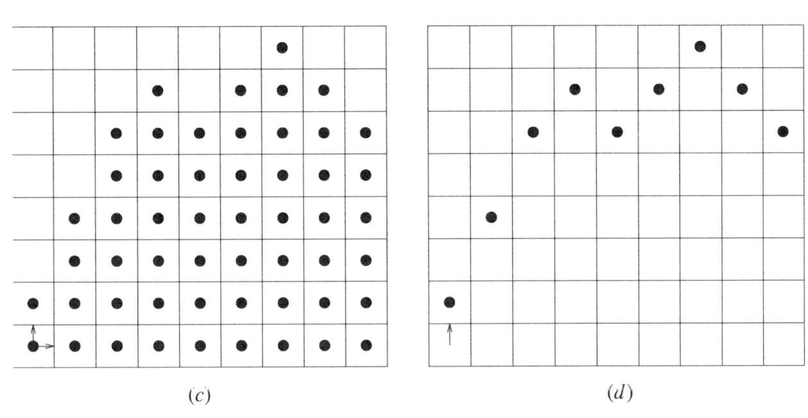

FIGURE 6.20 Second discretized one-dimensional function k defined as a domain consisting of three successive column positions and a finite portion of its umbra which lies on or below the function k: The dilation of the umbra of f and k and the surface of the dilation of the umbras of f and k are shown. (*a*) Function k; (*b*) umbra $U[k]$; (*c*) $U[f] \oplus U[k]$; (*d*) $f \oplus k = T[U[f] \oplus U[k]]$.

Let $F, K \subseteq E^{N-1}$ and $f: F \rightarrow E$ and $k: K \rightarrow E$. The *erosion* of f by k is denoted by $f \ominus k, f \ominus k : F \ominus K \rightarrow E$, and is defined by

$$f \ominus k = T[U[f] \ominus U[k]], \tag{6.52}$$

$$(f \ominus k)(x) = \max\{f(x+z) - k(z) \mid z \in K\}. \tag{6.53}$$

Figure 6.21 shows the erosion of the umbra of f and k and the surface of the erosion of the umbras of f and k.

Properties

1. The basic relationship between the surface and umbra operations is that they are, in a certain sense, inverses of each other. More precisely, the surface operation will always undo the umbra operation. That is, the surface operation is an inverse to the umbra operation:

$$T[U[f]] = f, \tag{6.54}$$

where $F \subseteq E^{N-1}$ and $f: F \rightarrow E$.

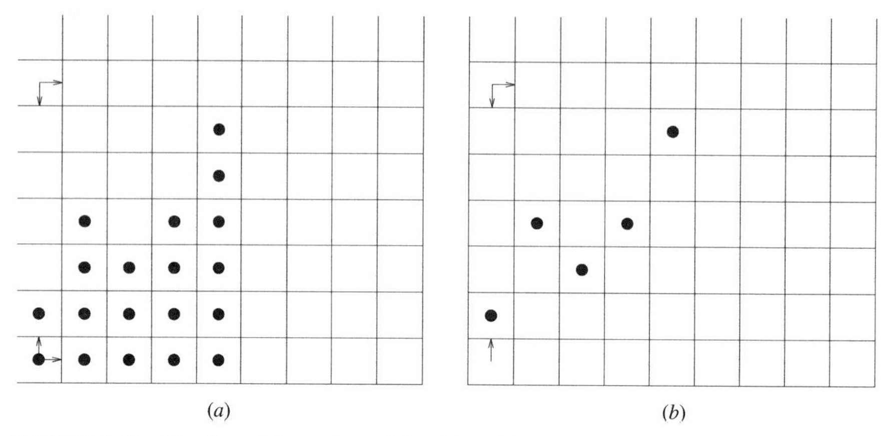

FIGURE 6.21 Erosion of the umbra of f and k and the surface of the erosion of the umbras of f and k: (a) $U[f] \ominus U[k]$; (b) $f \ominus k = T[U[f] \ominus U[k]]$.

2. The umbra operation is not an inverse to the surface operation. Without any constraints on the set A, the strongest statement that can be made is that the umbra of the surface of A contains A:

$$A \subseteq U[T[A]]. \tag{6.55}$$

3. When the set A is an umbra, then the umbra of the surface of A is A. In this case the umbra operation is an inverse to the surface operation.

$$\text{If } A \text{ is an umbra, then } A = U[T[A]].$$

4. Suppose that A and B are umbras. Then $A \oplus B$ and $A \ominus B$ are umbras.

6.3.2 Umbra Homomorphism Theorem

The umbra homomorphism theorem states that the operation of taking an umbra is a homomorphism from the gray-scale morphology to the binary morphology.

Theorem 6.3.1 *Let F, $K \subseteq E^{N-1}$ and $F : f \rightarrow E$ and $k : K \rightarrow E$. Then,*

1. $U[f \oplus k] = U[f] \oplus U[k]$
2. $U[f \ominus k] = U[f] \ominus U[k]$

Properties We state the commutativity and associativity of gray-scale dilation and the chain rule for gray-scale erosion:

1. $f \oplus k = k \oplus f$.
2. $k_1 \oplus (k_2 \oplus k_3) = (k_1 \oplus k_2) \oplus k_3$.
3. $(f \ominus k_1) \ominus k_2 = f \ominus (k_1 \oplus k_2)$.

6.3.3 Gray-Scale Opening and Closing

Gray-scale opening and closing are defined in an analogous way to opening and closing in the binary morphology and they have similar properties.

Let $f : F \to E$ and $k : K \to E$. The gray-scale *opening* of f by structuring element k is denoted by $f \circ k$ and is defined by

$$f \circ k = (f \ominus k) \oplus k. \tag{6.56}$$

The gray-scale *closing* of f by structuring element k is denoted by $f \bullet k$, and is defined by

$$f \bullet k = (f \oplus k) \ominus k. \tag{6.57}$$

Properties

1. Let $f : F \to E$ and $g : G \to E$. Suppose that $F \subseteq G$. Then $f \leq g$ if and only if $U[f] \subseteq U[g]$.
2. $g \leq f \ominus k$ if and only if $f \geq g \oplus k$.
3. Let $A \subseteq E^{N-1} \times E$ and $D \subseteq E^{N-1} \times E$. Then $A \subseteq D$ implies that $T[A](x) \leq T[D](x)$.
4. $(f \circ k)(x) \leq f(x)$ for every $x \in F \circ K$.
5. $f(x) \leq (f \bullet k)(x)$ for every $x \in F$.
6. Idempotency of opening is

$$(f \circ k) \circ k = f \circ k. \tag{6.58}$$

7. Idempotency of closing is

$$(f \bullet k) \bullet k = f \bullet k. \tag{6.59}$$

6.3.4 Duality

Let $f : F \to E$. The reflection of f is denoted by $\check{f}$, $\check{f} : \check{F} \to E$, and is defined by $\check{f}(x) = f(-x)$. Let $f : F \to E$ and $k : K \to E$. Then, for any $x \in (F \ominus \check{K}) \cap (F \oplus K)$,

$$\begin{aligned}
-(f \oplus k)(x) &= \max \{f(x-z) + k(z) \mid z \in K, x - z \in F\} \\
&= \min \{-f(x-z) - k(z) \mid z \in K, x - z \in F\} \\
&= \min \{-f(x-z) - \check{k}(z) \mid z \in (-K), x - z \in F\} \\
&= \min \{-f(x-z) - \check{k}(z) \mid z \in K\} \\
&= ((-f) \ominus \check{k})(x).
\end{aligned} \tag{6.60}$$

It follows immediately from the gray-scale dilation and erosion duality that there is a gray-scale opening and closing duality:

$$-(f \circ k) = (-f) \bullet \check{k}, \tag{6.61}$$

$$f \bullet k = -((-f) \circ \check{k}). \tag{6.62}$$

This means that we can think of closing like opening. To close f with a paraboloid structuring element, we take the reflection of the paraboloid, turn it upside down, and slide it over the top surface of f. The closing is the surface of all the lowest points reached by the sliding paraboloid.

6.4 BIBLIOGRAPHICAL REMARKS

Since this chapter is somewhat additional, we give the bibliographical remarks in brief. First of all, mathematical morphology was originated by G. Matheron in his study of porous materials connected to oil exploration in the late 1960s [1]. Since 1975 morphological analysis has been developing rapidly. Some good professional books are Serra's edited volumes [2, 3]. A highly readable book is written by C. R. Giardina and E. R. Dougherty [4]. It covers three fundamental areas of morphological analysis: (1) binary morphological algebra, (2) gray-scale morphology, and (3) morphological filtering of both image and signals. In all cases the Euclidean (or analog) theory is introduced and then discussion of digital implementation follows. It gives an intuitive perspective to a reader's understanding. As a tutorial paper, Haralick, Sternberg, and Zhuang's paper can be recommended in which the related historical work is introduced [5]. Serra's paper is also good for a beginner [6]. As a very recent work, Jackway and Deriche's work on multiscale morphology is interesting and related to scale-space filtering [7].

BIBLIOGRAPHY

[1] G. Matheron, *Random Sets and Integral Geometry,* New York: Wiley, 1975.

[2] J. Serra, *Image Analysis and Mathematical Morphology,* New York: Academic Press, 1983.

[3] J. Serra, *Image Analysis and Mathematical Morphology: Theoretical Advances,* vol. 2, New York: Academic Press, 1988.

[4] C. R. Giardina and E. R. Dougherty, *Morphological Methods in Image and Signal Processing,* Englewood Cliffs, NJ: Prentice Hall, 1987.

[5] R. M. Haralik, S. R. Sternberg, and X. Zhuang, "Image analysis using mathematical morphology," *IEEE Trans. Pattern Anal. Machine Intell.,* vol. PAMI-9, no. 4, pp. 532–550, 1986.

[6] J. Serra, "Introduction to mathematical morphology," *Comp. Vision, Graphics, Image Proc.,* vol. 35, pp. 283–305, 1986.

[7] P. T. Jackway and M. Deriche, "Scale-space properties of the multiscale morphological dilation-erosion," *IEEE Trans. Pattern Anal. Machine Intell.,* vol. 18, no. 1, pp. 38–51, January 1996.

CHAPTER SEVEN

Feature Extraction Using Linear Methods

We have already introduced feature extraction using linear methods in Chapter 3. Two different mathematical expressions were introduced, finite dimensional representation and infinite dimensional representation. More specifically, n features, ξ^k, $k = 1, 2, \ldots, n$ were extracted according to the formula

$$\xi^k = A_i^k x^i, \tag{7.1}$$

where the A_i^k were called measuring vectors $k = 1, 2, \ldots, n$ and x^i was a vector representation of an image on an observed plane. Here we used Einstein's notation for summation. On the other hand, ξ^k was also represented for a two-dimensional function $f(x, y)$, an image on an observed plane, as

$$\xi^k = \int A^k(x, y) f(x, y)\,dxdy, \tag{7.2}$$

where $A^k(x, y)$ were called measuring function, $k = 1, 2, \ldots, n$. From these representations we can see that feature vector ξ^k was constructed by linear combination of an observed image.

Here we use both expressions, but for the latter some mathematical background is now necessary. The latter is better than the former from a theoretical point of view.

First, some real examples of linear feature extraction will be presented as an introduction. We will then focus on two points: *invariant feature extraction* and *efficient feature extraction.*

We have also introduced measuring using either moment or trigonometric functions, both of which were introduced by spanning functions whose spaces admit some transformations. They are very popular in pattern recognition as well as other fields.

7.1 MOMENT

First, the Nth *moment* of an image $f(x, y)$ is represented as

$$M_{pq} = \int\int_R x^p y^q f(x, y) dx dy, \tag{7.3}$$

where $N = p + q$.

The range of integration is limited to $x \in (-1, 1)$ and $y \in (-1, 1)$ in order to prevent the divergence of the integral at higher orders of the moment. As an example of $f(x, y)$, a simple rectangle is taken whose center is located at the origin of the coordinate system. The rectangle is filled and has $2a$ length along the x-axis and $2b$ length along the y-axis.

The *central moment* is defined as

$$\mu_{ij} = \int\int_R (x - \bar{x})^i (y - \bar{y})^j f(x, y) dx dy \tag{7.4}$$

where $(\bar{x}, \bar{y})$ and center of the gravity of $f(x, y)$. For the rectangle, μ_{ij} is obtained as

$$\mu_{ij} = \begin{cases} \dfrac{4a^{i+1} b^{j+1}}{(i+1)(j+1)}, & i \text{ and } j \text{ are even,} \\ 0 & \text{otherwise,} \end{cases} \tag{7.5}$$

assuming that the rectangular has uniform density 1. In particular, the lower-order moments are

$$\mu_{00} = 4ab,$$
$$\mu_{01} = \mu_{10} = \mu_{11} = 0,$$
$$\mu_{20} = \frac{4a^3 b}{3}, \quad \mu_{02} = \frac{4ab^3}{3}. \tag{7.6}$$

Obviously μ_{00} indicates the area of the rectangle, and $\mu_{01} = \mu_{10} = \mu_{11} = 0$ indicates the symmetry of the rectangle. Finally μ_{20} and μ_{02} indicate the dispersion along x-axis and y-axis, respectively. Using μ_{00}, we can find a and b as $a = \sqrt{3\mu_{20}/\mu_{00}}$ and $b = \sqrt{3\mu_{02}/\mu_{00}}$. Thus the lower moment features show the dimensional and symmetrical properties. Therefore they are widely used as statistical interpretation of a distribution function.

As mentioned before, in the feature space we can easily normalize size. Specifically, the size factor λ is introduced, and a given image $f(x, y)$ is changed as $f(x/\lambda, y/\lambda)$. Thus the moment is represented as

$$\mu'_{ij} = \int\int_R f(x/\lambda, y/\lambda) x^i y^j \, dx \, dy \tag{7.7}$$

If we take the range R to include the transformed image, μ'_{ij} is further represented as

$$\mu'_{ij} = \lambda^{i+j+2} \int\int_R f(X, Y) X^i Y^j \, dX \, dY, \tag{7.8}$$

where $X = x/\lambda$.

Thus

$$\mu'_{ij} = \lambda^{2+i+j} \mu_{ij}, \tag{7.9}$$

and in particular

$$\mu'_{00} = \lambda^2 \mu_{00}. \tag{7.10}$$

Therefore the following equation holds:

$$\frac{\mu'_{ij}}{(\mu'_{00})} = \lambda^{i+j} \frac{\mu_{ij}}{\mu_{00}},$$

and so we can define the following new μ'_{ij} as a *size invariant moment,* if μ'_{00} is normalized so that $\mu'_{00} = 1$ holds:

$$\mu'_{ij} = \frac{\mu_{ij}}{{\mu_{00}}^{(i+j+2)/2}} \tag{7.11}$$

For the rectangle, μ'_{ij} is obtained as

$$\mu'_{ij} = \begin{cases} \dfrac{(a/b)^{(i-j)/2}}{4^{(i+j)/2}\,(i+1)(j+1)}, & i \text{ and } j \text{ are even,} \\ 0 & \text{otherwise.} \end{cases} \tag{7.12}$$

In particular, for μ'_{i0}, μ'_{0j}, and μ'_{ii}, they are calculated as

$$\mu'_{i0} = \frac{(a/b)^{i/2}}{4^{i/2}(i+1)},$$

$$\mu'_{0j} = \frac{(b/a)^{j/2}}{4^{j/2}(j+1)}, \tag{7.13}$$

$$\mu'_{ii} = \frac{1}{4^i(i+1)^2}.$$

We note here that $\mu'_{ij} \cdot (i+j)$ depend on the ratio between a and b, and so they are size invariant due to the normalization. On the other hand, there is no contribution of μ'_{ii} for differentiating a set of figures of rectangles. We can see the behavior of the set of rectangles on the feature plane (μ'_{20}, μ'_{02}) which is shown in Fig. 7.1. The features are constrained to the curve $\mu'_{20} = K/\mu'_{20}$, where the coefficient K is $1/4^2 3^2$. (However, it is taken as 1 in the figure.) Thus the features are represented continuously on the feature plane, which is what we expected.

So far we have examined a set of rectangles. Now let us consider a set of ellipsoids, whose contour is represented as

$$\frac{x^2}{a^2} + \frac{y^2}{b^2} = 1. \tag{7.14}$$

The ellipsoid is filled with uniform density 1, as were the rectangles. The central moments are

$$\mu^e_{ij} = \begin{cases} \dfrac{2(b/a)^{j+1}}{(j+1)\int_{-a}^{a} x^i(\sqrt{a^2 - x^2})^{j+1}dx}, & j = \text{even}, \\ 0 & \text{otherwise}. \end{cases} \tag{7.15}$$

Specifically, the lower moments are given as

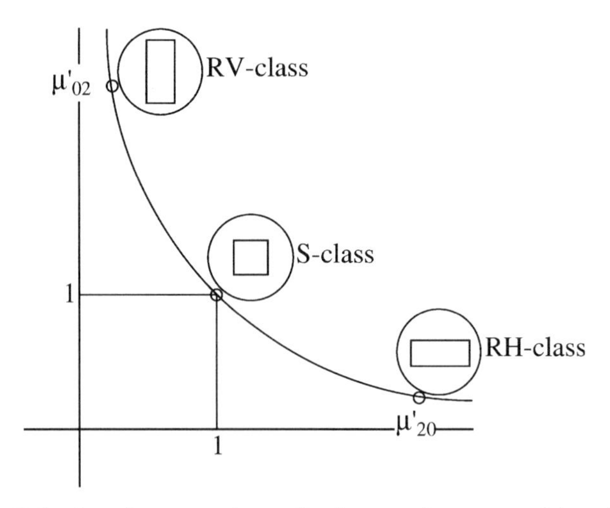

FIGURE 7.1 Behavior of a rectangle on the feature plane spanned by μ'_{20}–μ'_{02} feature moments, where S-class, RH-class, and RV-class denote square class, rectangle class elongated horizontally, and rectangle class elongated vertically, respectively.

$$\mu^e_{00} = \pi ab,$$
$$\mu^e_{10} = \mu^e_{01} = \mu^e_{11} = 0,$$
$$\mu^e_{20} = \frac{\pi a^3 b}{4}, \tag{7.16}$$
$$\mu^e_{02} = \frac{\pi a b^3}{4}.$$

We can see that they are very similar to the corresponding moments of the rectangles. Therefore the differentiation between rectangles and ellipsoids seem to be difficult. The key features distinguishing them are corner and roundness. We can see this difficulty in the simulation experiment of developing a rectangle moments, which is shown in Fig. 7.2. The figures are reconstructed rectangles using Nth-order moments. These figures show that in order to recover the original rectangle, the moments must be greater than 10th order. Tenth order means that we need 66 moments. [Generally, Nth order requires $(N+1)(N+2)/2$ moments.] The type of moments we use here are called *Zernike* moments; they will be explained later. Another example is shown in Fig. 7.3 in which Legendre polynomials were used. We see that a 16th order is necessary to recover the original shape of the letter "E." Thus it tells that in order to differentiate between "○" and "□," we need to construct a very high (>66) dimensional feature space. However, we note that the infinite dimension of the functional space is reduced to a finite one, mathematically speaking, and this finiteness (i.e., convergence) is guaranteed. In this regard some mathematical background will be introduced next.

7.2 SOME MATHEMATICAL BACKGROUND OF FUNCTIONAL ANALYSIS

So far we considered two sets of shapes, rectangles and ellipsoids. In general, any set of shapes can be considered to satisfy the following condition, where a shape in the set is denoted as $f(x, y)$:

$$\iint_R f(x, y)^2 dxdy < \infty. \tag{7.17}$$

It is obvious that any character image has this property. In general, functions that satisfies the above condition are called *square summable.* A set of all the square-summable functions is called an L_2 *space.* For simplicity, functions on one-dimension are considered. Specifically, $L_2[a, b]$ denotes L_2 space in which the functions are defined on the domain $[a, b]$. All the continuous functions on the domain are included in $L_2[a, b]$. In this space the distance between any two element functions can be defined as

$$d(f, g) = \|f - g\|, \tag{7.18}$$

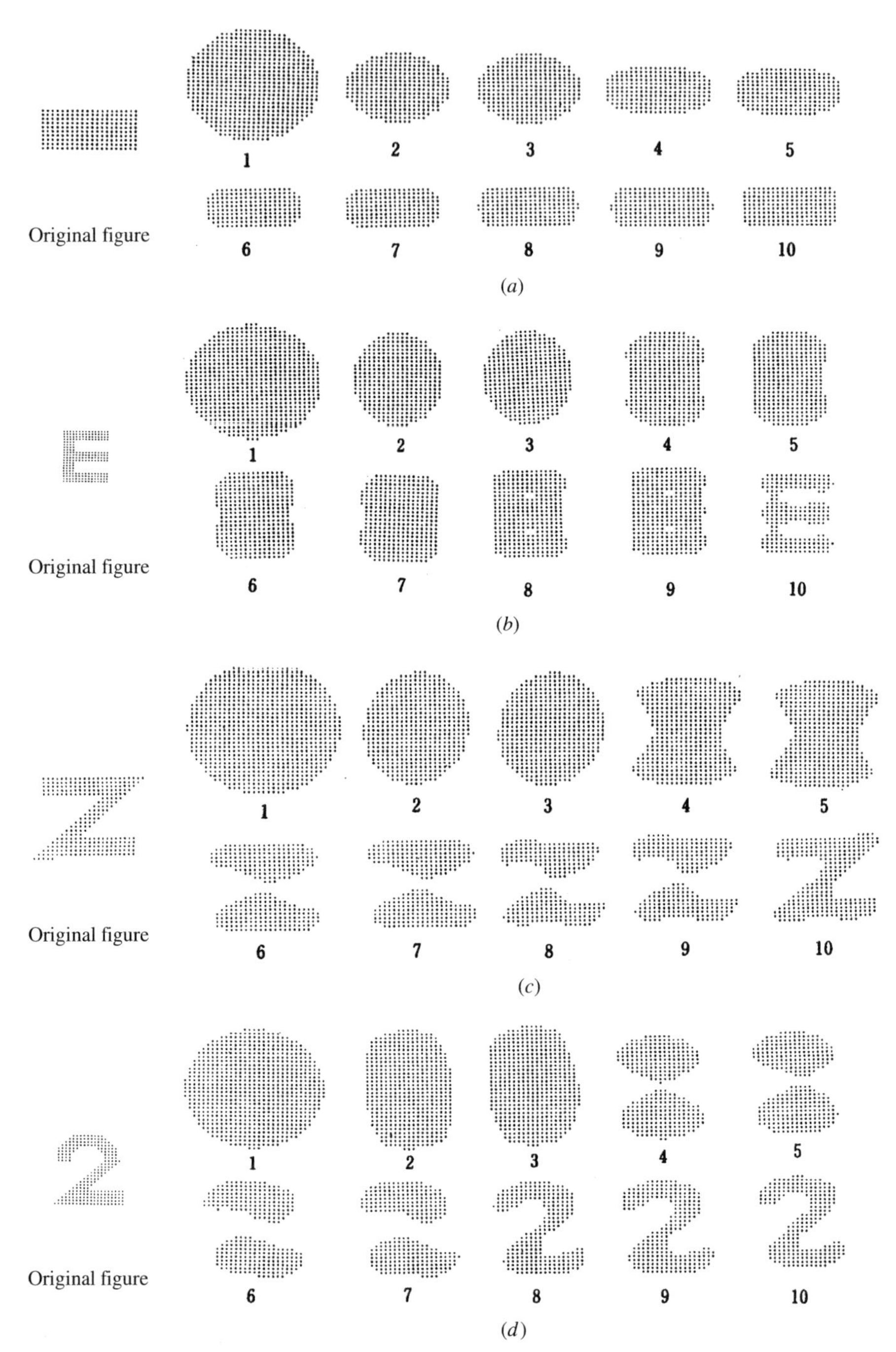

FIGURE 7.2 Some reconstruction experiments based on the Zernike moment. The number below each reconstructed shape indicates the order of the moment used for the reconstructions (*a*), (*b*), (*c*), and (*d*) which correspond, respectively, to shapes of the original rectangle, "E," "Z," and "2." From reference [55], © 1997, Ohm.

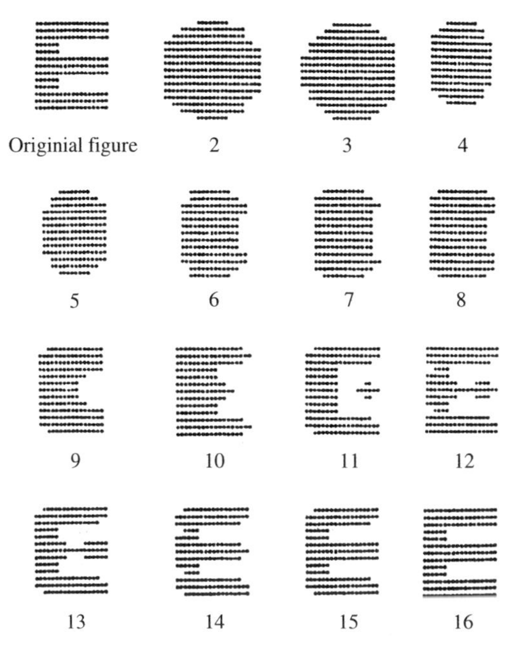

FIGURE 7.3 Reconstruction experimental results using Legendre polynomials; the original image is "E" shown at the top left and the number below each reconstructed shape is the order used for that shape reconstruction. From reference [2], © 1998, *J. Opt. Soc. Am.*

where $\|\cdot\|$ is a *norm,* which is defined by inner product,

$$\langle f, g\rangle \equiv \int_a^b f(t)g(t)dt, \tag{7.19}$$

$$\|f\| \equiv \sqrt{\langle f, f\rangle}. \tag{7.20}$$

Notice that on this space $\|f\| = 0$ does not necessarily imply that $f = 0$, since f may be nonzero on a set of measure zero. The metric space is *complete;* in other words, every Cauchy sequence in the space converges. The sequence $\{f_n\}$ in a metric space $X = (X, d)$ is said to be Cauchy if for every $\varepsilon > 0$ there is an $N = N(\varepsilon)$ such that

$$d(f_m, f_n) < \varepsilon \quad \text{for every} \quad m, n > N.$$

In general, a space in which an inner product is defined and complete is called a *Hilbert space,* denoted as H. Later we will present a beautiful structure of Hilbert space. Here we focus on a remarkable property of Hilbert space, the existence of a total orthogonal sequence, except when $H \neq \{0\}$. Generally, a *total set* in a normed space X is a subset $M \subset X$ whose span is dense in X. That is, M is *total* in X if and only if

$$\overline{spanM} = X.$$

The bar denotes closure. An orthogonal sequence in an inner product space X which is total in X is called a *total orthogonal sequence.* If M is total in X, then there does not exist a nonzero $x \in X$ that is orthogonal to every element of M; that is, if $x \perp M$, then $x = 0$. Another important criterion for totality is that in the *Bessel inequality,* equality holds, namely

$$\sum_k |\langle f, e_k \rangle|^2 = \|f\|^2, \tag{7.21}$$

which is called a *Parseval relation,* where $\{e_k\}$ is an orthogonal sequence. Furthermore, if H contains an orthogonal sequence that is total in H, then H is *separable.* X is said to be separable if it has a countable subset that is dense in X. If H is separable, every orthogonal sequence in H is countable. Thus we can consider the orthogonal sequence as an orthogonal basis for H, which makes infinite-dimensional H analogous to finite-dimensional Hilbert space. The sequence, $1, x, x^2, \ldots$ belongs to $L_2[-1, 1]$ and is total but not orthogonal. However, it is easily orthogonalized by the *Gram-Schmitt process.* The resulting orthogonal sequence is well-known as *Legendre polynomials.* The most popular orthogonal sequences are

$$\sin n\pi x, \quad \cos n\pi x \quad (n = 1, 2, \ldots)$$

and

$$e^{in\pi x} \quad (n = 0, \pm 1, \pm 2, \ldots),$$

which belong to $L^2[0, 1]$. We notice here that the above $L^2[0, 1]$ is extended to complex space to include $e^{in\pi x}$ in which the inner product is defined as

$$\langle f, g \rangle = \int_0^1 f(x)\overline{g(x)}dx. \tag{7.22}$$

7.3 INVARIANT FEATURES AND ZERNIKE MOMENTS

Moments are invariant to shift transformations, as we have already shown for the central moment, μ_{ij}. The explicit relations between μ_{ij} and M_{pq} are

$$\mu_{ik} = \sum_{r=0}^{j} \sum_{s=0}^{k} {}_jC_r \, {}_kC_r (-\bar{x})^{j-r} (-\bar{y})^{k-s} M_{rs}, \tag{7.23}$$

where

$$\bar{x} = \frac{M_{10}}{M_{00}}, \quad \bar{y} = \frac{M_{01}}{M_{00}}. \tag{7.24}$$

That is, normalization is performed in the feature space.

We have already mentioned invariant moments for size-transformation. However, another invariant size moment is also used, which is given as

$$\mu'_{ij} = \frac{\mu_{ij}}{(\mu_{20} + \mu_{02})^{(i+j+2)/4}}. \tag{7.25}$$

This can be derived in the same manner as the derivation of (7.11) based on (7.9) and (7.10), rotation and reflection transformation. We can construct a very nice orthogonal sequence of *Zernike moments* [1], such as Teague used to construct *invariant features for rotation and reflection* [2]. We will introduce his work. As mentioned before, the basic function series of moments $\{x^i y^i\}$, $i, j = 0, 1, 2, \ldots$ is total but not orthogonal. Legendre polynomials were generated to be orthogonal. However, to make moments rotation invariant, it is appropriate to change the coordinate system to polar coordinates. In this way the resultant moments are orthogonalized. Thus *Zernike polynomials* $V_{nl}(x, y)$ are defined as

$$\begin{aligned} V_{nl}(x, y) &= V_{nl}(r\cos\theta, r\sin\theta) \\ &= R_{nl}(r)\exp(il\theta), \end{aligned} \tag{7.26}$$

where θ is the angle between the radial direction and y-axis, and

$$R_{nl}(r) = \sum_{s=0}^{(n-|l|)/2} (-1)^s \cdot \frac{(n-s)!}{s!\,([(n+|l|)/2]-s)!\,([(n-|l|)/2]-s)!}\, r^{n-2s} \tag{7.27}$$

The $V_{nl}(x, y)$ satisfy the following orthogonal relations:

$$\iint_R dxdy[V_{nl}(x, y)]^* V_{mk}(x, y) = \frac{\pi}{n+1}\,\delta_{mn}\delta_{kl}, \tag{7.28}$$

where R is the region $x^2 + y^2 \leq 1$.

An image $f(x, y)$ is expanded using the Zernike polynomials, namely

$$f(x, y) = \sum_n \sum_l A_{nl} V_{nl}(r, \theta), \tag{7.29}$$

where $n = 0, 1, 2, \ldots$ and l is an integer satisfying

$$|l| \leq n \quad \text{and} \quad (n - |l|) \quad \text{even}.$$

The coefficients A_{nl} are called Zernike moments and given by the following formula:

$$\begin{aligned} A_{nl} &= \frac{n+1}{\pi} \iint_R dxdy\, f(x, y) V^*_{nl}(r, \theta) \\ &= \frac{n+1}{\pi} \iint_R rdrd\theta\, f(r, \theta) R_{nl} e^{-il\theta}. \end{aligned} \tag{7.30}$$

According to (7.27), $R_{n,l} = R_{n,-l}$ and $f(x, y)$ is a real valued function.
Therefore

$$A_{nl} = (A_{n,-l})^* \tag{7.31}$$

holds.

Naturally the Zernike moments are related to the ordinary ones. For the lower orders, A_{nl} are shown below:

$$\begin{aligned}
A_{00} &= \frac{\mu_{00}}{\pi} = \frac{1}{\pi}, \\
A_{11} &= A_{1,-1} = 0, \\
A_{22} &= \frac{3(\mu_{02} - \mu_{20} - 2i\mu_{11})}{\pi}, \\
A_{20} &= \frac{3(2\mu_{20} + 2\mu_{02} - 1)}{\pi}.
\end{aligned} \tag{7.32}$$

Now let us see the behavior of the Zernike moments under rotation transformation. Suppose that $x - y$ coordinates are rotated by θ_0 counterclockwise. Then an image $f(r, \theta)$ is transformed to $f(r, \theta - \theta_0)$. Therefore the transformed A_{nl}, denoted as A'_{nl}, is given by

$$\begin{aligned}
A'_{nl} &= \frac{n+1}{\pi} \iint_R r dr d\theta \, f(r, \theta - \theta_0) R_{nl}(r) \exp(-il\theta) \\
&= A_{nl} \exp(-il\theta_0).
\end{aligned} \tag{7.33}$$

This shows the simple relation between A_{nl} and A'_{nl}. That is, A'_{nl} is obtained by rotating original A_{nl} by θ_0 inversely in complex plane.

For the reflection, a simple relation can be obtained. The point $P(r, \theta)$ in $r - \theta$ coordinate system is reflected about the axis that passes through the origin and angle θ_0 counterclockwise to the y-axis as shown in Fig. 7.4. The resultant point is denoted $P'(r', \theta')$. Then

$$r' = r, \quad \theta' = 2(\theta_0 - \theta) + \theta = 2\theta_0 - \theta,$$

hold as illustrated in the figure.

The new Zernike moment A'_{nl} is given by

$$\begin{aligned}
A'_{nl} &= \frac{n+1}{\pi} \iint_R dx dy \, f(x, y) R_{nl} \exp[-il(2\theta_0 - \theta)] \\
&= A^*_{nl} \exp(-i2l\theta_0).
\end{aligned} \tag{7.34}$$

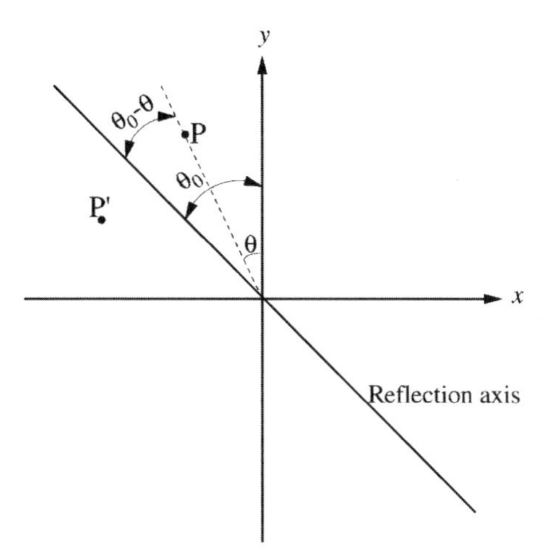

FIGURE 7.4 Reflection and related rotation.

According to the simple relationships between A_{nl} and A'_{nl}, we can derive invariant features easily. A_{nl} is a complex number, so it is represented as

$$A_{nl} = |A_{nl}| \exp(i\phi_{nl}). \tag{7.35}$$

First of all, A_{00}, $A_{11} \times A_{1,-1} = |A_{11}|^2$ are constants regardless of the given image, as seen from (7.32), and so they can be neglected. From now on the invariant features for both rotation and reflection are denoted as S_i and found from the lower-order moments. Now A_{20} and A^*_{20} are invariant for rotation and reflection, respectively, but A_{20} is equal to A^*_{20} according to (7.32). Thus the invariant feature S_1 is obtained as

$$S_1 = A_{20}. \tag{7.36}$$

In general, $A'_{nl} \times A'_{n,-l}$ cancel to each other in their phase parts:

For rotation: $A'_{nl} \times A'_{n,-l} \rightarrow A_{nl} \times A_{n,-l}$.
For reflection: $A'_{nl} \times A'_{n,-l} \rightarrow A^*_{nl} \times A^*_{n,-l}$.

However, these two values are the same according to (7.31), which is $|A_{nl}|^2$. Thus the following invariant features are derived:

$$S_2 = A_{2,2} \times A_{2,-2} = |A_{2,2}|^2,$$
$$S_3 = A_{3,3} \times A_{3,-3} = |A_{3,3}|^2,$$
$$S_4 = A_{3,1} \times A_{3,-1} = |A_{3,1}|^2. \tag{7.37}$$

On the other hand, $A'_{3,3} \times (A'_{3,-1})^3$ can be canceled in their phase parts:

For rotation: $A'_{3,3} \times (A'_{3,-1})^3 \rightarrow A_{3,3} \times (A^*_{3,1})^3$.
For reflection: $A'_{3,3} \times (A'_{3,-1})^3 \rightarrow A^*_{3,3} \times (A_{3,1})^3$.

They are related by a complex conjugate. Therefore their real part can be invariant, namely

$$S_5 = A_{3,3}(A^*_{3,1})^3 + \mathrm{CC}$$
$$= 2|A_{3,3}||A_{3,1}|^3 \cos(\phi_{33} - 3\phi_{31}), \tag{7.38}$$

where CC denotes a complex conjugate.

In the same manner, higher-order invariant features can be obtained. The invariant features can be represented by ordinal moments, which are shown below:

$$S_1 = A_{20} = \frac{3\{2(\mu_{20} + \mu_{02}) - 1\}}{\pi}$$

$$= |A_{2,2}|^2 = \frac{9\{(\mu_{20} - \mu_{02}) + 4(\mu_{11})^2\}}{\pi^2},$$

$$S_3 = |A_{3,3}|^2 = \frac{16\{(\mu_{03} - 3\mu_{21})^2 + (\mu_{30} - 3\mu_{12})^2\}}{\pi^2},$$

$$S_4 = |A_{3,1}|^2 = \frac{144\{(\mu_{02} + \mu_{21})^2 + (\mu_{30} + \mu_{12})^2\}}{\pi^2},$$

$$S_5 = (A_{3,3})^* \times (A_{3,1})^3 + \mathrm{CC}$$

$$= \frac{13824}{\pi^4}[(\mu_{03} - 3\mu_{21})(\mu_{03} + \mu_{21})\{(\mu_{03} + \mu_{23})^2$$

$$- 3(\mu_{30} + \mu_{12})^2\} - (\mu_{30} - 3\mu_{12})(\mu_{30} + \mu_{12})$$

$$\times \{(\mu_{30} + \mu_{12})^2 - 3(\mu_{03} + \mu_{21})^2\}],$$

$$S_6 = (A_{3,1})^2 A^*_{22} + \mathrm{CC}$$

$$= \frac{864}{\pi^3}[(\mu_{02} - \mu_{20})\{(\mu_{03} + \mu_{21})^2 - (\mu_{30} + \mu_{12})^2\}$$

$$+ 4\mu_{11}(\mu_{03} + \mu_{21})(\mu_{30} + \mu_{12})]. \tag{7.39}$$

For further invariant features see the referenced paper [2]. A simulation experiment was performed to confirm the above results for a 20×10 mesh rectangle rotated by 0, 45, and 90 degrees. Results are shown in Table 7.1.

TABLE 7.1 Calculated Values of Invariant Features $S(i)$, $i = 1, 2 \cdots 11$ of a Rectangle of Width 10 and Length 20 for Rotations, 0°, 45°, and 90°

Rotation Angle	0°	45°	90°
S(1)	–0.55864	–0.55864	–0.55864
S(2)	0.01424	0.01424	0.01424
S(3)	0.00000	0.00000	0.00000
S(4)	0.00000	0.00000	0.00000
S(5)	0.00000	0.00000	0.00000
S(6)	–0.00000	–0.00000	–0.00000
S(7)	0.00033	0.00033	0.00033
S(8)	0.09023	0.09024	0.09023
S(9)	0.24341	0.24340	0.24341
S(10)	0.00329	0.00329	0.00329
S(11)	–0.07171	–0.07171	–0.07171

Source: From reference [55], © 1997 Ohm.

Note: The unit of width and length is one pixel.

7.3.1 Pseudo-Zernike Moments

We can construct a variant of Zernike moments defined as

$$R_{nl}(r) = \sum_{s=0}^{n-|l|} (-1)^l \cdot \frac{(2n+1-S)!}{s!(n-|l|-s)!(n+|l|+1-s)!} r^{n-s}, \tag{7.40}$$

where $n = 0, 1, 2, \ldots, \infty$ and l takes on positive and negative integer values subject to $|l| \leq n$ only [3]. The Zernike moments become *pseudo-Zernike moments* as we replace its radial polynomials $\{R_{nl}(r)\}$ in (7.27) with those of (7.40). The pseudo-Zernike polynomial are also a complete set of orthogonal functions on the unit disk. Therefore a given image $f(x, y)$ is expanded as shown in (7.29), and on eliminating the condition $n - |l| = even$. Then $\{A_{nl}\}$ and $\{V_{nl}(x, y)\}$, become the pseudo-Zernike moments and the pseudo-Zernike polynomials, respectively. It has been shown that the pseudo-Zernike moments are less sensitive to image noise than the conventional Zernike moments [4]. Actually a comparison study has showed that pseudo-Zernike moments give better recognition results than the conventional Zernike moments [5]. Next we will demonstrate this.

7.3.2 Experiment Using Rotation Invariant Features

Suppose that we want to construct a rotation invariant recognition OCR system that can be used in a robot [5]. The characters labeled on the objects can be rotated by any number of angles. In the experiment, three kinds of moments were used, Zernike moments (ZM), pseudo-Zernike moments (PM), and *rotational moments* (RM). ZM and PM were described before. The RM of order n is defined as

$$D_{nl} = \int_0^{2\pi} \int_0^{\infty} r^n e^{-il\theta} f(r\cos\theta, r\sin\theta) r dr d\theta, \tag{7.41}$$

where $n = 0, 1, 2, \cdots, \infty$ and l takes on any positive and negative integer values. For details, see the Teh and Chin [4].

The moments used are as follows:

1. For ZM, $n = 2, \ldots, 7$, $n \geq l \geq 0$, $n - l =$ even.
2. For PM, $n = 2, \ldots, 5$, $n \geq l \geq 0$.
3. For RM, $n = 2, \ldots, 5$, $n \geq l \geq 0$.

The character set used has numerals of four kinds of fonts which are almost the same in their shape and size. The character images were sampled by 64×64 and binarized.

For each standard character image, called NORM, some deformations were performed. That is, skew, expansion, and vertical expansion by 8%, respectively. These deformed images are abbreviated as SKEW, HORI, and VETI. The 8% deformation corresponds to four pixels on the sampled plane. Each deformed image was rotated randomly as follows:

NORM $\rightarrow$ 50°, 330°, 55°.
SKEW $\rightarrow$ 290°, 230°, 150°.
HORI $\rightarrow$ 270°, 110°, 120°.
VETI $\rightarrow$ 20°, 280°, 285°.

For numeral "0," the deformed and rotated images are shown in Fig. 7.5. There are also additive noise images generated. That is, the SNRs are 50 db and 25 db, which correspond to 0.32% and 5.6% white-black reversals, respectively. For the learning, no rotation images are used, 4×10 kinds. The rest were used as test samples.

The discrimination was performed by a conventional neural network of three layers. This neural network can be regarded as a nonlinear discrimination system. The experimental results are shown in Table 7.2 and Table 7.3. From the experiment we can conclude that a *rotation invariant character recognition system* can be constructed in a practical sense for printed character sets that are not very different from each other. In general, it also shows the importance of feature extraction. The recognition system gives a good example of a combined feature extraction and neural network. It would be very difficult to construct such an invariant recognition system using only a neural network.

7.4 FOURIER EXPANSION

The most popular and important function sequence is the trigonometric function sequence, which belongs to L_2. *Fourier expansion* has very broad application areas,

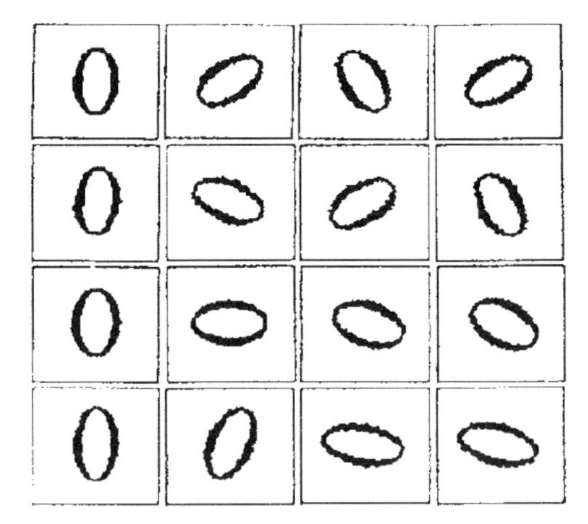

FIGURE 7.5 Sixteen images of character "0." From reference [5].

and there are many excellent books on the subject. Therefore we focus on the feature extraction application to character recognition, skipping the mathematical background. In image processing, area filters are constructed based on Fourier expansion. In the area of character recognition, feature extractions based on Fourier expansion are known as *Fourier descriptors*. One reason why Fourier expansion is used in the feature extraction of shape is that the boundary is closed. The trigonometric function

TABLE 7.2 Correct Recognition Rates (%) for ZM, PM, and RM Feature Extraction Methods Using a Conventional Neural Network

Order of Moments	Zernike Moments	Pseudo-Zernike Moments	Rotational Moments
2	63.3	77.5	
3	79.2	100.0	
4	92.5	100.0	
5	96.6	100.0	87.5

Source: From reference [5].

TABLE 7.3 Effects of Noises on the Recognition Rates for ZM, PM, and RM Feature Extraction Methods Using a Conventional Neural Network

Noise Level	Zernike Moments	Pseudo-Zernike Moments	Rotational Moments
No noise	100.0	100.0	87.5
50 dB	100.0	99.2	40.0
25 dB	78.3	98.3	13.3

Source: From reference [5].

sequence is cyclic and very appropriate to describe a closed boundary efficiently. As will be shown later, this can be compared with moment expansion. However, there is a direct application to the two-dimensional plane, which will be discussed first.

7.4.1 Circular Harmonic Expansion

The polar representation of an image is denoted $f(r, \theta)$. It can be expanded as

$$f(r, \theta) = \sum_{M=-\infty}^{\infty} f_M(r) \exp(iM\theta), \tag{7.42}$$

$$f_M(r) = \frac{1}{2\pi} \int_0^{2\pi} f(r, \theta) \exp(-iM\theta) d\theta. \tag{7.43}$$

As seen above, $f(r, \theta)$ is expanded by keeping r constant as a function of θ. Therefore the Fourier coefficients are a function of r. Such expansion is called *circular harmonic expansion.* The merit of the expansion is that the magnitudes of Fourier coefficients are invariant to the rotation [6]. That is, the clockwise rotated image $f(r, \theta + \alpha)$ by α, is represented as

$$f(r, \theta + \alpha) = \sum_{M=-\infty}^{\infty} f_M(r) \exp(iM\alpha) \exp(iM\theta), \tag{7.44}$$

so the Fourier coefficients of $f(r, \theta + \alpha)$ are

$$f_M(r, \alpha) = f_M(r) \exp(iM\alpha). \tag{7.45}$$

Therefore the magnitudes of the coefficient are the same as the original ones.

Figure 7.6 shows the results of character recognition for related printed characters. The image "E" is rotated by 45°, –90°, and 180°. The rotated images can be recognized. The top half of the figure shows input characters and the bottom half shows the recognition system's output. The strength of the light intensity of each dot shows the certainty of its recognition.

7.4.2 Fourier Descriptor

The application of Fourier expansion to character recognition began early. However, systematic research was first done by Zahn and Roskies in 1972; their result is called the Z&R method [7]. In 1986 another approach taken by Persoon and Fu produced the P&F method [9]. In the Z&R method at each point of a boundary of a given image, a tangential line is drawn which makes an angle with the x-axis and is taken as a dependent variable. However, it becomes discontinuous at the corner points of a polygonal shape, and this makes for slow convergence of the sequence. The P&F method improved this point. Nevertheless, Z&R method gave excellent theoretical perspective on the Fourier descriptor. So we will describe both methods comparatively.

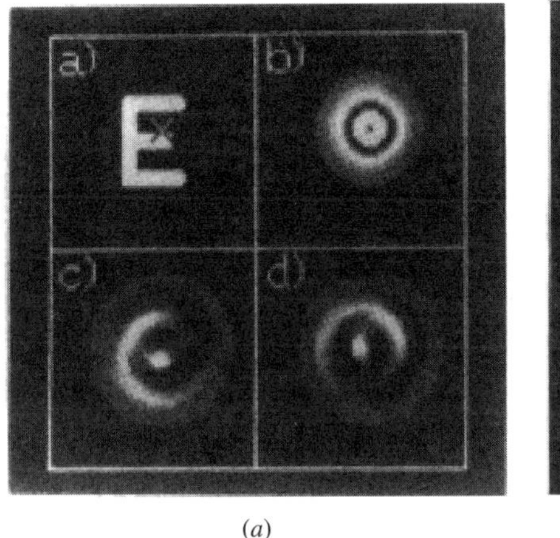

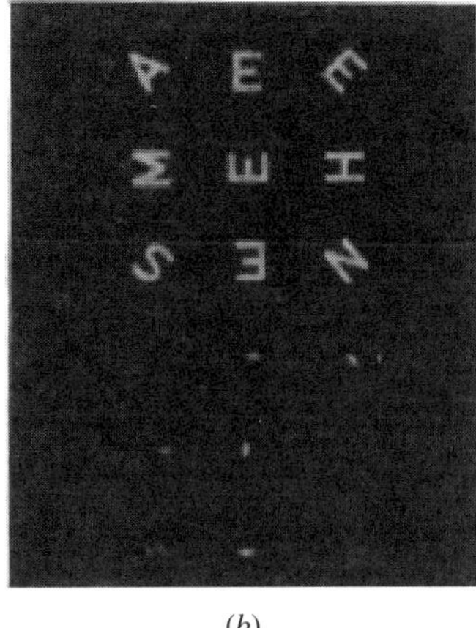

(*a*) (*b*)

FIGURE 7.6 Experimental results of character recognition based on circular harmonic expansion: (*a*) Expansion of an image "E"; (*b*) demonstration of its rotation invariant feature, where the rotated "E" by –45°, 90°, and 180° are recognized as shown in corresponding light marks on the output plane. From reference [6], © 1998, *J. Opt. Soc. Am.*

Shape Representations

Z&R Method Taking a starting point on a boundary of an image, let the length l measured from the starting point to any point on the boundary be an independent variable. Let $z(l)$ be the point at distance l along the boundary. Let $\theta(l)$ be the angle between the x-axis and the line tangent to the curve at $z(l)$, as illustrated by Fig. 7.7. The coordinate of the the boundary point $z(l)$ is a complex number expressed as

$$z(l) = z(0) + \int_0^l \exp\,(i\theta(\lambda))d\lambda, \tag{7.46}$$

which will be derived later.

Since the boundary is a closed curve, $\theta(l)$ is a periodic function whose period is the total length L. The Z&R method is used to expand the periodic function as a Fourier series. Instead of $\theta(l)$, a normalized function is defined first as

$$\phi(l) = \theta(l) - \theta(0), \tag{7.47}$$

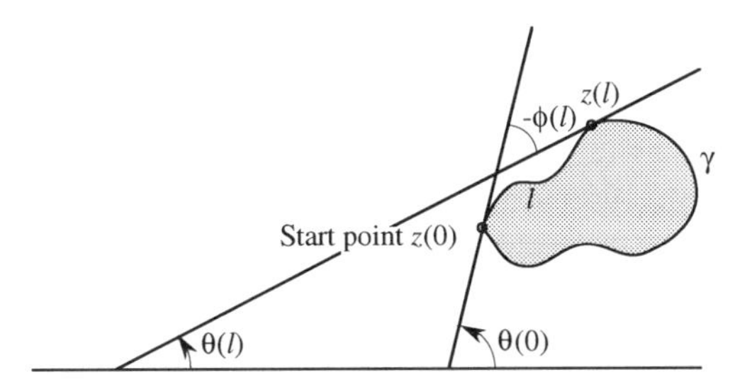

FIGURE 7.7 Notation for shape representation based on the Z&R method.

where

$$\phi(0) = 0,$$
$$\phi(L) = -2\pi. \tag{7.48}$$

So $\phi(l)$ is not periodic. To make $\phi(l)$ periodic, a function $\phi^*(t)$ is defined as

$$\phi^*(t) = \phi(l) + t,$$

where l is set to $Lt/2\pi$ so that

$$\phi^*(t) = \phi\left(\frac{Lt}{2\pi}\right) + t, \qquad t \in [0, 2\pi]. \tag{7.49}$$

Thus $\phi^*(0) = \phi(2\pi) = 0$ as desired.

Consider a circle whose peripheral length is L. Its radius is $L/2\pi$, and so the relation $l = Lt/2\pi$ can be interpreted as the relation between the center angle t and its arc length l spanned by t. If the boundary is a circle, $\phi(l) = -t$ as $\phi^*(t)$ becomes 0. Thus $\phi^*(t)$ can be interpreted as the difference in shape between a given boundary and a circle. The $\phi^*(t)$ depends on the shape of the boundary and is invariant during the shift, rotation, and dilatation/contraction of a given image. However, we need to be careful with this invariance, as will be mentioned later.

The Fourier sequence of $\phi^*(t)$ is given as

$$\phi^*(t) = \mu_o + \sum_{k=1}^{\infty} \{a_k \cos(kt) + b_k \sin(kt)\} \tag{7.50}$$

$$= \mu_o + \sum_{k=1}^{\infty} A_k \cos(kt - \alpha_k), \tag{7.51}$$

$$\mu_o = \frac{1}{2\pi}\int_0^{2\pi} \phi^*(t)dt$$

$$a_k = \frac{1}{\pi}\int_0^{2\pi} \phi^*(t)\cos(kt)dt$$

$$= A_k \cos \alpha_k,$$

$$b_k = \frac{1}{\pi}\int_0^{2\pi} \phi^*(t)\sin(kt)dt$$

$$= A_k \sin \alpha_k. \tag{7.52}$$

Now the boundary curve is γ. $\{A_k, \alpha_k\}$ is defined as the Fourier descriptor of the γ. Thus $\{A_k, \alpha_k\}$ is the polar coordinate expression of the Fourier descriptor $\{a_k, b_k\}$, and the components are called the kth harmonic amplitude and kth phase, respectively.

P&F Method Let the coordinate of a boundary point of a given image at a distance l from a starting point along the boundary and in a clockwise direction be $u(l)$. Then $u(l)$ is expressed on the complex plane as

$$u(l) = x(l) + iy(l). \tag{7.53}$$

It is obvious that $u(l)$ is cyclic. The P&F method is based on a direct Fourier expansion of $u(l)$ in which there is no discontinuous point so long as l is continuous.

Since $u(l)$ is a complex function, its Fourier expansion is expressed by the complex representation

$$u(l) = \sum_{k=-\infty}^{\infty} c_k \exp\left(ik\frac{2\pi}{L}l\right), \tag{7.54}$$

$$c_k = \int_0^L u(l) \exp\left(-i\frac{2\pi}{L}kl\right)dl, \tag{7.55}$$

where L is the period of the $u(l)$.

Curve Reconstruction It is important for a given image to be represented by a small number of Fourier series terms. Fast convergence is required for efficient information compression and feature extraction. Therefore we will give the reconstruction method of γ from the Fourier descriptor so that we can observe the degree of resemblance between the original and reconstructed ones.

Z&R Method Here we give $z(l)$ of (7.46) as a function of $\theta(l)$ first in terms of $\phi(l)$ and then $\phi^*(l)$. That is, $z(l)$ is finally given by the Fourier descriptor.

Figure 7.8 illustrates the differential geometry of two closed points on a curve. The two points are denoted as P and Q and are expressed as $z(l) = x(l) + iy(l)$ and

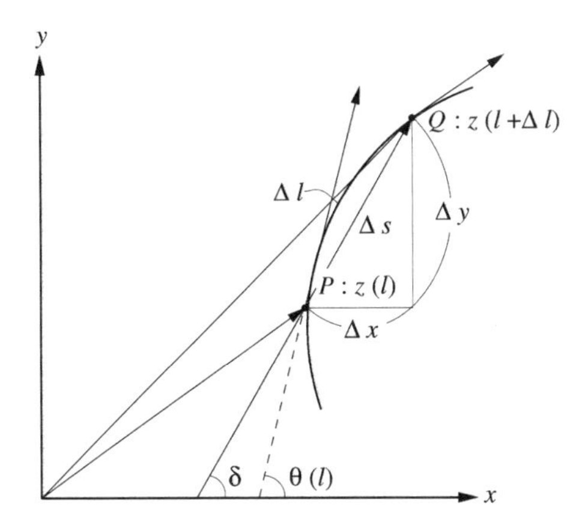

FIGURE 7.8 Differential geometry of a tangential line and a curve.

$z(l + \Delta l)$, respectively, where Δl is the distance from P to Q along the curve. Therefore the distance between P and Q denoted as Δz is expressed as

$$\begin{aligned} \Delta z &= z(l + \Delta l) - z(l) \\ &= \Delta x + \Delta y. \end{aligned} \tag{7.56}$$

The length of the line segment $\overline{PQ}$ is given by

$$\Delta s = \sqrt{(\Delta x)^2 + (\Delta y)^2}. \tag{7.57}$$

Now let the angle between $\overline{PQ}$ and x-axis (positive direction) be δ. The following relations hold:

$$\begin{aligned} \Delta x &= \Delta s \cos \delta, \\ \Delta y &= \Delta s \sin \delta. \end{aligned} \tag{7.58}$$

Taking $\Delta l \to 0$, Δ is rewritten by d so that $ds = dl$, and $\delta = \theta(l)$:

$$\begin{aligned} dx &= x(l + dl) - x(l) = dl \cdot \cos \theta(l), \\ dy &= y(l + dl) - y(l) = dl \cdot \sin \theta(l). \end{aligned} \tag{7.59}$$

Then dx and dy are integrated as

$$\begin{aligned} x(l) - x(0) &= \int_0^l \cos \theta(\lambda) d\lambda, \\ y(l) - y(0) &= \int_0^l \sin \theta(\lambda) d\lambda. \end{aligned} \tag{7.60}$$

Thus we reach the (7.46) which is just a complex representation of the above equations,

$$z(l) = z(0) + \int_0^l \exp(i\theta(\lambda)) d\lambda. \tag{7.61}$$

From the (7.47) and (7.49) we obtain the following relations:

$$\begin{aligned} \theta(l) &= \phi(l) + \theta(0) \\ &= \phi^*(t) - t + \theta(0). \end{aligned} \tag{7.62}$$

Substituting the second equation above for $\phi^*(t)$ in (7.51), changing the independent variable l to t, and noting that $t = (2\pi/L)l$, the following final reconstruction formula is obtained:

$$z(l) = z(0) + \frac{L}{2\pi} \int_0^{2\pi l/L} \exp i\left[-t + \theta(0) + \mu_0 + \sum_{k=1}^{\infty} A_k \cos(kt - d_k)\right] dt. \tag{7.63}$$

P&F Method In this case the reconstruction of a curve γ is simple. Since $u(l) = x(l) + iy(l)$, the following relations are given immediately by (7.54) and (7.55):

$$x(l) = \sum_{k=-\infty}^{\infty} a_k \cos \frac{2\pi kl}{L},$$

$$y(l) = \sum_{k=-\infty}^{\infty} a_k \sin \frac{2\pi kl}{L}. \qquad (7.64)$$

Usually the original closed curve γ is approximated by some finite number of Fourier series terms, say $k \leq N$. In the case of the Z&R method, the reconstructed curve fails to be closed if the number of the terms is not enough. In contrast, in the P&F method the reconstructed curve is always closed regardless of Fourier convergence because the approximated sequence always has period L regardless of N in the direct expansion of boundary coordinates.

Some General Relations between Shape and Fourier Descriptors Now we will mention the relations between the shape of a closed curve γ and its Fourier descriptor.

Z&R Method Suppose that we have a pair of simple closed curves γ and γ' whose shapes are the same but only their position $z(0)$, rotation angle $\theta(0)$, and L are different. Then the relations

$$A_k' = A_k, \quad \alpha_k' = \alpha_k \qquad (7.65)$$

hold. This means that the Fourier descriptor is invariant for shift, rotation, and size of a given image. However, we must note that there is a subtle condition: No change of relative starting point. This is not easily satisfied in practice. Therefore we investigate the case where the relative starting points are different by Δl for the same curves γ and γ'. Then the following relations can be derived:

$$A_k' = A_k$$

$$\alpha_k' = \alpha_k + k\Delta\alpha,$$

$$\Delta\alpha = \frac{-2\pi\Delta l}{L}$$

$$\mu_0' = \mu_0 + \theta(0) - \theta'(0) + \Delta\alpha. \qquad (7.66)$$

These relations tell us that the phase α_k depends on not only the shape but also on the starting point, although the amplitudes are kept the same. This result meets our intuition for trigonometric functions.

To obtain stable features that are invariant of the starting point, we need to define another appropriate phase feature as follows:

$$F_{kj}[\gamma] = j^*\alpha_k - k^*\alpha_j, \tag{7.67}$$

where j^* and k^* are given as

$$j^* = \frac{j}{\gcd(j,\, k)}, \quad k^* = \frac{k}{\gcd(j,\, k)}. \tag{7.68}$$

Here $\gcd(j, k)$ denotes greatest common divisor of j and k. The invariance of $F_{kj}[\gamma]$ is given as

$$\begin{aligned} F_{kj}[\gamma'] &= j^*\alpha'_k - k^*\alpha'_j \\ &= \frac{j(\alpha_k + k\Delta\alpha) - k(\alpha_j + j\Delta\alpha)}{\gcd(j,\, k)} \\ &= \frac{j\alpha_k - k\alpha_j}{\gcd(j,\, k)} \\ &= j^*\alpha_k - k^*\alpha_j \\ &= F_{kj}[\gamma]. \end{aligned}$$

Now we look into the case of symmetric curves. First, let us consider two curves γ and γ' that are mirror images and have identical starting points, as shown in Fig. 7.9. The following relations hold:

$$A'_k = A_k, \quad \alpha'_k + \alpha_k \equiv \pi \pmod{2\pi}, \quad \mu'_0 + \mu_0 = -\Delta\phi_0, \tag{7.69}$$

where $\Delta\phi_0$ is the bend at the starting point.

Next we move to investigate the effect on Fourier descriptors of the axial and rotational symmetries of a single curve. For the axial symmetry, we have two starting points that are a mirror image to each other, for which $\alpha'_k = \alpha_k + n\Delta\alpha$ holds in general. We can still apply the relation $\alpha'_k + \alpha_k = \pi$, so the following relation holds for a single curve:

$$2\alpha_k = \pi - k\Delta\alpha. \tag{7.70}$$

We can normalize the expression of the phase shift of the starting point by $\beta = \Delta\alpha/2 = (\pi - 2\alpha_k)/(2k)$ using the above equation, and we can substitute β for $\Delta\alpha$ in the rotation of $\alpha'_k = \alpha_k + k\Delta\alpha$ in (7.66). The resulting normalized $\{\alpha'_k\}$ has the following simple property:

$$\alpha_k \equiv \frac{\pi}{2} \pmod{\pi}. \tag{7.71}$$

Here the meaning of normalization is that the starting point lies on an axis of symmetry; see Fig. 7.8.

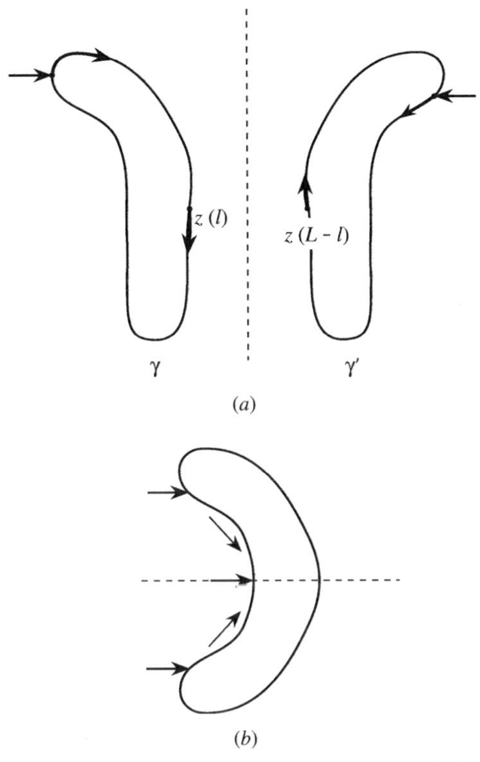

FIGURE 7.9 Starting point and symmetry in curves: (*a*) γ and γ' are reflections of one another and have the same relative starting point as shown by the arrows. (*b*) When the curve has a symmetry axis (dotted line), the starting point can be moved to a point on the axis.

Concerning rotational symmetry, a curve having k-fold rotational symmetry has the following property:

$$A_n = 0 \quad \text{for} \quad n \not\equiv 0 \pmod{k}; \tag{7.72}$$

that is, the harmonic amplitudes of the curve become zero save those where the harmonic members are just multiple of k, and this meets our intuition.

P&F Method We can expect fast convergence of the Fourier series when $u(l)$ is continuous. The reconstruction of a γ that is described by the Fourier descriptor is simple. However, $u(l)$ is complex, and therefore $c_k^* \neq c_{-k}$. We need to note that

$$du = dx + idy = \cos\theta(l)dl + i\sin\theta(l)dl \tag{7.73}$$

holds, so

$$\left|\frac{du}{dl}\right| = \frac{du}{dl} \cdot \frac{d^*u}{dl} = 1 \qquad (0 \le l \le L) \tag{7.74}$$

must be satisfied. This sets limitations on $\{c_k\}$; that is, a partial sum of $u(l)$

$$\sum_{k=-M}^{M} c_k \exp\left(ik\frac{2\pi}{L}l\right) \tag{7.75}$$

must be substituted in the equation above. Then we find that except for $k \neq 0$, all the rest of the c_k's are zero, which means that the equation above is only satisfied by a circle. Therefore we reconsider l as a parameter to avoid this problem.

In this regard we will be freer if the curve γ is constructed by two independent vectors that are a function of parameter t,

$$\mathbf{V}(t) = \begin{Bmatrix} x(t) \\ y(t) \end{Bmatrix}.$$

An example of this expression is given later.

For a linear image that neither loops nor crosses over itself, simple relations using the Fourier descriptor can be derived. As shown in Fig. 7.10, such line necessarily has end points, one of which is denoted E. A starting point S is located counterclockwise away from E by $\alpha/2$, as shown in the figure. Let the point distance l from E be P and the counter point of P be Q. When the width of the line becomes zero, P and Q coincide. We will derive the property of the line only in the ideal case where the width is zero. For the other case where the width is not zero, refer to the paper [7]. For the ideal case, then, the coordinates of P and Q are equal, so

$$u(l) = u(L - l + \alpha) \tag{7.76}$$

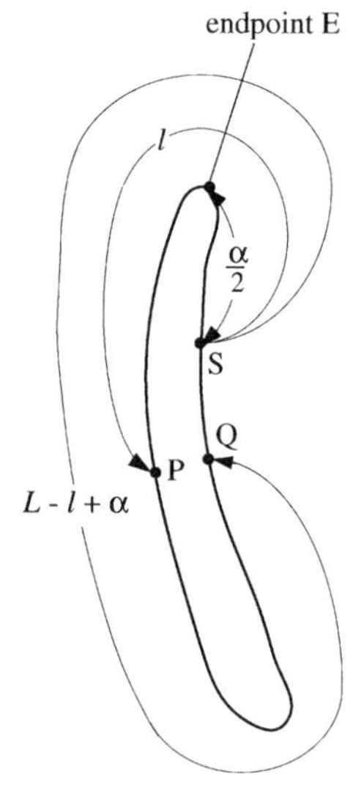

FIGURE 7.10 Notation used to find features of a line based on a Fourier descriptor.

holds and the Fourier descriptor c_k can be given by

$$c_k = c_{-k} \exp\left(-ik\frac{2\pi}{L}\alpha\right). \tag{7.77}$$

If the starting point is taken at the end point, then $\alpha = 0$ and the following simple relation is derived:

$$c_k = c_{-k}. \tag{7.78}$$

Fourier Descriptors for Polygonally Approximated Curves An image can be polygonally approximated as we will describe later in Chapter 8. Here we give the calculations of the Fourier descriptors for both feature extraction methods.

Z&R Method A given polygonal curve is assumed to have m vertices $V_0, V_1, \ldots, V_{m-1}$, and each edge (V_{i-1}, V_i) has length Δl_i. The change in angular direction at vertex V_i is $\Delta\phi_i$ as shown Fig. 7.11. By these definitions, $\phi(l)$ can be represented as

$$\phi(l) = \begin{cases} \sum_{i=1}^{k} \Delta\phi_i & \text{for } \sum_{i=1}^{k} \Delta l_i \le l < \sum_{i=1}^{k+1} \Delta l_i, \\ 0 & \text{for } 0 \le l < \Delta l_1, \\ -2\pi & \text{for } l_m \le l < l_m + \varepsilon, \end{cases} \tag{7.79}$$

where ε is a very small constant. Now we have

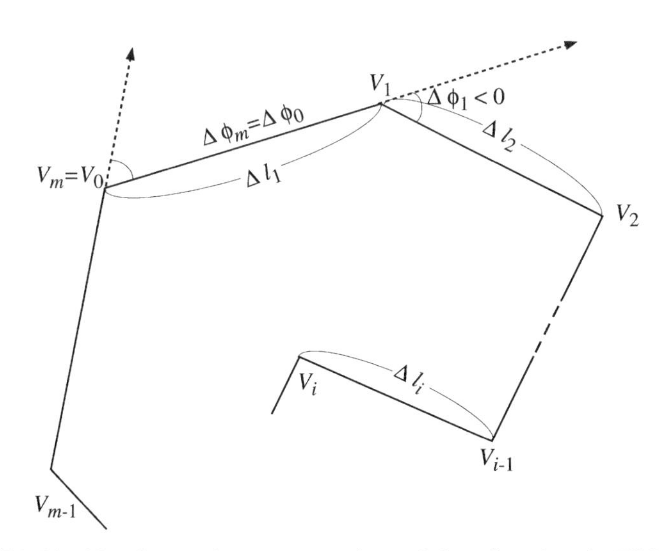

FIGURE 7.11 Notation used to extract a polygonal shape based on the Z&R method.

$$l_k = \sum_{i=1}^{k} \Delta l_i,$$

$$l_m = L,$$

$$\sum_{i=1}^{m} \Delta \phi_i = -2\pi. \tag{7.80}$$

Then, setting l by $\lambda = Lt/2\pi$, we calculate μ_0, a_n, and b_n for the polygonal curve as follows:

$$\begin{aligned}
\mu_0 &= \frac{1}{2\pi} \int_0^{2\pi} \phi^*(t)dt \\
&= \frac{1}{L} \int_0^{L+\varepsilon} \phi(\lambda)d\lambda + \pi \\
&= \frac{\int_{l_1}^{l_2} \phi(\lambda)d\lambda + \int_{l_2}^{l_3} \phi(\lambda)d\lambda + \cdots + \int_{l_m}^{l_{m+\varepsilon}} \phi(\lambda)d\lambda}{L} + \pi, \\
&= \frac{(\{\Delta\phi_1(l_2 - l_1) + (\Delta\phi_1 + \Delta\phi_2)(l_3 - l_2) + \cdots + (\Delta\phi_1 + \cdots + \Delta\phi_m)(l_{m+\varepsilon} - l_m)\})}{L} + \pi, \\
&= \frac{-\Delta\phi_1 l_1 - \Delta\phi_2 l_2 - \cdots - \Delta\phi_m l_m - 2\pi(L+\varepsilon)}{L} + \pi, \\
&= -\frac{1}{L} \sum_{k=1}^{m} l_k \Delta\phi_k + \pi, \qquad \text{setting } \varepsilon \to 0; \\
a_n &= \frac{1}{\pi} \int_0^{2\pi} \phi^*(t) \cos(nt)dt, \\
&= \frac{1}{\pi} \int_0^{L} \left(\phi(\lambda) + \frac{2\pi\lambda}{L} \right) \cos \frac{2\pi n\lambda}{L} d\lambda, \\
&= \frac{-1}{n\pi} \sum_{k=1}^{m} \Delta\phi_k \sin \frac{2\pi n l_k}{L}; \\
b_n &= \frac{1}{\pi} \int_0^{2\pi} \phi^*(t) \sin(nt)dt, \\
&= \frac{2}{L} \int_0^{L} \left(\phi(\lambda) + \frac{2\pi\lambda}{L} \right) \sin \frac{2\pi n\lambda}{L} d\lambda, \\
&= \frac{1}{n\pi} \sum_{k=1}^{m} \Delta\phi_k \cos \frac{2\pi n l_k}{L}.
\end{aligned} \tag{7.81}$$

P&F Method The necessary notation is shown in Fig. 7.12. The following definitions apply:

$$l_k = \sum_{i=1}^{k} |\mathbf{V}_i - \mathbf{V}_{i-1}| = \sum_{i=1}^{k} r_i \qquad (k \neq 0),$$

$$l_0 = 0,$$

$$b_k = \frac{\mathbf{V}_{k+1} - \mathbf{V}_k}{|\mathbf{V}_{k+1} - \mathbf{V}_k|}$$

$$= \frac{\mathbf{r}_{k+1}}{r_{k+1}}$$

$$= e^{i\theta_{k+1}}$$

$$l_m = L. \tag{7.82}$$

Then $u(l)$ is described separately as

$$u(l) = V_0 + (l - l_0)e^{i\theta_i} \qquad (0 \leq l < r_1),$$

$$u(l) = V_{m-1} + (l - l_{m-1})e^{i\theta_m} \qquad \left(\sum_{i=1}^{m-1} r_i \leq l < \sum_{i=1}^{m} r_i\right). \tag{7.83}$$

These $u(l)$ representations are substituted for $u(l)$ in (7.54), and the Fourier descriptor a_n is calculated as follows:

$$c_n = \left\{ \int_0^{\gamma_1} (V_0 + (l - l_0) \exp(i\theta_1)) \exp\left(-i\frac{2\pi}{L}nl\right) dl + \cdots \right.$$

$$\left. + \int_{\gamma_1 + \cdots + \gamma_{m-1}}^{\gamma_1 + \cdots + \gamma_m} (V_{m-1} + (l - l_{m-1}) \exp(i\theta_m)) \exp\left(-i\frac{2\pi}{L}nl\right) dl \right\} / L$$

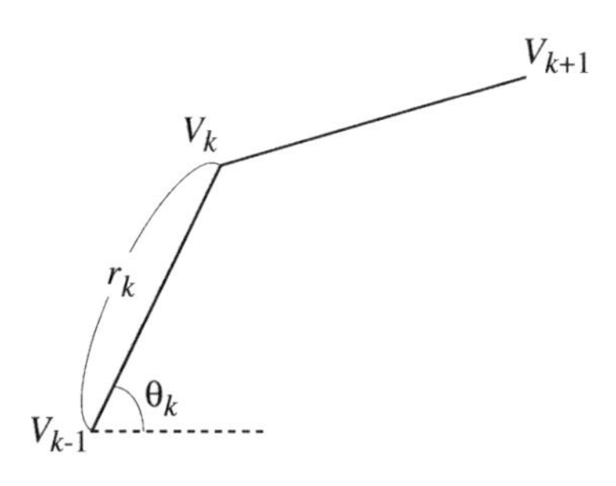

FIGURE 7.12 Notation used to describe polygonal shape based on the P&F method.

$$= \frac{\sum_{k=1}^{m} (b_{k-1} - b_k) \exp\left(-in \dfrac{2\pi}{L} l_k\right)}{L(2\pi n/L)^2},$$

$$c_0 = \frac{1}{L} \int_0^L u(l)dl$$

$$= \left\{ \int_0^{\gamma_1} (V_0 + (l - l_0) \exp(i\theta_1))dl + \cdots \right.$$

$$\left. + \int_{\gamma_1 + \cdots + \gamma_{m-1}}^{\gamma_1 + \cdots + \gamma_m} (V_{m-1} + (l - l_{m-1}) \exp(i\theta_m))dl \right\} / L$$

$$= \frac{\sum_{k=1}^{m} (b_{k-1} - b_k) l_k^2 + b_0 L}{2L} + \frac{\sum_{k=1}^{m} (l_k V_{k-1} - l_{k-1} V_k)}{L}. \tag{7.84}$$

The quantity c_0 is not relevant to the polygon's shape and only related to its position, but it is necessary for its reconstruction.

Reconstruction Experiments So far we have described the theoretical aspects of Fourier descriptors. Now we compare the two methods by reconstruction experiments in which closed curves are approximated by the Fourier descriptors of up to N terms. The comparisons are given visually. Figure 7.13 shows the results of reconstructions by the two methods Z&R and P&F for $N = 5$, 10, and 50. The original images are binary patterns on a 64×64 plane which are represented by polygonal curves. Obviously the P&F method is better than Z&R in terms of convergence. In the P&F method, the difference between "2" and "Z" becomes clear with $N \geq 5$. On the other hand, as expected, some closed curves are not closed even if $N = 10$ for the Z&R method. With $N = 50$, the reconstructions are complete for the case of the P&F method. The Z&R method is 1.5 times faster than P&F, using the polygonal reconstruction formulas. The polygonal reconstruction formula is very efficient and so is about one order of magnitude faster than using the ordinal integral formulas.

Overall, we can see that the essential features of the simple curves in the examples cited are described by lower-frequency terms. Compared with the method of moments, convergence is dramatically improved. This point will be further described quantitatively next.

Elliptic Fourier Descriptors Elliptic Fourier descriptors are directly related to Freeman encoding, and this relation is intuitive, as will be seen below. We will describe a practical method devised by F. P. Kuhl and C. R. Giardina [10].

A closed contour $(x(t), y(t))$, $t \in [0, T]$, where T is the length of the contour, has an Nth-order approximation

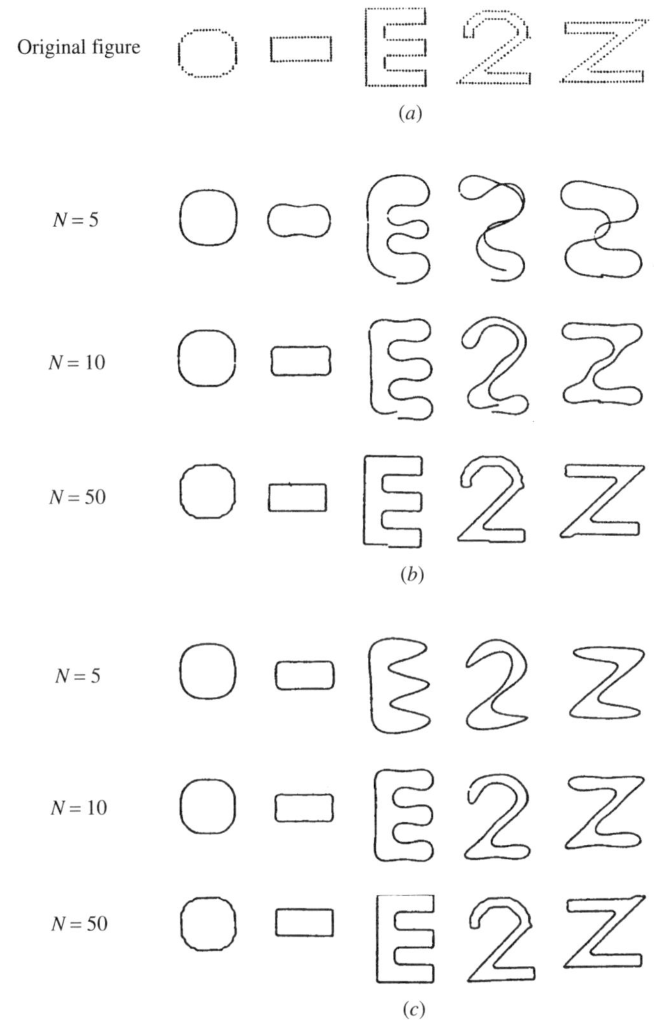

FIGURE 7.13 Some reconstruction experiments using two kinds of Fourier descriptors: (*a*) Original images; (*b*) Z&R; (*c*) P&F. From reference [55], © 1997, Ohm.

$$x(t) = A_0 + \sum_{n=1}^{N} \left[a_n \cos \frac{2n\pi t}{T} + b_n \sin \frac{2n\pi t}{T} \right], \tag{7.85}$$

$$y(t) = C_0 + \sum_{n=1}^{N} \left[c_n \cos \frac{2n\pi t}{T} + d_n \sin \frac{2n\pi t}{T} \right]. \tag{7.86}$$

In Freeman encoding we need to represent a series of discrete points of the contour. However, in Freeman encoding the directions are essentially limited to either horizontal/vertical or diagonal. Therefore, if discrete version of $(x(t), y(t))$ is represented by (x_i, y_i), then x_i, y_i can be expressed as

$$x_i = \sum_{i=1}^{i} \Delta x_i, \tag{7.87}$$

$$y_i = \sum_{i=1}^{i} \Delta y_i, \tag{7.88}$$

where $\Delta x_i = x_i - x_{i-1}$ and $\Delta y_i = y_i - y_{i-1}$. The digitized parametric expression is written as

$$t_i = \sum_{i=1}^{i} \Delta t_i, \tag{7.89}$$

where $\Delta t_i = \sqrt{\Delta x_i^2 + \Delta y_i^2}$.

The Fourier coefficients of the nth harmonic a_n and b_n are most easily found by noticing that $x(t)$ is piecewise linear and continuous for all time (parameter t). This means that the derivative $\dot{x}(t)$ is very easily derived because the time derivative $\dot{x}(t)$ consist of the sequence of piecewise constant derivatives $\Delta x_i/\Delta t_i$ in the interval $t_{i-1} < t < t_i$. Since the time derivative is also periodic with period T, it can be represented by the Fourier series

$$\dot{x}(t) = \sum_{i=1}^{\infty} \left[\alpha_n \cos \frac{2n\pi t}{T} + \beta_n \sin \frac{2n\pi t}{T} \right], \tag{7.90}$$

where

$$\alpha_n = \frac{2}{T} \int_0^T \dot{x}(t) \cos \frac{2n\pi t}{T}\, dt, \tag{7.91}$$

$$\beta_n = \frac{2}{T} \int_0^T \dot{x}(t) \sin \frac{2n\pi t}{T}\, dt. \tag{7.92}$$

Then

$$\begin{aligned} \alpha_n &= \frac{2}{T} \sum_{i=1}^{N} \frac{\Delta x_i}{\Delta t_i} \int_{t_{i-1}}^{t_i} \cos \frac{2n\pi t}{T}\, dt \\ &= \frac{2}{T} \sum_{i=1}^{N} \frac{\Delta x_i}{\Delta t_i} \left(\sin \frac{2n\pi t_i}{T} - \sin \frac{2n\pi t_{i-1}}{T} \right). \end{aligned} \tag{7.93}$$

Likewise β_n is given as

$$\beta_n = -\frac{2}{T} \sum_{i=1}^{N} \frac{\Delta x_i}{\Delta t_i} \left(\cos \frac{2n\pi t_i}{T} - \cos \frac{2n\pi t_{i-1}}{T} \right). \tag{7.94}$$

Then $\dot{x}(t)$ is also obtained directly from its continuous representation of

$$x(t) = A_0 + \sum_{n=1}^{\infty} \left[a_n \cos \frac{2n\pi t}{T} + b_n \sin \frac{2n\pi t}{T} \right], \tag{7.95}$$

as

$$\dot{x}(t) = \sum_{n=1}^{\infty} \left[-\frac{2n\pi}{T} a_n \sin \frac{2n\pi t}{T} + \frac{2n\pi}{T} b_n \cos \frac{2n\pi t}{T} \right]. \tag{7.96}$$

Now we have two expressions of $\dot{x}(t)$, and by equating them, the coefficients are obtained as

$$a_n = \frac{T}{2n^2\pi^2} \sum_{i=1}^{N} \frac{\Delta x_i}{\Delta y_i} \left[\cos \frac{2n\pi t_i}{T} - \cos \frac{2n\pi t_{i-1}}{T} \right], \tag{7.97}$$

$$b_n = \frac{T}{2n^2\pi^2} \sum_{i=1}^{N} \frac{\Delta x_i}{\Delta t_i} \left[\sin \frac{2n\pi t_i}{T} - \sin \frac{2n\pi t_{i-1}}{T} \right]. \tag{7.98}$$

Likewise the coefficients of $y(t)$ are obtained as

$$c_n = \frac{T}{2n^2\pi^2} \sum_{i=1}^{N} \frac{\Delta y_i}{\Delta t_i} \left[\cos \frac{2n\pi t_i}{T} - \cos \frac{2n\pi t_{i-1}}{T} \right], \tag{7.99}$$

$$d_n = \frac{T}{2n^2\pi^2} \sum_{i=1}^{N} \frac{\Delta y_i}{\Delta t_i} \left[\sin \frac{2n\pi t_i}{T} - \sin \frac{2n\pi t_{i-1}}{T} \right]. \tag{7.100}$$

The DC components, A_0 and C_0, can be obtained similarly, but they are more complex and not so important. Therefore they are omitted here.

The truncated Fourier approximation to a contour is expressed as

$$x(t) = A_0 + \sum_{n=1}^{N} X_n, \tag{7.101}$$

$$y(t) = C_0 + \sum_{n=1}^{N} Y_n, \tag{7.102}$$

where the components of the projections X_n, Y_n $(1 \le n \le N)$ are represented as functions of parameter t as

$$X_n(t) = a_n \cos \frac{2\pi nt}{T} + b_n \sin \frac{2\pi nt}{T}, \tag{7.103}$$

$$Y_n(t) = c_n \cos \frac{2\pi nt}{T} + d_n \sin \frac{2\pi nt}{T}. \tag{7.104}$$

The points (X_n, Y_n) all have elliptic loci which are described in detail by C. Giardina and F. Kuhl [10]. This means that the Fourier approximation to the original contour can be viewed as the addition in proper phase relationship of rotating phasers. See the original paper [10] for a detailed illustration. There are many examples of Fourier approximations.

The phase shift from the first major elliptic axis is given by

$$\theta_1 = \frac{1}{2}\tan^{-1}\frac{2(a_1b_1 + c_1d_1)}{a_1^2 + b_1^2 + c_1^2 + d_1^2}. \tag{7.105}$$

Then the coefficients can be rotated to zero phase shift and normalized as

$$\begin{pmatrix} a_n^* & b_n^* \\ c_n^* & d_n^* \end{pmatrix} - \begin{pmatrix} \cos n\theta_1 & -\sin n\theta_1 \\ \sin n\theta_1 & \cos n\theta_1 \end{pmatrix}\begin{pmatrix} a_n & b_n \\ c_n & d_n \end{pmatrix}, \qquad n = 1, 2, \ldots. \tag{7.106}$$

The normalized coefficients are used as descriptors for the recognition of characters of known rotation. These coefficients are all independent of translation, which affects only 0th-order coefficients. To normalize them further for scaling, each coefficient must be divided by an appropriate factor, say the square sum of the magnitudes of the two semi-axes of the first ellipse:

$$D = \sqrt{a_1^2 + b_1^2 + c_1^2 + d_1^2}. \tag{7.107}$$

Rotation invariant descriptors are given by Lin and Hwang [11]. After selecting a random starting point, following the pixels clockwise along an outer boundary of a character image, and naming the semi-axis lengths of the k ellipse A_k and B_k, respectively, the following descriptors are defined:

$$I_k = A_k + B_k, \qquad k = 1, 2, \ldots, \tag{7.108}$$

$$J_t = abs(A_tB_t), \qquad t = 1, 2, \ldots \tag{7.109}$$

where $4\pi J_t$ is the area of the tth ellipse and

$$K_{1,j} = (A_1^2A_j^2 + B_1^2B_j^2)\cos\theta_{1,j} + (A_1^2B_j^2 + A_j^2B_1^2)\sin\theta_{1,j}, \qquad j = 1, 2, \ldots, \tag{7.110}$$

where $\theta_{1,j}$ is the angle between the major axis of the first and jth ellipse. The $K_{1,j}$ is rotation invariant. These coefficients are related to the elliptic Fourier coefficients mentioned above by the formulas

$$I_k = a_k^2 + b_k^2 + c_k^2 + d_k^2, \tag{7.111}$$

$$J_t = a_td_t - b_tc_t, \tag{7.112}$$

$$K_{1,j} = (a_1^2 + b_1^2)(a_j^2 + b_j^2) + (c_1^2 + d_1^2)(c_j^2 + d_j^2) \tag{7.113}$$

$$+ 2(a_1c_1 + b_1d_1)(a_jc_j + b_jd_j). \tag{7.114}$$

Recognition Experiments We introduce a character recognition experiment based on Fourier descriptors, which was conducted by Taxt, Olafsdottir, and Daehlen [12]. The data set is a mixed one of handprinted numerals and lowercase Roman alphabet, where 0 and *O* were not counted. The size of the data tested is 3554 characters, approximately 100 for each class. The quality is not described explicitly, but there are given some examples from which we can infer that the quality of the data was loosely constrained. In general, to the Japanese the lowercase Latin alphabet seems to be sloppy. Notice that both "2" and "Z" are included.

For a discrimination method, these authors used a generalized Mahalonobis distance given by Hjort [13]. They calculated the distance between two classes ω_i and ω_j in feature space according to the formula

$$M_{i,j} = (\delta^2 + \gamma^2)^{1/2},$$

where

$$\delta^2 = (\bar{\mu}_i - \bar{\mu}_j) \sum^{-1} (\bar{\mu}_i - \bar{\mu}_j)^T,$$

$$\gamma^2 = 4 \log \frac{\left|\sum^{-1}\right|}{\left|\sum_i\right|^{1/2} \left|\sum_j\right|^{1/2}},$$

and

$$\sum = \frac{1}{2}\left(\sum_i + \sum_j\right).$$

Here $\bar{\mu}_i$ and $\bar{\mu}_j$ are the mean vectors and $\sum_i$ and $\sum_j$ are the covariance matrices of the classes ω_i and ω_j, respectively.

They used 12 features and compared elliptic Fourier and Zahn and Roskies's Fourier descriptors. The elliptic Fourier descriptors which were independent of the character rotation gave a satisfactory recognition rate of 98.6% (no rejection). The normalized descriptors gave slightly higher correct classification rates than the non-normalized descriptors. The rotation-independent elliptic Fourier descriptors were clearly superior to the Fourier descriptors of Zahn and Roskies. This is true for both normalized and nonnormalized features. This was expected because the convergence of Zahn and Roskies's Fourier descriptors is not good, as mentioned earlier.

Granlund's Fourier Descriptor We introduce briefly a Fourier descriptor using a complex plane as in E. Persoon and K. S. Fu's work; this approach was taken earlier by G. H. Granlund in 1972 [14]. He gave an invariant form of the Fourier descriptor as

$$b_n = \frac{a_{1+n} a_{1-n}}{a_1^2}, \tag{7.115}$$

where $a_n b_n = 0, 1, 2, \ldots$ are the Fourier coefficients. They are defined as usual:

$$a_n = \frac{1}{T} \int_0^T u(t) \exp\left[\frac{-in2\pi t}{T}\right] dt, \tag{7.116}$$

where $u(t)$ is a complex number function of a parameter t.

To show the invariant nature of b_n, we select an arbitrary starting point on a contour and express it as

$$u(t) = u^{(0)}(t). \tag{7.117}$$

Then any function that differs from $u(t)$ by only a starting point is expressed as

$$u(t) = u^{(0)}(t + \tau). \tag{7.118}$$

The Fourier coefficients of $u(t)$ are denoted using $u^0(t)$ as

$$\begin{aligned} a_n &= \frac{1}{2\pi} \int_0^{2\pi} u^{(0)}(t + \tau) \exp[-int] dt \\ &= \exp[in\tau] \cdot \frac{1}{2\pi} \int_0^{2\pi} u^{(0)} \exp[-int] dt \\ &= \exp[in\tau] a_n^{(0)}. \end{aligned} \tag{7.119}$$

Thus a_n differ only by phase factor from $a_n^{(0)}$, which is naturally an intuitive result. Concerning rotation, we can expect the same result as

$$a_n = \exp[j\phi] a_n^{(0)}, \tag{7.120}$$

where the center of the gravity is positioned at the origin and the contour is rotated in an angle ϕ in a positive direction around the origin. Similarly it can be shown that the dilatation of a contour with a factor R results in the following expressing:

$$a_n = R a_n^{(0)}. \tag{7.121}$$

Therefore, in general, the Fourier coefficients of a contour generated by transformation, rotation, and dilatation of its arbitrary starting point are given by

$$a_n = \exp[in\tau] R \exp[j\phi] a_n^{(0)}. \tag{7.122}$$

We notice here that

$$a_n = \begin{cases} a_n^{(0)} & \text{for } n \neq 0, \\ a_0^{(0)} + Z & \text{for } n = 0, \end{cases} \tag{7.123}$$

where Z is a complex vector of the translation. Using the general form of Fourier coefficients, we can prove the invariance of b_n easily as

$$
\begin{aligned}
b_n &= \frac{a_{1+n}a_{1-n}}{a_1^2} \\
&= \frac{a_{1+n}^{(0)} \exp[i(1+n)z]R \exp[i\phi] a_{1-n}^{(0)} \exp[i(1-n)\tau]R \exp[i\phi]}{[a_1^{(0)} \exp(i\tau)R \exp(i\phi)]^2} \\
&= \frac{a_{1-n}^{(0)} a_{1-n}^{(0)}}{[a_1^{(0)}]^2} = b_n^{(0)}.
\end{aligned} \tag{7.124}
$$

Granlund also gave another form of the invariant Fourier descriptor using small-size handprinted characters of the Roman alphabet. His experiment obtained good results.

Accuracy of Curve Approximation by Fourier Series As stated before, a curve can be represented by a vector with a parameter, say t, as

$$
\mathbf{V}(t) = \begin{Bmatrix} x(t) \\ y(t) \end{Bmatrix}. \tag{7.125}
$$

For a closed curve $x(t)$ and $y(t)$ are better approximated by Fourier series than other orthogonal series:

$$
x(t) = A_0 + \sum_{n=1}^{\infty} a_n \cos \frac{2n\pi t}{T} + b_n \sin \frac{2n\pi t}{T}, \tag{7.126}
$$

$$
y(t) = B_0 + \sum_{n=1}^{\infty} c_n \cos \frac{2n\pi t}{T} + d_n \sin \frac{2n\pi t}{T}. \tag{7.127}
$$

The merit of this representation is that we can separate the two-dimensional representation into two independent one-dimensional ones very simply. This simplification allows us to determine a formula for the approximation of a curve. To determine its accuracy, some measure of closeness between two curves is defined. We take

$$
\|\mathbf{V}(t) - \mathbf{V}_n(t)\| = \max \left[\sup_{0 \le t \le T} |x(t) - x_n(t)|, \sup_{0 \le t \le T} |y(t) - y_n(t)| \right]. \tag{7.128}
$$

A graphic interpretation of $\|\mathbf{V}(t) - \mathbf{V}_n(t)\| < \varepsilon$, $(\varepsilon > 0)$ is possible if the x-y plane with a grid of squares with sides ε partitioned as shown in Fig. 7.14. As the figure shows, it meets our intuition.

Now we present a formula that tells how many terms are necessary for a specified value of ε [8]. To understand the formula, we need to introduce a concept of total variation in an interval $[0, T]$, denoted as $V_0^T(\cdot)$. As stated before, we treat a set of continuous closed curves. However, the derivatives of $x(t)$, $y(t)$ of a curve denoted as $\dot{x}(t)$, $\dot{y}(t)$ are not necessarily continuous at a finite number of points.

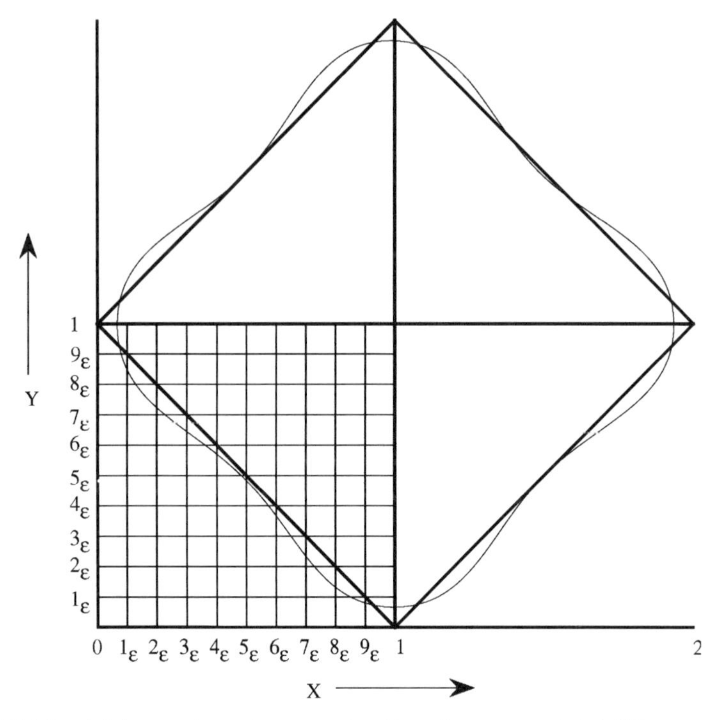

FIGURE 7.14 A 3-harmonic vector approximation of a square rotated 45° on a grid. From reference [8], © 1998, Academic Press.

Since the curve treated is continuous, the quantities $V_0^T(\dot{x}(t))$ and $V_0^T(\dot{y}(t))$ are bounded in practice. Strictly speaking, it is known that there exist continuous functions that are nondifferentiable at every point. Recently such functions have been used in practice, so caution should be exercised. Then $V_0^T(\dot{x}(t))$ can be calculated by taking the sum of absolute changes in height of $\dot{x}(t)$ over the total period. Once these total variations are obtained, it is proved that for a specified $\varepsilon > 0$,

$$\|\mathbf{V}_n(t) - \mathbf{V}(t)\| < \varepsilon, \tag{7.129}$$

provided that $n > K$ where

$$K = \frac{T}{2\pi^2\varepsilon} \max\,[V_0^T(\dot{x}(t)),\ V_0^T(\dot{y}(t))]. \tag{7.130}$$

For a proof refer to the literature cited [8].

In the illustration of moments, a simple example of expanding a rectangular image was given. To compare with it, an expansion of a square will be given. The square is placed as shown in Fig. 7.14. Then $x(t)$ and $y(t)$ are simply triangular wave functions, and so the derivatives $\dot{x}(t)$ and $\dot{y}(t)$ can be easily obtained. The $x(t)$ and $\dot{x}(t)$, and $y(t)$ and $\dot{y}(t)$, are shown in Fig. 7.15 (a) and (b), respectively. For example, $V_0^T(\dot{x}(t))$ is

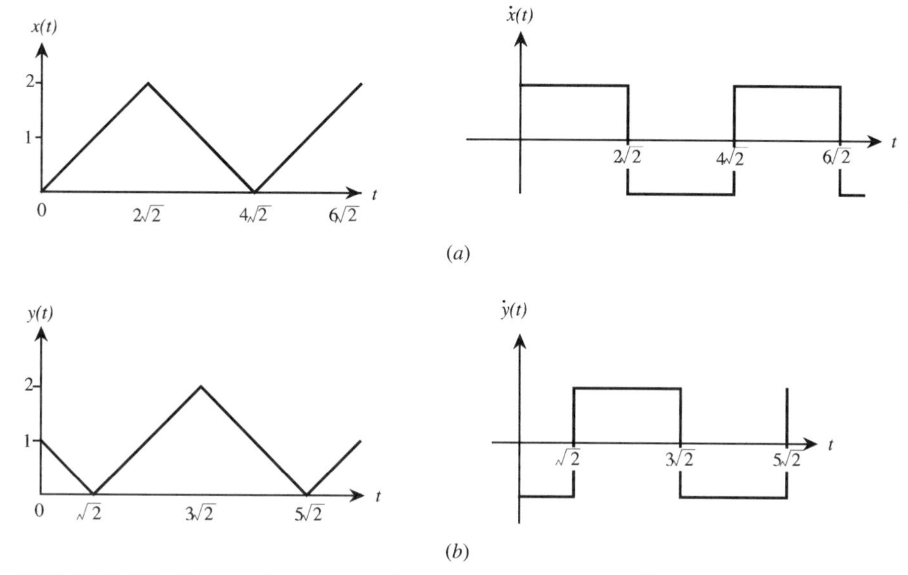

FIGURE 7.15 Parametric representation (*a*) and its derivatives (*b*) of a square rotated 45°. From reference [8], © 1998, Academic Press.

obtained as the sum of high $\dot{x}(t)$ jumps at points 0, $2\sqrt{2}$, and $4\sqrt{2}$, namely, $1/\sqrt{2}$, $2/\sqrt{2}$, and $1/\sqrt{2}$, respectively. Therefore

$$\overset{T}{\underset{0}{V}}(\dot{x}(t)) = \overset{T}{\underset{0}{V}}(\dot{y}(t)) = 2\sqrt{2}.$$

Then

$$K = \frac{T}{2\pi^2\varepsilon} \max\left[\overset{T}{\underset{0}{V}}(\dot{x}(t)), \overset{T}{\underset{0}{V}}(\dot{y}(t)) \right] = \frac{4\sqrt{2}}{2\pi^2\varepsilon}\, 2\sqrt{2} = \frac{8}{\pi^2\varepsilon}.$$

For example, taking $\varepsilon = 0.1$, nine terms in the approximation is enough. In this simple case we can find each known term analytically. That is,

$$x(t) = 1 - \frac{8}{\pi^2} \sum_{n=0}^{\infty} \frac{1}{(2n+1)^2} \cos \frac{2\pi t(2n+1)}{T}, \tag{7.131}$$

$$y(t) = 1 - \frac{8}{\pi^2} \sum_{n=0}^{\infty} (-1)^n \frac{1}{(2n+1)^2} \sin \frac{2\pi t(2n+1)}{T}, \tag{7.132}$$

where $T = 4\sqrt{2}$.

In each series, even harmonics are absent, so the first five nonzero terms up to $2n + 1 = 9$ are all that are required to ensure that

$$\|\mathbf{V}(t) - \mathbf{V}_n(t)\| < 0.1.$$

As shown in Fig. 7.14, five nonzero terms suffice. Therefore, compared with the moments method, we see that using the Fourier descriptor is a dramatically more efficient way in feature extraction for a closed curve.

7.4.3 Experiments Using Fourier Descriptors

Persoon and Fu conducted recognition experiments for handprinted numerals based on Fourier descriptors. In our discussion of their method, first a primitive introduction for the discrimination scheme on a feature space is given. Thus we begin by defining distance on a feature space. Let $\{a_n\}$ and $\{b_n\}$ be FDs of a curve α and FDs of a curve β, respectively. We assume that M harmonics are used. However, the positional information does not contribute to the recognition, so it is removed. Specifically we set $a_0 = b_0$. Therefore the distance between the two curves is given as

$$d(\alpha, \beta) = \left[\sum_{\substack{n=-M \\ n \neq 0}}^{M} |a_n - b_n|^2 \right]^{1/2} \tag{7.133}$$

On the other hand, we have to consider the normalization of scale, rotation, and starting point. The normalization of the position is not necessary due to the nature of FDs; it is implicitly done by neglecting the zeroth term. Here an ideal normalization scheme is described first. The approach is based on normalization in feature space, and the normalization parameters are chosen so that the distance becomes minimized, as described in Chapters 1 and 3. Specifically the $\{b_n\}$ of the curve β is multiplied by $Se^{i(n\alpha + \phi)}$ where the parameters S, ϕ, and α are for scaling, rotation, and starting point, respectively. These parameters are chosen to minimize

$$\sum_{\substack{n=-M \\ n \neq 0}}^{M} |a_n - Se^{i(n\alpha + \phi)} b_n|^2 . \tag{7.134}$$

The optimum value of S, ϕ, and α are found by solving for the roots of a periodic function. This is the traditional method of least squares and is given in P&F's paper.

P&F tried two kinds of the normalizations. One is suboptimal and the other optimal. Recognition experiments using the two normalizations were conducted.

Suboptimal Case As seen before, P&F FDs have very good convergence. Therefore we can expect that the first-order coefficient contains much of a character's image. Accordingly the normalization parameters S, ϕ, and α are chosen such that a_1 and a_{-1} become pure imaginary numbers and such that $|a_1 + a_{-1}| = 1$. All the numerals are normalized for this rule; that is, $a_n (n \neq 0)$ is multiplied by $Se^{i(\phi + n\alpha)}$. The normalization is uniform to all the FDs used. Thus the operative features are as follows:

$$
\begin{aligned}
x_1 &= |a_1 - a_{-1}|, \\
x_2 &= R_e(a_2), \\
x_3 &= I_m(a_2), \\
x_4 &= R_e(a_{-2}), \\
x_5 &= I_m(a_{-2}), \\
x_6 &= R_e(a_3), \text{ etc.}
\end{aligned}
\tag{7.135}
$$

In order to classify an unknown numeral, the distance between that numeral and each numeral in the training set is computed. The distance used is

$$
\left[\sum_{i=1}^{29} (x_i - x_i')^2 \right]^{1/2}, \tag{7.136}
$$

where $\{x_i\}$ and $\{x'_i\}$ are the features corresponding to the numerals between which the distance is computed. The unknown numeral is then given the same class label as the closest numeral in the training set.

For the training set 160 samples were used. Therefore the number of masks per class is 16 on average. There were 500 test samples. The result is shown in Table 7.4, known as a confusion table. From the results we can observe that numerals "4" and "8" are the most difficult characters. The basic reason is the difficulty of normalization. Some examples on starting points are illustrated in Fig. 7.16. A starting point is the first black point encountered when scanning a character image from top left to bottom right. This is the usual scanning convention.

TABLE 7.4 Classification Result (Suboptimal Procedure)

w_k w_i	0	1	2	3	4	5	6	7	8	9
0	39				6		1		3	
1	2	50		1	3	3	2		10	1
2	1		47						2	
3				46						
4	1				27				2	1
5						47			2	
6	1		1		2		47		2	
7	1		2	1	1			50		2
8	4				4				26	1
9	1			2	7				3	45

Source: From reference [9], © 1998, IEEE.

Note: w_i = original class, w_k = class classified, error rate = no. misclassified/total sample size = 76/500 = 15.4%.

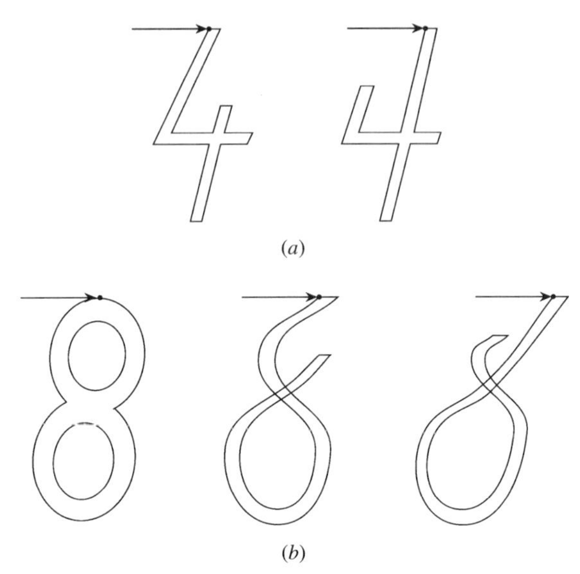

FIGURE 7.16 Difficulty of normalization of the starting point: (*a*) Examples for character "4"; (*b*) examples for character "8," where the dot indicated by an arrow shows the starting point.

Optimal Case For the optimal case the same training samples were used and the same test samples were tried. The recognition result is shown in Table 7.5 as a confusion table. We can observe a great improvement for the character "4."

The improvement is due to the sophisticated normalization technique. However, in the case of "8," the improvement is not so conspicuous. This is because only the outer boundary is used. Therefore the confusion between "1" and "8" is very large. In other words, the loop feature was not detected. Taking this into consideration, the correct recognition rate of 89.4% is not bad. However, the processing time was 14 s/c

TABLE 7.5 Classification Result (Optimal Procedure)

w_k w_i	0	1	2	3	4	5	6	7	8	9
0	42				4		1		1	
1	1	50			1	2	2		10	1
2	1		47						2	
3				47						
4	1				43	1	1		3	2
5						47			2	
6			1				46		1	
7			2	1				50		2
8	4								31	1
9	1			2	2					44

Source: From reference [9], © 1998, IEEE.

Note: w_i = original class, w_k = class classified, and error rate = no. misclassified/total sample size = 53/500 = 10.6%.

using a CDC 6500 (a famous scientific computer at that time). For suboptimal case the processing time was 150 m s/c.

Lai and Suen's Experiment More recently Lai and Suen [15] noticed the weakness of using only the outer boundary in the FDs and added the a loop feature as well as other key features such as curvature and line direction to the FDs. They constructed a hybrid recognition system for handprinted numerals and achieved a very high correct recognition rate of the order of 99% for a large test sample set of 8653 characters. Ten harmonics were used in the FDs, and they used a binary tree classification scheme. This tells us that FDs are essentially global features, but that quasi-local and local features are also important in character recognition.

7.5 KARHUNEN-LOÈVE EXPANSION

So far we have described linear feature extraction based on a built-in total set of functions. There is also a property of invariance to some part of the affine transformations. Thus the next step is to find an efficient feature extraction base so that a given set of shapes is effectively represented by the base. Since the base depends on a given set of characters, we suppose a full set of characters that belongs to a given alphabet and its statistical nature. We will give two mathematical approaches. One is based on linear algebra in a vector space and the other is on functional analysis in a continuous functional space. The former was given by Otsu [26], and the latter was given by Iijima [17].

7.5.1 Linear Algebra Approach on a Vector Space

The essence of this approach is that a high-dimensional space F_i is constructed by an invariant feature extraction base, so it is efficient to reduce the high-dimensional space to a lower-dimensional space so that an efficient discrimination is performed. Thus a feature vector $\mathbf{x} \in F_i \subset V^M$ is linearly mapped by a matrix U^t to a lower-dimensional vector $\mathbf{y} \in F_d \subset V^N$. That is,

$$\mathbf{y} = U^t\mathbf{x}, \quad U = [\mathbf{u}_1, \ldots, \mathbf{u}_N], \quad \mathbf{u}_i \in V^M : V^M \to V^N \ (N < M), \tag{7.137}$$

where $\mathbf{u}_i$, $i = 1 \sim N$ is an orthogonal base. Therefore $U^tU = [\mathbf{u}_i^t \mathbf{u}_j] = [\delta_{ij}] = I_N$ holds.

Multiplying U to the equation $\mathbf{y} = U^t\mathbf{x}$,

$$U\mathbf{y} = UU^t\mathbf{x} \equiv P_U\mathbf{x} \equiv \hat{\mathbf{x}}, \tag{7.138}$$

where P_U is a *projection operator/projector* to the *range space* $R(U)$, which is spanned by the orthogonal base. We note that in general $\mathbf{x}$ is not equal to $\hat{\mathbf{x}}$ which is an approximation of $\mathbf{x}$. Only if $N = M$, $\hat{x} = \mathbf{x}$; that is, $(U^t)^{-1}$ exists because U becomes a $M \times M$ orthogonal matrix. When $N < M$, $\hat{\mathbf{x}}$ is called the projection vector on $R(U)$. The norm $\|\hat{\mathbf{x}} - \mathbf{x}\|$ gives the minimum distance to $R[U]$ which is a well-known property of Hilbert space, as explained at the end of this chapter. On the other hand, the projection operator P_U is defined as follows:

$$P_U = U(U^tU)^{-1}U^t. \tag{7.139}$$

Noting $U^tU = I_N$, we have

$$U(U^tU)^{-1}U^t = UU^t. \tag{7.140}$$

It is easily proved that $U(U^tU)^{-1}U^t$ is an idempotent matrix as $U(U^tU)^{-1}U^tU(U^tU)^{-1}U^t = U(U^tU)^{-1}U^t$. Therefore

$$P_U^2 = P_U. \tag{7.141}$$

Now let $\mathbf{x}, \mathbf{y} \in V^M$, $\mathbf{x}$ and $\mathbf{y}$ be resolved to $\mathbf{x} = \mathbf{x}_1 + \mathbf{x}_2$, and $\mathbf{y} = \mathbf{y}_1 + \mathbf{y}_2$, where $\mathbf{x}_1, \mathbf{y}_1 \in R(U)$ and $\mathbf{x}_1 = P_U\mathbf{x}$ and $\mathbf{y}_1 = P_U\mathbf{y}$ hold. Therefore

$$(\mathbf{x}_2, P_U\mathbf{y}) = (\mathbf{y}_2, P_U\mathbf{x}) = 0, \tag{7.142}$$

and then

$$\begin{aligned}(\mathbf{x}, P_U\mathbf{y}) &= (\mathbf{x}_1 + \mathbf{x}_2, P_U\mathbf{y}) \\ &= (P_U\mathbf{x} + \mathbf{x}_2, P_U\mathbf{y}) = (P_U\mathbf{x}, P_U\mathbf{y}) \\ &= (P_U\mathbf{x}, P_U\mathbf{y} + \mathbf{y}_2) = (P_U\mathbf{x}, \mathbf{y}) = (\mathbf{x}, P_U^t\mathbf{y}).\end{aligned}$$

The above equation holds for any vector $\mathbf{x}$ and $\mathbf{y}$. Therefore

$$P_U^t = P_U. \tag{7.143}$$

A projector that satisfies the both $P^t = P$ and $P^2 = P$ is called an *orthogonal projector.* Now we consider a total set of data characters, denoted as $X = \{\mathbf{x} \in V^M\}$, and the best approximating orthogonal base U of X in the meaning of the norm. That is, the formulation of the above problem is given by

$$\varepsilon^2(U) = E_X \|\hat{\mathbf{x}} - \mathbf{x}\|^2, \tag{7.144}$$

where $E_X(\cdot)$ denoted an expected value of $(\cdot)$ on the set of X. For simplicity it is denoted as E. As shown above, the best approximation of $\mathbf{x}$; that is, $\hat{\mathbf{x}}$ is given by $\hat{\mathbf{x}} = UU^t\mathbf{x} = P_U\mathbf{x}$ when U is given. Therefore the equation above is expressed as

$$\varepsilon^2(U) = E \| UU^t\mathbf{x} - \mathbf{x}\|^2. \tag{7.145}$$

Furthermore the right term of the equation above is expanded by the trace as

$$
\begin{aligned}
\varepsilon^2(U) &= \mathrm{tr}(UU^t E\mathbf{xx}^t UU^t) - 2\,\mathrm{tr}(UU^t E\mathbf{xx}^t) \\
&\quad + \mathrm{tr}(E\mathbf{xx}^t) \\
&= \mathrm{tr}R_X - \mathrm{tr}(U^t R_X U),
\end{aligned}
$$

where $R_X \equiv E(\mathbf{xx}^t)$, is called *autocorrelation matrix* of data X, and is a positive definite symmetric matrix. For the derivation of the equation above, see the comment on the trace at the end of this chapter. Actually, using the formula $\mathrm{tr}(AB) = \mathrm{tr}(BA)$,

$$
\mathrm{tr}(UU^t E\mathbf{xx}UU^t) = \mathrm{tr}(UU^t UU^t E\mathbf{xx}^t)
$$

is derived. On the other hand, UU^t is projection matrix. Therefore $UU^tUU^t = UU^t$, which is substituted to the above equation to obtain

$$
\mathrm{tr}(UU^t E\mathbf{xx}^t UU^t) = \mathrm{tr}(UU^t E\mathbf{xx}^t). \tag{7.146}
$$

The same derivation is applied to $\mathrm{tr}(UU^t E\mathbf{xx}^t)$, and we obtain

$$
\mathrm{tr}(UU^t E\mathbf{xx}^t) - \mathrm{tr}(U^t E\mathbf{xx}^t U).
$$

The $\varepsilon^2(U)$ is a function of U, so we want to minimize $\varepsilon^2(U)$ concerning U, but we need to note that there is a constraint on U, namely $U^tU = I_N$. Therefore the Lagrange unknown multiple coefficient method is employed introducing diagonal matrix $\Lambda = \mathrm{diag}(\lambda_1, \cdots \lambda_N)$. That is, the following $J(U)$ is minimized:

$$
J(U) = \varepsilon^2(U) + \mathrm{tr}[(U^tU - I_N)\Lambda]. \tag{7.147}
$$

Differentiating by U and setting the equation to zero, we derive

$$
\frac{\partial J(U)}{\partial U} = -2R_X U + 2U\Lambda = 0. \tag{7.148}
$$

Here the following formula is used:

$$
\frac{\partial}{\partial A}\,\mathrm{tr}\,(A^t SA) = 2SA. \tag{7.149}
$$

Thus the minimized U is obtained as

$$
R_X U = U\Lambda. \tag{7.150}
$$

The normal orthogonal base U is obtained by the eigenvectors of the real symmetrical matrix R_X. The ε^2 is calculated using this result,

$$
\varepsilon^2(U) = \mathrm{tr}\,R_X - \mathrm{tr}\,(U^tU\Lambda) = \mathrm{tr}\,R_X - \mathrm{tr}\,\Lambda. \tag{7.151}
$$

The tr R_X is the sum of eigenvalues of R_X, and tr Λ is the sum of eigenvalues of the adopted eigenvector of U. Therefore, in order to minimize $\varepsilon^2(U)$, it is enough to choose, say, N eigenvectors whose eigenvalues are greater than or equal to those of the other eigenvectors. Then the error is given by

$$\varepsilon_N^2 \equiv \min_U \varepsilon^2(U) = \sum_{i=1}^{M} \lambda_i - \sum_{i=1}^{N} \lambda_i = \sum_{i=N+1}^{M} \lambda_i. \tag{7.152}$$

If each member of data X is canonicalized, namely its average value is zero, the same procedure can be applied to the data X. Setting $\mu \equiv E\mathbf{x}$, $\mathbf{x} - \mu$ is substituted to $\mathbf{x}$ in the above formulation and derivation, which results in the following formula:

$$\Sigma_X U = UA, \tag{7.153}$$

where Σ_X is defined as $E_X[(\mathbf{x} - \mu)(\mathbf{x} - \mu)^t]$ and is called a *covariance matrix*.

7.5.2 Functional Analysis Approach on a Function Space

The essence of this approach is that all the data X are normalized, so for each category of a set $J^{(r)}$, $r = 1, 2, \cdots, L$, of a given alphabet only one normalized and standard shape, $g^r(x)$, corresponds to $J^{(r)}$. However, the normalizations considered are limited to linear and selected nonlinear ones. They are, by far, not enough to capture all the variations of shape. Therefore we consider all the normalized character shapes that belong to some of the $\{J^{(r)}\}$, denoted as $\mathcal{D}$. The number of the elements of the set $\mathcal{D}$ is tremendous, so for each member of the set $\mathcal{D}$ a real number can be assigned in a one-to-one manner. Thus a normalized character shape is expressed as $g(x;\, \alpha)$ using a real parameter α; if it belongs to $J^{(r)}$, then it is denoted as $g^{(r)}(x;\, \alpha)$. Here we assume that the set of parameter α, denoted as $\mathcal{D}$, forms a compact and continuous domain. Furthermore we assume that each member of the set $\mathcal{D}$ is *canonicalized* and thus denoted as $\mathcal{D} = \{h(x;\, \alpha)\}$. The canonicalization is nothing but a linear transformation of $g(x;\, \alpha)$ so that the average taken on $\mathcal{D}$ is zero. A formal description will be given later. For convenience a one-dimensional pattern is considered here. Thus we construct the set of functions, but we need to introduce a structure to the set. In other words, we need a norm and a distance to make the set a functional space, which is done by introducing inner product so that a Hilbert space is constructed. Our aim is to find an orthogonal base in the space in the sense that the orthogonal base is the most efficient to use to extract features of the functional space made of a given alphabetic character set. The formalization of this scenario is given by

$$J(\varphi(x)) = \frac{\int_{\mathcal{D}} w(\alpha)(h(x;\, \alpha), \varphi(x))^2 d\alpha}{\|\varphi(x)\|^2}, \tag{7.154}$$

where $w(\alpha)$ is an appearance probability of indexed character $h(x;\, \alpha)$. $J(\varphi(x))$ is a functional and function of the function $\varphi(x)$. The feature extraction is formalized such that the best function $\varphi(x)$ gives the maximum value over the domain $\mathcal{D}$. Therefore let us consider maximizing the functional $J(\varphi)$ with respect to $\varphi(x)$, which can be carried out putting $\varphi(x) \Rightarrow \varphi(x) + \varepsilon\eta(x)$; that is, a variation method is employed, where $\eta(x)$ is an arbitrarily continuous function and ε is a parameter. The following derivation is based on Noguchi's work [19]. Equation (7.154) is changed to

$$J[\varphi(x) + \varepsilon\eta(x)] \cdot \int (\varphi(x) + \varepsilon\eta(x))^2 dx$$

$$= \int_{\mathcal{D}} w(\alpha)\left[\int h(x;\, \alpha)(\varphi(x) + \varepsilon\eta(x))dx\right]^2 d\alpha \tag{7.155}$$

Then the equation above is differentiated by ε,

$$\left.\frac{\partial J}{\partial \varepsilon}\right|_{\varepsilon=0} = 0 \tag{7.156}$$

in order to find the best function $\varphi(x)$. The equation derived is

$$\int \eta(x)\left[J[\varphi(x)]\varphi(x) - \int\left(\int_{\mathcal{D}} \omega(\alpha)h(x;\, \alpha)h(x';\, \alpha)d\alpha\right)\varphi(x')dx'\right] dx = 0. \tag{7.157}$$

The function $\eta(x)$ is arbitrary, so $[\cdot]$ must be set to zero. Thus

$$J[\varphi(x)]\varphi(x) = \int K(x, x')\varphi(x')dx', \tag{7.158}$$

$$K(x, x') \equiv \int_{\mathcal{D}} \omega(\alpha)h(x;\, \alpha)h(x';\, \alpha)d\alpha. \tag{7.159}$$

Equation (7.158) is a Fredholm integral equation of the second kind with a kernel $K(x, x')$ that is also symmetric and real. In this case it is well-known that there exists an orthogonal eigenfunction system. However, the real problem is how to solve it. The formulation above has the intrinsic problem that it is impossible to obtain the set $\mathcal{D}$ expressing all the variation of character shapes. In other words, the model above is conceptual, although it gives a basic framework of the linear feature extraction theory.

In practice, the kernel must be reduced to a drastically simpler one; that is, (7.158) and (7.159) are simplified to

$$\lambda\varphi(x) = \int K(x, x')\varphi(x')dx', \tag{7.160}$$

$$K(x, x') = \sum_{r=1}^{r} \omega_r h_r(x)h_r(x'), \tag{7.161}$$

where $\lambda \equiv J(\varphi(x))$. Here h_r represents the rth class of a given alphabet. The domain $\mathcal{D}$ is reduced to an isolated finite point set. Of course the $h_r(x)$ can be obtained by taking the average of the sampled data that belong to the rth class, or the size of the class can be increased when the variation is large. However, the simplest way of absorbing the small variation is blurring, as mentioned before. In practice, ω_r is taken as $1/L$, so the statistical nature is virtually eliminated in the formulation above. On the contrary, the structure of the space introduced by inner product dominates, as noted before.

The kernel of the equation above is called the Pinchele-Coursat kernel, or degenerated kernel, and it is easily converted to an eigenvalue problem as shown below. See [18] for a discussion.

First, we put

$$C_r = \int \sqrt{\omega_r} h_r(x')\varphi(x')dx', \qquad r = 1, 2, \cdots, L. \tag{7.162}$$

We notice that the defined coefficients include $\varphi(x')$, so they are unknown coefficients which we can solve within a framework of simultaneous equations. Next, using these coefficients, the integral equation is rewritten as

$$\lambda\varphi(x) = \sum_{r=1}^{L} \sqrt{\omega_r} C_r h_r(x). \tag{7.163}$$

This shows that $\lambda\varphi(x)$ must coincide with a suitable linear combination of the function $h_r(x)$. To obtain simultaneous equations, the equation above is multiplied by $\sqrt{\omega_s} h_s(x)$; $s = 1, 2, \ldots, L$, and is integrated. Then

$$\lambda \int \sqrt{\omega_s} h_s(x)\varphi(x)dx = \sum_{r=1}^{L} \sqrt{\omega_r}\sqrt{\omega_s} C_r \int h_r(x)h_s(x)dx. \tag{7.164}$$

Looking at (7.164), we note that the left term is an unknown coefficient C_s multiplied by λ and that the right term includes a known coefficient denoted as a_{rs},

$$a_{rs} = \int \sqrt{\omega_r}\sqrt{\omega_s} h_r(x)h_s(x)dx. \tag{7.165}$$

Thus we obtain simultaneous equations of the unknown coefficient C_i, in general, $i = 1, 2, \ldots, l$:

$$\lambda C_s = \sum_{r=1}^{L} a_{rs} C_r = \sum_{r=1}^{L} a_{sr} C_r. \tag{7.166}$$

The $a_{rs} = a_{sr}$ comes from the definition of a_{rs} which makes the solution realistic as mentioned below. The equation above can be represented neatly by using a vector and a matrix. That is, taking a column vector $\mathbf{c} = (C_1, C_2, \ldots, C_L)^t$ and an $L \times L$ matrix A with an element in the sth row, rth column being a_{sr}, we can rewrite the equation above as

$$\lambda \mathbf{c} = A\mathbf{c}. \tag{7.167}$$

This is the eigenvalue equation with respect to the matrix A. Because of canonicalization, the rank of the matrix A is less than or equal to $L - 1$. A detailed description of canonicalization will be given at the end of this section. As noted earlier, A is a symmetric matrix. Therefore the equation above has a nontrivial solution, and all the eigenvalues are real. Let us consider the eigenvalue problem of matrix A. The eigenvalues of matrix A are defined as the roots of a characteristic equation

$$|A - \lambda E| = 0. \tag{7.168}$$

The roots of the equation are arranged in descending order: $\lambda_1 \geq \lambda_2 \geq \cdots \geq \lambda_{L-1}$. Let the L row column vector $\mathbf{c}_m = (C_{1m}, C_{2m}, \ldots, C_{Lm})^t$ be the eigenvector corresponding to the mth eigenvalue λ_m with respect to the matrix A. We can assume that these eigenvectors are normalized. That is,

$$\mathbf{c}_n^t \mathbf{c}_m = \delta_{nm}, \qquad m, n = 1, 2, \ldots, L - 1. \tag{7.169}$$

Thus the eigenfunction $\varphi_m(x)$ corresponding to the eigenvalue λ_m can be written as

$$\varphi_m = \sum_{r=1}^{L} \frac{1}{\sqrt{\lambda_m}} \sqrt{\omega_r} C_{rm} h_r(x), \tag{7.170}$$

which is slightly different from (7.163). This is because the eigenfunctions are normalized:

$$(\varphi_m(x), \varphi_n(x)) = \delta_{mn}, \qquad m, n = 1, 2, \ldots, L - 1. \tag{7.171}$$

Thus we have obtained the optimal orthogonal base for feature extraction for a given data set of characters. Naturally each canonicalized standard pattern can be represented by a linear combination of the normalized eigenfunctions:

$$h_r(x) = \sum_{n=1}^{L-1} \beta_{rm} \varphi_m(x), \tag{7.172}$$

$$\beta_{rm} = \int h_r(x) \varphi_m(x) dx, \qquad r = 1, 2, \ldots, L. \tag{7.173}$$

The set of the coefficients β_{rm} has the following properties:

$$\sum_{r=1}^{L} \omega_r \beta_{rm} = 0. \tag{7.174}$$

To prove the equation, let us explain the *canonicalized standard patterns.* A standard pattern $g_i(x)$, $i = 1, 2, \dots, L$, is selected and the average pattern $\tau(x)$ is defined as

$$\tau(x) = \frac{\sum_{i=1}^{L} \omega_i g_i(x)}{\left\| \sum_{i=1}^{L} \omega_i g_i(x) \right\|}. \tag{7.175}$$

Then a canonicalized standard pattern of the ith class is defined using $\tau(x)$ as

$$h_i(x) = g_v(x) - d_i \tau(x), \tag{7.176}$$

$$d_i = (g_i(x), \tau(x)). \tag{7.177}$$

The canonicalized patterns must satisfy the following properties:

$$\sum_{i=1}^{L} \omega_2 h_i(x) = 0, \tag{7.178}$$

$$\int_{-l}^{l} h_i(x) dx = 0, \qquad i = 1, 2, \dots, L. \tag{7.179}$$

That is, $h_i(x)$, $i = 1, \dots, L$, are not linear independent because all ω_i cannot be zero. Therefore the property of (7.174) is nothing but a reflection of the property of (7.178). The second property of the set β_{rm} is as follows:

$$\sum_{r=1}^{L} \omega_r \beta_{rm} \beta_{rn} = \lambda_m \delta_{mn}, \qquad m, n = 1, 2, \dots, L-1. \tag{7.180}$$

This is due to the nature of the normalized and orthogonal base, φ_m, $m = 1, \dots, L-1$. Finally we note that an arbitrary observed and normalized pattern can be represented in terms of the eigenfunctions as

$$h(x) = \sum_{m=1}^{L-1} \beta_m \varphi_m(x) + \varepsilon(x), \tag{7.181}$$

$$\beta_m = (h(x), \varphi_m(x)), \tag{7.182}$$

where $\varepsilon(x)$ is the orthogonal to $\tau(x)$ and $\{\varphi_m(x)\}$. This shows that an arbitrary data character $h(x)$ includes the component that lies outside of the space denoted W. In other words, the $\varepsilon(x)$ belongs to $W^{\perp}$. So long as $h(x) \cong h_i(x)$, $i \in L-1$, in the sense of the norm, there is no problem, but this is not always true. This point will be further discussed later.

7.5.3 Factor Analysis and Karhunen-Loève Expansion

So far we have introduced two kinds of Karhunen-Loève expansion methods. Eventually the linear algebra method results in solving the eigenvector/value problem of a covariance matrix, which is known as factor analysis in the field of multivariate analysis. However, the factor analysis approach is used to treat high-dimensional matrices, namely of patterns on the order of $30 \times 30 = 900$. The latter pattern recognition method concerns only the L dimension, namely the number of classes of a given alphabet. In principle, however, both methods treat the same number of standard patterns which are linearly independent, so they must have the same number of orthogonal functions vectors, which is equal to or less than L. For the canonicalized case $L \Rightarrow L - 1$. Here we give the relation between them. First we construct the $L - 1$ vectors:

$$\varphi_m = \frac{1}{\sqrt{L}} \sum_{r=1}^{L} \frac{1}{\sqrt{\lambda_m}} C_{rm} \mathbf{h}_r, \qquad m = 1, 2, \ldots, L - 1, \tag{7.183}$$

where λ_m is a set of the eigenvalues of the matrix A whose elements are defined in (7.165) and C_{rm} is as before. We show that these vectors are the eigenvectors of the following covariance matrix:

$$S = \frac{1}{L} \sum_{r=1}^{L} \mathbf{h}_r \mathbf{h}_r^t, \tag{7.184}$$

and φ_m, φ_n, $m \neq n$ are also orthogonal. The $\mathbf{h}_r$ is a set of canonicalized standard vectors whose dimension is, say $N \times N$, the number of pixels of an observed plane. Actually $\mathbf{h}_r$ is given as a matrix $[h_r(i, j)]$, so a one-dimensional $N^2 \times 1$ vector $\mathbf{h}_r$ is constructed by column or row scanning as

$$\mathbf{h}_r = \begin{bmatrix} h_r(1, 1) \\ h_r(2, 1) \\ \\ h_r(N, 1) \\ h_r(1, 2) \\ \vdots \\ h_r(N, 2) \\ \vdots \\ h_r(1, N) \\ \vdots \\ h_r(N, N) \end{bmatrix}. \tag{7.185}$$

This representation of a two-dimensional picture matrix by a vector somewhat goes against our intuition, but that problem will be discussed later. At any rate, $\mathbf{h}_r$ is interpreted as a set of L samples from a statistical population with the mean vector 0.

We can write the following equations according to (7.167):

$$\lambda_m \mathbf{c}_m = A\mathbf{c}_m, \tag{7.186}$$

$$a_{ij} = \frac{1}{L} (\mathbf{h}_i, \mathbf{h}_j) = \frac{1}{L} \mathbf{h}_i^t \mathbf{h}_j. \tag{7.187}$$

Multiplying the covariance matrix S and φ_m and simplifying it using the relations as shown above, we reach the desired relation between S and φ_m:

$$
\begin{aligned}
S\varphi_m &= \frac{1}{L}\sum_{r=1}^{L} \mathbf{h}_r\mathbf{h}_r^t \frac{1}{\sqrt{L}}\sum_{t=1}^{L} \frac{1}{\sqrt{\lambda_m}} C_{tm}\mathbf{h}_t \\
&= \sum_{r=1}^{L} \mathbf{h}_r \frac{1}{\sqrt{L}} \frac{1}{\sqrt{\lambda_m}} \sum_{t=1}^{L} C_{tm}\ 1/L\ \mathbf{h}_r^t\mathbf{h}_t \\
&= \frac{1}{\sqrt{L}} \frac{1}{\sqrt{\lambda_m}} \sum_{r=1}^{L} \mathbf{h}_r \sum_{t=1}^{L} a_{rt}C_{tm} \\
&= \frac{1}{\sqrt{L}} \lambda_m \sum_{r=1}^{L} \frac{1}{\sqrt{\lambda_m}} C_{rm}\mathbf{h}_{rm} \\
&= \lambda_m\varphi_m.
\end{aligned}
\tag{7.188}
$$

That is, φ_m and λ_m are the eigenvector and the eigenvalue of the covariant matrix S. On the other hand, using (7.183)

$$
\begin{aligned}
(\varphi_m, \varphi_n) &= \frac{1}{L} \frac{1}{\sqrt{\lambda_m}} \frac{1}{\sqrt{\lambda_n}} \left(\sum_{r=1}^{L} C_{rm}\mathbf{h}_r, \sum_{r=1}^{L} C_{tm}\mathbf{h}_t \right) \\
&= \frac{1}{L} \frac{1}{\sqrt{\lambda_m}} \frac{1}{\sqrt{\lambda_n}} \sum_{r=1}^{L}\sum_{t=1}^{L} C_{rm}C_{tn}a_{rt} \\
&= \frac{1}{L} \frac{\sqrt{\lambda_n}}{\sqrt{\lambda_m}} \frac{1}{\sqrt{\lambda_m}} \sum_{r=1}^{L} C_{rm}C_{rm}
\end{aligned}
\tag{7.189}
$$

and also (7.169), we obtain the desired relation

$$
(\varphi_m, \varphi_n) = \delta_{mn}. \tag{7.190}
$$

That is, the set of φ_m is normally orthogonal.

7.5.4 Karhunen-Loève Expansion of a Set of Closed Lines

Fourier series expansion is an effective feature extraction method, as mentioned before. Here we show that a Fourier series expansion is indeed a Karhunen-Loève expansion. In both approaches an optimum function series in the sense of the least mean square for a set of closed lines. A set of closed lines can be very broad however, and we need to limit the range of the set in such a way that each element of the set has equal length. In practice, this restriction can be relaxed a little because the real condition is that the same number of sampling points of each element of the set of closed lines is the same.

If this condition is satisfied, each element of the set can be represented by a vector of the same dimension. To infer a general form of the covariance matrix of the set, let us consider that a line is represented by a parameter l on a Gaussian coordinate $Z(l)$. That is, for the four-dimensional case it is expressed as

$$\begin{pmatrix} Z(l_1) \\ Z(l_2) \\ Z(l_3) \\ Z(l_4) \end{pmatrix} \rightarrow \begin{pmatrix} Z_1 \\ Z_2 \\ Z_3 \\ Z_4 \end{pmatrix},$$

where $l_{i+1} - l_i = \Delta l$ (const):

$$\begin{pmatrix} Z_1 \\ Z_2 \\ Z_3 \\ Z_4 \end{pmatrix}, \begin{pmatrix} Z_2 \\ Z_3 \\ Z_4 \\ Z_1 \end{pmatrix}, \begin{pmatrix} Z_3 \\ Z_4 \\ Z_1 \\ Z_2 \end{pmatrix}, \begin{pmatrix} Z_4 \\ Z_1 \\ Z_2 \\ Z_3 \end{pmatrix}.$$

For the set the covariance matrix is calculated as

$$\begin{vmatrix} \sum_{i=1}^{4} z_i^2 & \sum_{i=1}^{3} z_i z_{i+1} + z_4 z_1 & 2z_1 z_3 + 2z_2 z_4 & \sum_{i=1}^{3} z_i z_{i+1} + z_4 z_1 \\ \sum_{i=1}^{3} z_i z_{i+1} + z_4 z_1 & \sum_{i=1}^{4} z_i^2 & \sum_{i=1}^{3} z_i z_{i+1} + z_4 z_1 & 2z_1 z_3 + 2z_2 z_4 \\ 2z_1 z_3 + 2z_2 z_4 & \sum_{i=1}^{3} z_i z_{i+1} + z_4 z_1 & \sum_{i=1}^{4} z_i^2 & \sum_{i=1}^{3} z_i z_{i+1} + z_4 z_1 \\ \sum_{i=1}^{3} z_i z_{i+1} + z_4 z_1 & 2z_1 z_3 + 2z_3 z_4 & \sum_{i=1}^{3} z_i z_{i+1} + z_4 z_1 & \sum_{i=1}^{4} z_i^2 \end{vmatrix}.$$

Looking at the result, we can infer the case of N circular vectors to be

$$C = \begin{vmatrix} C_0 & & C_1 & & C_2 & & & & C_2 & C_1 \\ & C_0 & & C_1 & & C_2 & & & & C_2 \\ C_1 & & \ddots & & \ddots & & \ddots & & & \\ & C_1 & & \ddots & & \ddots & & \ddots & & \\ C_2 & & \ddots & & \ddots & & \ddots & & \ddots & \\ & C_2 & & \ddots & & \ddots & & \ddots & & C_2 \\ & & \ddots & & \ddots & & \ddots & & \ddots & \\ & & & \ddots & & \ddots & & \ddots & & C_1 \\ C_2 & & & & \ddots & & \ddots & & C_0 & \\ C_1 & C_2 & & & & C_2 & & C_1 & & C_0 \end{vmatrix}. \tag{7.191}$$

The symmetry and beautiful structure of the matrix can be decomposed further to a simpler form, employing a *cyclic matrix* P defined as

$$P = \begin{bmatrix} 0 & 1 & 0 & & & & 0 & 0 \\ 0 & 0 & 1 & 0 & & & & 0 \\ 0 & 0 & 0 & 1 & 0 & & & \\ & \ddots & & \ddots & & \ddots & & \\ & & \ddots & & \ddots & & \ddots & \\ & \ddots & & \ddots & & \ddots & & 0 \\ 0 & & \ddots & & \ddots & & \ddots & 1 \\ 1 & 0 & & & & & & 0 \end{bmatrix}, \tag{7.192}$$

$$P \begin{bmatrix} Z_1 \\ Z_2 \\ \vdots \\ \vdots \\ Z_n \end{bmatrix} = \begin{bmatrix} Z_2 \\ Z_3 \\ \vdots \\ \vdots \\ Z_n \\ Z_l \end{bmatrix}.$$

Using the cyclic matrix, the covariant matrix is expressed as

$$C = C_0E + C_1P + C_2P^2 + \cdots + C_1P^{N-1}. \tag{7.193}$$

Here C is a polynomial function of P, so the eigenvector/eigenvalue problem is replaced by that of P according to the theorem of Frobenius. That is, if a matrix V is U^n, then the eigenvector of V is the same as U, and its eigenvalue is given by λ^n, where λ is the eigenvalue of the matrix U. Thus the eigenvalue of C denoted as C_λ is given as

$$C_\lambda = C_0 + C_1\lambda^1 + C_2\lambda^2 + \cdots + C_1\lambda^{N-1}. \tag{7.194}$$

The eigenvalue of P is easily obtained, noting that

$$P^N = E. \tag{7.195}$$

That is, let λ be the eigenvalue of P. The eigenvalue of matrix $P^N - E$ is $\lambda^N - 1$, and it must be zero by (7.195). Thus λ must be an N multiple root. Letting $e(2\pi i/N) = \omega$, they are $1, \omega, \omega^2, \ldots, \omega^N$. For each the corresponding eigenvector is given as

$$\begin{pmatrix} 1 \\ 1 \\ 1 \\ \vdots \\ 1 \end{pmatrix}, \begin{pmatrix} 1 \\ \omega \\ \omega^2 \\ \vdots \\ \omega^{n-1} \end{pmatrix}, \begin{pmatrix} 1 \\ \omega^2 \\ \omega^4 \\ \vdots \\ \omega^{2(n-1)} \end{pmatrix}, \ldots, \begin{pmatrix} 1 \\ \omega^{n-1} \\ \omega^{2(n-1)} \\ \vdots \\ \omega^{(n-1)^2} \end{pmatrix} \tag{7.196}$$

and denoted as $\mathbf{W}(k)$, $k = 0, 1, 2, \ldots, N-1$ for simplicity. Its eigenvalue $\lambda(k)$ is expressed as

$$\lambda(k) = \sum_{j=0}^{N-1} C_j \exp\left(\frac{i2\pi}{N} kj\right), \qquad k = 0, 1, 2, \ldots, N-1. \tag{7.197}$$

Then we construct the matrix

$$W = [W(0), W(1), \ldots, W(n-1)],$$

$$\{W\}_{kj} = \exp\left(\frac{i2\pi}{N} kj\right). \tag{7.198}$$

The inverse matrix of $\mathbf{W}$ is given as

$$\{W^{-1}\}_{kj} = \frac{1}{N} \exp\left(-\frac{i2\pi}{N} kj\right) \tag{7.199}$$

Therefore the matrix C is diagonized by

$$C = WDW^{-1}, \tag{7.200}$$

$$D_{kk} = \lambda(k). \tag{7.201}$$

Thus we can consider the space spanned by W^{-1} instead of W, which is a *discrete Fourier transformation;* that is, a given vector $\mathbf{f}$ can be expanded by W^{-1} as

$$W^{-1}f = \frac{1}{N} \sum_{j=1}^{N-1} f(j) \exp\left(-\frac{i2\pi}{N} kj\right), \qquad k = 0, 1, \ldots, N-1. \tag{7.202}$$

7.5.5 Experiment

The Karhunen-Loève expansion has been applied to different character sets such as numerals, Roman alphabet, Katakana, and Kanji. Among the many research examples, we introduce here an experiment conducted by Noguchi [19] which is very systematic and consistent with the theory.

The character set he used for the experiment is printed style Katakana 46-character alphabet standardized as JE1DA-5 by the Japan Electronic Industry Development Association Standard, as shown in Fig. 7.17. The size of each character was 2.73×1.89 mm, with some surrounding marginal space included for segmentation. The characters were binarized, segmented, and placed on a 31×43 grid frame. Then they were normalized in position and brightness. For example, taking $e(x_i, y_i)$ to be the character normalized, the process proceeds as follows:

$$f(x_i, y_i) = e(x_i + \lfloor x_d \rfloor; y_i + \lfloor y_d \rfloor) - K_0, \tag{7.203}$$

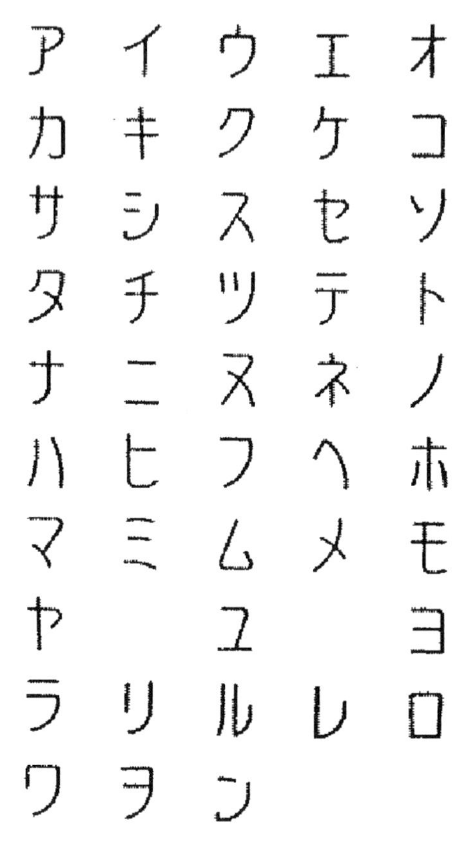

FIGURE 7.17 Katakana character set (Japan Electric Industry Development Association Standard).

where x_d, y_d, K_0 are normalizing parameters and $\lfloor x \rfloor$ means the greatest integer less than or equal to x. This is necessary for space quantization. These normalizing parameters were decided according to the conditions

$$\sum_{j=1}^{43} \sum_{i=1}^{31} e(x_i + x_d, y_j) \sin\left(\frac{\pi x_i}{2l_x}\right) = 0, \tag{7.204}$$

$$\sum_{j=1}^{43} \sum_{i=1}^{31} e(x_i, y_j + y_d) \sin\left(\frac{\pi y_i}{2l_y}\right) = 0, \tag{7.205}$$

$$K_0 = \frac{1}{4l_x l_y} \sum_{j=1}^{43} \sum_{i=1}^{31} e(x_i, y_i), \tag{7.206}$$

where $2l_x = 4.00$ mm, $2l_y = 5.00$ mm. As mentioned before, the position normalization is conducted so that its average value is equal to zero. Therefore the preceding position normalization is a little different from the conventional one, but sin (x) is almost equal to x in the neighborhood of the original point, so it is not essential in this context.

The normalized character is further preprocessed by blurring and canonicalization. For example, let $g(x_i, y_i)$ be a blurred pattern transformed from $f(x_i, y_i)$ so that

$$g(x_i, y_i) = \sum_{t=-3}^{3} \sum_{s=-3}^{3} \frac{c}{\sigma} \exp\left(-\frac{(x_i - u_{i+s})^2 + (y_j - v_{j+t})^2}{2\sigma^2}\right) f(u_{i+s}, v_{j+t}), \qquad (7.207)$$

where $\sigma = 0.177$ and c is a constant. Then the weighting normalization is performed:

$$\sum_{t=-3}^{3} \sum_{s=-3}^{3} \frac{c}{\sigma} \exp\left(-\frac{(u_{i+s})^2 + (v_{j+t})^2}{2\sigma^2}\right) = 1 \qquad (7.208)$$

for all i, j, and $1 < i < 11$, $1 \leq j < 15$. The Gaussian weights are shown in Fig. 7.18. The sampling interval of the blurring process is 0.36 mm on the frame, so the final dimensions of the frame were reduced to 11×15. That is, let $\mathbf{g}_r$ be a blurred standard pattern corresponding to the rth pattern class of the Katakana alphabet with $L = 46$

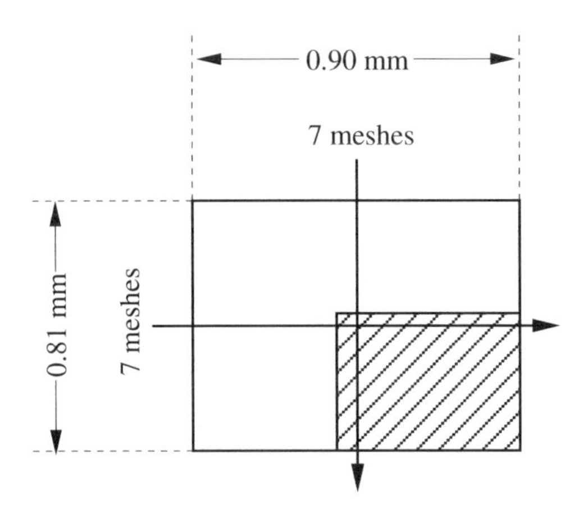

Gaussian weights

0.0784	0.0601	0.0271	0.0072
.0633	.0485	.0218	.0058
.0331	.0254	.0114	.0030
.0112	.0086	.0039	.0010

Sampled weights at original mesh points

0.0979	0.0997	0.0997	0.0979
.1051	.1071	.1071	.1051
.1051	.1071	.1071	.1051
.0979	.0997	.0997	.0979

FIGURE 7.18 Partial list of Gaussian weights. From reference [19].

classes. Let τ be the average pattern of the set of standard blurred patterns. Then it is defined as follows:

$$\tau = \frac{\sum_{r=1}^{L} \mathbf{g}_r}{\left\| \sum_{r=1}^{L} \mathbf{g}_r \right\|}. \tag{7.209}$$

Thus the canonicalized pattern of rth pattern class was obtained using the τ above as

$$\mathbf{h}_r = \mathbf{g}_r - \alpha_r \tau, \tag{7.210}$$

$$\alpha_r = (\mathbf{g}_r, \tau). \tag{7.211}$$

According to the experiment $\{\|\mathbf{g}_r\|\}$ ranges between 0.316 and 0.461 and $\{\alpha_r\}$ between 0.165 and 0.379. On the other hand, $\{\|h_r\|\}$ ranges between 0.228 and 0.335. Therefore we see that the canonicalized patterns are distributed in size almost uniformly in the space. The matrix $A(a_{ij})$, $a_{ij} = 1/L\,(\mathbf{h}_i, \mathbf{h}_j)$ was constructed and the eigenvalue equation

$$\|A - \lambda E\| = 0 \tag{7.212}$$

was solved, so the eigenvectors were given according to the earlier formula (7.183) as

$$\varphi_m = \sum_{r=1}^{L} \frac{1}{\sqrt{\lambda_m}} \frac{1}{\sqrt{L}} c_{rm} \mathbf{h}_r, \qquad m = 1, 2, \cdots, L-1, \tag{7.213}$$

where the vector $C_m = (C_{1m}, C_{2m}, \ldots, C_{Lm})^t$ is the eigenvector corresponding to the mth eigenvalue λ_m with respect to the matrix A, and the eigenvalues are labeled in the descending order, $\lambda_1 \geq \lambda_2 \geq \cdots \geq \lambda_{L-1}$. The extraction ratio is defined as

$$\eta_m = \frac{\sum_{i=1}^{m} \lambda_i}{\sum_{i=1}^{L-1} \lambda_i}. \tag{7.214}$$

Eigenvalues and extraction ratios are shown in the Table 7.6, and the graph of extraction ratios in Fig. 7.19 shows our expectation that the character set would be described by much smaller dimension compared with L and $K = 11 \times 15$ in this case. The distribution of the standard canonicalized patterns in the two-dimension subspace constructed as φ_1 and φ_2 is shown in Fig. 7.20. As expected, the distribution is broad and uniform in the space.

Remarks on the K-L Expansion An image can be represented by an $n \times m$ matrix. In general, an $n \times m$ unit vector can be generated as

TABLE 7.6 Eigenvalues and Extraction Ratios

i	Eigen-value	Extraction Ratio	i	Eigen-value	Extraction Ratio	i	Eigen-value	Extraction Ratio
1	0.01289	0.16545	16	0.00096	0.90560	31	0.00015	0.99060
2	0.01085	0.30468	17	0.00092	0.91741	32	0.00012	0.99219
3	0.00927	0.42366	18	0.00082	0.92791	33	0.00011	0.99366
4	0.00616	0.50268	19	0.00073	0.93723	34	0.00010	0.99497
5	0.00552	0.57356	20	0.00062	0.94525	35	0.00007	0.99587
6	0.00455	0.63200	21	0.00059	0.95284	36	0.00006	0.99665
7	0.00432	0.68747	22	0.00049	0.95917	37	0.00005	0.99734
8	0.00340	0.73115	23	0.00044	0.96481	38	0.00005	0.99796
9	0.00293	0.76875	24	0.00039	0.96981	39	0.00004	0.99834
10	0.00235	0.79890	25	0.00034	0.97423	40	0.00003	0.99881
11	0.00203	0.82490	26	0.00031	0.97821	41	0.00003	0.99916
12	0.00147	0.84382	27	0.00023	0.98119	42	0.00003	0.99950
13	0.00142	0.86210	28	0.00022	0.98406	43	0.00002	0.99974
14	0.00123	0.87795	29	0.00020	0.98661	44	0.00001	0.99990
15	0.00119	0.89325	30	0.00016	0.98865	45	0.00001	1.00001

Source: From reference [19].

$$\mathbf{e}^1 = \begin{pmatrix} 1 \\ 0 \\ \vdots \\ \vdots \end{pmatrix}, \quad \mathbf{e}^2 = \begin{pmatrix} 0 \\ 1 \\ \vdots \\ \vdots \end{pmatrix}, \quad \mathbf{e}^3 = \begin{pmatrix} 0 \\ 0 \\ 1 \\ \vdots \end{pmatrix} \cdots, \tag{7.215}$$

each of which has dimension $n \times m$. Then the following equation holds:

$$(\mathbf{e}^i, \mathbf{e}^j) = \delta_{ij}. \tag{7.216}$$

That is, the vectors generated are orthogonal and independent. Thus a Hilbert space is constructed. This result gives us a strange feeling because the distance between $\mathbf{e}^i$ and $\mathbf{e}^i$ $i \neq j$ is equal to $\sqrt{2}$ according to the distance definition (7.18). Now, e^1 and e^2, for example, are very close, and the distance between e^1 and e^2 is very short. However, this is due to confusion between two types of spaces. One is Hilbert space and the other is two-dimensional Euclidean space. Nevertheless, the result is strange. To bridge the two spaces, a correlation among the vectors is employed. That is, if the correlation between $\mathbf{e}^1$ and $\mathbf{e}^2$ is strong, then it can be merged to a one-dimensional vector. This idea can be uniformly applied to all the space.

In the example shown in Fig. 7.21, it is assumed that there is a strong correlation in the neighborhood around each point. The size of the neighborhood is regarded as a disk whose radius is one pixel. Because of the strong correlation inside the neighborhood disk, it is collapsed to a one-dimensional space. Therefore an 8×8 dimensional space is reduced to a 4×4 dimensional space. Effectively, by this reduction,

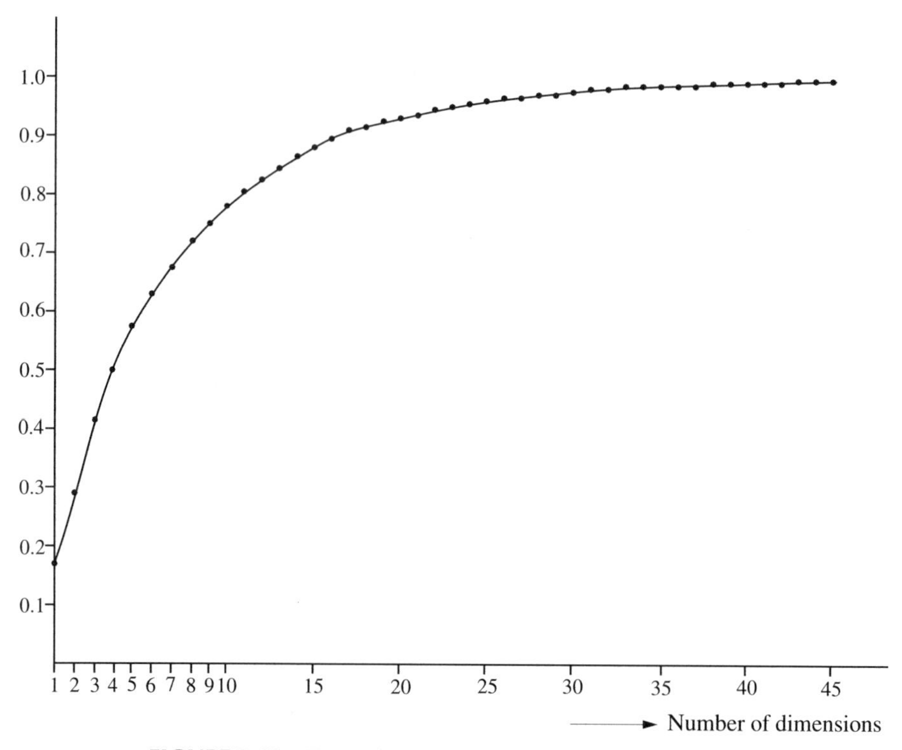

FIGURE 7.19 Extraction ratios graphed. From reference [19].

vectors $\mathbf{e}^i$, $i = 1 \sim 16$, on the 4×4 dimensional space behave as almost noncorrelated vectors and can well represent a given image. The meaning of rough sampling also can be understood from this point of view.

Note on Trace Let $\mathbf{x}$ be a N-dimensional vector. Then its square norm $\|\mathbf{x}\|^2$ (i.e., an inner product $\mathbf{x}^t\mathbf{x}$) is equal to the trace of the matrix $\mathbf{x}\mathbf{x}^t$,

$$\mathrm{tr}(\mathbf{x}\mathbf{x}^t) = \sum_{i=1}^{N} [\mathbf{x}\mathbf{x}^t]_{ii} = \sum_{i=1}^{N} x_i x_i = \mathbf{x}^t\mathbf{x}. \tag{7.217}$$

On the other hand, an averaging operation E can be exchanged with the trace operation because of a linear nature of a trace, $E\,\mathrm{tr}(\mathbf{x}\mathbf{x}^t) = \mathrm{tr}(E(\mathbf{x}\mathbf{x}^t))$. Furthermore the following equations hold:

$$\begin{aligned} \mathrm{tr}(A^t) &= \mathrm{tr}(A), \\ \mathrm{tr}(A + B) &= \mathrm{tr}(A) + \mathrm{tr}(B), \\ \mathrm{tr}(\alpha A) &= \alpha\,\mathrm{tr}(A). \end{aligned} \tag{7.218}$$

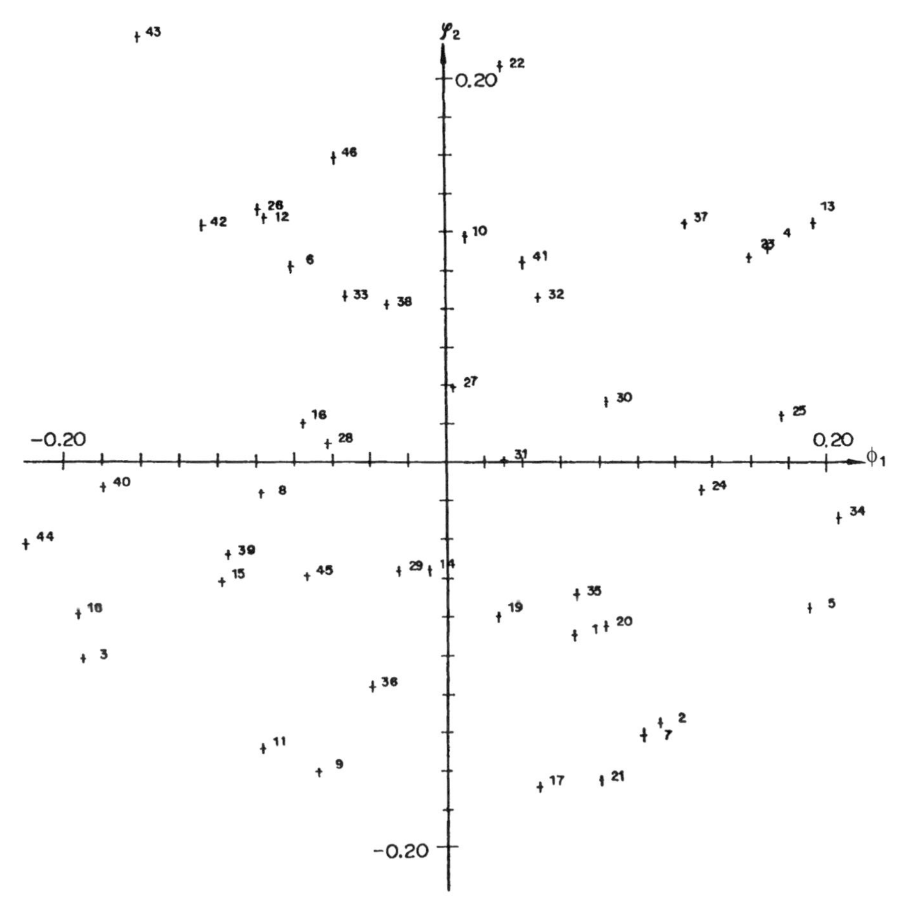

FIGURE 7.20 Distribution of the standard canonicalized patterns in 2D subspace of the feature space. From reference [19].

Let A be an $n \times m$ matrix, and let B be an $m \times n$ matrix, then the following relation holds:

$$\operatorname{tr}(AB) = \operatorname{tr}(BA). \tag{7.219}$$

Also the following derivatives of the trace hold:

$$\begin{aligned} \frac{\partial}{\partial A} \operatorname{tr}(A^t B) &= B, \\ \frac{\partial}{\partial A} \operatorname{tr}(A^t R A) &= (R + R^t)A, \\ \frac{\partial}{\partial A} \operatorname{tr}(A^t R A) &= A^t(R + R^t). \end{aligned} \tag{7.220}$$

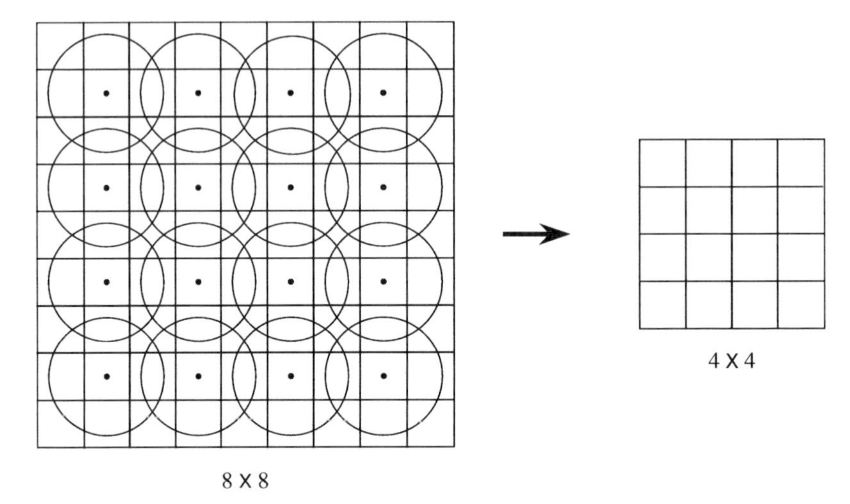

FIGURE 7.21 Dimension reduction by taking the correlation among neighborhoods into account.

Note on Hilbert Space Let us consider a Hilbert space X and its nonempty subset M and $\mathbf{x}$ in X. The distance between $\mathbf{x}$ and M is defined as

$$\delta = \inf_{\tilde{\mathbf{y}} \in M} \|\mathbf{x} \in \tilde{\mathbf{y}}\|. \tag{7.221}$$

The problem is whether there is a $\mathbf{y} \in M$ that is closest to the given $\mathbf{x}$, and if such an element exists, whether it is unique. This is an existence and uniqueness problem. In a Hilbert space, the following theorem is known.

Theorem 7.5.1 *Let X be an inner product space and $M \neq \phi$ a convex subset that is complete. Then for every given $\mathbf{x} \in X$ there exists a unique $\mathbf{y} \in M$ such that*

$$\delta = \inf_{\tilde{\mathbf{y}} \in M} \|\mathbf{x} - \tilde{\mathbf{y}}\| = \|\mathbf{x} - \mathbf{y}\|. \tag{7.222}$$

Furthermore the following properties hold:

Lemma 1 (Orthogonality) *Let M be a complete subspace Y and $x \in X$ fixed. Then $\mathbf{z} = \mathbf{x} - \mathbf{y}$ is orthogonal to Y.*

Lemma 2 (Direct Sum) *Let Y be any closed subspace of a Hilbert space H. Then*

$$H = Y \oplus Z, \qquad Z = Y^{\perp}. \tag{7.223}$$

Therefore for every $\mathbf{x} \in H$ there is a $\mathbf{y} \in Y$ such that

$$\mathbf{x} = \mathbf{y} + \mathbf{z}, \qquad \mathbf{z} \in Z = Y^{\perp}, \tag{7.224}$$

where $\mathbf{y} = P\mathbf{x}$, *called the (orthogonal) projection of* $\mathbf{x}$ *on* Y. *The distance between* $\mathbf{x}$ *and* Y, $\delta = \|\mathbf{x} - \mathbf{y}\|$ *is minimum distance* $\mathbf{x}$ *to* Y.

For the details, see Kreyszig [20] for example.

7.6 BIBLIOGRAPHICAL REMARKS

Invariant Feature Extraction

A systematic comparative study was conducted for various moment invariant feature extraction methods by S. O. Belkasim, M. Shridhar, and M. Ahmadi [21]. The methods compared were as follows:

1. Zernike moment invariants (ZMI).
2. Pseudo-Zernike moment invariants (PZMI).
3. Normalized Zernike moment invariants (NZMI).
4. Normalized pseudo-Zernike moment invariants (NPZMI).
5. Teaque-Zernike moment invariants (TZMI).
6. Hu moment invariants (HMI).
7. Bamieh moment invariants (BMI).
8. Regular moment invariants (RMI).

They used unconstrained handprinted numerals and military data on aircrafts. The sample sizes were 320 and 288. The number of classes in numeral data set was 10, and 18 were used for the aircraft data set. The discrimination method employed was the K-nearest-neighbor rule. The result for the numeral data using moments up to a sixth-order set is shown in Table 7.7.

In the table, BMI is omitted because its performance was comparatively poorer than that of the other methods. That BMI is poorer than the other methods of recognition is true for all orders less than or equal to seventh order. As seen in the table,

TABLE 7.7 Arabic-Numeral Recognition Using Moment Invariants up to Sixth Order

Type	Error (%)	Rejection (%)	Discrimination Factor
NPZMI	8.44	0.63	91.50
NZMI	5.00	1.25	94.94
ZMI	6.88	2.50	92.95
RMI	5.94	0.94	94.01
TZMI	8.44	2.50	91.35
PZMI	12.19	1.88	87.58
HMI	11.88	5.94	87.37

Source: From reference [21].

NZMI had the best performance among all the invariants. For example, for the aircraft data set the correct recognition rate and error rates were 99.30% and 0.69%, respectively, for both fifth and sixth orders. Among experiments for noisy data, the best result was due to NZMI as well.

Theoretical Feature Extraction

For a book that focuses on invariant feature extraction for planer objects, see [56] which treats invariance to projective transformation based on algebraic invariants.

Concerning a theoretical feature extraction based on functional analysis or statistical approach, there is a systematic survey paper by H. Hamamoto [22]. Since it is written in Japanese, we have included some parts of it in this chapter. Recall that theoretical feature extraction is divided into two classes: nonlinear and linear ones.

The nonlinear method introduced by J. W. Sammon tried to keep a structure on an observed plane (space) in its feature space [23]. The error between the distances on a Euclidean plane and its feature space was minimized by the *descending method.* A. K. Jain and J. Mao applied the method to a neural network [24]. However, both approaches were based on a descending principle, so there is no guarantee that the minimum value was obtained by their methods. This aspect was improved by R. W. Klein and R. C. Dubes using a simulated annealing method [25]. However, this method requires enormous computation time. The theoretical views of N. Otsu, and K. Fukunaga and S. Ando, proposed nonlinear methods based on a posteriori probability, [26, 27] and [28]. Unfortunately, a posteriori probability is not known in general and is difficult to estimate it from finite number of samples. Therefore, because of these difficulties, there is no effective method established so far.

By contrast, for the linear method, an analytical approach can be taken and its execution is easy. So-called K-L expansion was proposed by Iijima [29] and Watanabe [30] independently. This was generalized by Y. T. Chien and K. S. Fu [31]. Its properties were studied by T. Y. Young [32]. A rigorous treatment of its optimality was given by H. Ogawa [33]. His claim was that the K-L expansion is not a necessary condition but only a sufficient condition for the best approximation problem. That is, what is important is not the K-L eigenvectors themselves but the subspace spanned by those K-L eigenvectors. The relation between K-L expansion and neural network was investigated theoretically by K. Funahashi [34].

A good discussion of the subspace approach is given by Oja [35]. The subspace approach based on K-L expansion was proposed independently by S. Watanabe, P. E. Lambert, and C. A. Kulikowski [36] and by Iijima et al. [37]. However, in both works the subspace of each class is represented independently of its other classes, so they are not necessarily optimum in terms of discrimination. This point was improved by S. Watanabe and N. Pakvasa [38], by Iijima's compound similarity method, and by K. Fukunaga and W. L. G. Koontz [39]. Recently a new approach was taken by S. Yamashita and H. Ogawa based on the *relative K-L operator* [40].

There is another approach to linear feature extraction method, namely discrimination analysis, which was not included in this book because it would have shifted the focus of the text. Briefly, in the discrimination analysis approach, features that maximize Fisher's estimation function are extracted, and defined by a ratio of dispersion between classes to that within a class. The extracted feature vectors are not orthogonal, although they are independent and so generate an orthogonal system. This is a kind of pretransformation was introduced by Y. Hamamoto, T. Kanaoka, and S. Tomita [41], and by T. Okada and S. Tomita who studied the relationship between K-L expansion and discrimination analysis theoretically [42].

A problem with discrimination analysis is that only $m - 1$ features can be obtained from an m class problem [43]. This is a big constraint on the designer of a pattern recognition system. So far, to cope with this problem, some methods were proposed that divide it roughly into three classes.

The first approach is to add another feature(s) that maximizes *Fisher's estimation function* under the orthogonal condition. This was done first by J. W. Sammon [44], who treated it as a two-class problem. D. H. Foley and J. W. Sammon [45] gave a successive method to obtain other features but still limited it to two classes. T. Okada and S. Tomita [48] proposed a normalized orthogonal discrimination vector method to solve a multi-class problem. Further development was done by Y. Hamamoto et al. [46].

The second approach is to extend Fisher's estimation function so that the limitation of the number of features is solved. In many linear feature extraction methods, features are extracted, assuming normal distribution, from the difference among pattern distributions. It is well-known that a normal distribution is described by the average vectors and the covariance matrices. Then the extension of Fisher's estimation function is done such that discrimination information is obtained from a covariance matrix. W. Malina proposed a new method along this line [47]. He added to an ordinal Fisher's estimation function the difference among covariance matrices, called *dispersion difference.* T. Okada and S. Tomita proposed a way of generating an orthogonal system from Malina's estimation function [48]. M. Aladjem and I. Distein tried to improve Malina's estimation function [49, 50]. However, J. Fehlauer and B. A. Eisenstein [51] proposed a new feature estimation function that represents the difference among covariance matrices in terms of the dispersion ratio. This is that when a given pattern set is projected on a feature axis, the distance among average vectors and dispersion of another classes are maximized under the condition of constant dispersion of a noticed class. E. S. Gelsema and G. Eden [52] improved Fehlauer feature estimation function and studied the several properties of the estimation function. Nevertheless, Fehlauer's feature estimation function has a problem that the reliability to unknown samples is low, since features whose variation is large are obtained due to maximization of the dispersion. Therefore Y. Hamamoto et al. defined a feature estimation function that takes into account the constrained condition of the dispersion and so maximizes this feature estimation function by imposing an orthogonal system construction. This is just a mixture of the first and the second approaches.

The third approach is to solve the constrained problem of number of features through nonparameterization of covariance matrices. The number of the features is basically limited by the rank of the between classes covariance matrices which constitute Fisher's feature estimation function. Therefore K. Fukunaga and J. M. Mantock [53] obtained regular between classes covariance matrices by nonparameterization based on which the discrimination analysis was done. However, in the nonparameterization, many parameters are necessary, for which some experimental results were reported by M. E. Aladjem [54]. Thus in the discrimination analysis various improvements have been done to solve the constrained problem of the feature number limitation.

BIBLIOGRAPHY

[1] M. Born and E. Wolf, *Principle of Fourier Optics,* New York: Pergamon, 1975.

[2] M. R. Teague, "Image analysis via the general theory of moments," *J. Opt. Soc. Am.,* vol. 70, no. 8, pp. 179–187 August 1980.

[3] A. B. Bhatia and E. Wolf, "On the circle polynomials of Zernik and related orthogonal sets," *Proc. Camb. Phil. Soc.,* vol. 50, pp. 40–48, 1954.

[4] C. H. Teh and R. T. Chin, "On image analysis by the methods of moments," *IEEE Trans. Pattern Anal. Machine Intell.,* vol. 10, no. 4, pp. 496–513, 1988.

[5] Y. Morita, K. Fukurotani, and M. Mikkaichi, "Numeric character recognition using rotation invariant features." *Trans. IECE Japan,* vol. J23-D-11, no. 11, pp. 1906–1909, 1990.

[6] Y. Hsu, H. H. Arsenault, and G. April, "Rotation-invariant digital pattern recognition using circular harmonic expansion," *Appl. Optics,* vol. 21, no. 22, pp. 4012–4015, November 1982.

[7] C. T. Zarn and R. Z. Roskies, "Fourier descriptor for closed curves," *IEEE Trans. Comp.,* vol. C-21, no. 3, pp. 269–281, March 1972.

[8] C. R. Giardina and F. P. Kuhl, "Accuracy of curve approximation by harmonically related vectors with elliptical loci," *Comp. Graphics Image Proc.,* vol. 6, pp. 277–285, 1977.

[9] E. Persoon and K. Fu, "Shape discrimination using Fourier descriptors," *IEEE Trans. Syst., Man Cybern.,* vol. SMC-7, no. 3, pp. 170–179, March 1977.

[10] F. P. Kuhl and C. R. Giardina, "Elliptic Fourier features of a closed contour," *Comp. Vision, Graphics Image Proc.,* vol. 18 pp. 236–258, 1982.

[11] C. S. Lin and C. L. Hwang, "New forms of shape invariants from elliptic Fourier descriptors," *Pattern Recogn.,* vol. 20, pp. 535–545, 1987.

[12] T. Taxt, J. B. Olafsdottir, and M. Daehen, "Recognition of handwritten symbols," *Pattern Recogn.,* vol. 23, no. 11, pp. 1155–1166, 1990.

[13] N. L. Hjort, "Notes on the theory of statistical symbol recognition," Norweigian Computing Center Report, no. 778, 1986.

[14] G. H. Granlund, "Fourier preprocessing for hand print character recognition," *IEEE Trans. Compt.,* pp. 195–201, February 1972.

[15] M. T. Y. Lai and C. Y. Suen, "Automatic recognition of characters by Fourier descriptors and boundary line encodings," *Pattern Recogn.,* vol. 14, no. 1–6, pp. 383–393, 1981.

[16] N. Otsu, "Mathematical studies on feature extraction in pattern recognition," Researches of the Electrotechnical Laboratory, no. 818, July 1981.

[17] T. Iijima, "Basic theory of feature extraction for visual pattern," *J. IECE Japan,* vol. 46, no. 11, November 1963.

[18] J. Kondo, *Integral Equations,* Oxford Applied Mathematics and Computing Science Series, Oxford: Clarendon Press, 1991.

[19] Y. Noguchi, "Pattern recognition systems—On the basics of the feature extraction technique," Researches of the Electrotechnical Laboratory, no. 739, April 1973.

[20] E. Kreyszig, *Introductory Functional Analysis with Applications,* New York: Wiley, 1989.

[21] S. O. Belkasim, M. Shridhar, and M. Ahmadi, "Pattern recognition with moment invariants: A comparative study and new results," *Pattern Recogn.,* vol. 24, no. 12, pp. 1117–1138, 1991.

[22] Y. Hamamoto, "Recent trends in the theory of pattern recognition," *J. IECE Japan,* vol. 77, no. 8, pp. 853–864, 1994.

[23] J. W. Sammon, "A nonlinear mapping for data structure analysis," *IEEE Trans.,* vol. C-18, no. 5, pp. 401–409, May 1969.

[24] A. K. Jain and J. Mao, "Artificial neural network for nonlinear projection of multivariate," *Proc. Int. Joint Conf. Neural Networks,* Baltimore, pp. III335–III340, 1992.

[25] R. W. Klein and R. C. Dubes, "Experiments in projection and clustering by simulated annealing," *Pattern Recogn.,* vol. 22, no. 2, pp. 213–220, 1989.

[26] N. Otsu, "An optimal nonlinear transformation based on variance criterion for pattern recognition I. 1st derivation," *Bull. Electrotechn. Lab.,* vol. 36, no. 12, pp. 815–830, December 1972.

[27] N. Otsu, "An optimal nonlinear transformation based on variance criterion for pattern recognition II. Its properties and experimental confirmation," *Bull. Electrotechn. Lab.,* vol. 37, no. 3, pp. 283–295, March 1973.

[28] K. Fukunaga, and S. Ando, "The optimum nonlinear features for a scatter criterion in discriminant analysis," *IEEE Trans.,* vol. IT-23, no. 4, pp. 453–459, July 1977.

[29] T. Iijima, "Theory of pattern recognition," *J. IECE Japan,* vol. 46, no. 11, pp. 1582–1591, November 1963.

[30] S. Watanabe, "Karhunen-Loeve expansion and factor analysis: Theoretical remarks and applications," *Trans. 4th Prague Conf. Inf. Theory,* pp. 635–660, 1965.

[31] Y. T. Chien and K. S. Fu, "On the generalized Karhunen-Loeve expansion," *IEEE Trans.,* vol. IT-13, no. 3, pp. 518–520, July 1967.

[32] T. Y. Young, "The reliability of linear feature extractors," *IEEE Trans.,* vol. C-20, no. 9, pp. 967–971, September 1971.

[33] H. Ogawa, "Karhunen-Loeve subspace," *Proc. 11th Int. Conf. Pattern Recogn.,* The Hague, pp. 75–78, 1992.

[34] K. Funahashi, "On the approximate realization of identity by three-layer neural networks," *Trans. IECE Japan,* vol. J73-A, no. 1, pp. 139–145, January 1990.

[35] E. Oja, *Subspace Methods of Pattern Recognition,* Tokyo: Research Studies Press, 1983.

[36] S. Watanabe, P. E. Lambert, C. A. Kulikowski, J. L. Buxton, and R. Walker, "Evaluation and selection of variables in pattern recognition," in J. T. Tou, ed., *Computer and Information Sciences II,* New York: Academic Press, pp. 91–112, 1967.

[37] T. Iijima, H. Genchi, and K. Mori, "A theoretical study of pattern identification by matching method," *Proc. 1st USA-Japan Comp. Conf.,* pp. 42–28, 1972.

[38] S. Watanabe and N. Pakvasa, "Subspace method in pattern recognition," *Proc. 1st Int. Joint Conf. Pattern Recogn.,* Washington DC, pp. 25–32, 1973.

[39] K. Fukunaga and W. L. G. Koontz, "Application of the Karhunen-Loeve expansion to feature selection and ordering," *IEEE Trans.,* vol. C-19, no. 4, pp. 311–318, April 1970.

[40] S. Yamashita and H. Ogawa, “Relative Karhunen-Loeve operator,” Technical Report IECI Japan, vol. PRU92-50, pp. 39–44, November 1992.

[41] Y. Hamamoto, T. Kanaoka, and S. Tomita, “Orthogonal discriminant analysis for interactive pattern analysis,” *Proc. 10th Int. Conf. Pattern Recogn.,* Atlantic City, pp. 424–427, 1990.

[42] T. Okada and S. Tomita, “Optimum discriminant analysis for display systems,” *Trans. IECE Japan,* vol. J65-A, no. 4, pp. 277–280, April 1982.

[43] A. K. Jain, “Advance in statistical pattern recognition,” in P. A. Devijver and J. Kittler, eds., *Pattern Recognition Theory and Applications,* New York: Springer-Verlag, pp. 1–19, 1987.

[44] J. W. Sammon, “An optimal discriminant plane,” *IEEE Trans.,* vol. C-9, no. 9, pp. 826–829, September 1970.

[45] D. H. Foley and J. W. Sammon, “An optimal set of discriminant vectors,” *IEEE Trans.,* vol. C-24, no. 3, pp. 281–289, March 1975.

[46] Y. Hamamoto, Y. Matsuura, T. Kanaoka, and S. Tomita, “A note on the orthonormal discriminant vector method for feature extraction,” *Pattern Recogn.,* vol. 24, no. 7, pp. 681–684, 1991.

[47] W. Malina, “On an extended Fisher criterion for feature selection,” *IEEE Trans.,* vol. PAMI-3, no. 5, pp. 611–614, September 1981.

[48] T. Okada and S. Tomita, “An extended Fisher criterion for feature extraction—Marina’s method and its problems,” *Trans. IECE Japan,* vol. J67-A, no. 3, pp. 159–165, March 1984.

[49] M. Aladjem, “Parametric and nonparametric linear mappings of multi-dimensional data,” *Pattern Recogn.,* vol. 24, no. 6, pp. 543–553, 1991.

[50] M. Aladjem and I. Distein, “A multiclass extension of discriminant mappings” *Proc. 11th Int. Conf. Pattern Recogn.,* The Hague, pp. 101–104, 1992.

[51] J. Fehlauer and B. A. Eisenstein, “A declustering criterion for feature extraction in pattern recognition,” *IEEE Trans.,* vol. C-27, no. 3, pp. 261–266, March 1978.

[52] E. S. Gelsema and G. Eden, “Mapping algorithms in ISPAHAN,” *Pattern Recogn.,* vol. 12, no. 3, pp. 127–136, 1980.

[53] K. Fukunaga and J. M. Mantock, “Nonparametric discriminant analysis,” *IEEE Trans.,* vol. PAMI-5, no. 6, pp. 671–678, November 1983.

[54] M. E. Aladjem, “A statistical technique for evaluating the significant of control parameters of mapping procedures,” *Pattern Recogn. Lett.,* vol. 14, no. 8, pp. 631–636, August 1993.

[55] S. Mori and T. Sakakura, *Fundamentals of Image Recognition II,* Tokyo: Ohm, 1990.

[56] T. H. Reises, *Recognizing Planar Objects Using Invariant Image Features,* New York: Springer-Verlag, 1991.

CHAPTER EIGHT

Feature Extraction Based on Structural Analysis

Feature extraction is a central component of an OCR system. In particular, for feature extraction based on structural analysis, much work has been done to date. We try to give a systematic description of feature extraction methods based on structural analysis in this and succeeding chapters. First a basic consideration is given, and then methods are classified according to how shapes are viewed. This chapter begins with an introduction to feature extraction based on structural analysis.

8.1 PRIMITIVE, FEATURE, AND DESCRIPTION

Before proceeding to an explanation of features, it is useful and necessary to introduce another term, *primitive,* which is close to *feature* in its meaning. The concept of primitive is derived from atomism, which was proposed by the Greek philosopher Democritus. It is well-known that this idea has a reality in substances found in nature. On the other hand, this idea is also applied to linguistics. That is, language consists of sentences, each of which consists of words, each of which consists of characters. As a result the atoms of language are characters. Therefore it is very natural to consider such a scheme in shape description. However, primitives are not unique in shape, which is different from atoms in nature. On the contrary, primitives are chosen subjectively. That is, they depend on and determine how a shape is decomposed.

This is illustrated in Fig. 8.1. A typical character image "A" is shown with three kinds of primitives in panels (*b*), (*c*), and (*d*). Panel (*b*) gives the most basic primitive in image recognition. It is an edge that has two attributes of direction and strength. In character recognition the attribute of direction is important and fully used in practice. Panel (*c*) is based on run length encoding; namely each run is regarded as

FIGURE 8.1 Variety of primitives and their levels: (*a*) Character image "A"; (*b*) edge primitive and its integration of line, hole, and concavity; (*c*) run length coding and its integration of convex blobs; (*d*) chain encoding and its integration of line segments, corner, end points, and branching points.

a primitive whose attribute is width/length and position. In panel (*d*) some thinning preprocessing is assumed. After that the thinned lines are coded by chain encoding. That is, the primitive is chain encoded, and this is also a typical primitive for description of thinned lines and contours of an image. Thus, for the same image, there is a variety of primitives depending on how a given image is described. In addition there are other primitives, as we will see later. So far we have seen that a primitive is likely to be similar to the coding element from which an image is constructed, but it is not always true, as is illustrated later.

The primitives shown above are "primitive" in the sense that they must be integrated to give a global feature. For example, a series of edges whose directions are almost the same constructs a straight line segment as shown in panel (*b*) on the right. We note that such a line can be regarded as a primitive at a higher level. On the other hand, the straight line segment can be seen as a feature of the image of "A." Another example can be given. The edges located along the inside lines of the image "A" can be propagated to their ground (white region), and they collide with each other approximately at the center point, which can be regarded as a primitive representing a hole. This is also a higher-level primitive, which is in turn seen as a feature of the image. More examples are shown in panels (*c*) and (*d*) on the right. Thus we can say that a higher-level primitive is also a feature. These terms are used interchangeably in the literature and also in this book. One significant nuance is that a primitive is used for structural description and a feature is used for differentiating character images, as seen later. So far it has been easy to choose an appropriate primitive that has a physical reality. However, it is not so easy to find a suitable primitive in general, as in Fig. 8.2. In panels (*a*) and (*b*) a pixel whose value is black can be a primitive, but it is too simple to be a constructing element. Therefore it is hard to proceed to the higher level based on such a simple primitive. Although a point is the simplest primitive, it alone has no primitive shape element in whatever the view point. In the extreme case, panel (*d*), we cannot identify the primitive because there are many kinds of elements. Thus it is fundamentally questionable whether we can find a primitive as a physical element of shape. In other words, we must abandon a strictly physical world to an abstract world. In Fig. 8.2 the common primitive is a group of points. In human perception, grouping is a well-known capability, such as was discovered by Gestalt psychologists a long time ago. In the field of pattern recognition, grouping is known as *clustering,* on which some research work has been done but no effective and generally applicable method is known [1]. Specifically, for the case of a series of dot points there exists an effective way of clustering, known as the Hough transformation [2], in which a line on the R^2 plane is represented by a parametric linear equation as

$$\rho = x \cos \theta + y \sin \theta, \tag{8.1}$$

where ρ and θ are parameters, and (x, y) is a coordinate of a point. The above equation represents a line model and also can be regarded as a primitive representation of

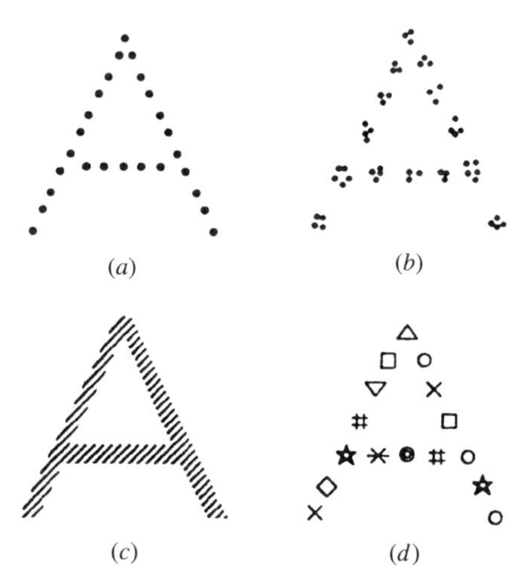

FIGURE 8.2 Some examples of a shape composed of unusual primitives.

a line. Based on the model, a series of dot points are integrated to a real primitive. This is one example of a primitive that is an abstract concept for humans, although it is not applicable to the general cases shown in panels (*b*), (*c*), and (*d*). Elementary particles are also described in very abstract ways that are interesting but of little practical use. However, the time seems to be coming when engineers in the field of pattern recognition will have to consider seriously such a problem. We don't pursue it further here. The objects we treat are *regular* in the sense that physical primitives can be found easily. Even for regular image data, there remain many problems. Some detailed discussion about the definition of a regular shape is mentioned in Pavlidis's paper [3]. Actually there are so many variations in shape that belong to a category that it is not easy to find primitives that effectively represent the variations. Therefore some rules or criteria to select primitives are necessary. Pavlidis offered some useful guidelines which we quote below [3]:

1. Primitives must conform with our intuitive notions of "simpler" components of a "complex" picture.
2. Primitives must have a well-defined mathematical characterization.
3. Primitives and characterization must be subject independent.
4. The shape description by primitive has the same complexity as the raw data in terms of information.[1]
5. When a full set[2] is reduced to a thin set by some type of limiting processes, the "regular" shaped components should be reduced to "regularly" shaped thin components.

Of course it is difficult to satisfy these conditions simultaneously. In particular, condition 3 roughly pertains to the idea mentioned above that character images can be divided into two classes. One is printed character images, and the other is hand-printed or handwritten character images. It is difficult to find primitives that are applicable to both classes because the former is rich in noise and the latter is rich in variations. Therefore, even for a class of hand-printed characters, for example, two kinds of primitives must usually be used if there is noise. We might also consider that multidimensional viewpoints can be more effective sometimes than depending on one kind of primitive.

We continue our examination of the rules to obtain a strategy for seeking primitives. For this purpose, let us consider the examples shown in Fig. 8.3. In panel (*a*) the primitive is a vector whose attribute is length and direction. This is certainly a first candidate for line figures or figure contours. That is, it is intuitive and can be expressed mathematically and thus may be generally applicable. However, the description using this primitive requires many primitives along a line that is concave or convex, as shown in panel (*a*). On the other hand, the primitive shown in panel (*b*) is the monotone function, which has only four kinds of primitives. The primitive is intuitive and also mathematically strictly defined. It can be applied to both fine line

[1] The description differs a little from Pavlidis's.

[2] A subset of the Euclidean plane is full if its boundary is normal and it has a nonempty interior [3].

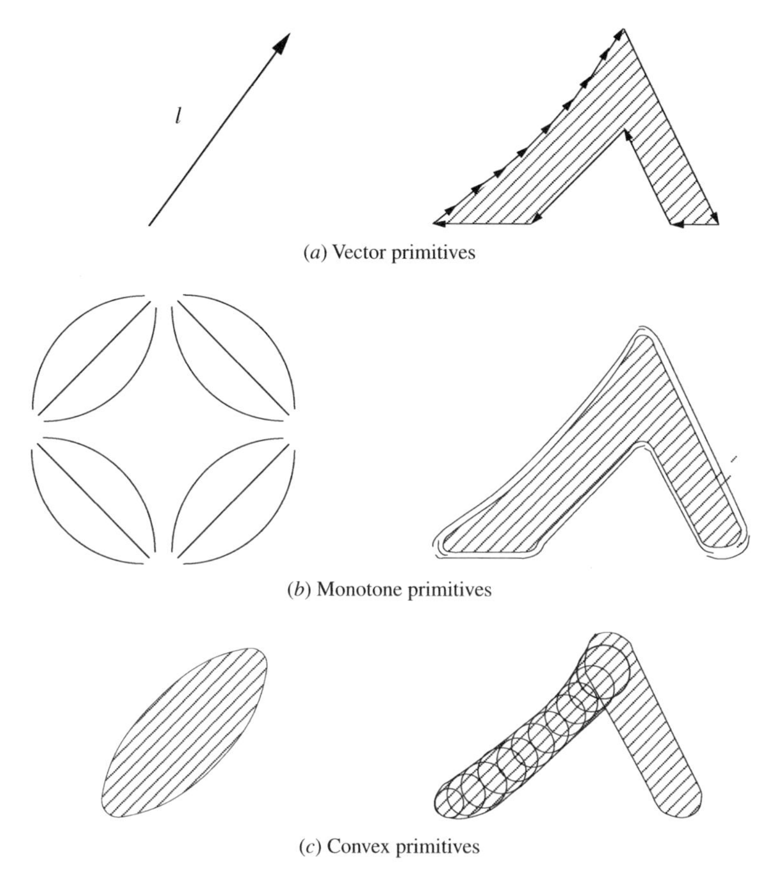

(*a*) Vector primitives

(*b*) Monotone primitives

(*c*) Convex primitives

FIGURE 8.3 Some primitives and their applications. From reference [99], © 1997, Ohm.

and blotchy figures. Here convexity and concavity can be disregarded, since the figure is very simply described. Actually it is too simple for us to discriminate between convex and concave lines, so it runs counter to condition 4. Since the primitive description is hierarchically based, as mentioned before, the monotone primitive is a level higher than a vector primitive. The monotone primitive enables a rough segmentation of line or contours. If a more exact description is needed after the segmentation is done, the segmented line can be described by another primitive (e.g., the vector). Thus condition 4 is satisfied by this hierarchical description, although it necessitates another primitive. A good combination of primitives is required. On the other hand, from this hierarchical point of view, good preprocessing is also important. A thinning process can be used to provide a first approximation of lines. In particular, a graphic representation of lines is an effective way to further describe certain primitives. The thinning process disregards certain detailed information so that an effective and simple primitive can be derived. In this regard a polygonal approximation of an image can be effective if the degree of its approximation is appropriate.

The vertices and edges of a polygonal approximation can be good primitives. For example, the shallow concavity of panel (*a*) can be represented by three vertices and two edges. A vertex has the important attribute of an angle, as we will fully illustrate in the polygonal approximation used in this chapter.

The last example is the primitive of convexity shown in panel (*c*). The same argument can be applied. That is, the convex primitive is intuitive and can be mathematically expressed, but it is not so simple for a description of convex or concave blobs as can be seen in panel (*c*). Therefore a polygonal approximation may be appropriate for a given blob in this case too. Then, besides the mathematical characterization of primitives, the engineering point of view is important. Usually a character is scanned sequentially from top left to bottom right. Therefore, in considering primitives, it is efficient to conform with scanning convention. For example, one row can be taken as a primitive in which run length encoding can be adopted. The crossing method is a simple case of that. That is, it counts only the total number of black runs in each row, disregarding the geometric feature. Furthermore it does not make any connection of black runs to the neighboring rows. The continuity of runs in the vertical direction, which is very important, can be easily obtained by a thinning method based on the LAG, as mentioned earlier. Thus continuity is extracted from consecutive rows, and from this higher primitives are constructed dynamically. Such a scheme was tried historically and is being used in current OCR systems. It is called the *stream-following method,* which suggests its dynamic nature, and we will explain the method fully in this chapter. Now that we have finished presenting a general picture of primitives, feature, and description, we can move on to describing the details of some typical primitives and feature extraction methods.

8.2 CONVEX DECOMPOSITION

As mentioned in the previous section, convexity is a good candidate for a primitive on shape description. The examples that we use here are based on Pavlidis's work [4]. This work is interesting more from a theoretical than a practical point of view. At any rate we began with some mathematical background which is interesting in itself. The arrangement of lines on a Euclidean plane has some interesting characteristics. Figure 8.4 (*a*) shows an example in which four lines are placed in a construct of convex and half-infinite regions. The total number of regions is easily calculated by the number of lines and multiple crossing points. For $\mathbf{x} \in R^2$ let $\lambda(\mathbf{x})$ be number of lines that pass point $\mathbf{x}$. When $\lambda(\mathbf{x}) = 1$, it is called a usual point. When $\lambda(\mathbf{x}) = 2$, the point is called a double point, and so on. Let $M(L) = (p_1, p_2, \ldots, p_s)$ be a set of multiple crossing points. Then we have the following theorem:

Theorem 8.2.1 *Let l_i be ith line arranged on the plane, and let there be n lines. Let $L = l_1 \cup l_2 \cup \ldots \cup l_n$. Then L divides R^2 into the following number of convex regions:*

$$f = 1 + n + \sum_{p_i \in M(L)} (\lambda(p_i) - 1).$$

Actually in panel (*a*), for example, $n = 4$, the number of double points is 3 and the number of triple points is 1. Therefore $f = 1 + 4 + 3 \times (2 - 1) + 1 \times (3 - 1) = 10$ as

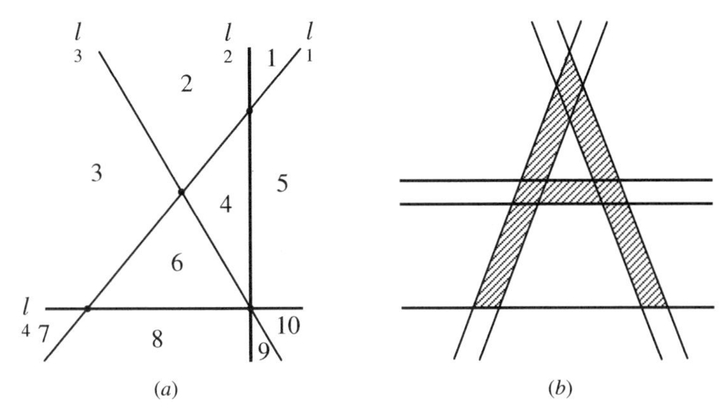

FIGURE 8.4 Nature of lines on a plane: (*a*) Example of Theorem 1; (*b*) example of polygonal description.

shown in panel (*a*). If a shape is approximated by a polygon, then its edges construct lines that divide the polygon into convex regions according to Theorem 8.2.1. This is illustrated in Fig. 8.4 (*b*). Therefore convex decomposition is mathematically guaranteed. However, we need to identify the primitive here. For any further polygonal description it is appropriate to take a *half-plane* as the primitive as shown in Fig. 8.5(*a*). It is convenient to assume that each half-plane generated by a line has a positive or negative sign, such that a line $\vec{ab}$ connects point a to point b with the left side being the positive side and the right side the negative side. We call the plus side a half-plane generated by a line. Now, if we follow a given polygon along its edges counterclockwise, then the polygon is represented by the intersections of half-planes as shown in Fig. 8.5 (*b*), in which $\triangle ABC$ is constructed as an intersection of half-planes, h_1, h_2, and h_3. Notice that on encountering a hole, our direction is changed to clockwise. This is better seen if we follow the polygon's edge on its interior, staying on the left-hand side. Let us describe a systematic convex decomposition method based on the half-plane primitive which considers following *Q-sequences*:

$$Q = \bigcap_{i=1}^{n} h_i, \quad Q_j = \bigcap_{i=1,\, i \neq j}^{n} h_i, \quad Q_{jk} = \bigcap_{i=1,\, i \neq j,k}^{n} h_i, \ldots.$$

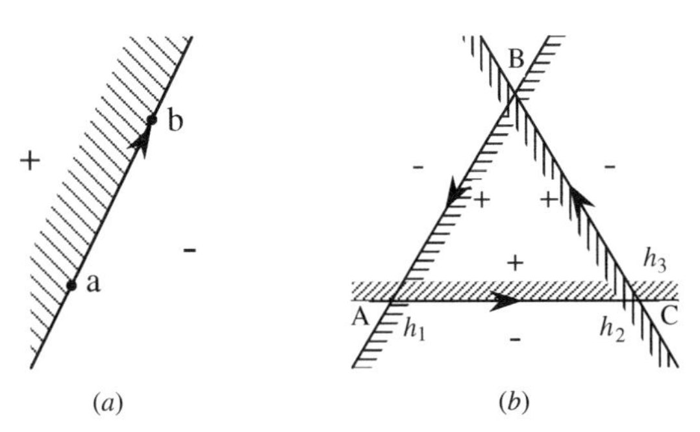

FIGURE 8.5 Some primitives: (*a*) Definition of half-plane; (*b*) example of polygon description by half-planes.

We explain the Q-sequence using the example shown in Fig. 8.6 (*a*). First Q is constructed by the intersections of all the half-planes whose boundaries include the sides of a given polygon. In panel (*a*) we have a polygon with six sides, and $Q = \cap_{i=1}^{6} h_i$, which is $\triangle BHG$. Q_j is constructed from Q by removing the jth half-plane. For example, Q_1 and Q_2 are now the trapezoids *CDEH* and *JIGF*. Next Q_{kj} is constructed from Q_k by removing the jth half-plane. For example, Q_{12} is rectangle *JDEF*. However, this rectangle *JDEF* is no longer a subset of the polygon. In the same manner all the Q_{1j} ($j = 2, 3, \ldots, 6$) are not subsets of the polygon. Therefore we turn back to Q_1, which is a maximal subset of the polygon generated by the Q-sequence. In this sense, the Q_1 is a convex primitive. Let us follow another Q-sequence, for example, Q_{23} is $\triangle AGF$ and a subset of the polygon. Therefore we try Q_{23i} ($i = 1, 4, 5, 6$), in which Q_{23i} ($i = 1, 5, 6$) all are not finite figures and Q_{234} is the same as Q_{23}. Therefore Q_{23} is a convex primitive. In Fig. 8.6 (*b*) the process of constructing convex decomposition is shown as a tree structure, called a Q-tree. The root of the Q-tree is Q, $\triangle BHG$, for example, where $(\bar{1}\,\bar{2}\,3\,4\,\bar{5}\,6, \phi)$ means that to construct $\triangle BHG$ all the half-planes are used. But $\triangle BHG$ is actually constructed by only upper barred half-planes, and ϕ means that no half-plane is excluded to construct $\triangle BHG$. That is, a list of half-planes need to be excluded in order to construct $\triangle BHG$. So we have two sets of half-planes to consider: The set of the former is called the formative list (F.List); the set of the latter is

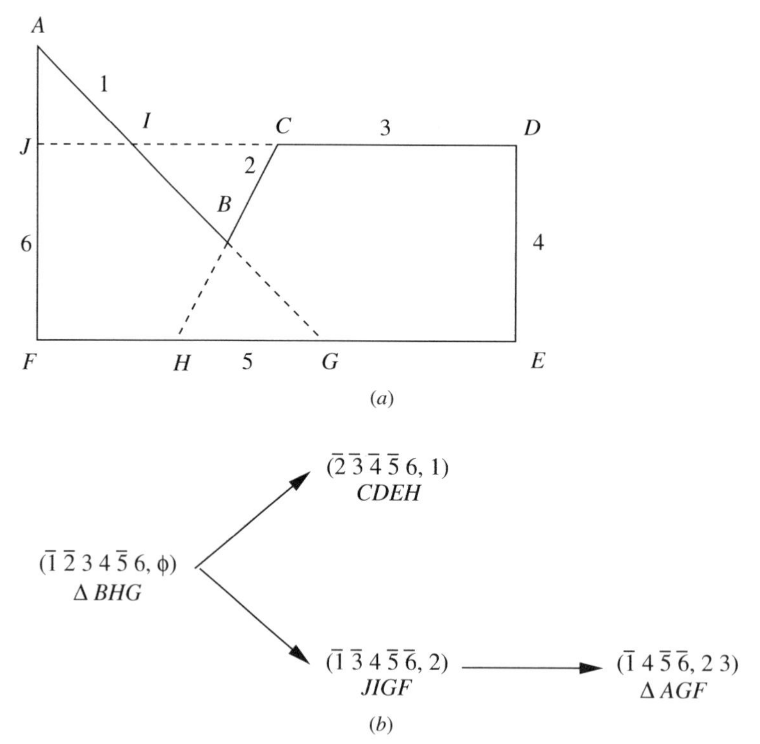

FIGURE 8.6 Convex decomposition: (*a*) Given polygon; (*b*) process of decomposition for the polygon shown in (*a*), expressed as a Q-tree.

called the exclusion list (E.List). For some more exact descriptions, let us introduce some definitions:

Definition 8.2.1 *A nonempty member of a Q-sequence which also is a subset of polygon P is called a nucleus of P if all the previous elements of the Q-sequence are empty.*

Definition 8.2.2 *A nonempty member of a Q-sequence which is also a subset of P is called a primary (convex) subset of P if all the subsequent members of the Q-sequence are not subsets of P.*

Then the following fundamental theorem holds:

Theorem 8.2.2 *The union of all the primary subsets of P equals P.*

Proof For the proof refer to Pavlidis [4].

Although we said that the primary convex subsets in our example consist of primitive convex subsets, it is unfortunately generally not true. That is, in many cases, but not in all, the primary subsets form a minimal cover of *P*. Such an example is shown in Fig. 8.7. The primary subsets are rectangles, *ACRL, NEGJ, BCHJ,* and *MEFL,* while only the first two rectangles form a minimal cover. It is an open question whether there is a way to reduce the representation to a minimal form. The problem can sometimes be solved ad hoc. Actually we can exclude those primary subsets that are contained in the union of two or more other subsets. However, this procedure is not applicable in general. For us humans to find the minimal cover may be intuitively clear, so we might look for some additional criteria for primitive convex subsets, which makes this an interesting open problem. Aside from this problem, we can proceed to describe representation by the convex primitives. Some simple examples are shown in Fig. 8.8, in which convex primitives are represented by vertices and the relationship of connection between them is represented by edges as shown in the middle panel. However, as shown $\triangle$ cannot be distinguished from Y. For this reason the nuclei are introduced, as defined earlier. Intuitively the nuclei are connecting polygons that are common relative to the convex primitives shown in panel (*a*) as shaded polygons. By assigning a vertex to each nucleus, more exact graphic representations are obtained as shown in panel (*c*). Now the degeneration of $\triangle$ and Y is resolved. Thus, by introduction of nuclei, we see that for some figures degeneration can be resolved. However, we pursue this problem further in examining Fig. 8.9. Strictly speaking, we are looking at figures such as $\{\angle, \neg\}$, $\{\top, \bot\}$, and $\{\times, +\}$. All the shapes have the same graphic representations despite the apparent differences in their structures. Intuitively these differences are due to connections of nuclei with convex primitives. For example, in the L-shaped figures shown at the top, the corner nucleus connects with the two convex primitives at only one side. However, in the T-shaped figures shown in the middle of the figure, the nucleus connects with the horizontal convex primitive at two sides and with the vertical convex primitive at one side. Such connection relations can be represented by

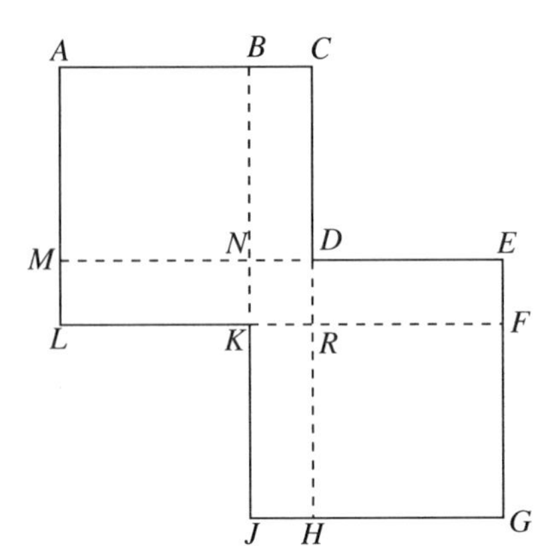

FIGURE 8.7 Polygon with a nonminimal representation by primary subsets. From reference [4], © 1998, Pergamon Press.

the number of edges between vertexes as shown in the figure on the right. Quantitatively this connection number is evaluated by the following formulation:

$$T(P, N) = |E_P \cap F_N|,$$

where F_N is the N nucleus F.List and E_P is the P convex primitive E.List. The notation $|L|$ denotes the number of elements included in L. In the case of panel (a), $T(P.N) = 1$ in both convex primitives. In the case of panel (b), the E.List of the horizontal convex primitive, say, P is $E_P = (h_5, h_6)$ and the F.List of the nucleus, say, N is $F_N = (h_1, h_2, h_5, h_6)$. Therefore $E_P \cap F_N = (h_5, h_6)$, and so $|E_P \cap F_N| = 2$ as expected. This is explained intuitively as follows: The two half-planes h_5 and h_6 of the nuclei are just the excluding half-planes of the horizontal convex primitive. This means that the horizontal convex primitive is separated by the two half-planes above. Thus by the multiple edge relation the degeneration of each pair is resolved. On the other hand, among the pairs shown in the figure we cannot resolve the degeneration. This problem is not serious, and it concerns the capability of the structural representation of a graph. In other words, so far the graphic representation is invariant to position and rotation. If we want to differentiate a pair further, we need the appropriate coordinates on a Euclidean plane. Eventually we will need these values in character recognition, where higher hierarchical representation is required. We do not pursue this problem here further, since it will be discussed fully later.

Here we notice another essential point of this scheme of convex decomposition. This is illustrated by the example shown in Fig. 8.10. Real data are not ideal, and there are some redundant or nonessential convex primitives appearing as a result of the convex decomposition such as c and a in the figure. They should be merged to the key convex primitives as $c \rightarrow d$ and $a \rightarrow b$. Of course then we violate the condition for the "convex primitive" being the merged primitive, $a \cup b$. However, this is necessary in practice if the algorithms of the convex decomposition are to be useful. That is, some "roughness" must be introduced to the primitive that does not affect the

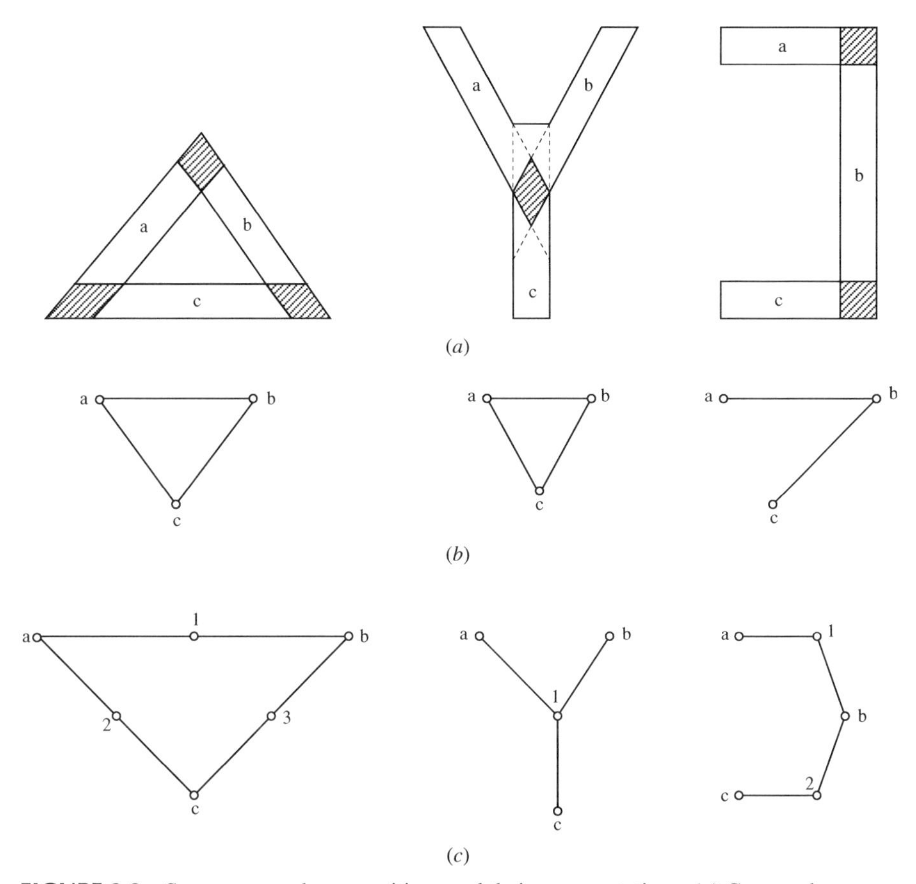

FIGURE 8.8 Some convex decompositions and their representations: (*a*) Convex decomposition with nucleus shaded; (*b*) simple graph representations not using nucleus; (*c*) more precise graph representation using nucleus. From reference [99], © 1997, Ohm.

mathematical characteristics. This is likely to become a general principle in pattern recognition. Although convexity is very simple as a primitive and the theory is beautifully constructed based on a polygonal approximation that itself has no problem in practice, there is still an essential problem in the descriptive capability of "roughness" or "flexibility." This is a good lesson. The reader can refer to related work mentioned in the Bibliographical Remarks at the end of the chapter.

8.3 STREAM-FOLLOWING METHOD

We introduced the stream-following method in the previous section and earlier in Chapter 5 in our discussion of thinning methods. Actually graphic representations like the run length graph (RLG) are a main product of the stream-following method. Here we describe how to construct such graphic representations systematically, in

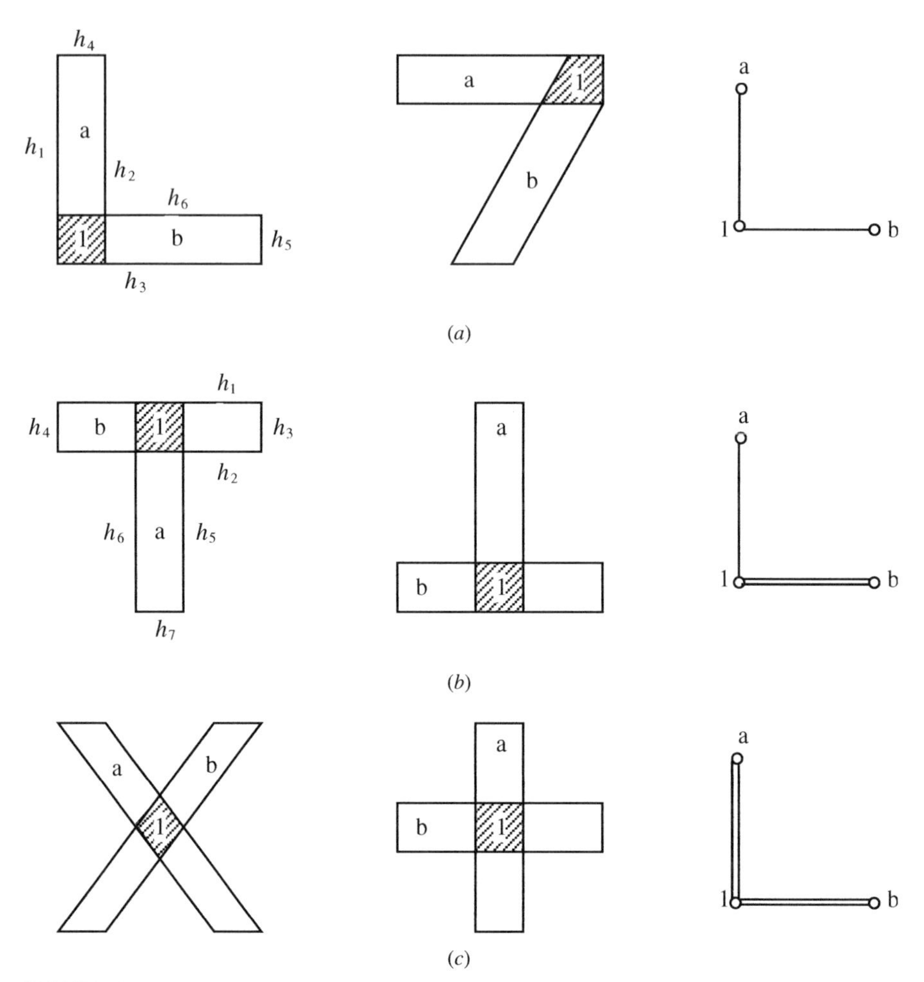

FIGURE 8.9 Some convex decompositions that need multiple edges between vertices: (*a*) Figures without multiple edges in their graphic representations; (*b*) figures with one multiple edge; (*c*) figures with double edges. From reference [55], © 1996, Ohm.

terms of their expansion and their graphic properties from a character recognition viewpoint.

8.3.1 Nadler's Quasi-topological Code Generator

Nadler's approach [6] to the stream-following method is very simple. The lowest-level primitive is nothing but a 2×1 window, as shown in Fig. 8.11. As mentioned before, stream following is dynamic in nature, so the scanning order is important. The 2×1 window is scanned from the left bottom to the top. Then the next scanning takes place overlapped one column to obtain information of connectivity of a given image. Our objective is to construct a graphic representation as shown in Fig. 8.12 (bottom). The graph has four kinds of vertices, S (Start), E (End), C (Close), and O

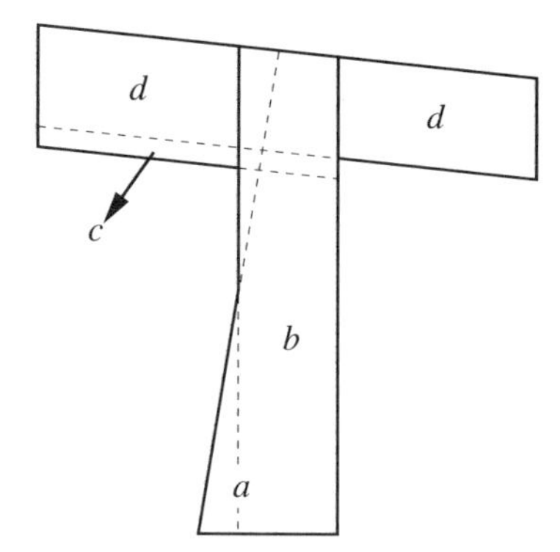

FIGURE 8.10 Example of the rigid property of convex decomposition, which is a frequent occurrence in practice. From reference [99], © 1997, Ohm.

(Open). These are primitives of a given image. For coding purposes one more primitive N (Null) is used. The naming of these primitives reflects the dynamic method of feature extraction based on the stream-following approach. Start and End are obvious. Close and Open denote that the streams following boundaries are closed and opened, respectively. Null mean no feature, only a connection.

Now we will describe how to obtain such a graphic representation using the 2×1 window. It is done by constructing an automaton which was named the *quasi-topological code generator* by Nadler. The naming of "quasi-topological" is somewhat confusing viewing from a strict mathematical viewpoint. However, the term *quasi-topology* has been widely used in the field of character recognition. At any rate we begin with the illustration of the automaton using a sample shown in Fig. 8.12. First we notice that since a 1×2 window is used, only four kinds of input patterns are assumed, (0, 0), (0, 1), (1, 0), and (1, 1). The outputs are five kinds, S, E, C, D, and N. The binary image shown in panel (*a*) is scanned by the 1×2 window from the left bottom to the top. The first input pattern is (0, 0) of course, and we assume that the initial state is labeled as "0." The input series of (0, 0) does not change the initial state of the automaton, which is shown in Fig. 8.13. Thus the 1×2 window proceeds to the tip of the image over the whiteground. That is, (0, 1) is input to the "0" state, and the state of the automaton changes its state to "1." To this state there is a possibility that some (0, 1) patterns are input to the "1" state, but the automaton must keep its "1" state. Therefore a loop is necessary to the state. After a series of (0, 1), the (0, 0) pattern appears as shown at the left top. Then the automaton outputs "S," and the state returns to "0." This behavior of the automaton is indicated in Fig. 8.13 by (○), (,), and [·] which denote state, input, and output, respectively. After some further sequential (0, 0), the 2×1 window reaches the top side of the frame and moves to the right bottom by one column. The first input is (1, 0) which is input to the "0" state. This is a new state, and we assume that the automaton moves next to state "2." A series of (1, 0) patterns can occur, so a loop is necessary. Next in our example, the input is (1, 1), which is a new state, so we assume that the automaton transits to state "3." Again a loop is necessary for the series (1, 1) to move to a new state. Here (0, 1) is input, and since it is a new state,

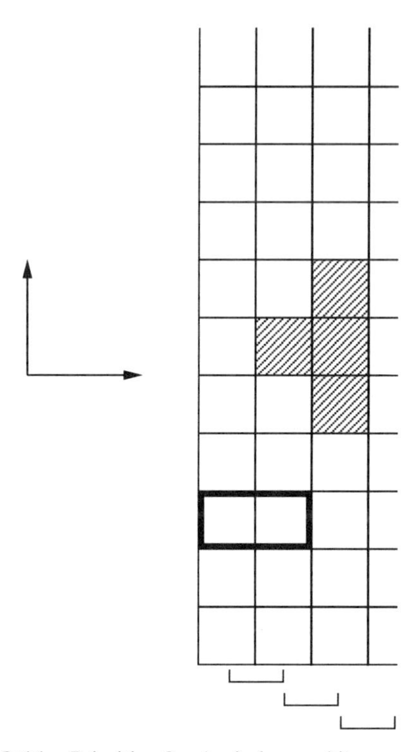

FIGURE 8.11 Primitive 2×1 window and its scanning order.

we assume that the automaton transits to state "5." The labeling of each state is arbitrary, but here we follow Nadler's labeling. Finally (0, 0) is input, and the automaton returns to "0" state and outputs N, which denotes nothing but a connection. Until this point it follows the transition diagram shown in Fig. 8.13. In a similar manner a complete transition diagram can be constructed.

The final result is shown in Table 8.1, which gives a transition table. In all, seven states are necessary but very simply described. Notice that only five states were described in Nadler's original transition table. This is because he disregarded the patterns of the more complex cases as shown in Fig. 8.14. In other words, his cases exclude complex boundaries and focus on a *simply connected boundary.* If we can assume some thinning of lines preprocessing, then the assumption of a simply connected boundary would be satisfied, although some side effects might be expected. Therefore we will show how to construct a graphic representation from outputs of the automaton, using such an example.

Let S be output so that a vertex S is generated at column 1. Then, by the next columnwise scanning numbered as 2, N is output, and a dummy vertex is generated which is necessarily connected to the S vertex. Likewise the outputted N's generate dummy vertices which are connected as shown in Fig. 8.12 (bottom). The 1×2 windows reaches the second tip of the image, and a new vertex S is generated. In the next columnwise scanning at column 6, a vertex 0 is generated which is connected to the new vertex S in the order of occurrences of the output. The dummy vertex for N in

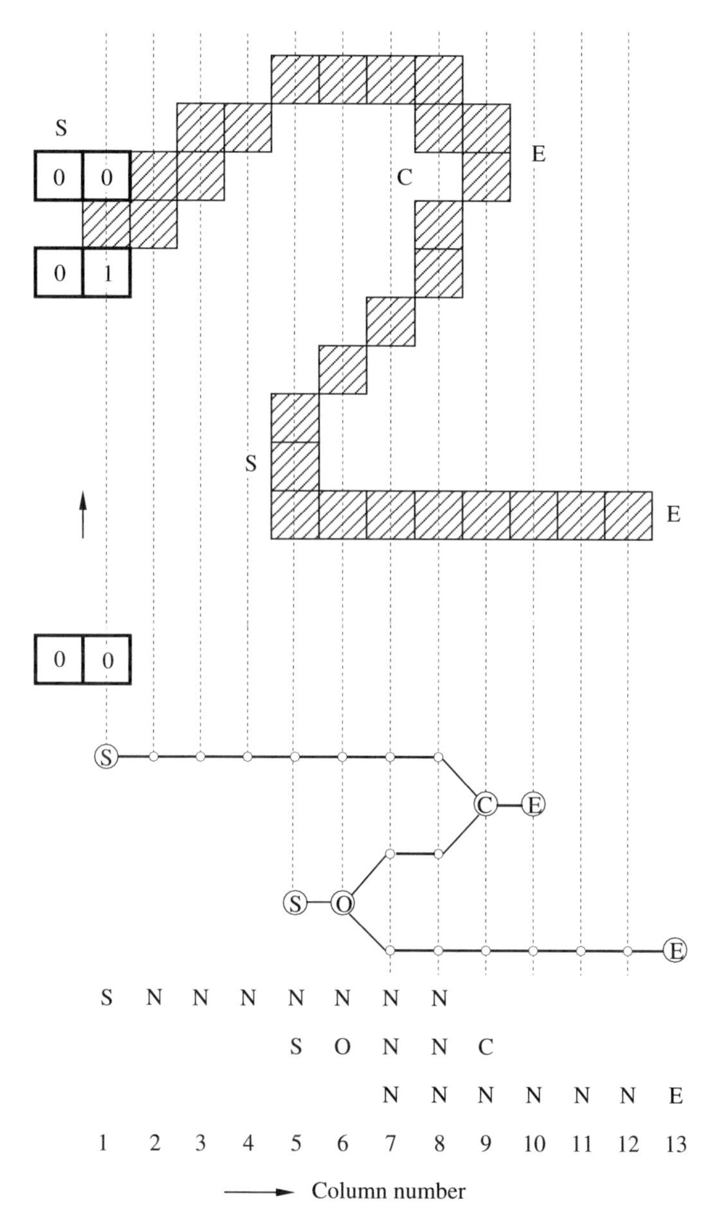

FIGURE 8.12 Construction of a LAG-like graph.

the same columnwise scanning is connected to the previous N vertex, respecting the order. In the scanning at column 7, three dummy vertices are generated. The first two vertices are connected to the 0 vertex. Usually a 0 vertex requires a branch, meaning two N's. On the contrary, at column 9 a new vertex C is generated; then the previous two N vertices are connected to the C vertex. That is, the two streams are merged at vertex C. A new vertex E generated at column 10 is connected to the C vertex fol-

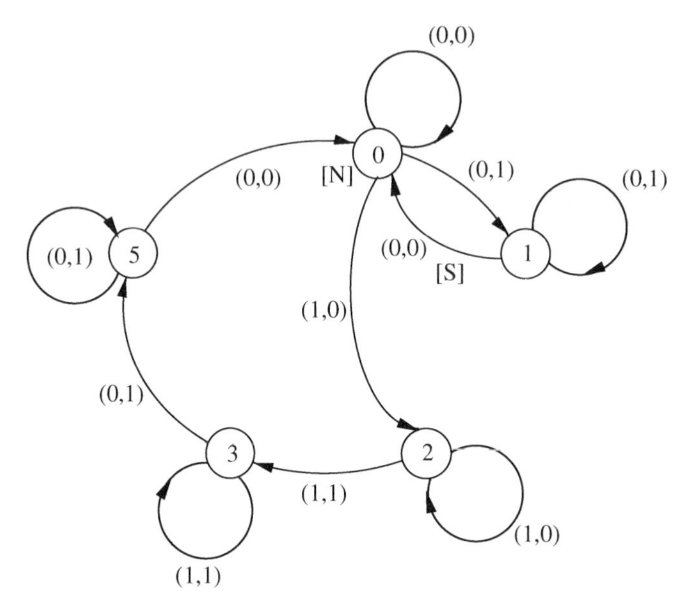

FIGURE 8.13 Nadler's quasi-topological code generator applied to the sample shown in Fig. 8.12 (top); this is not a complete transition diagram of the automaton.

lowing the established order. But we need a more precise algorithm to construct such a graph for a more complex image pattern.

8.3.2 Representation of Patterns Based on the Stream-Following Method

In the previous section we showed a way of coding of an image based on the stream-following method, such as was illustrated by the graph in Fig. 8.12. In this section we

TABLE 8.1 Transition Table of Nadler's Quasi-topological Code Generator

State	Input			
	00	01	10	11
0	0	1	2	3
1	0[S]	1	4	3
2	0[E]	5	2	3
3	0[N]	5	4	3
4	0[N]	6[O]	4	3[O]
5	0[N]	5	7[C]	3[O]
6	0[N]	6	7[C]	7[C]
7	0[N]	6[O]	7	6[O]

Source: From reference [5] with corrections.

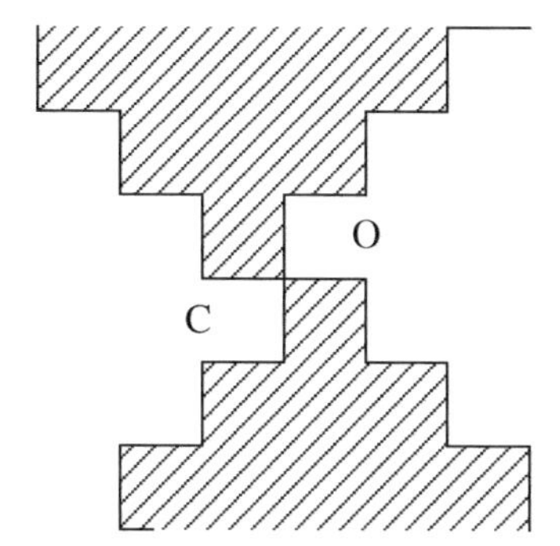

FIGURE 8.14 Complex boundary such as is disconnected if 4-connectness is adopted.

will discuss the coding in some detail. As will become clear, the purpose of this section is not only to discuss coding but also to introduce the basic problem of pattern recognition of "equivalence" relations.

It should be obvious that the coding described in Fig. 8.12 is redundant; the coding can be represented by a graph such that all the dummy N vertices are reduced to make a line. However, in general, metric information is disregarded in a graph, so it is too flexible. In other words, too many types of patterns can be collapsed into a graph. In this sense we need some intermediate representation in pattern recognition. Let us look first at simple coding based on the stream-following method. Rather than a formal presentation, we will consider an example of *topo-coding,* which is shown in Fig. 8.15 (*a*). Comparing with the redundant coding shown in Fig. 8.12 (bottom), we note that every column includes some elements of the set {S, E, O, C}. That is, a column that consists of only N's is contracted. We call such coding *the first vertical quasi-topological coding.* However, this coding method is not sufficient to represent a pattern compactly. Actually, although the semi-graphical patterns shown in Fig. 8.16 have a common "similar" pattern, all the patterns differ. For example, consider the numeral "3," and notice the end points located at the left side of the pattern. These end points are regarded as starting points from the stream-following perspective. Their related projective positions on the x-axis do not matter in terms of "3." Therefore we need a more compact coding so that these variations can be absorbed. Such coding is shown in the Fig. 8.15 (*b*). Informally this coding is constructed as

$$\text{S, SN, NSN} \rightarrow \text{sss},$$

$$\text{NNC, NC} \rightarrow \text{cc},$$

$$\text{NE, E} \rightarrow \text{ee}.$$

That is, further contraction of N's is performed, which is called *the second vertical quasi-topological coding.* According to this coding, all the variations shown in Fig. 8.15 are represented by one coding as

sss, non, cc, ee.

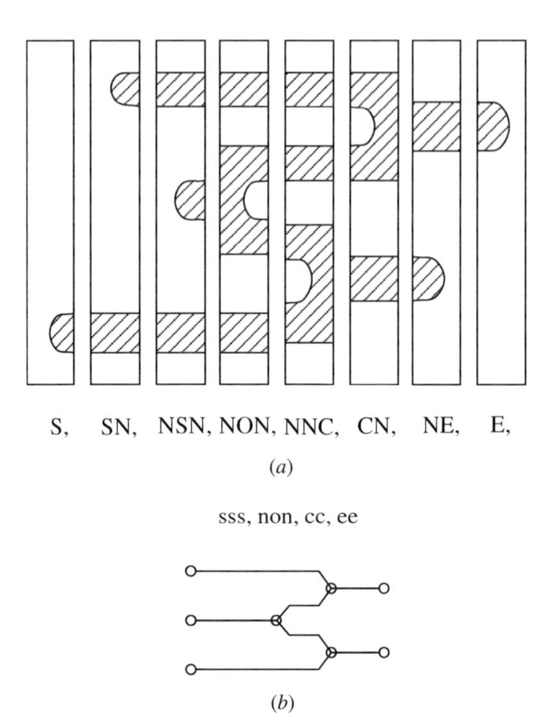

FIGURE 8.15 Making the redundant graph compact: (*a*) Intermediate representation, called topo-coding; (*b*) a compact representation.

In order to obtain the coding exactly, we need to extract hidden relations among vertices over the ground (white region), which is called the *dependent relation graph.* One example for the pattern of (sss, non, cc, ee) is shown in Fig. 8.17. We note that each vertex has its own two white sides, except for two concave types of vertices: o (open) and c (closed). Obviously such two vertices have only one white region by the nature of their connectivity. Taking a dynamic view, all the vertices have either one input edge and three outlet edges or three input edges and one outlet edge. For vertex s_1, there are one white edge input from the ground and three outlet edges in which one is black ordinal one and the other two outlets are spanned over the two grounds separated by the black stroke edge. For example, the vertex s_1 has a "0" input edge, a "2" black outlet edge, and "1" and "3" white outlet edges. The "1" white outlet edge is connected over the ground to the nearest vertex s_2 in its ground region. On the other hand, the "3" outlet white edge is connected to the vertex e_1 which is the nearest vertex on the opposite side. Note that the "3" cannot be connected to c_2 because it must have only one white input edge "9" in its bay(cavity) constructed by the edges "2" and "12." By these illustration we think that the reader can understand the manner of construction of the dependent relation graph. A formal description is omitted. This graph is interesting for us because it describes the relations over the ground.

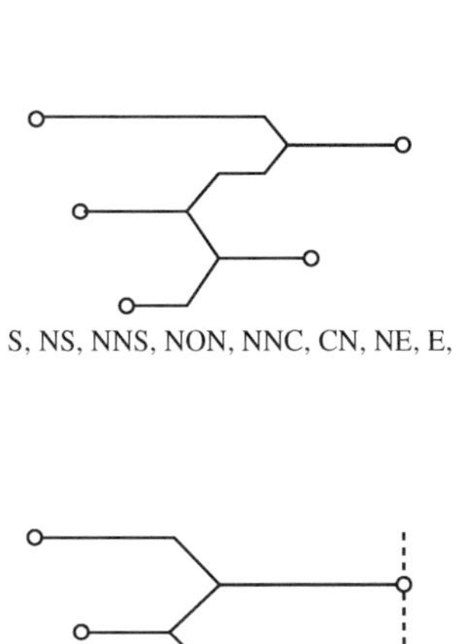

S, NS, NNS, NON, NNC, CN, NE, E,

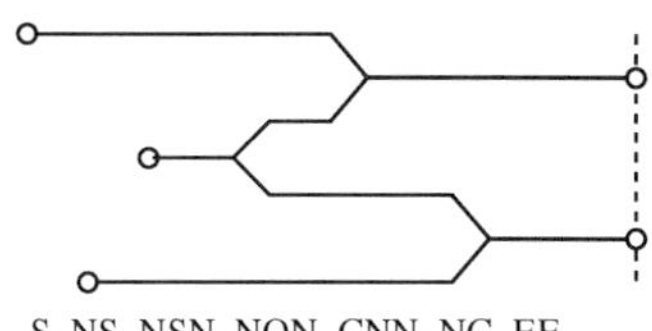

S, NS, NSN, NON, CNN, NC, EE,

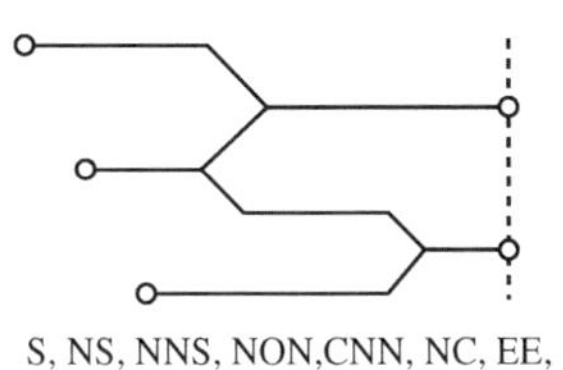

S, NS, NNS, NON,CNN, NC, EE,

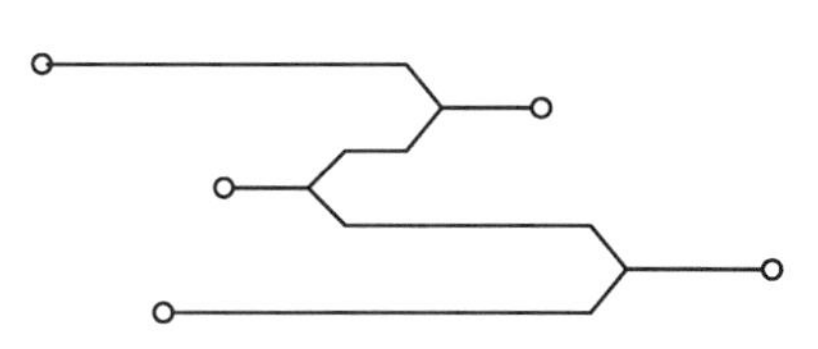

S, NS, NSN, NON, CNN, ENN, C, E,

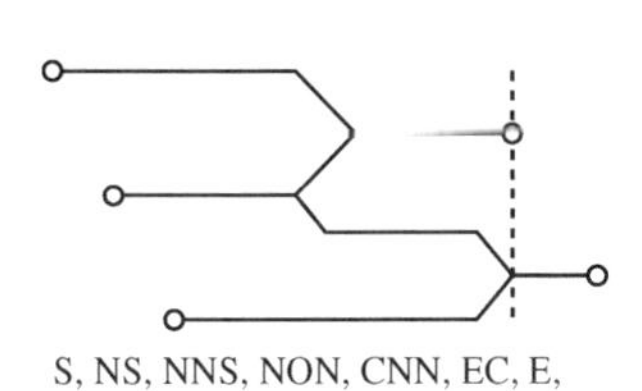

S, NS, NNS, NON, CNN, EC, E,

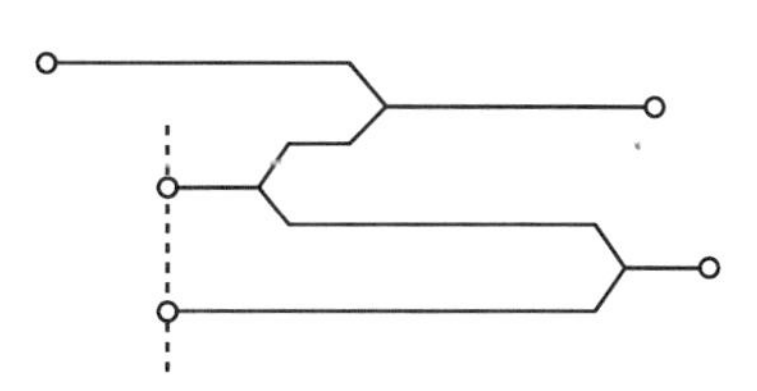

S, NSS, NON, CNN, NC, EN, E

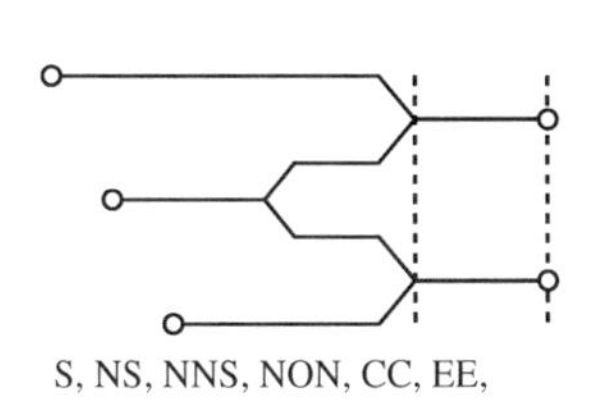

S, NS, NNS, NON, CC, EE,

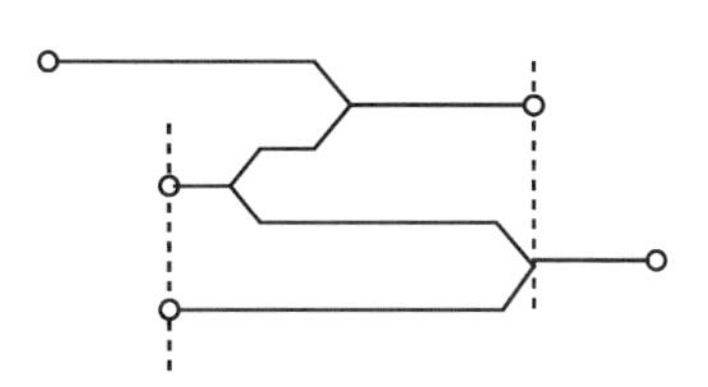

S, NSS, NON, CNN, EC, E,

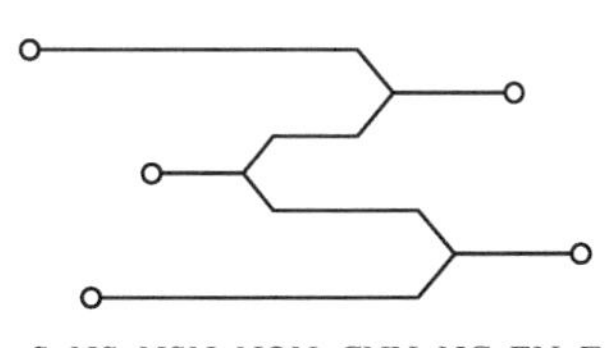

S, NS, NSN, NON, CNN, NC, EN, E,

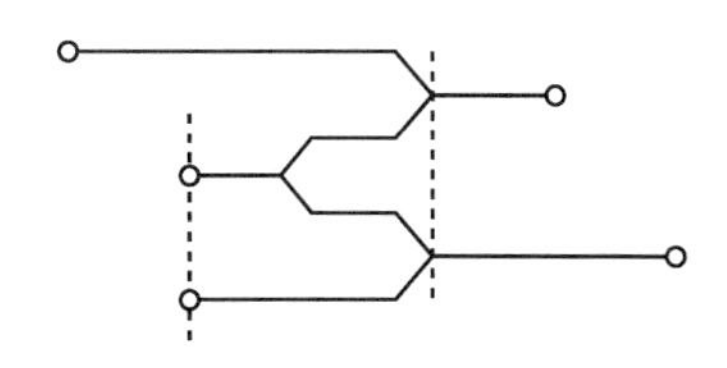

S, NSS, NON, CC, EN, E,

FIGURE 8.16 Examples of top-coding of “3” that can be regarded as the same pattern.

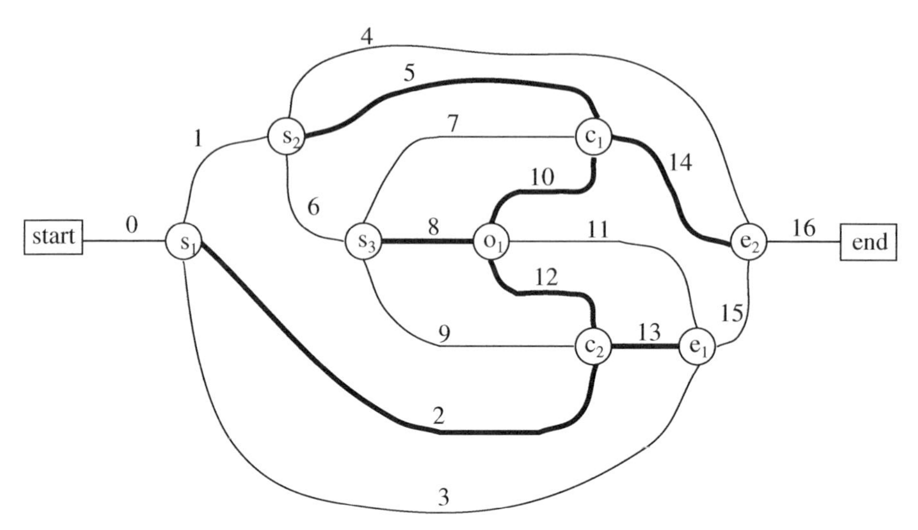

FIGURE 8.17 Dependent relation graph for the coded pattern (sss, non, cc, ee).

Now we look at the usefulness of this dependent relation graph using some examples. These examples are drawn in Fig. 8.18 (a') and represented a one-directional graph shown in panel (b'). However, it is reasonable to differentiate three kinds of shapes in their symbol representation. For each of these configurations, corresponding dependent relation graphs are drawn in panel (c'). Their second kind quasi-topological codings are all different as we intended, in which the (ss, no, nen, c, e) shown across the configurations in panel (c') need a dependent relation graph. That is, scanning from the left the first vertex s is connected to the second s vertex by a white output edge, which is represented as (ss, . . .). Other coding can be obtained by informal contraction of N as mentioned before. However, it is a delicate and important problem to what degree we contract a given image into a simple symbol representation. Too much contraction causes too much degeneration of shape. We have to consider some trade-off between simple coding and too much degeneration. We now move to the degeneration problem of the stream-following method.

Degeneration Problem and Its Resolution There is an intrinsic degeneration problem in the stream-following approach because the stream following is only in one direction such as vertical or horizontal. A very typical example is shown in Fig. 8.19 in which four kinds of characters, "9," "6," "4," and "0," have the same coding as (S, O, C, E) with high probabilities, as can be guessed by looking at their shapes. Note that we consider horizontal quasi-topological codings here. From the coding (S, O, C, E), we imagine "0," but in that coding are hidden three other character shapes.

In order to resolve this degeneration, we must use a metric length as shown in Fig. 8.20 (a) and (b) for five functions defined by using lengths a, b, c, and h. Based on these functions, the degeneration is resolved by a tree structure as is shown in panel

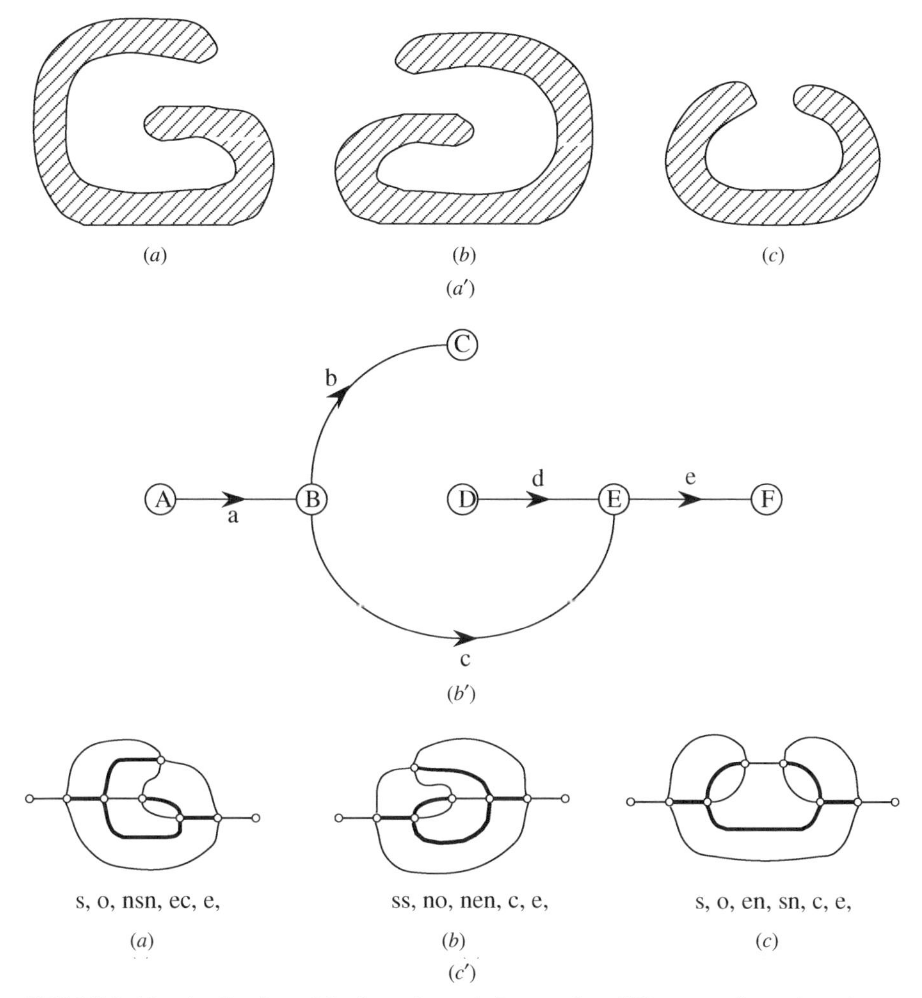

FIGURE 8.18 Application of the dependent relation graph to different configurations shown in (*a*′); (*b*′) the directional graph; (*c*′) the dependent relation graphs.

(*c*). The tree is generated by a function f_0 at its root. The filtered input data to the root are divided into three classes: (4, 9) *class,* (0, 4) *class,* and (6) *class;* and these correspond to $class < 85$, $85 \le class < 125$, and $125 \le class$, respectively. The classifications are very stable. That is, they are robust with respect to small changes of thresholds for the 13,000 samples of data (handprinted numerals). The merit of the stream-following approach is in its high-processing speed. An actual OCR system, called the Recognition Filter, which was created by Nakajima achieved about 260 cps on a SPARC station IPX, 32-bit SPARC CPU, 40 MHz, 28.5 MIPS, using the tree structure mentioned above [8].

FIGURE 8.19 Different numeral images coded to the same (S, O, C, E).

In sum, multidirectional quasi-topological codings have proved effective in resolving degenerations. Usually two directions, vertical and horizontal, are used. In addition the two diagonal directions can be useful, and this has been tried by some researchers mentioned in the Bibliographical Remarks to this chapter.

8.4 POLYGONAL APPROXIMATION

We have seen that polygonal approximation provides an important means to boundary decomposition. It can be done by either keying on the edge/side or keying on the angle/curvature of a polygon. In this section we focus on the first approach. Another later section will be devoted to the second approach.

8.4.1 Mathematical Representation of Approximation Criterion

Let us begin with some mathematical background for polygonal approximations. For convenience we use a one-dimensional wave form $f(x)$, which is assumed to be continuous on some interval $[a, b]$ on the real axis, although it is given by discrete points in practice. Despite this simple representation, there is no loss in generality, and rather the description becomes more transparent. That is to say, the problem of polygonal approximation can be restated as the problem of approximating $f(x)$ by some simple functions.

The first point of the approximation problem is selection of the "simple" functions. The selection itself, however, cannot be done mathematically. This is divided appropriately considering the purpose of the approximation. For example, we have used a trigonometric function as a Fourier descriptor. The nth-order polygonal function is typical for simplicity. For the purpose of obtaining a smooth function, the cubic spline function is generally used in graphics. Later, when we describe curvature, we will introduce this function. The second point of the approximation is the metric of the approximation: That is, how do we decide the distance between $f(x)$ and its approximation $F(x)$. Viewing this problem in general, it can be rephrased as a problem on how to decide a scalar value for a given function $g(x)$, namely a functional $\Phi(g)$. For example, the most frequently used functional is

$$\Phi(g) = \left(\int_a^b g(x)dx \right)^{1/2}. \tag{8.2}$$

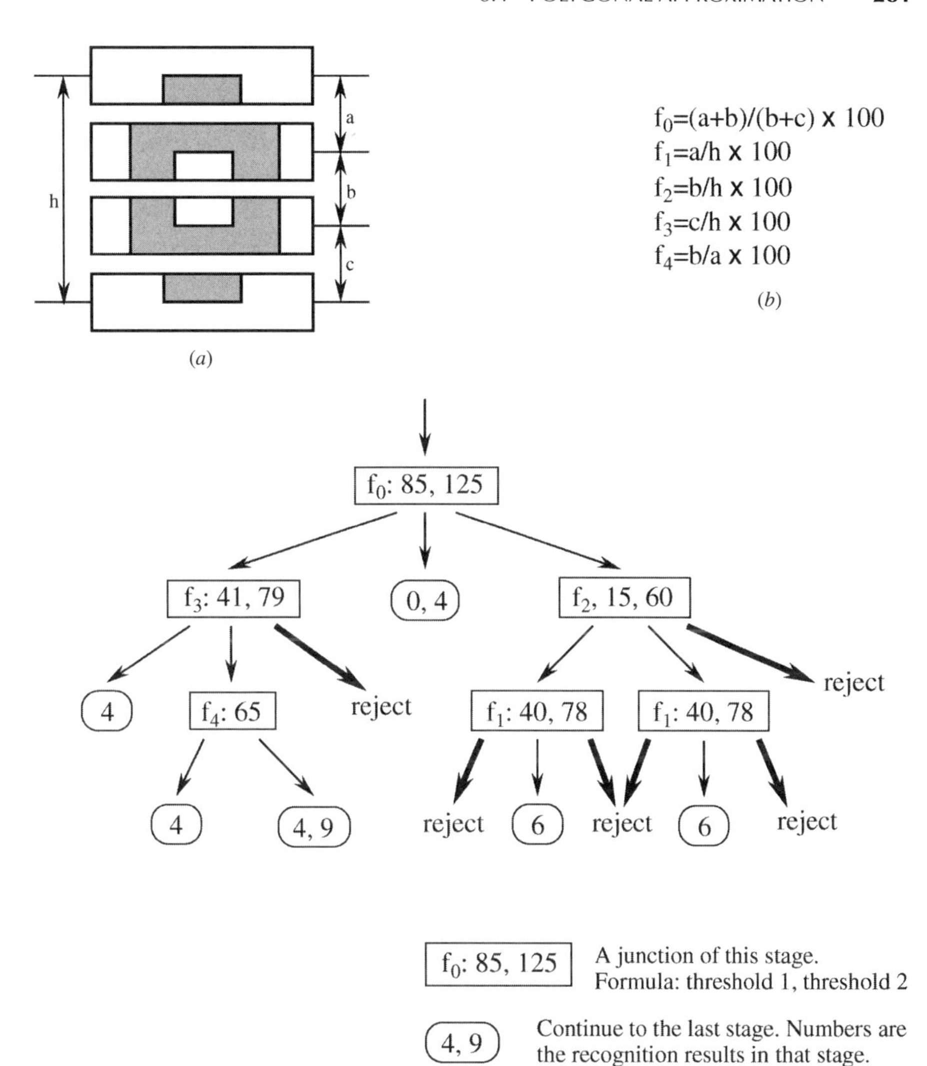

FIGURE 8.20 Classification using Euclidean metrics: (*a*) Vertical connection representation; (*b*) simple formulas based on the Euclidean metric; (*c*) subtree for the classification of the degenerated shapes. From reference [8], © 1998, IEEE.

Let us assume that $\Phi(g)$ is finite so that g is what is called a square-summable function, as mentioned before. The set of all functions that are square summable is usually designated by the symbol L_2. $\Phi(g)$ is denoted as $\|g\|$, called the *norm*. The following properties hold:

1. $\|g\| \geq 0$ and $\|g\| = 0$ if and only if $g(x) \sim 0$.
2. $\|kg\| = |k| \cdot \|g\|$, where k is an arbitrary scalar value, in particular, $\|-g\| = \|g\|$.
3. $\|f + g\| \leq \|f\| + \|g\|$.

Obviously there is strong analogy with the absolute value $|x|$ of real number x. The utility of the absolute value lies in its giving a distance such that the distance between x and y, $\rho(x, y)$, is given by $\rho(x, y) = |x - y|$. Applying the analogy, we can define the following distance between f and g, $\rho'(f, g)$:

$$\rho'(f, g) = \|f - g\|. \tag{8.3}$$

Thus, in taking a norm, we obtain a definition of distance. We notice, however, that $\rho'(f, g)$ must satisfy the three conditions of distance for the analogy to hold. These conditions are those just described above: conditions 1 to 3, to which we add the following fourth condition:

$$\rho'(f, g) = \rho'(g, f). \tag{8.4}$$

The relation of this equation to our distance definition (8.3) is obvious.

Given this preliminary background, let us now formalize the approximation of a function $f(x)$ more rigorously. First we need to consider the representation of an approximating function $F(x)$, which is given a priori. The representation must have a form that is easy to adjust without changing it. For example, an nth-order polynomial function can be represented as

$$F(x) = a_0 + a_1x^1 + a_2x^2 + \cdots + a_nx^n. \tag{8.5}$$

Therefore the problem of its approximation is how to adjust a set of parameters, $A = (a_0, a_1, \ldots, a_n)$, where each parameter a_i is a real number. The A is included explicitly in $F(x)$ as $F(A, x)$, and the problem of approximation of a function $f(x)$ by $F(A, x)$ is stated as follows: Find A^* satisfying the inequality

$$\|F(A^*, x) - f(x)\| \leq \|F(A, x) - f(x)\| \qquad \text{for every } A. \tag{8.6}$$

If we can find such an A^*, it is an optimal solution of the approximation problem in the sense of the norm defined above. However, we need to think hard about the "optimization" because it usually requires a lot of computing time to obtain an optimal solution. Rather, if we give up on a rigorous optimal solution and satisfy ourselves with a near-optimal solution, then the computing time will be shortened drastically. Therefore it is important to consider the trade-off between optimal and near-optimal solutions. Furthermore we should consider whether an optimal solution is really critical to our recognition needs.

The approximating function $F(x)$ is represented in practice as

$$F(A, x) = \sum_{i=0}^{n} a_i\phi_i(x), \tag{8.7}$$

where $\phi_i(x)$, $i = 0, \sim n$ are linearly independent. Therefore, by the linear independence imposed on $F(A, x)$, we have for an arbitrary x that $F(A, x) = 0$ holds only when all the $a_i = 0$. In other words, so long as $\sum_{i=0}^{n} | a_i | > 0$, there exists x such that $F(A, x) \neq$

0. The nth-order polynomial function satisfies this condition, and the orthogonal function series (e.g., 1, cos x, sin x, cos $2x$, sin $2x$, . . .) in $[-\pi, \pi]$ satisfy the condition as Fourier descriptors, as we have mentioned much before. Finally the problem of approximating functions is whether there exists a solution and whether it is unique if it exists. Here we need not be so nervous in practice; the solution is guaranteed for a piecewise linear function which is what we are dealing with. In general, for approximating a continuous function $f(x)$ by a polynomial form, there exists the famous theorem of Weierstrass:

Theorem 8.4.1 (Weierstrass Approximation Theorem) *Let $f(x)$ be a continuous function given on an interval $[a, b]$. Then for a given $\varepsilon > 0$ there exists an n such that there is an nth-order polynomial function $P_n(f, x)$ that satisfies the condition*

$$|P_n(f, x) - f(x)| < \varepsilon, \qquad x \in [a, b]. \tag{8.8}$$

Proof The basic strategy follows from *the theorem,* for it is to be proved that there exists a piecewise linear function $F(f, x)$ approximating $f(x)$ as mandated by the theorem. The interval $[a, b]$ is divided into N subintervals as

$$a = x_0 < x_1 < x_2 < \cdots < x_N = b,$$

and on each interval $F(f, x)$ is given as

$$F(f, x) = \frac{(x_{k+1} - x)y_k + (x - x_k)y_{k+1}}{x_{k+1} - x_k}, \tag{8.9}$$

where $y_k = f(x_k)$, $x_k < x < x_{k+1}$.

The function takes values of $f(x)$ at the two end points of subinterval $[x_k, x_{k+1}]$, and between them it is a line connecting points (x_k, y_k) and (x_{k+1}, y_{k+1}). It is intuitively obvious that on increasing N, the absolute difference between $f(x)$ and $F(f, x)$ will enter in the range of ε eventually. For an exact proof refer to [9]. ■

We proceed to consider the meaning of norm. For convenience we denote a norm in L_2 space as $L_2(f)$, and we generalize it as

$$L_2(f) = \left[\int_a^b |f(x)|^2 dx\right]^{1/2}, \tag{8.10}$$

$$L_p(f) = \left[\int_a^b |f(x)|^p dx\right]^{1/p}, \tag{8.11}$$

where $p \geq 1$. $L_p(f)$ has properties of the norm, which is proved by using Minkowski's inequality: $1/p$

$$\sqrt[1/p]{\int_a^b |f(x) + g(x)|^p dx} \leq \sqrt[1/p]{\int_a^b |f(x)|^p dx} + \sqrt[1/p]{\int_a^b |g(x)|^p dx}. \tag{8.12}$$

Then for optimization the following form is used in practice:

$$\int_a^b |F(A, x) - f(x)|^p dx. \tag{8.13}$$

The calculation of $\sqrt[1/p]{\cdot}$ is nothing more than changing resolution, or normalization in some sense, so essentially the form described above is enough.

Now let us consider how the situation changes when increasing p. Let $|F(A, x) - f(x)| = \varepsilon(x)$. This means that a large weight is imposed for large $\epsilon(x)$. For $p = 1$, uniform weight is taken for all the $\varepsilon(x)$. What happens when p is extremely large? It means that we must take $\max_{x \in [a,b]} |F(A, x) - f(x)|$. In the limit, this is called *Chebyshev's norm* and denoted as L_∞. For Chebyshev's norm, the existence of optimal solution is proved in the sense of the formulation of (8.6). We can observe the difference between $L_2(f)$ and $L_\infty(f)$ in Fig. 8.21. In panel (*a*) notice that for $L_2(f)$ the existence of a sharp spike around the center of the wave segment is almost ignored. In contrast, for $L_\infty(f)$ the piecewise approximating function is raised upward due to the pulse. In this case notice that the interval is approximated by only one linear function. In fact one difficult problem in approximations using piecewise linear functions lies in finding an appropriate number of linear functions automatically. Humans probably see it as one linear function on which there exists a spike. Therefore humans see it from the perspective of both $L_2(f)$ and $L_\infty(f)$ metrics, and it is difficult to say which one is correct. For a detailed description of the approximation of functions, see the excellent book by Rice [1].

8.4.2 Piecewise Linear Approximation Based on $L_2(f)$

One reason $L_2(f)$ is favored is that an optimal function is easily obtained analytically by fixing the number of subintervals. Here we show it. First we give a formalization of the approximation. We use the piecewise linear function given in (8.9) but change the notation as shown in Fig. 8.22. A given function $y(t)$ whose domain is an interval $[t_1, t_N]$ on the t-axis is approximated by $N - 1$ linear function $x(t)$. That is, the square of the norm $L_2(y, x)$ is given by

$$L_2^2(y, x) = \int_{t_1}^{t_N} (y(t) - x(t))^2 dt, \tag{8.14}$$

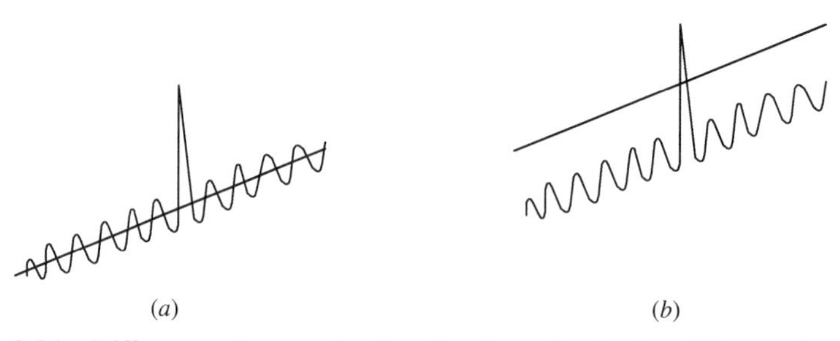

FIGURE 8.21 Difference of approximation functions due to the difference between the norms: (*a*) Norm L_2; (*b*) norm L_∞.

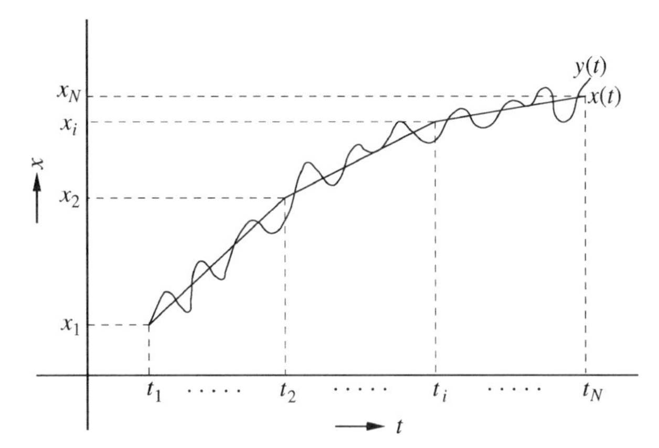

FIGURE 8.22 Notations used for the piecewise linear approximation based on $L_2(f)$; $y(t)$ is approximated by $x(t)$.

where $x(t)$ is given in $t_i \leq t \leq t_{i+1}$, $i = 1, 2, \ldots, N-1$, by

$$x(t) = \frac{x_{i+1}(t - t_i) + x_i(t_{i+1} - t)}{t_{i+1} - t_i}. \tag{8.15}$$

Therefore the $L_2^2(y, x)$ is expressed as

$$L_2^2(y, x) = \sum_{i=1}^{N-1} \int_{t_i}^{t_{i+1}} (y(t) - x(t))^2 dt. \tag{8.16}$$

There are two ways to obtain an optimal solution based on this formulation. One was given by Cantoni [6], and the other by Bellman [7] based on his dynamic programming. Here we introduce the former because of its educational value in that the approximation process is clear and neat. Of course, in pattern recognition dynamic programming is an important technique too, as is fully explained in Chapter 13.

Let us start with the simple case in which a given $y(t)$ is approximated by two $x(t)$'s. That is, $N = 3$, and so the interval $[t_1, t_3]$ is divided into two subintervals, $[t_1, t_2]$ and $[t_2, t_3]$, where t_2 can be regarded as a parameter. The optimization is done by moving this parameter t_2. As a first step, t_2 is fixed, and under this condition optimum $x(t)$ is found such that the $L_2^2(y, x)$ takes minimum value. Because of the nature of piecewise linearity, the optimum $x(t)$ is equivalent to finding $\hat{x}_1$, $\hat{x}_2$, and $\hat{x}_3$ in this simple case. $L_2^2(y, x)$ is formulated as

$$L_2^2(y, x) = \sum_{i=1}^{2} \int_{t_i}^{t_{i+1}} \left[y^2(t) - 2y(t) \cdot \left(\frac{x_{i+1}(t - t_i) + x(t_{i+1} - t)}{t_{i+1} - t_i} \right) \right.$$
$$\left. + \left(\frac{x_{i+1}(t - t_i) + x_i(t_{i+1} - t)}{t_{i+1} - t_i} \right)^2 \right] dt \tag{8.17}$$

We set each coefficient as

$$B(t_i, t_{i+1}) = \int_{t_i}^{t_{i+1}} \frac{y(t)(t - t_i)}{t_{i+1} - t_i} dt, \tag{8.18}$$

$$B'(t_i, t_{i+1}) = \int_{t_i}^{t_{i+1}} \frac{y(t)(t_{i+1} - t)}{t_{i+1} - t_i} dt, \tag{8.19}$$

$$A(t_i, t_{i+1}) = \int_{t_i}^{t_{i+1}} \frac{(t_{i+1} - t)^2}{(t_{i+1} - t_i)^2} dt, \tag{8.20}$$

$$A'(t_i, t_{i+1}) = \int_{t_i}^{t_{i+1}} \frac{(t - t_i)^2}{(t_{i+1} - t_i)^2} dt, \tag{8.21}$$

$$A''(t_i, t_{i+1}) = \int_{t_i}^{t_{i+1}} \frac{(t - t_i)(t_{i+1} - t)}{(t_{i+1} - t_i)^2} dt, \tag{8.22}$$

$$d = \int_{t_1}^{t_3} y(t)^2 dt. \tag{8.23}$$

Then the equation (8.17) is expressed as

$$\begin{aligned} L_2^2(y, x) = &-2[x_1 B'(t_1, t_2) + x_2(B(t_1, t_2) + B'(t_2, t_3)) \\ &+ x_3 B(t_2, t_3)] + [x_1^2 A(t_1, t_2) + x_2^2(A'(t_1, t_2) + A(t_2, t_3)) \\ &+ x_3^2 A'(t_2, t_3) + 2x_1 x_2 A''(t_1, t_2) + 2x_2 x_3 A''(t_2, t_3)] + d \end{aligned} \tag{8.24}$$

This form is second order concerning unknown variables x_1, x_2, and x_3. Therefore we can find the optimum values of x_1, x_2, and x_3 by solving only linear equations. For this purpose $L_2^2(y; x_1, x_2, x_3)$ is partially differentiated in each variable and set to zero. Thus the linear equation is expressed as

$$\begin{bmatrix} A(t_1, t_2) & A''(t_1, t_2) & 0 \\ A''(t_1, t_2) & A'(t_1, t_2) + A(t_2, t_3) & A''(t_2, t_3) \\ 0 & A''(t_2, t_3) & A'(t_2, t_3) \end{bmatrix} \begin{bmatrix} \hat{x}_1 \\ \hat{x}_2 \\ \hat{x}_3 \end{bmatrix} = \begin{bmatrix} B'(t_1, t_2) \\ B(t_1, t_2) + B'(t_2, t_3) \\ B(t_2, t_3) \end{bmatrix}. \tag{8.25}$$

In the equation above the coefficient matrix A is symmetric and constructed by the coefficients of the second-order terms of (8.24). Therefore, for an arbitrary vector variable,

$$\mathbf{X} = \begin{bmatrix} x_1 \\ x_2 \\ x_3 \end{bmatrix},$$

$$\mathbf{X}^T A \mathbf{X} = \sum_{i=1}^{2} \int_{t_i}^{t} \left(\frac{x_{i+1}(t - t_i) + x_i(t_{i+1} - t)}{(t_{i+1} - t_i)} \right)^2 dt, \tag{8.26}$$

holds. That is,

$$\left.\begin{array}{ll}\mathbf{X}^T A\mathbf{X} > 0, & \text{for } \mathbf{X} \neq \mathbf{0} \\ \mathbf{X}^T A\mathbf{X} = 0, & \text{for } \mathbf{X} = \mathbf{0}\end{array}\right\} \tag{8.27}$$

holds. Thus A is positive definite, and by the theory of matrices, $|A|$ is positive, so the linear equation has a unique solution. This solution gives a minimum globally.

The preceding derivation can be easily generalized. Let X be set as follows fixing N:

$$\mathbf{X} = \begin{bmatrix} x_1 \\ x_2 \\ \vdots \\ x_N \end{bmatrix}.$$

Then $L_2^2(y, x)$ is expressed as

$$L_2^2(y, x) = -2B^T\mathbf{X} + \mathbf{X}^T A\mathbf{X} + d, \tag{8.28}$$

where

$$d = \int_{t_1}^{t_N} y^2(t)dt. \tag{8.29}$$

Now B is an $N \times 1$ column matrix. For $j = 2, 3, \ldots, N-1$, its elements are given as

$$b_j = B(t_{j-1}, t_j) + B'(t_j, t_{j+1}), \tag{8.30}$$

$$b_1 = B'(t_1, t_2), \tag{8.31}$$

$$b_N = B(t_{N-1}, t_N). \tag{8.32}$$

On the other hand A is $N \times N$ matrix, and its elements are given as

$$a_{11} = A(t_1, t_2), \tag{8.33}$$

$$a_{NN} = A'(t_{N-1}, t_N). \tag{8.34}$$

For the other suffixes i and j,

$$\begin{aligned} a_{ij} = {} & A''(t_{j-1}, t_j)\delta_{i,j-1} + A''(t_j, t_{j+1})\delta_{i,j+1} \\ & + [A'(t_{j-1}, t_j) + A(t_j, t_{j+1})]\delta_{i,j}. \end{aligned} \tag{8.35}$$

Clearly A is a symmetric matrix, that is $a_{ij} = a_{ji}$, and it is positive definite. In general, the necessary and sufficient condition that form

$$L_2^2(y, x) = -2B^T\mathbf{X} + \mathbf{X}^T A\mathbf{X} + d \tag{8.36}$$

must have a minimum value at $x = \hat{x}$ is that

$$\left.\frac{\partial L_2^2(y, x)}{\partial x}\right|_{x=\hat{x}} = 0 \tag{8.37}$$

and that A is positive definite. Then the solution is given by

$$\hat{\mathbf{X}} = A^{-1}B. \tag{8.38}$$

In the derivation above, t_i, $i = 2, 3, \ldots, N-1$, are fixed. In practice, these are not fixed, for the numbers are randomly selected, and in each t_i's distribution $\hat{\mathbf{X}}$ is calculated using (8.38) for finding the maximum of these results. Also, while the number of segments are fixed, in practice, N is changed somewhat in order to see the stability of N's variation. In Fig. 8.23 an approximation of $y(t) = 100 \sin(t)$ is shown. The solid line shows the result explained so far and the dashed line shows the result when an equal interval is used. In terms of their curve, the difference between them is not so obvious, as is shown.

8.4.3 Piecewise Linear Approximation Based on $L_\infty(f)$

There are many methods based on Chebyshev's norm,

$$\max_{a \le x \le b} |f(x) - g(x)|,$$

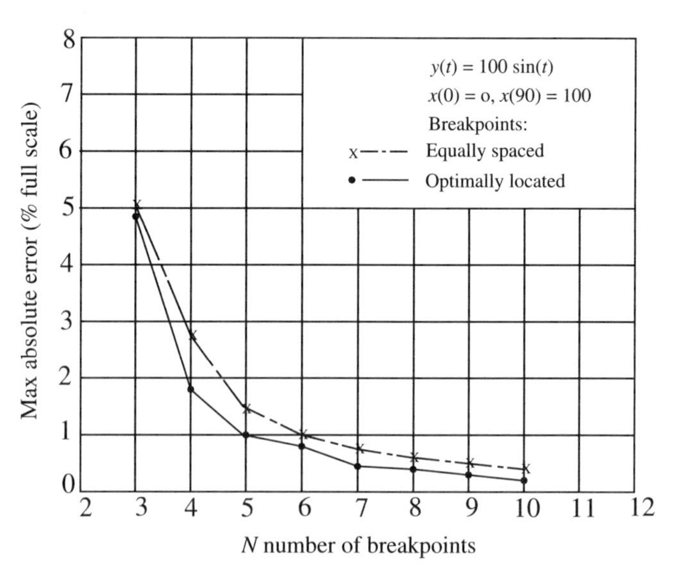

FIGURE 8.23 Result of simulation based on Cantoni's method. From reference [6], © 1998, IEEE.

where $f(x)$ is a given function and $g(x)$ is an approximating piecewise linear function. The advantage of this approach is that it is sensitive to spikes which are sometimes an important signal. Another advantage is the simplicity in calculating $L_\infty(f)$, since it is nothing more than a maximizing operation. However, we need to regard a little the optimality implicit in this approach. Thus, given a piecewise linear function that is approximated, we let S_k be the kth segment of the approximating function. The required optimization then is stated as follows: Find the minimum number of S_k that satisfy a specified criterion function

$$f(S_k) \leq \alpha, \tag{8.39}$$

where $\alpha = \text{const}$. Then $f(S_k)$ is defined as follows:

$$f(S_k) \equiv \max\ (\text{distance}(\mathbf{x}, S_k)), \tag{8.40}$$

where $\mathbf{x}$ is an arbitrary point on the approximated curve whose interval is given by the two end points of the segment, and distance $(\mathbf{x}, S_k)$ means the distance between $\mathbf{x}$ and line segment S_k.

However, since the optimum requirement makes the problem unduly hard, we need to relax the requirement and "choose" a minimum number of S_k. Ramer solved this problem iteratively in a very simple way [11]. In our experience, Ramer's method is very practical and effective. Therefore we introduce his method, although there are many methods based on Chebyshev's norm. Ramer's algorithm is illustrated by the example shown in Figs. 8.24 and 8.25. In Fig. 8.24 an open curve $\widehat{AB}$ is shown, which is approximated by a piecewise linear function. First, both end points A and B are connected by a line segment, $\overline{AB}$. Then the distance d is connected to any $\mathbf{x}$ according to the definition (8.40). From Fig. 8.24 it is obvious that

$$\max_{\mathbf{x} \in \widehat{AB}} d = dc. \tag{8.41}$$

Then the point C is checked to see whether it is a vertex of the approximating function. That is, some specified threshold value α is set, and if $dc > \alpha$, $\widehat{AB}$ is divided

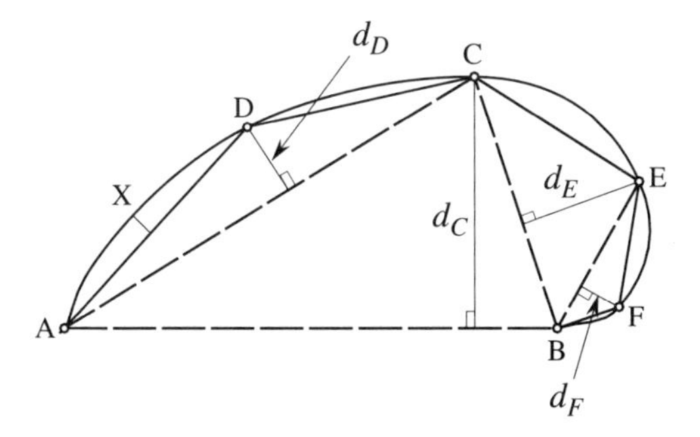

FIGURE 8.24 Ramar's method. From reference [11], © 1998, Academic Press (modified).

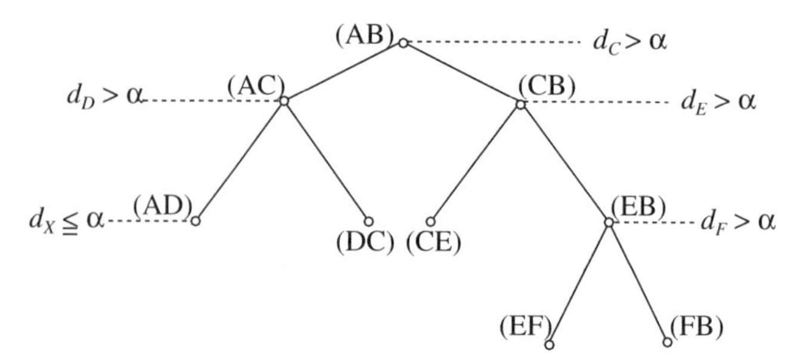

FIGURE 8.25 Binary tree representation of the example in Fig. 8.24. From reference [11], © 1998, Academic Press (modified).

into two open curves $\widehat{AC}$ and $\widehat{CB}$. Otherwise, no division is performed. For each line segment $\overline{AC}$ and $\overline{CB}$, the same procedure is carried out as shown and thus the given open curve $\widehat{AB}$ is approximated precisely. This process is represented by a tree structure as shown in Fig. 8.25. Ramer's advantage lies also in this simple data structure.

Now let us explain his algorithm more precisely using the tree structure. Ramer also used two stacks of first-in and last-out vertices, which he called CLOSED and OPEN. The stacking process is shown in Table 8.2. The CLOSED stack accumulates chosen vertices in order of increasing distance from the point A. In contrast, the order of the OPEN stack vertices is not yet decided. First, point A is regarded as a starting vertex and shifted to CLOSED. The vertex B, however, is not yet decided in its order from A, and so is moved to OPEN. Now the examined curve segment is $\widehat{AB}$, which is specified by the two highest vertices of the stacks. $\widehat{AB}$ is examined and a new vertex C is found. Since C's order is not yet decided, it is moved into OPEN. Then the two highest vertices of the stacks are A and C, so $\widehat{AC}$ is examined. As a new vertex D is found, and its order is not yet decided, it is moved to OPEN. Now the two highest vertices of the stacks are A and D, so $\widehat{AD}$ is examined, but no new vertex is found. With D's order known (i.e., it is next to A), D is popped out of OPEN and moved to CLOSED. Then the highest vertices of the stacks are D and C, so $\widehat{DC}$ is examined,

TABLE 8.2 Stack Representation of the Approximating Process of Ramer's Method by the Binary Tree in Fig. 8.25

CLOSED	OPEN	Segment	Vertex
A	B	(AB)	C
A	B, C	(AC)	D
A	B, C, D	(AD)	—
A, D	B, C	(DC)	—
A, D, C	B	(CB)	E
A, D, C	B, E	(CE)	—
A, D, C, E	B	(EB)	F
A, D, C, E	B, F	(EF)	—
A, D, C, E, F	B	(FB)	—
A, D, C, E, F, B			

Source: From reference [11], © 1998, Academic Press.

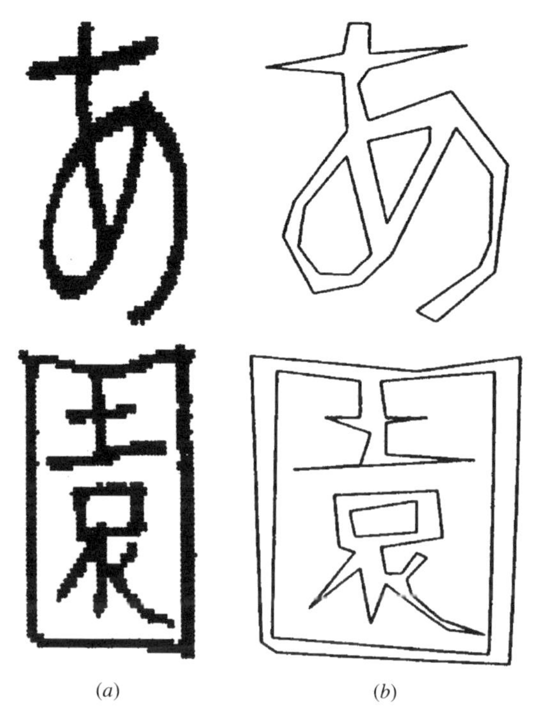

FIGURE 8.26 Some polygonal approximations based on Ramer's method. From reference [99], © 1997, Ohm.

but no new vertex is not found. Then *C*'s order is decided as next to *D,* and *C* is moved to CLOSED. And so it continues until finally all the ordered sequence for vertices of this approximating function are found in CLOSED.

Some experimental results are given in Fig. 8.26. As seen, the method works quite well for ordinal figures. One point should the mentioned here on the use of a closed curve. In that case we need to choose two points/vertices that must be efficient. Usually the two points chosen are the extreme points; one is a maximum and the other is a minimum, taking the *x-axis* as a projection line, as shown in Fig. 8.27. We notice,

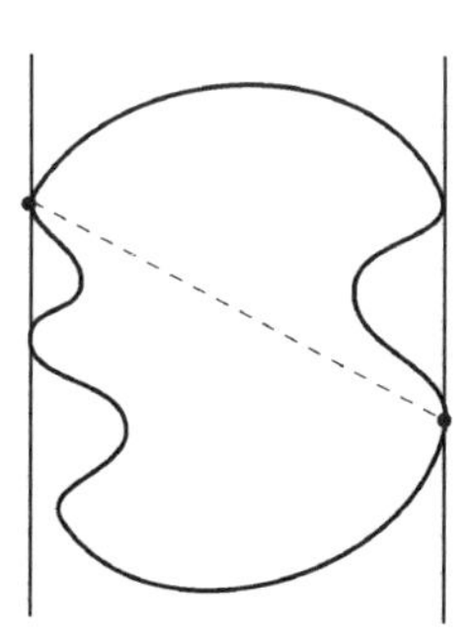

FIGURE 8.27 The figure illustrates one important point of taking an initial line when Ramer's method is applied.

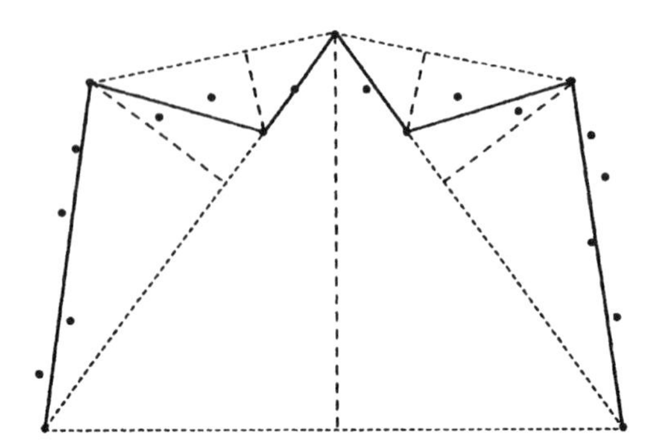

FIGURE 8.28 Weakness of Ramer's method.

however, that Ramer's method is weak for noisy images. An example is shown in Fig. 8.28. As noted before, by the nature of its metric, it is sensitive to spikelike noise, which sometimes results in a distorted polygon (open polygon) as shown. To the human eye it appears as a polygon of three edges, one flat horizontally and the two others almost vertical.

8.5 SPLIT-AND-MERGE METHOD

The split-and-merge method is a practical answer to optimization problems that consume much computing time. It was given by Pavlidis [12]. The idea here is useful not only in character recognition but also in image processing, in general. There are two approaches to the optimization problem. One is that bound by a given error, a minimum number of intervals/segments m is sought. The other is that under a given m, a minimum error is sought. The former has complexity of $N \log m$ or N; the latter has complexity of $N^a (a > 2)$. Therefore the desired practical method is to solve the problems while keeping complexity close to N.

8.5.1 Local Optimal Piecewise Linear Approximation Method

To illustrate the split-and-merge method, we need to introduce a local optimization piecewise linear approximation method. This is for the second approach, that is, for us to seek the minimum error of the approximation while keeping the number of intervals constant; this is, however, a local approximation and not global. The objective of split-and-merging is to obtain a global approximation which changes the number of segments/intervals.

The basic strategy of locally optimal piecewise linear approximation is to shift little by little from an arbitrary initial approximation. Specifically, in one dimension each vertex is taken along the x-axis such that u_i, $i = 0 \sim m$, and $e_1, e_2, \ldots, e_i$, is the error norms of intervals, $[u_0, u_1], [u_1, u_2], \ldots, [u_{i-1}, u_i]$. Let each step of the approximation process be denoted by k, which is attached to the variable as a suffix at the right. We assume that the approximated function is continuous and smooth so that the error norm is a nondecreasing function of the length of the interval of continuous approximation. If the goal is to minimize the maximum error norm on each interval, then a balanced error is the optimal state, which occurs when the error norms on each interval are equal.

The shifting method is formulated as

$$u_i^{k+1} = u_i^k + \delta(e_{i+1}^k - e_i^k), \tag{8.42}$$

where δ is a sufficiently small constant. The application of this approximation procedure is illustrated in Fig. 8.29, which is self-explanatory. At the kth step, $e_2^k - e_1^k$ approaches zero, so $e_2^k = e_1^k$; that is, the solution has a balanced error. However, for discrete data it does not work this way. Therefore some other measure is necessary. We cannot pursue this topic further here. The reader who wants to study it can refer to Pavlidis's paper [12]. However, one important point must be mentioned. The maximum error norm cannot both increase and remain the same at any step. The procedure stops when any change/shift in dividing points will increase the error.

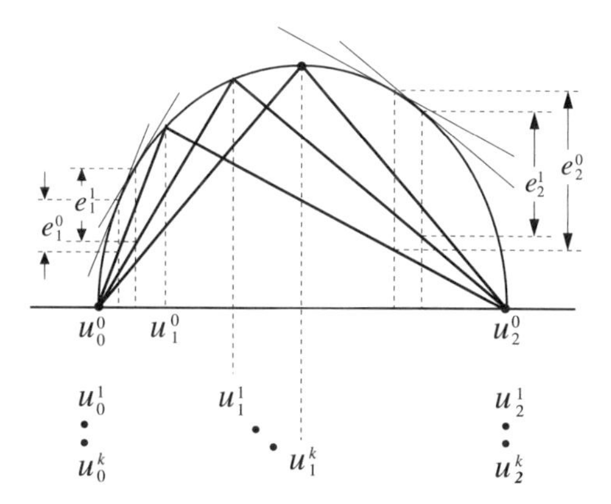

$$u_1^1 = u_1^0 + \delta(e_2^0 - e_1^0) \qquad [u_0^0, u_1^0] \dashrightarrow e_1^0$$

$$u_1^2 = u_1^1 + \delta(e_2^1 - e_1^1) \qquad [u_0^1, u_1^1] \dashrightarrow e_1^1$$

$$\vdots \qquad \bullet$$

$$u_1^k = u_1^{k-1} + \delta(e_2^k - e_1^k) \qquad [u_1^0, u_2^0] \dashrightarrow e_2^0$$

$$[u_1^1, u_2^1] \dashrightarrow e_2^1$$

$$\bullet$$

FIGURE 8.29 Illustration of locally optimal piecewise linear approximation, where the norm is taken at the maximum difference along the y-axis between the smooth approximated function and approximating polygon.

8.5.2 Split-and-Merge Algorithm

As mentioned before, the objective in using the split-and-merge algorithm is to make good initialization unnecessary and to allow for a variable number of segments. In this section we will give a sketch of this idea using a simple example. Before proceeding, however, we need to review the initial locally optimal piecewise linear approximation property for the example used, which is shown in Fig. 8.30. The locally optimal piecewise linear approximation was performed after the selection of a random dividing point, u_1^0, and it reached point A after some iterations. As noticed before, point A does not move to point B which is optimal in global sense. This is because further movement of A, say to A', results in an increase of the error norm, as is shown by $e_1 < e_1'$ and $e_2 < e_2'$. Here we use the L_∞ norm measured along the y-axis. That is, in both intervals the error norms increase, so the iteration process is stopped at A. Despite the large error norm remaining, further improvement cannot be made. Therefore split-and merge is necessary, which is schematically shown in Fig. 8.31 using the same example including the L_∞ norm. Naturally we assume that the error norm e_2 is greater than some specified bound, say, $E_{\max}$. Then the interval $[E_A, E_2]$ which gives the largest error norm is divided into the two intervals at the point that takes the largest error norm. The dividing point is denoted as E_B. In each divided interval, the norm error is measured and compared with $E_{\max}$. If the error norm is less than $E_{\max}$, then the process of splitting is stopped. Otherwise, the splitting is continued. In this case we assume that no further division is necessary, namely $e_2' < E_{\max}$. However, we want the number of intervals to be as small as possible, keeping the total error at least the same as before. For this reason merging is tried in the neighboring intervals. In this case intervals $[E_1, E_A]$ and $[E_A, E_B]$ are merged, and the error norm is measured over the $[E_1, E_A] \cup [E_A, E_B]$ which is $[E_1, E_B]$. As seen from panels (*b*) and (*c*), $e_2' > e_1$, so the error is decreased by the merging. We assume that the error norm in this new interval is less than $E_{\max}$, and therefore the split-and-merge process stops. As seen in the figure, a global optimization has been performed.

Now let us formalize the scenario mentioned above. First we restate the problem. Given an ordered point set $S = \{(x_i, y_i) | i = 1, 2, \ldots, N\}$, the problem is to divide the S into intervals, $S_1, S_2, \ldots, S_m$, with the number of m kept at a minimum and at the same time the error norm being less than $E_{\max}$ when each interval is approximated with a piecewise linear function. Here we notice that the PL approximation is inde-

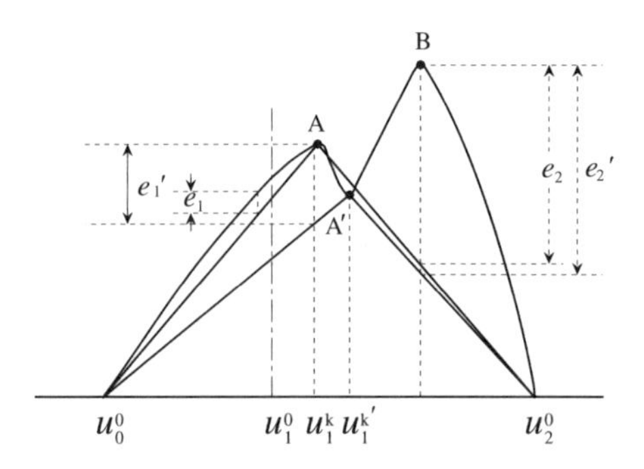

FIGURE 8.30 Illustration of optimization. From reference [99], © 1997, Ohm.

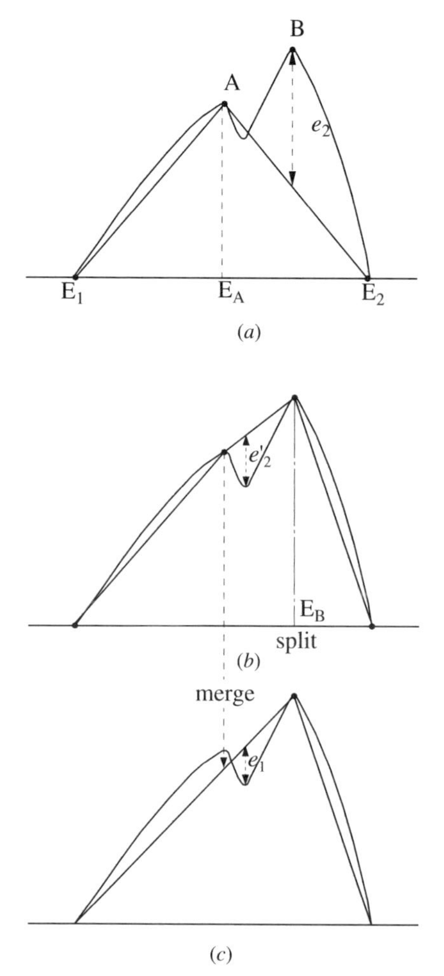

FIGURE 8.31 Split-and-merge method: (*a*) Initial division and local optimization; (*b*) split; (*c*) split and merge. From reference [99], © 1997, Ohm.

pendently given in each interval. This implicitly means that the continuity of the *PL* function at both end points of the interval with neighboring *PL* functions is not taken into account. Cantoni's method includes this continuity, which makes the calculation so complex. In our case, which is free from any continuity condition, the calculation is simplified, so split-and-merge is a useful method in practice.

To solve the above problem, we need first to decide how to take the error norm. Let $S_1, S_2, \ldots, S_{m_r}$ be the segments at the rth iteration and $E^1, E^2, \ldots, E^{m_r}$ the respective error norms. Two cases are considered. One is to have $E^i \leq E_{\max}$ for all segments. The other is to place the error norm E over all segments less than or equal to $E_{\max}$. In case of $L_\infty(f)$,

$$E_\infty = \max_i E_\infty^i, \tag{8.43}$$

and therefore the two cases coincide. For the $L_2(f)$ they differ, since

$$E_2 = \sum_{i=1}^{m_r} E_2^i. \tag{8.44}$$

In the split-and-merge algorithm, a locally optimal *PL* approximation is used to decide each end point and thus to adjust an interval such that on each iteration it cannot reverse any inequality of the form $E^i \leq E_{max}$ or $E \leq E_{max}$ as noticed before, which is called *R*. The following is the split-and-merge procedure for the first case, which is based on Pavlidis's paper [12].

STEP 1: For $i = 1, 2, \ldots, m_r$, if E^i exceed E_{max}, split the *i*th interval into two and increment m_r. The dividing point is determined by Rule A below. Calculate the error norms at each new interval.

STEP 2: For $i = 1, 2, \ldots, m_r - 1$, merge segments S_i and S_{i+1} provided that this will result in a new segment with $E^i \leq E_{max}$. Then decrease m_r by one, and calculate the error norm on the new interval.

STEP 3: Do one iteration of algorithm *R*.

STEP 4: If no changes in the segments have occurred in Steps 1–3, then terminate. Or else, go to Step 1.

Rule A: If two or more points are known where the pointwise error is greater than E_{max}, then use the average point among them as a dividing point. Otherwise, divide S_i in half.

We omit the second case because it is similar to the first case.

8.5.3 Orthogonal Least Squares Piecewise Linear Approximation

Next we look at how to measure error norms, in particular, for $L_2(f)$ which is natural in two-dimensional or higher-dimensional planes and neat in a mathematical sense. For a one-dimensional wave the error is estimated at each point.

Let y_1 be an approximated point series and $p(x_i)$ be the PL approximating function. As is usual, we take the error estimate e_i at point x_i as

$$e_i = |y_i - p(x_i)|. \tag{8.45}$$

However, we need to generalize this expression to the two-dimensional plane, as shown in Fig. 8.32. That is, we need to measure distance between a point (x_i, y_i) and a line. The line is represented as

$$\sin\varphi \cdot x + \cos\varphi \cdot y = d, \tag{8.46}$$

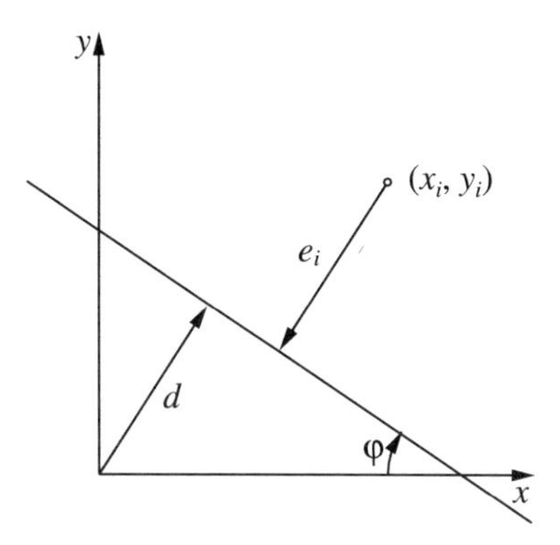

FIGURE 8.32 Error between point (x_i, y_i) and a line segment.

where φ and d are the parameters shown in the figure. Then the distance between the point (x_i, y_i) and the line is expressed as

$$e_i = |\sin \varphi \cdot x_i + \cos \varphi \cdot y_i - d|. \tag{8.47}$$

Equation (8.47) can be understood as a parallel line that passes through (x_i, y_i) and has the same angle as (8.46). Let d' be the parameter of the line passing (x_i, y_i). Then $d' - d$ is the distance required.

Now in the case of $L_2(f)$, the integral squared error is

$$E_2 = \sum e_i^2, \qquad (x_i, y_i) \in S_k. \tag{8.48}$$

In case of $L_\infty(f)$, the maximum error is

$$E_\infty = \max (e_i), \qquad (x_i, y_i) \in S_k, \tag{8.49}$$

where S_k is the set of points to be approximated.

The latter case is intuitive, but its optimization is complex. Actually it is a mathematical programming problem, whereas, for the former, there exists an explicit formula for E_2 whether e_i is given by (8.45) or (8.47). The parameters d and φ that minimize E_2 can be obtained by partially differentiating about φ and d and substituting e_i (8.47) into (8.48) after setting it to zero. The results are as follow:

$$\sin 2\varphi(V_{xx} - V_{yy}) + 2 \cos 2\varphi V_{xy} = 0, \tag{8.50}$$

$$d = \sin \varphi \cdot Vx + \cos \varphi \cdot V_y. \tag{8.51}$$

Let N_k be the number of points in S_k, then the constants in (8.50) and (8.51) are expressed as

$$V_x = \frac{1}{N_k} \sum_{S_k} x_i, \quad V_y = \frac{1}{N_k} \sum_{S_k} y_i,$$

$$V_{xx} = \sum_{S_k} (x_i - V_i)^2, \quad V_{yy} = \sum_{S_k} (y_i - V_y)^2,$$

$$V_{xy} = \sum_{S_k} (x_i - V_x)(y_i - V_y). \tag{8.52}$$

The error norm, E_2, is given by

$$E_2 = \sin^2 \varphi \cdot V_{xx} + \cos^2 \varphi \cdot V_{yy} + \sin^2 \varphi \cdot V_{xy}. \tag{8.53}$$

Here we notice that during an iteration where the set S_k varies, one can update the quantities given by (8.52) without having to recalculate the sums over each S_k. This is done by simply adding or subtracting the contributions of the added or removed points.

Furthermore we will see that only the quantity E_2 for each S_k is needed during the computation stage and that φ and d need to be calculated explicitly only at the end. Eliminating φ between (8.50) and (8.53), we obtain

$$E_2 = \frac{1}{2} \{V_{xx} + V_{yy} - [(V_{xx} - V_{yy})^2 + 4V_{xy}^2]^{1/2}\}. \tag{8.54}$$

Note on Orthogonal Least Squares We can obtain the result mentioned above more elegantly by solving an eigenvalue problem. In this sense this method is called *orthogonal least squares line fitting* or PL *approximation*. The framework is given by Duda and Hart in their book [1]. In this case e_i is given by (8.45), extending it to the two-dimensional case. For simplicity, suppose that the average point of the approximated points lies at the origin of the coordinates on the plane, and the PL approximation line's normal direction is denoted as **N**. The data are denoted as $\mathbf{x}_i = (x_i, y_i)$. Then the distance $d_i(N)$ from the point $\mathbf{x}_i$ to the line **N** is expressed as

$$d_i^2(N) = (\mathbf{N} \cdot \mathbf{x}_i)^2 = (\mathbf{N}^t \cdot \mathbf{x}_i)^2. \tag{8.55}$$

Therefore the total squared error is given as

$$\begin{aligned} d^2(\mathbf{N}) &= \sum_{i=1}^{n} d_i(\mathbf{N}) \\ &= \mathbf{N}^t \sum (\mathbf{x}_i \mathbf{x}_i^t) \mathbf{N} \\ &= \mathbf{N}^t R \mathbf{N}. \end{aligned} \tag{8.56}$$

Under the condition of $\|\mathbf{N}\| = 1$, namely $\mathbf{N}'\mathbf{N} = I$, we can construct the following minimizing problem employing the method of Lagrange multipliers:

$$J(\mathbf{N}) = \mathbf{N}'R\mathbf{N} + (\mathbf{N}'\mathbf{N} - I)\Lambda, \tag{8.57}$$

where Λ is a diagonal matrix. Thus

$$\frac{\partial J(\mathbf{N})}{\partial N} = -2R\mathbf{N} + 2\mathbf{N}\Lambda. \tag{8.58}$$

Setting (8.58) to zero, we obtain

$$R\mathbf{N} = \mathbf{N}\Lambda. \tag{8.59}$$

That is, we reach an eigenvalue problem as

$$\det\,(R - \lambda I) = 0. \tag{8.60}$$

On the other hand, we can find R as

$$R = \begin{pmatrix} \sum x_i^2 & \sum x_i y_i \\ \sum x_i y_i & \sum y_i^2 \end{pmatrix}.$$

Thus

$$\begin{vmatrix} \sum x_i^2 - \lambda & \sum x_i y_i \\ \sum x_i y_i & \sum y_i^2 - \lambda \end{vmatrix} = 0, \tag{8.61}$$

$$\lambda^2 - \left(\sum x_i^2 + \sum y_i^2\right)\lambda + \sum x_i^2 \sum y_i^2 - \sum (x_i y_i)^2 = 0. \tag{8.62}$$

The smaller solution of (8.62) gives E_2:

$$E_2 = \frac{1}{2}\left\{\sum x_i^2 + \sum y_i^2 - \left[\left(\sum x_i^2 - \sum y_i^2\right)^2 + 4\left(\sum x_i y_i\right)^2\right]^{1/2}\right. . \tag{8.63}$$

This is the same as (8.54).

8.5.4 Experimental Results

By experimental results we will show the effectiveness of this method. The test pattern is a cell outline which was used as a standard test pattern for various line-fitting methods. The original outline was digitized into 124 points, and E_{max} was set to 30 in all the experiments. First, Ramer's method was applied to the test pattern, and the result is shown in Fig. 8.33 (*a*) where some digital display errors can be observed. An orthogonal least squares PL approximation was applied to the segmentation obtained by Ramer's method, which was shown in Fig. 8.33 (*b*). The result is too rough, so to this result the split-and-merge method was applied, which was shown in Fig. 8.33 (*c*). As expected, compared with panels (*a*) and (*b*), an intuitively appealing result is obtained, although there are conspicuous discontinuities such as marked by the A's.

As is clear, by sacrificing continuity at neighboring sides, more stable and reasonable PL approximation can be achieved. Despite the presence of discontinuities, PL approximation has much appeal in practice, in particular, for its effectiveness on noisy images.

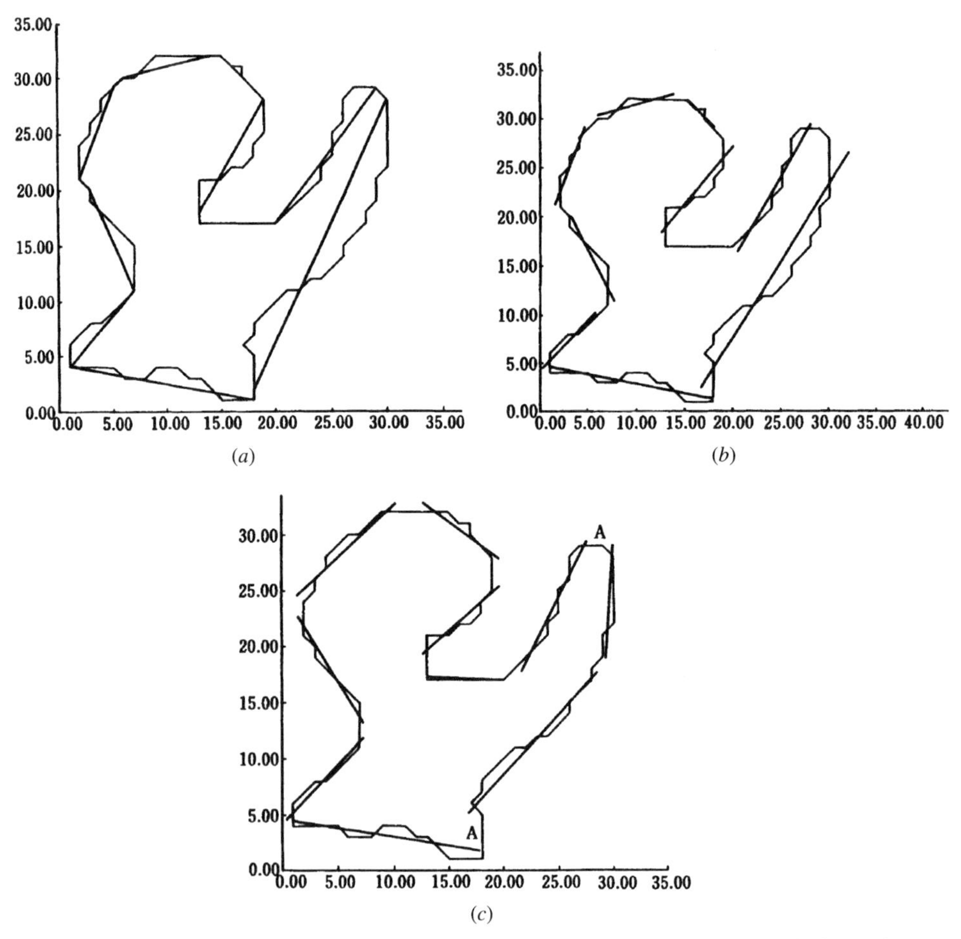

FIGURE 8.33 Three kinds of PL approximation methods: (*a*) Ramer's method; (*b*) application of orthogonal least square PL approximation to the segmented sides by Ramer's method; (*c*) split-and-merge method applied to (*b*). From reference [12], © 1998, IEEE.

8.6 CONTOUR-FOLLOWING SCHEMES

We mentioned in Section 8.3 on stream following that raster scanning sets up a natural order on a two-dimensional plane. Contour following also takes a suitable natural order that is invariant to rotation. In following the contours of an image, directional features such as can be easily detected and acute direction change points can be used as segmenting points of the contour so long as the image is not too noisy. Historically speaking, the first contour-following mechanism used an analog device called a flying spot scanner. In fact, originally, one of the most successful commercial OCR systems, IBM1287OCR, fully used this mechanism [13]. It operates as beam motion that is attenuated within a black region and draws a cycloid as it traverses the boundary. At the time it was first introduced, it was very fast and flexible, and another good point was that it could handle coarse boundaries. However, its computer counterpart was very slow and suffered from boundary noise. In the present computer generation, the situation has improved, and contour-following mechanisms can be realized with high speed as well as digital noise smoothing. Here we will mention some algorithms of feature detection based on contour following. First, an indirect method will be introduced in constructing a convex hull. Second, some direct methods will be described.

8.6.1 Outermost Points Method

Before proceeding to the details, we need to mention the *convex hull*. In doing so, we will define convexity. That is, on a plane, a region C is said to be convex if it always contains the line segment joining any two points in the region. A convex hull, then, can be briefly described as a minimum convex set of points that includes a convex shape. A convex hull can be easily constructed from a sheet of cardboard. First, draw a shape, and cut it out. Place it into a loop of an expanded rubber band. Then release the rubber band. The rubber band will form the desired convex hull. However, if the figure made of cardboard is not convex, then there will be gaps between the expanded rubber band and the cardboard shape. That is the feature we want. More specifically, this is what occurs for a simply connected concave figure. If a given figure is not simply connected, but connected, then there can be another feature region that is surrounded by the figure itself. Such feature is called, hole or lake (see Section 5.1). Such varying situations are shown in Fig. 8.34. The idea of the extraction of a con-

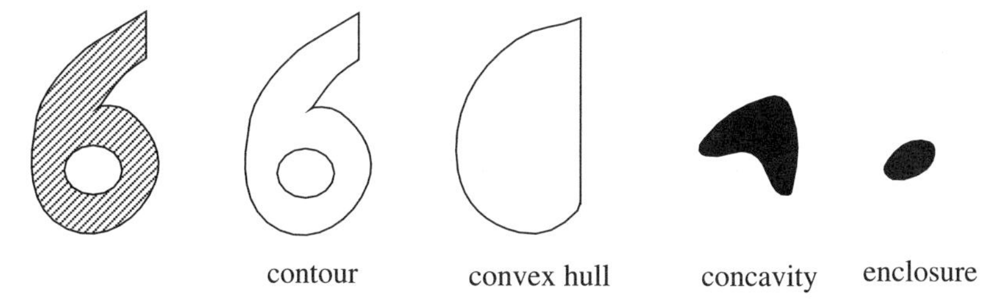

FIGURE 8.34 Schematic of Munson's idea. From reference [15], © 1998, IEEE.

cavity based on a convex hull was due to Munson [14]. However, he did not give a precise method for constructing a convex hull.

Now we proceed to the construction of a convex hull. While the definition of a convex hull is very simple, we cannot use it. We need a constructive definition. In the field of computational geometry, construction of a convex hull is one of the central themes; we refer the readers to the book by F. P. Preparata and M. I. Shamos [16]. Also a well-known simple algorithm is R. L. Graham's [17], and it is fully explained in the book cited above. Our position is that it is enough to form a convex hull approximately so that it can be fast. We rather introduce an efficient way of constructing a convex hull devised by Yamamoto [18], called the *outermost point method.* In passing, a constructive definition of a convex hull will be mentioned. The construction is shown in Fig. 8.35, where at every point on the boundary of the image a tangential line is drawn. Each tangential line divides the plane, on either half-plane of which the image is placed. Among these half-planes, we keep the tangential lines whose half-planes include the complete image, say, at the left-hand side, and call these supporting lines. The remaining tangential lines construct the enclosed line of the image, which forms the convex hull of the image. This is a constructive definition of a convex hull. The outermost point method is based on this constructive definition of a convex hull. More specifically, as shown in Fig. 8.36, the tangential line M is a supporting line but L is not. In what way can we differentiate these two tangential lines? To find the true tangential line, we set a vector **A** such that it is orthogonal to both M and L outside of a given image. Then we take the projection of vectors $\mathbf{P}_L$ and $\mathbf{P}_M$ to **A**. Here we notice that the relation $(\mathbf{P}_M \cdot \mathbf{A}) > (\mathbf{P}_L \cdot \mathbf{A})$ holds, where the inner product $(\mathbf{P}_M \cdot \mathbf{A})$ is a projection of $\mathbf{P}_M$ to **A**. This implies that there exist tangential and orthogonal lines on both sides of L. Therefore, when a base direction is fixed, in our case, vector **A**, an orthogonal supporting line can be located at a point, after tracing the boundary, that gives a maximum projection to **A**. As shown in the figure, the point that gives the minimum projection to **A** is also a supporting line. Ideally we need to take base directions infinitely; however, it is enough for our purpose to take eight directions. In order to use only the maximum criterion in the implementation of the scheme, an origin is chosen at the central gravity point

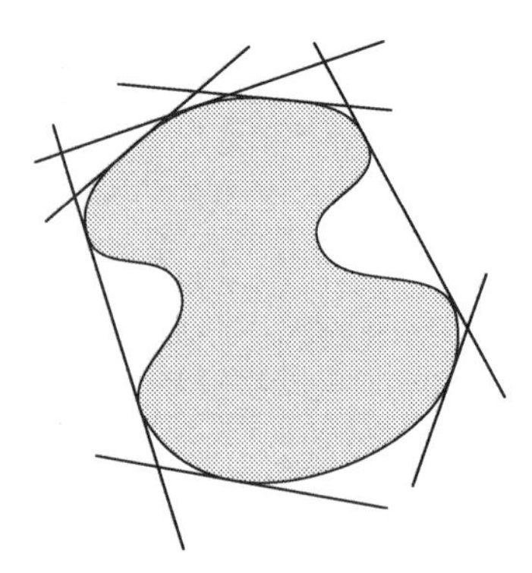

FIGURE 8.35 Constructive definition of a convex hull.

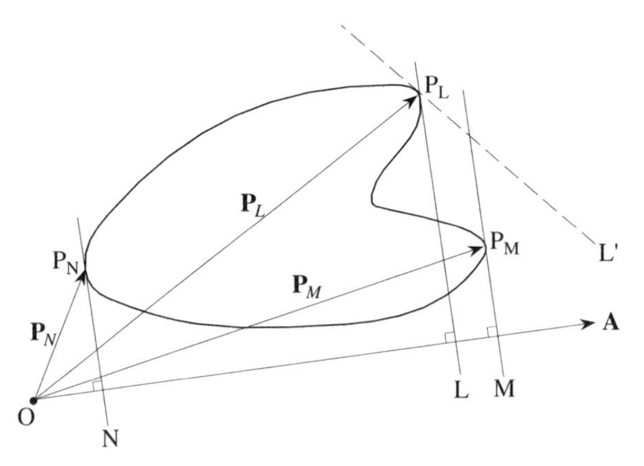

FIGURE 8.36 Principle of the outermost point method. Notice that point P_L seems to be an outermost point, but the real outermost point exists very close to P_L, just above P_L at which a supporting line L' is drawn (dotted line).

and $8 \times 2 = 16$ directions are used. Now let us formalize the procedure. Let $\mathbf{E}_i$ be an arbitrary point vector on a boundary, and let $\mathbf{A}_r$ ($r = 1 \sim 16$) be unit vectors having the basic directions chosen as Fig. 8.37, with $\mathbf{P}_r$ a point vector in the basic direction $\mathbf{A}_r$. Then formulation of the scheme is described as follows:

$$\bigcup_{r=1}^{16} \mathbf{P}_r = \bigcup_{r=1}^{16} (\mathbf{E}_{m(r)} | (\mathbf{A}_r \cdot \mathbf{E}_{m(r)}) = \max_{i=1}^{N} (\mathbf{A}_r \mathbf{E}_i)), \tag{8.64}$$

where N is the total number of the boundary points. From the procedure mentioned above, the name of the method "outermost point" may be understood. This method enables us to approximate a convex hull using a small number of base directions, and its sufficiency in practice has been proved by experience.

Now we can proceed to define contour following more rigorously. In general, there are two ways to consider: One is to trace the black boundary points; the other

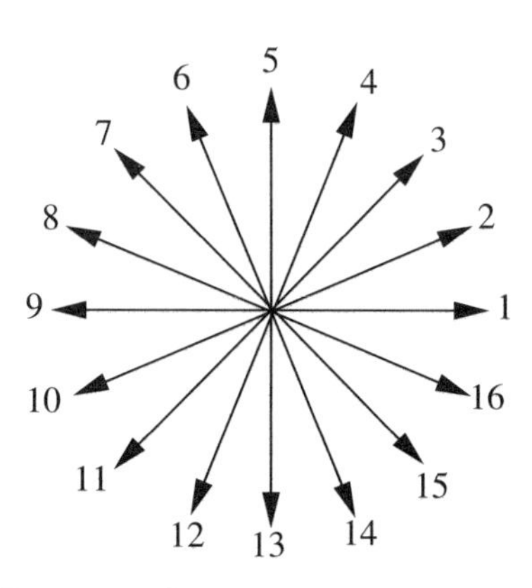

FIGURE 8.37 Definitions of direction $\mathbf{A}_r$, $r = 1 \sim 16$.

is to trace the white points which are contiguous to the black boundary points. In the following experiments the latter is used. The experimental results are shown in Fig. 8.38 (*a*). The image is a character "ア" of Katakana. The circle marks denote the outermost points, and characters ($1 \sim G$) within the circles indicate the basic directions. Connecting these points an approximated convex hull was constructed as seen in the figure. Here we notice that some directions are missing, 1, 3, for example, which have degenerated. Suppose an acute corner point at which we can draw any tangential lines. In this experiment a base direction was chosen that gives maximum projection value to the corresponding base direction vector compared with the maximum projection values to other base directions vectors.

Once a convex hull is found, it is an easy task to extract the concavity region. This is shown in Fig. 8.38 (*b*) as between the length of the adjacent outermost points L_2 and the length between the points along the boundary L_1. These lengths are compared, and if $L_1 - L_2 > \delta$, the boundary L_2 is segmented as concavity where δ is a threshold value. Otherwise, it is segmented as convex. Usually segmented convex boundaries are connected, so they are merged. The result is shown in panel (*b*), where the boundary line is segmented into five portions. The boundaries marked by 1 and 3, and by 2 and 4, are concavity and convex, respectively. The boundary marked by 5 is a hole, which is easily detected because the direction of the boundary following it is just opposite to the outer boundary. The attributes of concavity feature are also shown in panel (*b*). Point G is the center of gravity of the concavity region consisting of L_1 and L_2. The distance to the line L_2 from G is denoted as ρ, and the normal direction of the line L_2 is denoted as θ. Among these, θ, in particular, is the most important attribute.

As seen, the outermost point method gives a convex hull efficiently, and stable segmentation is based on that. However, it cannot handle shapes more complex than those of alphanumeric and Katakana complexity. This is shown in Fig. 8.39 for a character of Katakana "ス." In this case the left concavity region includes a convex portion that is marked by the concavity region. Therefore we proceed to improve this point by expanding the outermost point method. Later this will be easily solved by a direct contour-following method.

8.6.2 Extreme Point List Method

We showed above that a convex hull is sufficient to extract convex and concavity features of a contour. Only maximum points are taken in forming a convex hull. To extend the idea of the outermost list method, we trace relative/local maximum points [19]. Following a contour $Z(i)$, ($i = 1 \sim I$, numbered discrete points) projection to $\mathbf{e}_n - direction(n = 1, \sim I,$ numbered quantized direction $|\mathbf{e}_n|$) is calculated in the same way as we formed the convex hull. That is,

$$(Z(j) \cdot \mathrm{e}_n), \qquad n = 1, \sim I, \tag{8.65}$$

is the projection of $Z(i)$ to $\mathbf{e}_n$. In each direction relative/local maximum points are stored, and when one of the values exceeds some specified value, say ε, compared

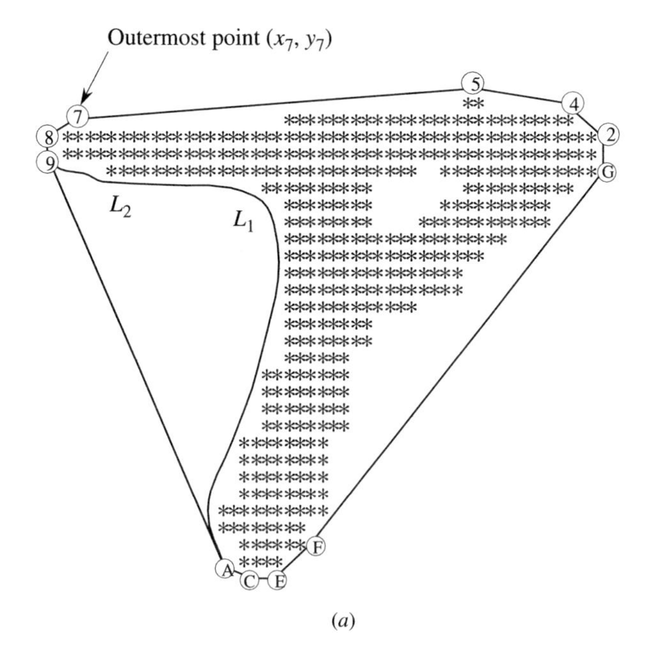

(*a*)

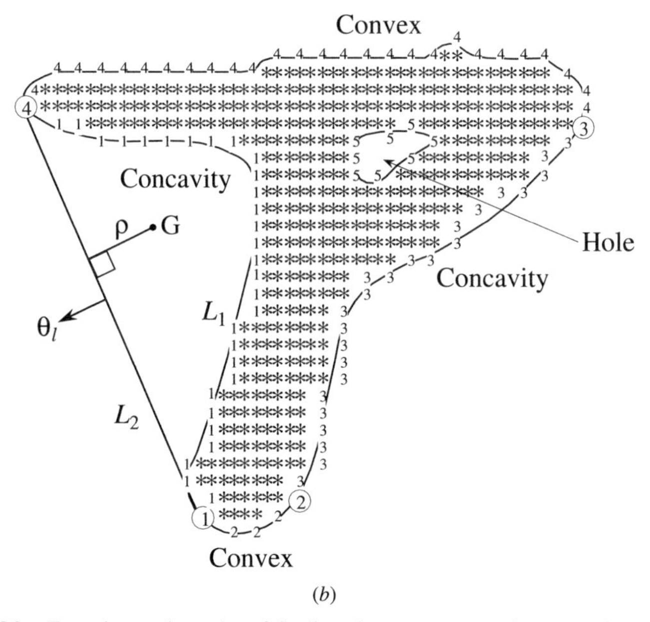

(*b*)

FIGURE 8.38 Experimental results of finding the outermost points (*a*); (*b*) segmentations based on those points. From reference [101].

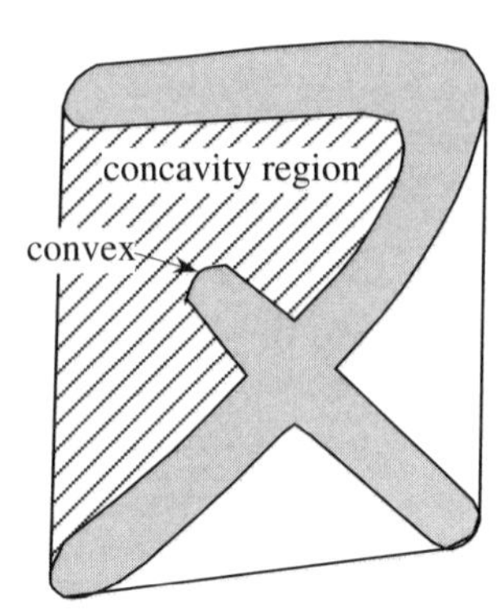

FIGURE 8.39 Drawback of outermost point method. From reference [101].

with the projected value of tracing point, this relative/local maximum point is listed as an "extreme point." Therefore this can be said to be on the idea of an expansion of the outermost list point method.

To assist the readers' understanding, this process is illustrated in Fig. 8.40 in which N is taken as 8 for simplicity. We start the tracing of the contour of a shape drawn at point P_1. The point P_1 is naturally both a maximum and a relative/local maximum point in the direction of $\mathbf{e}_7$. The tracing is done clockwise, so we reach point P_2. This point is also a maximum and a relative/local point in the direction $\mathbf{e}_8$. So they are stored and compared with the tracing point projection to $\mathbf{e}_7$ and $\mathbf{e}_8$. Now we reach point P_3, where the difference between $(Z(P_1) \cdot \mathbf{e}_7)$ and $(Z(P_3) \cdot \mathbf{e}_7)$ exceeds ε. So point P_1 is listed as an "extreme point." Then we reach point P_4, where the difference between $(Z(P_2) \cdot \mathbf{e}_8)$ and $(Z(P_4) \cdot \mathbf{e}_8)$ exceeds ε. So point P_2 is listed as above. In these two cases the maximum points coincide with the relative/local maximum points. Therefore we do not need to make a comparison, but this was explained as a unified procedure to illustrate maximum or relative/local maximum tracing. Then we reach point P_6 where the point P_6 is the relative/local maximum point in the direction of $\mathbf{e}_4$ and so it is stored. At point P_7, the difference between $(Z(P_6) \cdot \mathbf{e}_4)$ and $(Z(P_7) \cdot \mathbf{e}_4)$ exceeds ε, and so the point P_6 is listed as above. Therefore we see the that a concavity point was detected as expected.

This method includes the outermost point method by our setting $\varepsilon = \infty$ (greater than the size of the frame in which an character image is placed). On the other hand, if $\varepsilon = 0$, the result is nothing but the original shape. The changes according to ε's value are shown in Fig. 8.41.

8.6.3 Angle Change Analysis Method

Now we turn to a more intuitive method among contour-following schemes which was introduced by Yamamoto [101]. For simplicity we suppose a smooth continuous one-dimensional function, $y = f(x)$. As is well-known, concavity and convexity of such a function can be found by calculating the second derivative $f''(x)$ and by checking its sign. That is, if $f''(x) \leq 0$, then the gradient of the tangent line decreases, so this region is marked as convex (upward). If $f''(x) \geq 0$, then the gradient of the tangent

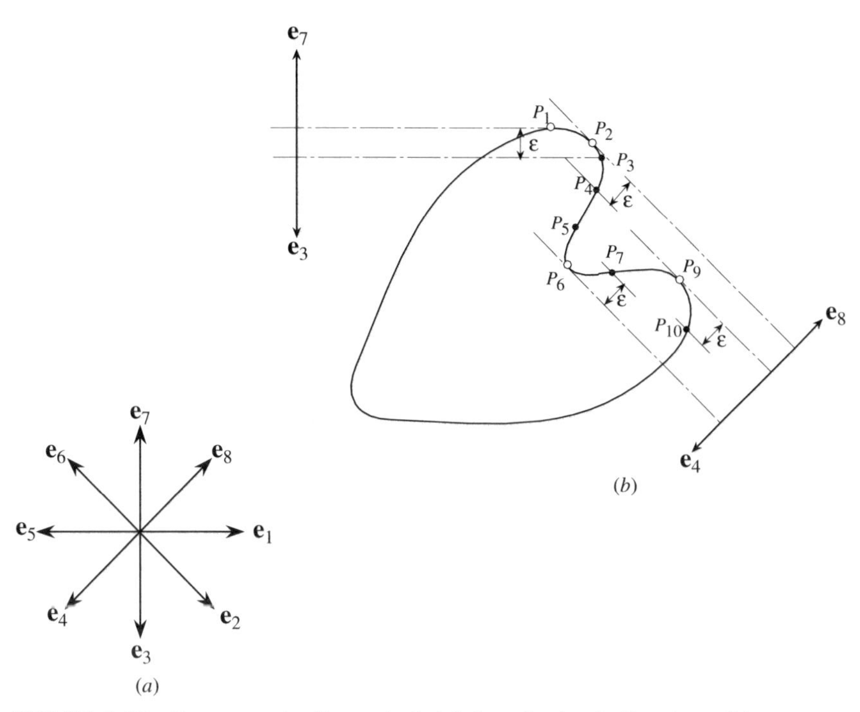

FIGURE 8.40 Extreme point list method: (*a*) Quantized unit directions; (*b*) extreme point list method.

line increases, so this region is marked as concave. Therefore the changing point of $f''(x)$'s sign is a segmenting point between convexity and concavity regions. In general, it is easy to obtain a tangent line of a boundary point, so the scheme can be applied to real data, but not so easily. This is illustrated in Fig. 8.42. In panel (*a*) a given wave shape is segmented according to the rule mentioned above. In this case both convexity and concavity regions are narrow and limited around to the tip and the bottom points. Other remaining parts are the portion where $f''(x) = 0$. However, this is counter to human intuition. Rather, by our intuition, the convex and concave portions overlap as shown in panels (*b*) and (*c*). This point will be formalized within the framework of the algebraic curve described in Section 9.2.

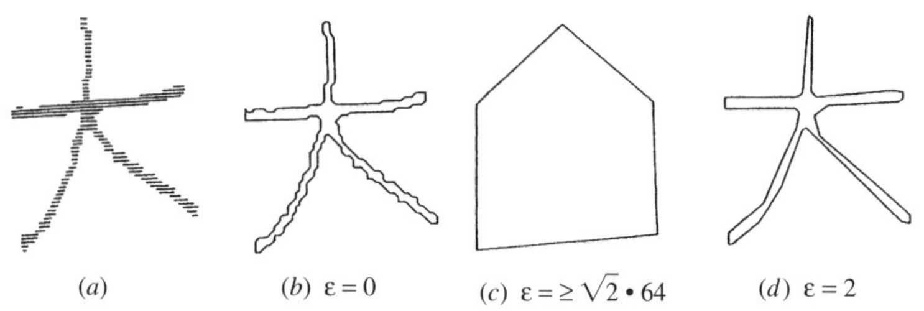

FIGURE 8.41 Change of polygonal approximation according to ε's values: (*a*) Original image. From reference [19].

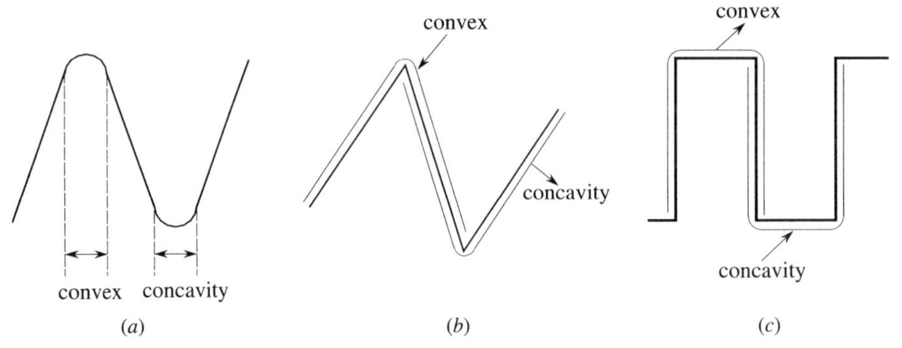

FIGURE 8.42 Problem of the convex and concavity segmentation based on second derivative (*a*) and the examples of intuitive convex and concavity segmentation (*b*) and (*c*). From reference [99], © 1997, Ohm.

Returning to the consideration above, we have a partly overlapped convex and concave segmentation of a contour. Therefore the contour tracing must be done twice. In the first tracing the starting points of concavity and convexity are found. In the second tracing, the end points of concavity and convexity using the known starting points are found. The detailed illustration of the algorithm is shown in Fig. 8.43. The first tracing is done clockwise looking at the shape from our right. At each boundary point i we take a normal direction θ_i of the tangent line, which is toward the foreground (inside of the shape). This angle is defined in Fig. 8.44.

Recall that raster scanning is performed from left to right and from top to bottom. So the starting point of the tracing, denoted as ST, is located first from which the tracing proceeds clockwise. For a while the angle θ_i decreases. The interval in which θ_i decreases is marked by a slightly thicker line. Outside the interval, θ_i begins to increase and continues increasing. This increasing interval is marked by a thin line. This increasing reaches a point, $\theta_i > \theta_l + \Delta\theta$, where θ_l is a minimum angle in the interval and $\Delta\theta$ is

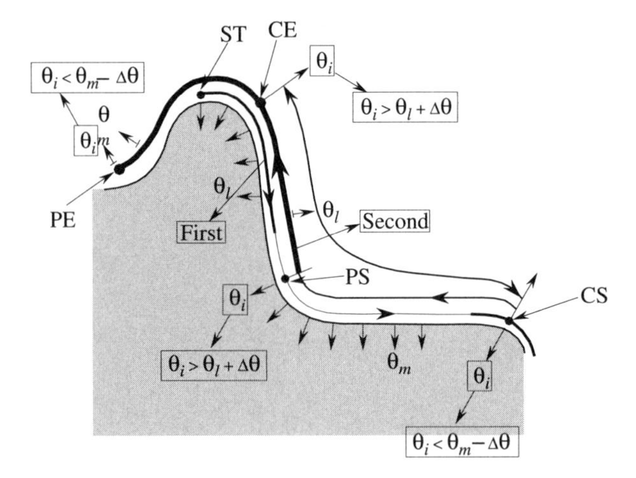

FIGURE 8.43 Concavity and convexity segmentation by detecting angle change.

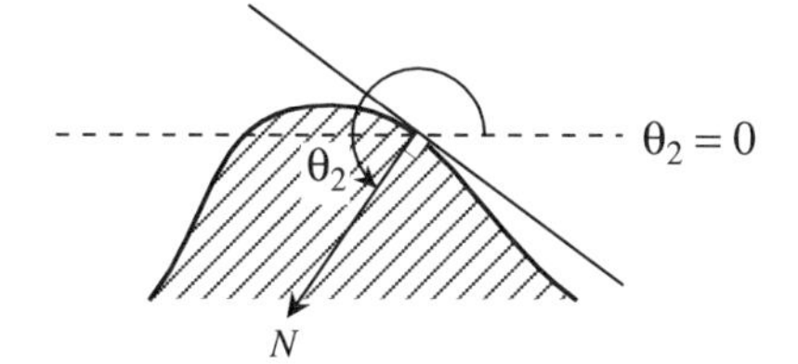

FIGURE 8.44 Definition of the angle at boundary.

a specified constant angle. This point is registered as the starting point of the convex segment, denoted as PS. It marks the start of the second tracing but the end point of the first trace. This idea will become clearer as we describe the second trace. Naturally, if we take $\Delta\theta$ large, then the overlapping of convex and concavity segments become large. In an experiment shown later, the $\Delta\theta$ was set at 45°. After finding PS, the θ_i will increase for a while but begin to decrease again, which is shown by a slightly thicker line (see Fig 8.43). We hold a maximum angle θ_m in the latest interval in which θ_i increases. Then θ_i reaches a point where $\theta_i < \theta_m - \Delta\theta$ holds. That point is registered as the starting point of a concave segment, denoted as CS. Thus PS and CS points are alternatively registered in the contour tracing until the trace reaches ST.

After reaching ST, the second tracing begins counterclockwise. The purpose of the second tracing is to find end points of convex and concave segments, since the start points of these segments have already been found in the first tracing. The starting points of the second tracing are the registered PS and CS points. The second tracing is also shown Fig. 8.43, and we begin our illustration from the CS point located at the right side. That is, we assume that the tracing traveled counterclockwise and reached the CS point. We notice that in the second tracing θ_i is taken as a normal of the tangent line to the background. The angle θ_i is decreasing and during this decreasing interval the minimum angle θ_l is held. When the angle turns to increase, we find the end point of the concavity segment such that $\theta_l > \theta_l + \Delta\theta$ begins to hold, denoted as CE. On the other hand, we encounter a PS point in this interval. So we start the tracing for finding the end point of the convex segment. Thus the two processes find end points of both concave and convex segment runs. In this convex case, θ_i increases, and within the increasing interval, the maximum angle θ_m holds. As the angle θ_i begins to decrease, a point is found such that $\theta_i < \theta_m - \Delta\theta$, which is registered as the end point of the convex segment, denoted as PE. This convex segment is marked by a thick line. We notice that the second tracing does not end at ST, in general, because the tracing must be done until finding all the pairs of PS-PE and CS-CE. Mathematically this is just a mapping from a contour to a Gauss map, which is introduced in Section 8.8.

Experiment An experimental result is shown in Fig. 8.45. The character used is Hirakana "あ." Both convex and concave portions are clearly extracted as seen. However, we notice that in some places the PS point coincides with the PE point in a neighboring convex portion. In that case only PE is marked, as seen at the top convex portion and the right convex portion of the short horizontal stroke near the top. Furthermore the

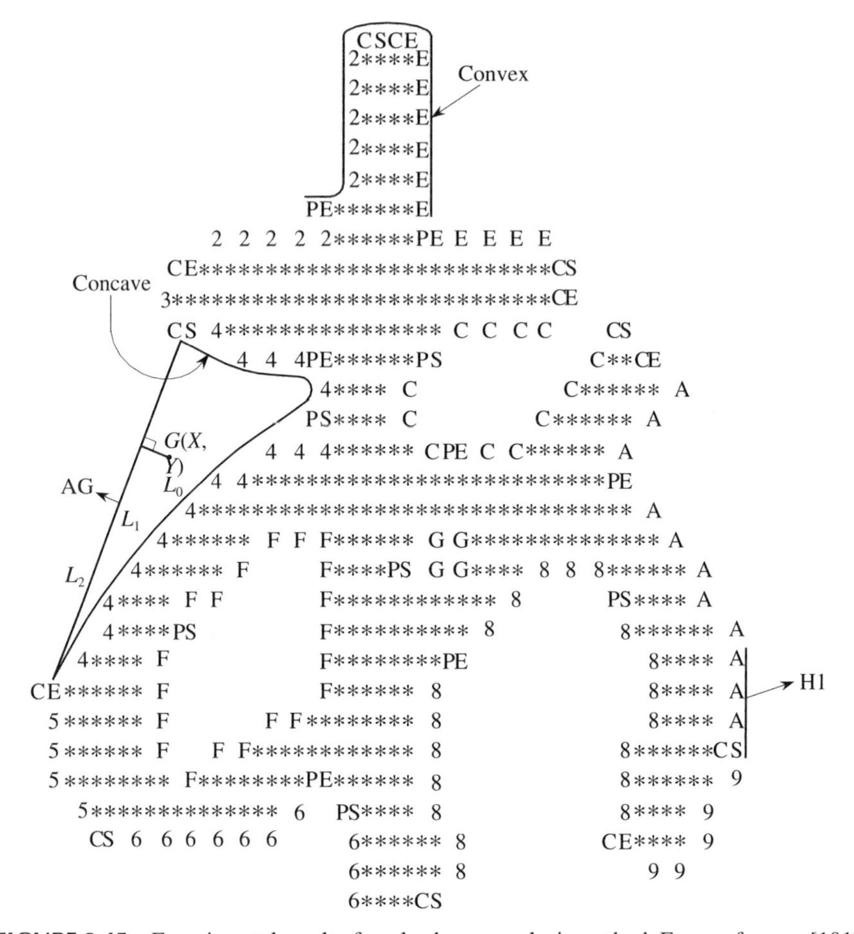

FIGURE 8.45 Experimental result of angle change analysis method. From reference [101].

segmentation IDs are marked for the concave segment when they are overlapped. Nevertheless, the intervals of convex segments can be seen looking at PE-PS pairs.

8.7 ANGLES AND CURVATURES

Corners are very important features, as can be seen in the drawing of Attneave's cat shown in Fig. 8.46 [20]. Much work has been done on corner detection so far. Polygonal approximation can be regarded as one approach to corner detection in a shape image, as described earlier. Here we introduce two other approaches. One is a digital or direct approach and the other is an analog or indirect approach. The former is classic in some sense but basic, and the latter is relatively modern, as will be seen in the next section.

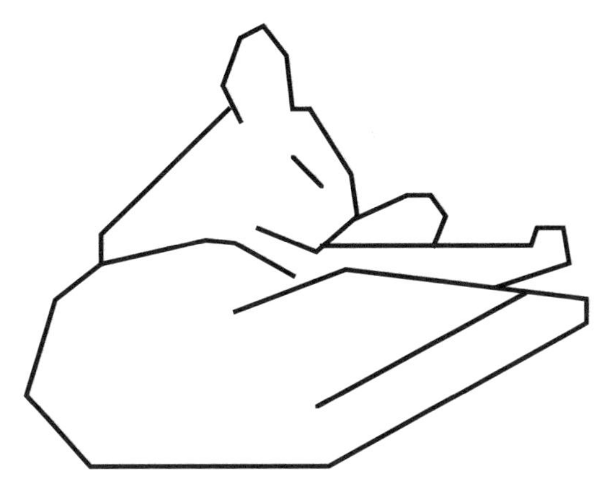

FIGURE 8.46 Atteneave's cat. From reference [20].

8.7.1 Definition of Angles

Before we begin discussing angles, we need to define a half straight line, $AB\infty$, that starts from A where vertex A passes a point B and extends to infinity.

The inverted half straight line of $AB * \infty$ is just opposite to $AB\infty$, which starts at A and passes through a point B^* on a straight line that is an extension of B^*A, as shown in Fig. 8.47. Next a two-sided shape is defined. Let $\vec{a} = OA\infty$ and $\vec{b} = OB\infty$ be different half straight lines and not inverted half straight lines to each other. Then $\vec{a} \cup \vec{b}$ is called a two-sided shape. Now we define an angle. Let X be an inner point of AB, where the points X and B lie on the same side of $\vec{b}$. The set of such points as X is called the interval portion of angle AOB, and the shape that consists of $\vec{a} \cup \vec{b}$ and all points of X is called an angle, denoted as $\angle AOB$, $\angle(\vec{a}, \vec{b})$, or $\angle O$. Each half straight line, $\vec{a}$ and $\vec{b}$, is called an angle side, respectively. The point O is called a vertex. The definition is illustrated in Fig. 8.48. When $\vec{a}$ and $\vec{b}$ are half straight lines to each other, one side of the straight line (i.e., the half-plane) is sometimes considered to be an angle. We notice that the exact definition of the measure of an angle remains to be defined, which is considerably complex.

The reason why the elementary term, angle, of geometry was introduced is to clarify the concept of angles. An angle is considered to be formed by sides that extend infinitely. For example, see the Fig. 8.49. The corner drawn in panel (a) is ambiguous as an angle, but that in panel (b) is clear. The figure illustrates the importance of sides to defining an angle [21].

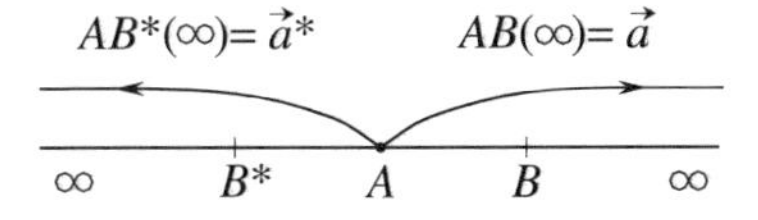

FIGURE 8.47 Definition of a half-line. From reference [21], © 1998, Iwanami.

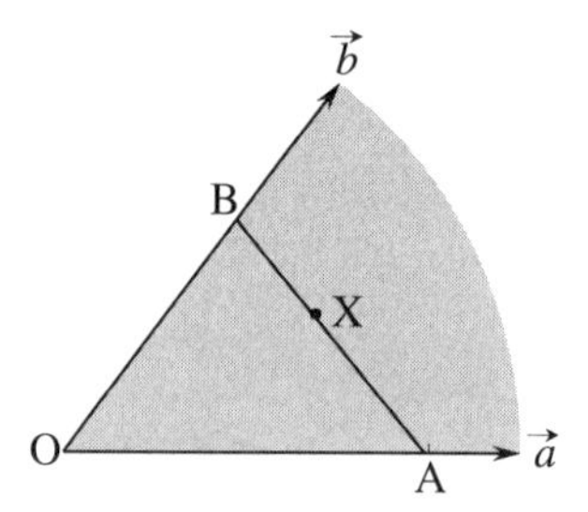

FIGURE 8.48 Definition of an angle: From reference [21], © 1998, Iwanami.

8.7.2 Digital Approach to Corner Detection

General algorithms of corner detection based on digital approaches have been developed. We will introduce the methods of is Rosenfeld-Johnston's algorithm [23] and Teh-Chin's algorithm [24]. The former is basic and comprehensive. The latter is regarded as the better one, and it addresses the difficulties of its forerunners.

Rosenfeld-Johnston's Algorithm For a continuous line, $y = f(x)$, it is well-known that curvature of the line is expressed in terms of derivatives as

$$\frac{d^2y/dx^2}{(1 + (dy/dx)^2)^{3/2}}. \tag{8.66}$$

However, it is not clear how to translate this formula to a digital curve, such as is defined as a sequence of integer-coordinate points, $P_1, \ldots, P_n$, where $P_i = (x_i, y_i)$. Actually the difficulty is that successive slope angles on the digital curve can only differ by a multiple of 45°, so small changes in slope are impossible. The first way to address this difficulty is to take the sides of a given point P_i, defining the derivative at P_i as

$$\frac{y_{i+k} - y_i}{x_{i+k} - x_i}, \tag{8.67}$$

where $k > 1$.

This k can be regarded as a smoothing factor as well as the length of one side. The second problem, however, is how to choose k. Before proceeding, let us look at some examples of angles spanned at point i (we took i instead of P_i for simplicity). These examples are shown in Fig. 8.50. In panel (a) we give a simple digital curve in which k

FIGURE 8.49 Important differences in the sides of a corner. From reference [23].

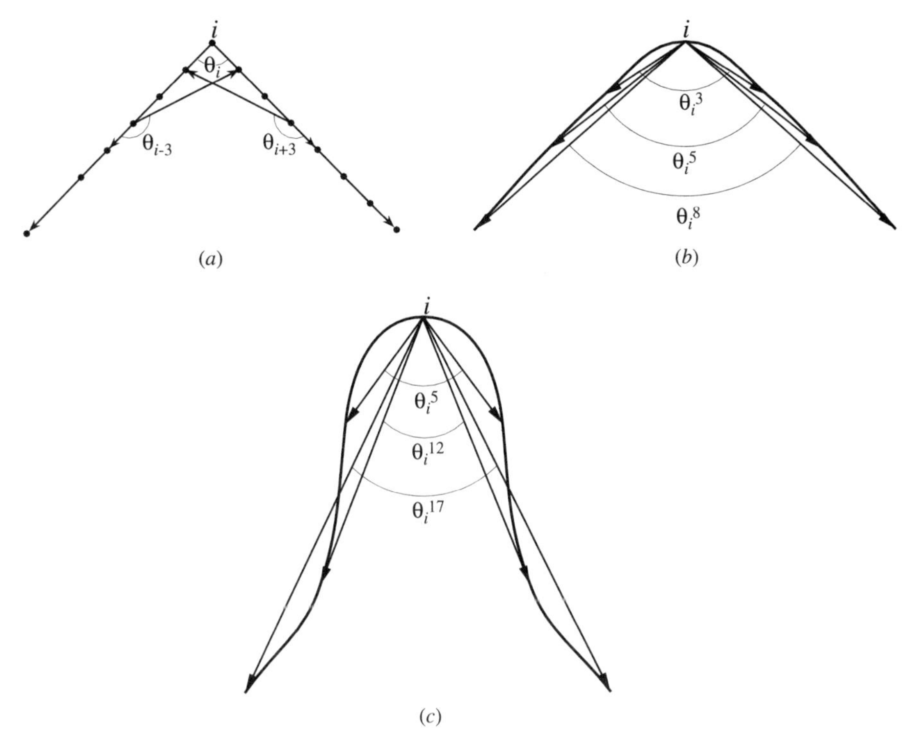

FIGURE 8.50 Some angles defined globally: (*a*) Acute case; (*b*) smooth case; (*c*) complex case.

is 4 and θ_{i-3}, θ_i, and θ_{i+3} are as marked. Obviously θ_i is most acute, so there is no problem even if $k = 1$ in this simple case. Therefore it is reasonable to take the acutest θ as a measure of curvature around point *i*. Other cases are shown in panels (*b*) and (*c*). In the case of (*b*), the θ spanned at *i* decreases as *k* increases. In the case of (*c*), the θ spanned at *i* decreases first and then increases, $\theta_i^5 > \theta_i^{12}$ and $\theta_i^{17} > \theta_i^{12}$. The latter two cases show the dependency θ on *k*. Therefore it is important to choose an appropriate *k*.

Now let us formalize the angle at point *i* as a curvature in a global sense. First we will find cos θ instead of θ because it is easier to handle and because θ and cosθ are related monotonically ($0 \le \theta \le 0$). We can easily obtain cosθ at point *i* as shown in Fig. 8.51. We have

$$\vec{a}_{ik} = \vec{x}_{i+k} - \vec{x}_i = (x_{i+k} - x_i,\ y_{i+k} - y_i), \tag{8.68}$$

$$\vec{b}_{ik} = \vec{x}_{i-k} - \vec{x}_i = (x_{i-k} - x_i,\ y_{i-k} - y_i). \tag{8.69}$$

Thus cos θ at *i*, called *k* – cosine, can be obtained as follows:

$$C_i^k = \frac{(\vec{a}_{ik} \cdot \vec{b}_{ik})}{|\vec{a}_{ik}| \cdot |\vec{b}_{ik}|}. \tag{8.70}$$

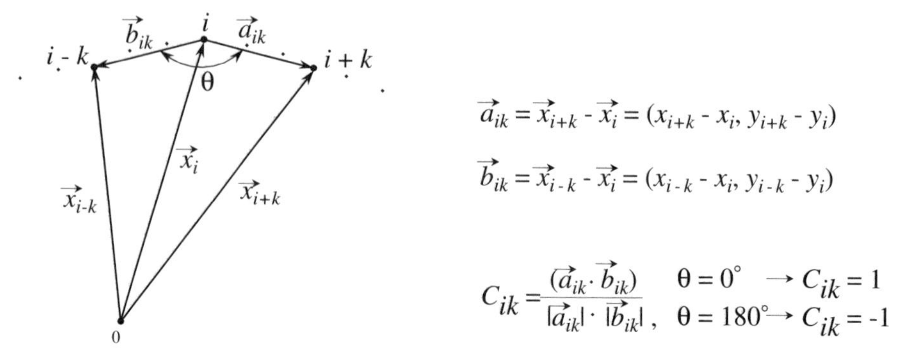

FIGURE 8.51 Illustration of k-cosine, C_i^k. From reference [99], © 1997, Ohm.

The most acute $\theta = 0°$ is given by $C_i^k = 1$ and no curve point $\theta = 180°$ is given by $C_i^k = -1$. Therefore, to seek the minimum of θ is equivalent to seek the maximum of C_i^k.

Now let us describe how to choose an appropriate value of k at each point i of a given curve. First, we take a window of an arbitrary length m and calculate $C_i^1, C_i^2, \ldots, C_i^m$. Let h be such that

$$C_i^m < C_i^{m-1} < \cdots < C_i^h \geq C_i^{h-1}. \tag{8.71}$$

We call C_i^h the cosine of at i and denote it by C_i. Finally we compare C_i, $C_{i\pm1}$, $C_{i\pm2}, \ldots$, and say that there is a curvature maximum at point i if $C_i \geq C_j$ for all j such that $|i - j| \leq h/2$. This is shown in Fig. 8.52. Rosenfeld and Johnston proposed to select m as $[n/10]$, where n is a total number of points (i.e., $[1/10]$) of the perimeter of the curve. This idea will be further discussed in the next section.

Notice A reader closely reading their paper may be a littled puzzled by the details. To help in understanding the numerical results, we use the image given in their paper, a chromosome shape, and their standard data for a comparison study. The data image we produced is shown in Fig. 8.53, and it will be compared with the results of other algorithms. The detected corner points are indicated by filled circles, some of which do not coincide with theirs. The numerical results are also shown in Table 8.3. The boundary points are numbered, starting from the double-circled point and proceeding clockwise in the figure. The parameter m is taken as 6. The corner points shown coincide exactly with the Teh and Chin's results shown in Fig. 8.60. Nevertheless, Rosenfeld and Johnston's algorithm is very simple, so readers can check this for themselves.

The algorithm can be easily understood from a visual inspection of Table 8.3. For example, the point numbered 32 is a typical case, where $C_i^6 < C_i^5 > C_i^4$ holds and the size h is determined as 5. The neighboring points that satisfy $|i - j| \leq h/2 = 2.5$ (i.e., points 30 through 32) were checked to see whether or not point 32 is an extreme point in the sense of a local maximum.

One important observation concerning the representation of the data is that as mentioned before, there are two methods of boundary representation of shape, that is,

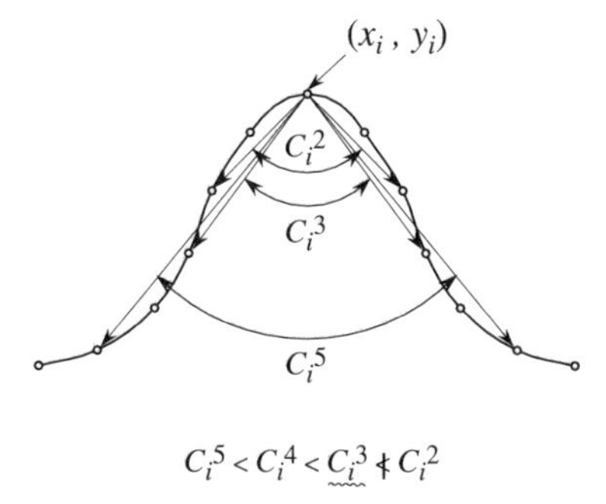

FIGURE 8.52 Selection of C_i^h at point (x_i, y_i) based on formula (8.69), where m is chosen as 5. From reference [99], © 1997, Ohm.

white and black boundaries. The chromosome shape is obviously a white boundary representation because, if it were not, the acute concave portion around the starting point would be filled by black points. However, a black boundary representation is more intuitive than a white one. For this reason we would use the black boundary representation if we want to compare the result with human intuition. This point will be discussed again when we examine the comparison study conducted by Teh and Chin.

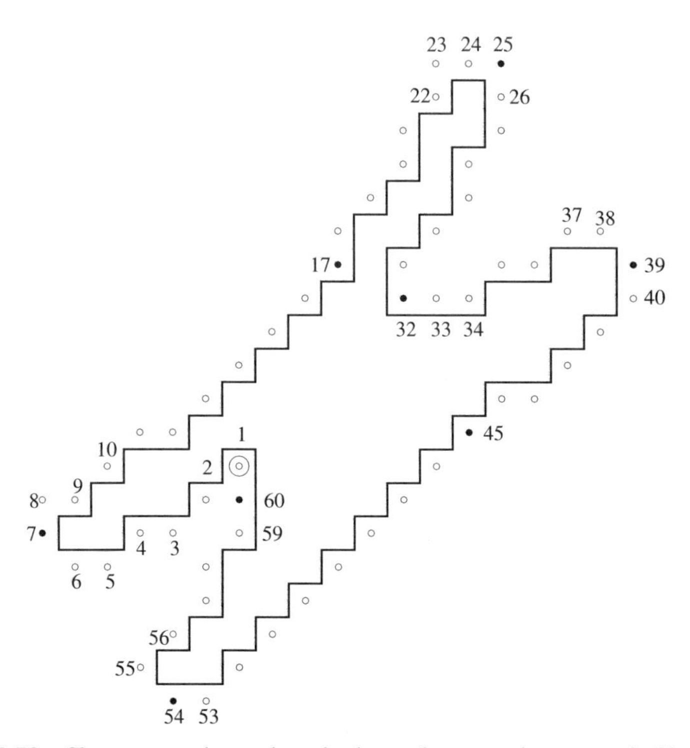

FIGURE 8.53 Chromosome shape where the detected corner points are marked by filled circles near the numbered boundary points. From reference [23] (modified) © 1998, IEEE.

TABLE 8.3 Numerical Results of Rosenfeld and Johnston's Algorithm Applied to the Data Shown in Fig. 8.50

Point Number	C_i^1	C_i^2	C_i^3	C_i^4	C_i^5	C_i^6
1	0.707	0.707	0.789	0.776	0.796	0.707
2	−1.000	−0.894	−0.196	0.447	0.196	0.243
3	−0.707	−0.949	−0.990	−1.000	−0.857	−0.707
4	−0.707	−1.000	−0.832	−0.800	−0.894	−0.800
5	−0.707	−0.600	−0.196	0.179	0.447	0.471
6	−0.707	0.000	0.316	0.707	0.844	0.854
7	−0.707	0.316	0.707	0.707	0.902	**0.938**
8	0.000	0.000	0.196	0.707	0.707	0.832
9	−0.707	−1.000	−0.555	−0.179	0.419	0.526
10	−1.000	−1.000	−0.981	−0.822	−0.625	−0.229
11	−0.707	−0.949	−1.000	−0.990	−0.908	−0.807
12	−0.707	−0.949	−0.981	−0.949	−0.990	−0.999
13	−1.000	−0.949	−0.981	−0.990	−0.937	−0.980
14	−1.000	−1.000	−0.981	−0.960	−0.976	−0.959
15	−1.000	−1.000	−0.981	−0.960	−0.976	−0.959
16	−1.000	−0.949	−0.981	−0.990	−0.937	−0.959
17	−0.707	−0.949	−0.981	−0.949	−0.970	**−0.916**
18	−0.707	−0.949	−1.000	−1.000	−0.991	−1.000
19	−1.000	−1.000	−1.000	−0.984	−1.000	−0.996
20	−0.707	−0.949	−0.965	−0.998	−0.994	−0.959
21	−0.707	−1.000	−0.981	−0.868	−0.759	−0.555
22	−0.707	−0.949	−0.868	−0.600	−0.077	0.496
23	0.000	−0.447	0.141	0.316	0.651	0.759
24	−1.000	0.000	0.316	0.832	0.800	0.888
25	0.000	0.000	0.707	0.740	0.919	0.949
26	−1.000	−0.316	−0.141	0.447	0.759	0.868
27	−0.707	−0.894	−0.496	−0.141	0.077	0.371
28	−0.707	−1.000	−0.965	−0.894	−0.844	−0.894
29	−0.707	−0.949	−0.965	−0.997	−1.000	−0.976
30	−1.000	−1.000	−0.832	−0.600	−0.083	−0.124
31	−0.707	0.000	0.124	0.600	0.514	0.614
32	0.000	0.447	0.789	0.651	**0.796**	0.707
33	−1.000	−0.316	0.316	0.707	0.585	0.519
34	−0.707	−0.894	−0.496	0.000	0.196	0.000
35	−0.707	−1.000	−1.000	−1.000	−0.976	−0.868
36	−0.707	−1.000	−0.949	−0.844	−0.707	−0.600
37	−0.707	−0.600	−0.196	0.141	0.371	0.385
38	−0.707	0.000	0.316	0.651	0.690	0.759
39	−0.707	0.000	0.555	0.600	**0.832**	0.814
40	−0.707	−0.316	0.000	0.569	0.606	0.768
41	−1.000	−0.949	−0.555	−0.316	0.110	0.284
42	−1.000	−0.949	−0.923	−0.776	−0.625	−0.364
43	−0.707	−0.949	−0.981	−0.960	−0.870	−0.778
44	−0.707	−0.949	−0.981	−0.990	−1.000	−0.970

TABLE 8.3 *(Continued)*

Point Number	C_i^1	C_i^2	C_i^3	C_i^4	C_i^5	C_i^6
45	−1.000	**−0.949**	−0.981	−0.990	−0.994	−1.000
46	−1.000	−1.000	−0.981	−0.990	−0.994	−0.996
47	−1.000	−1.000	−1.000	−0.990	−0.994	−0.996
48	−1.000	−1.000	−1.000	−1.000	−0.994	−1.000
49	−1.000	−1.000	−1.000	−1.000	−0.994	−0.973
50	−1.000	−1.000	−1.000	−0.990	−0.919	−0.857
51	−1.000	−1.000	−0.981	−0.857	−0.707	−0.316
52	−1.000	−0.949	−0.707	−0.316	0.316	0.447
53	−0.707	−0.316	0.316	0.707	0.707	0.832
54	−0.707	0.447	0.789	0.776	**0.870**	0.850
55	0.000	0.316	0.555	0.776	0.796	0.800
56	−1.000	−0.894	−0.496	0.000	0.371	0.471
57	−0.707	−0.949	−1.000	−0.970	−0.894	−0.800
58	−0.707	−1.000	−0.965	−0.970	−0.707	−0.707
59	−0.707	−0.894	−0.196	0.600	0.371	0.428
60	−1.000	0.447	0.707	0.707	**0.844**	0.646

Note: Arabic numbers and floating numbers indicate the detected corner points and their κ cosine values.

Teh and Chin Algorithm At the end of their paper, Rosenfeld and Johnston point out a basic problem with their algorithm. It is illustrated by Fig. 8.54. As the figure shows, if the point where the curve crosses itself is located some distance m from P, then C_i^m will be close to 1, and $C_i^{m-1} < C_i^m$. Therefore the procedure will detect a very sharp angle at P. This is an anomalous case, to be sure. This paradox would be instantiated by a curve that approaches itself at two points that are $n/10$ or closer. Rosenfeld and Johnston addressed this problem as follows: It is up to the user of the algorithm to pick an m appropriate to a class of curves rather than simply taking $m = n/10$ as was done here. In fact, if the user has special knowledge about the class of curves, he or she might well be advised to devise a segmentation procedure that takes this knowledge into account.

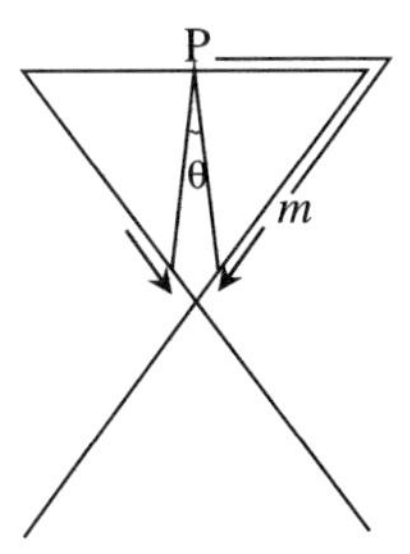

FIGURE 8.54 Abnormal curvature maximum at point *P*. From reference [23], © 1998, IEEE.

Actually this is a well-known fundamental problem of scale, as was first pointed out by Rosenfeld in his edge detection scheme [25]. Too large a window will smooth out the fine features of a curve, whereas a small window will generate a large number of redundant feature points. In other words, this is the problem of local versus global view. An analogous situation is the use of low-pass filtering to eliminate noise. Therefore it is very difficult to choose an appropriate smoothing parameter for all the scales. The choice of scale depends on the context, and this is a critical area in the development of current machine vision algorithms. In this respect, parameter setting is considered outmoded, and rather the idea is that the parameters should be selected according to the scale. Feature description then becomes hierarchical. This approach is also known as scale space filtering [26]. See the Bibliographical Remarks of at the end of the chapter.

Now we turn to an improvement in the algorithm above as was given by Teh and Chin [24], after several attempts by researchers. Their algorithm does not request an input smoothing parameter like m in the Rosenfeld and Johnston algorithm. They address the scale problem by making the window of each boundary point determined independently based on local properties. As will be shown later in an example, the reliability and accuracy of feature point detection depends not only on the accurate determination of discrete curvature, but primarily on the accurate determination of the smoothing factor or the region of support. Historically (see the Bibliographical Remarks) the feature point and the region of support have also been referred to as the dominant point and the domain $D(\cdot)$, respectively.

Now let us formalize the Ten-Chin algorithm. Let the sequence of n integer-coordinate points describe a closed curve C,

$$C = \{P_i = (x_i, y_i), i = 1, \dots, n\}, \tag{8.72}$$

where P_{i+1} is a neighbor of P_i (modulo n). The Freemen chain code of C consists of the n vectors

$$\vec{C}_i = \overline{P_{i-1}P_i}, \tag{8.73}$$

each of which can be represented by an integer

$$f = 0, \dots, 7, \tag{8.74}$$

where $f/4\pi$ is the angle between the x-axis and the vector. The chain of C is defined as $\{\vec{c}_i, i = 1, \dots, n\}$ and $\vec{c}_i = \vec{c}_{i \pm n}$.

To show and emphasize how important the determination of the region of support is, the following three different measures of significance were adopted, which correspond to different degrees of accuracy of discrete curvature measures:

1. *k-cosine measure* of the Rosenfeld-Johnston algorithm, C_i^k in (8.70).
2. *k-curvature measure,* which is the difference in mean angular direction of the k-vector (8.73) on the leading and trailing curve segment of the point P_i where the curvature is measured [27]:

$$\mathrm{CUR}_i^k = \frac{1}{k}\sum_{j=-k}^{-1} f_{i-j} - \frac{1}{k}\sum_{j=0}^{k-1} f_{i-j}. \tag{8.75}$$

3. 1-*curvature measure* ($k = 1$ of condition 2 above)

$$\mathrm{CUR}_i^1 = f_{i+1} - f_j. \tag{8.76}$$

Determination of Region of Support

STEP 1: Define the length of the chord joining the points P_{i-k} and P_{i+k} as

$$l_i^k = |\overline{P_{i-k}P_{i+k}}|. \tag{8.77}$$

Let d_i^k be the perpendicular distance of the point P_i to the chord $\overline{P_{i-k}P_{i+k}}$.

STEP 2: Start with $k = 1$. Compute l_i^k and d_i^k until

$$l_i^k \geq l_i^{k+1} \qquad \text{(condition 1)} \tag{8.78}$$

or

$$\frac{d_i^k}{l_i^k} \geq \frac{d_i^{k+1}}{l_i^{k+1}} \quad \text{for } d_i^k > 0 \qquad \text{(condition 2)}, \tag{8.79}$$

$$\frac{d_i^k}{l_i^k} \leq \frac{d_i^{k+1}}{l_i^{k+1}} \quad \text{for } d_i^k < 0. \tag{8.80}$$

Then the region of support of P_i is the set of points which satisfy either condition 1 or condition 2,

$$D(P_i) = \{(P_{i-k}, \ldots, P_{i-1}, P_i, P_{i+1}, \ldots, P_{i+k}) | \text{condition 1 or 2}\}. \tag{8.81}$$

Notice that the condition 2 is not used if $d_i^k = 0$, and it is related to Ramer's polygonal approximation mentioned before. We can show that the condition 1 alone is powerful enough to overcome the problems raised by Rosenfeld, Iohustar, Davis, and other researchers. The problems happen when the dominant points occur too close to one another. Figure 8.55 illustrates such a case. If the two dominant points i and j are very close as shown, then trouble occurs. The results in Fig. 8.55 tells us that size must be determined according to situation and not be constant for every point.

Figure 8.56 shows how the problem has been overcome by Teh and Chin's algorithm. In panel (*a*), the region of support of the point i is just the top horizontal segment, $k = 4$ as shown, because

$$l_i^1 < l_i^2 < l_i^3 < l_i^4 = l_i^5$$

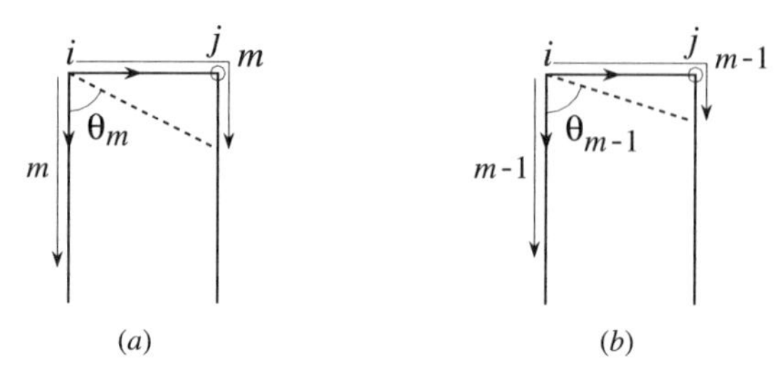

FIGURE 8.55 Difficulty of Rosenfeld size determination scheme. The corner marked by circle might be neglected by taking larger value of m. Notice at least the wrong angle $\theta_m < 90$ assigned to the i corner point. From reference [28], © 1998, IEEE.

holds, which is natural intuitively. The same thing is true at the close point j from symmetry. Panel (b) also gives thc shape discussed by Davis [28] as an example of the difficulty of setting an appropriate region of support of a corner point.

In this case condition 1 is also enough to choose the regions of support of the point i and j. In panel (c) every point on the circle is assigned a region of support of diameter l and cosine value of 0 using the condition 1 alone, resulting in every point being detected as a dominant point. As an analog or continuous model, it is reasonable. However, in the digital case the problem is that there are too many redundant dominant points. This point is discussed by Teh and Chin. The shape raised by Rosenfeld and Johnston, shown in Fig. 8.54, can be easily handled using condition 1 alone as well.

Further, in the Fig. 8.57 are given two other instances that call for condition 2. In panel ($a1$) at the point q,

$$l'_q < l_q^2 < l_q^3 \cdots < l_q^\infty$$

holds. That is, we cannot rely on only condition 1 in this case. Therefore condition 2 is checked as

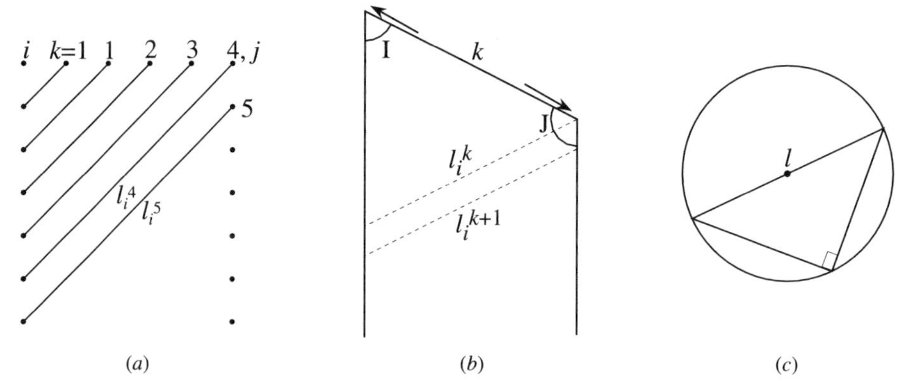

FIGURE 8.56 Some applications of the condition 1: (a)-(b) Davis's examples; (c) Teh and Chin's example.

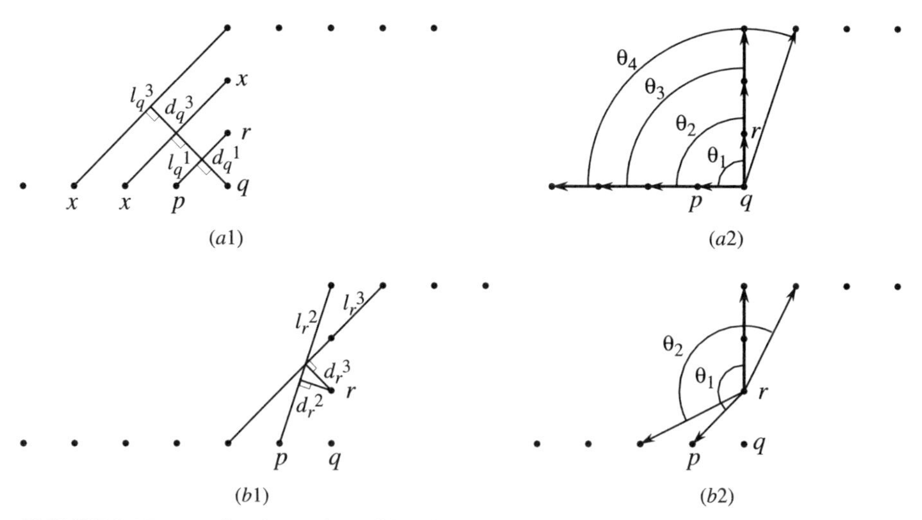

FIGURE 8.57 Applications of condition 2 and comparison with Rosenfeld-Johnston's idea of support.

$$\frac{d_q^1}{l_q^1} = \frac{d_q^2}{l_q^2} = \frac{d_q^3}{l_q^3},$$

which tells that the region of support of the point q is $k = 1$. This figure was raised and discussed in the Rosenfeld and Weska's paper [29]. They argued for a "best regions of support" behavior of 4, 3, and 2 for k's at p, q, and r points, respectively. Another example is shown in panel (b1), which gives $k = 2$ as the region of support at the point r. Concerning the discrepancy with Rosenfeld-Weska's consideration at point q, Teh and Chin claim in their paper that intuitively q should be assigned a best region of support of 1, since, if small k's give the same cosine value as large ones, the angle must be sharp, and then the smallest k (1, *not* 3) should be taken as the best region of support of q. This idea is somewhat controversial. Because k can be regarded as a measure of stability or reliability of the corner point. Figure 8.58 shows that the regions of support at the tips can intuitively be regarded as long sides. In the figure, k of both (a) and (b) is 1 according to Teh and Chin's algorithm. Figure 8.59 shows the problem raised by Davis [28] in which local and global angles coexist at point P_i. In these terms Teh-Chin's determination of the region of support can essentially be regarded as "local." At any rate, their algorithm works well, as will be shown later. We proceed to describe this algorithm in a complete manner.

Teh-Chin Algorithm

STEP 1: Determine the region of support of each point by the algorithm described above,

$$D(P_i) = \{P_{i-k_i}, \cdots, P_{i-1}, P_i, P_{i+1}, \cdots, P_{i+k_i}\}.$$

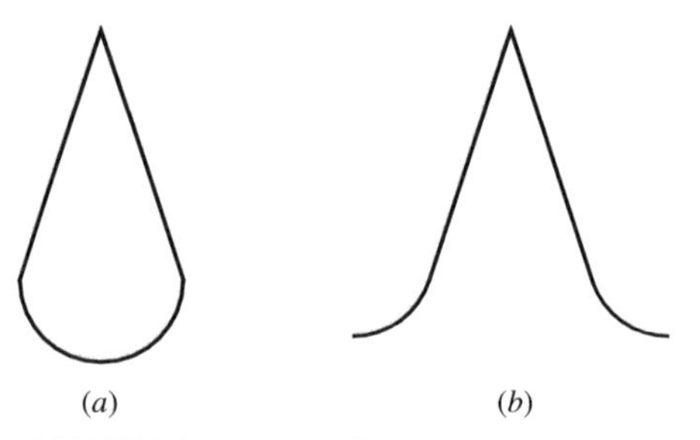

FIGURE 8.58 Reliability of a longer k corner.

STEP 2: Select a measure of significance (e.g., from one of the three measures defined before), and calculate $S(P_i)$.

STEP 3a: 1st pass: Perform nonmaxima suppression as follows; retain only those points P_i where

$$S(P_i) \geq S(P_j)$$

for all j such that

$$|i - j| \geq \frac{k_i}{2}.$$

STEP 3b: 2nd pass: Now suppress those points having zero 1-curvature ($\text{CUR}_i' = 0$).

STEP 3c: 3rd pass: For those points that survived the 2nd pass, if [k_i of $D(P_i)$] = 1 and if P_{i-1} or P_{i+1} still survive, then further suppress P_i if $S(P_i) \leq S(P_{i-1})$ or $S(P_i) \leq S(P_{i+1})$. If 1-curvature is selected as a measure of significance, then go to step 3d and do a 4th pass, else those points survived are the dominant points

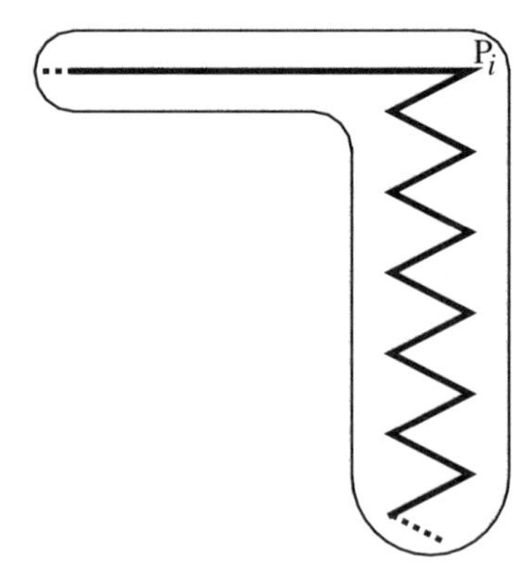

FIGURE 8.59 Global corner and local corners. From reference [28], © 1998, IEEE.

STEP 3d: 4th pass: For those groups of more than two points that still survived, suppress all points except the two end points of each group.
For those group of exactly two points that still survived,

if $(S(P_i) > S(P_{i+1}))$, then suppress P_{i+1}
else if $(S(P_i) < S(P_{i+1})$, then
suppress P_i
else if $(k_i > k_{i+1})$, then
suppress P_{i+1},
else suppress P_i.

Experimental Results and Comparative Studies Teh and Chin conducted systematic experiments and comparative studies. They compared their algorithm with other algorithms such as Rosenfeld-Johnston's, Rosenfeld-Weska's, Freeman-Davis [22], Sanker-Sharma's [30] and Anderson-Bezdek's [31]. One of the comparison studies is shown in Fig. 8.60 for a chromosome-shaped curve which has been widely used as a test shape in this field since Rosenfeld and Johnston used it. Panel (*a*) is the result of the Rosenfeld-Johnston's algorithm for $m = 6$, which was chosen according to the rule, $[n/10]$, and panel (*b*) is the result of the Rosenfeld-Weska's algorithm [29] which is an improved version of the Rosenfeld-Johnston's smoothing algorithm. Panels (*f*), (*g*), and (*h*) are the results of Teh-Chin algorithm on changing curvature measurement as *k*-cosine, *k*-curvature, and 1-curvature, as explained here earlier. In particular, it is surprising that a very simple measurement, 1-curvature, gives almost the same result as the other two. This shows the importance of the region of support argument, as mentioned before. To aid the reader's understanding, the numerical results for *k*-cosine measure are shown in Table 8.4, which can be used to check the Teh and Chin algorithm. In the table the "fail" in the selection column denotes a dropped-out point in the survival processing of their algorithm. The speeds in CPU time were also compared and were almost the same except for Sanker-Sharma's algorithm which was three times as slow as the rest. Sanker-Sharma's algorithm is the only other non-input-parameter method besides the Teh-Chin algorithm. We notice that the speeds are almost the same regardless of the three different curvature measurements. The details on these changing shape comparisons are discussed in their paper.

Notice As we noted earlier, our simulation result of Rosenfeld and Johnston's algorithm for the chromosome-shaped curve is exactly the same as that shown in Fig. 8.60 (*a*). Our simulation result of Teh and Chin's algorithm for the same shape obtained the same result shown in Fig. 8.60 (*f*). We need to make some observations based on our two results. If we look at the original data shown in Fig. 8.60 (*a*) intuitively, the result of Teh and Chin's algorithm seems to meet our intuition. As noticed before, the data points are given when the contour followed is the white boundary as shown in Fig. 8.53. If we follow along the black image boundary, then the result of Rosenfeld and Johnston's algorithm is reasonable except for points 1 and 59. Thus we are obliged to conclude that Teh and Chin's algorithm is better than Rosenfeld and Johnston's if we

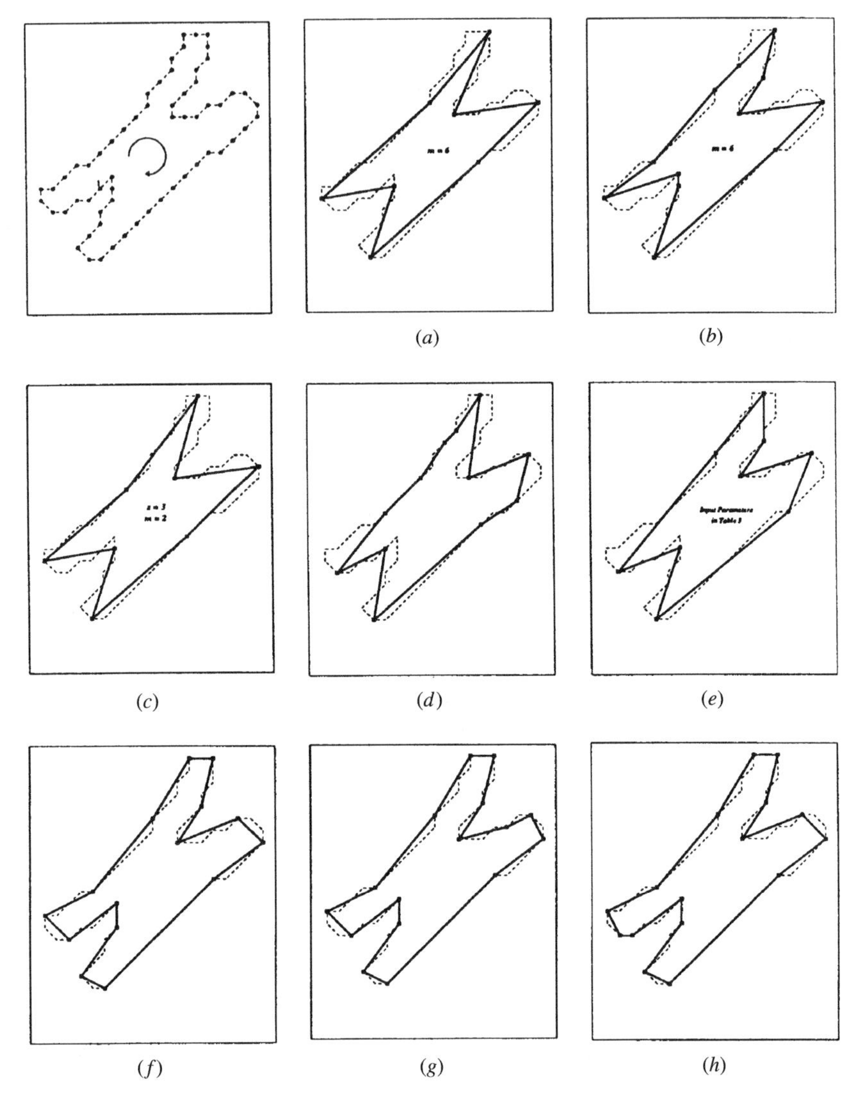

FIGURE 8.60 Chromosome-shaped curve. (*a*) Rosenfeld-Johnston algorithm. (*b*) Rosenfeld-Weszka algorithm. (*c*) Freeman-Davis algorithm. (*d*) Sankar-Sharma algorithm. (*e*) Anderson-Bezdek algorithm. (*f*) Teh-Chin algorithm (k-cosine). (*g*) Teh-Chin algorithm (k-curvature). (*h*) Teh-Chin algorithm (1-curvature). From reference [24], © 1998, IEEE.

regard the four end points of the data expressed in tracing the black boundary. However, this causes a problem around the acute corner point labeled to 1; that is, the concave region is buried. This was discussed previously. Furthermore, in Teh and Chin's original paper, an absolute value, $|S(p_i)|$, is used in their algorithm. It is very strange, as can be understood on examining the Table 8.4 for k-cosine. Therefore the symbol for absolute value, $|\cdot|$, has been removed in this text. This point will be further elaborated when we discuss "curvature" in the rigorous mathematical sense.

TABLE 8.4 Numerical Result of Teh and Chin's Algorithm Applied to the Chromosome Shaped Curve Using κ-cosine

Point Number	κ-cosine	κ	Selection	Point Number	κ-cosine	κ	Selection
1	0.707	1		31	0.124	3	Fail
2	−0.196	3	Fail	32	0.447	2	
3	−0.707	1	Fail	33	0.316	3	Fail
4	−0.707	1	Fail	34	−0.707	1	Fail
5	0.179	4		35	−0.707	1	Fail
6	0.000	2	Fail	36	−0.707	1	Fail
7	−0.707	1	Fail	37	0.385	6	
8	0.000	1		38	0.316	3	Fail
9	−0.707	1	Fail	39	0.000	2	Fail
10	−0.625	5	Fail	40	0.000	3	
11	−0.707	1	Fail	41	−0.316	4	Fail
12	−0.707	1		42	−0.625	5	Fail
13	−0.949	2	Fail	43	−0.707	1	Fail
14	−0.960	4	Fail	44	−0.707	1	
15	−0.960	4	Fail	45	−0.949	2	Fail
16	−0.949	2	Fail	46	−0.981	3	Fail
17	−0.707	1	Fail	47	−0.990	4	Fail
18	−0.707	1		48	−0.994	5	Fail
19	−0.984	4	Fail	49	−0.973	6	Fail
20	−0.707	1	Fail	50	−0.919	5	Fail
21	−0.707	1	Fail	51	−0.857	4	Fail
22	−0.707	1	Fail	52	−0.707	3	Fail
23	0.000	1		53	−0.316	2	
24	−1.000	1	Fail	54	−0.707	1	Fail
25	0.000	1		55	0.555	3	
26	−0.141	3	Fail	56	−0.496	3	Fail
27	−0.707	1	Fail	57	−0.707	1	Fail
28	−0.707	1	Fail	58	−0.707	1	Fail
29	−0.707	1		59	−0.707	1	
30	−0.600	4	Fail	60	−1.000	1	Fail

8.8 ANALOG APPROACH TO CORNER DETECTION

In this section we introduce an analog approach to detect a corner along a contour as an alternative to the digital approach which we described earlier. The basic idea is very simple. That is, a series of sampling points of a contour is approximated by a simple continuous function. Then we use a well-known formula of curvature from classic analytic geometry:

$$\frac{d^2y/dx^2}{(1 + (dy/dx)^2)^{3/2}}, \tag{8.82}$$

where $y = f(x)$ which represents a contour in continuous function. Historically, however, the digital approach appeared first, despite the fact that the formula of curvature was already known in the nineteenth century. The reason may be surmised from the fact that finding an appropriate continuous function representing a series of sampling points was not so easy. However, with the rapid development of computing power, a very nice approximating function was found in the field of computer graphics. This B-spline function, which is a kind of polynominal function, is easily derived. This is the idea behind a very simple method developed by G. Medioni and Y. Yasumoto [32].

8.8.1 Representation of a Curvature

We defined an angle as a quasi-global feature, whose lengths of its two opposite sides play an important role. In contrast, curvature is essentially locally defined. To make sure that this is well understood, the definition is given again below.

Definition of Curvature-I *For a point P on a curve γ another point Q on the γ is considered, which is set close to P, and the distance between P and Q along the γ is denoted as Δs. At the points P and Q tangential lines are drawn, then the angle between the two tangential lines is denoted as $\Delta\theta$. The curvature κ of the curve γ at point P is defined as*

$$\kappa = \lim_{\Delta s \to 0} \frac{\Delta\theta}{\Delta s} = \frac{d\theta}{ds}. \tag{8.83}$$

The notation is illustrated in Fig. 8.61, including the directions of Δs and $\Delta\theta$. That is, we need to know the curvature's sign. The plus and minus signs correspond to convexities down and up toward the top of the figure. Alternatively, we can give another definition of curvature that is more elegant and general than the definition mentioned above [33]. A curve γ is represented parametrically as

$$\mathbf{p} = \mathbf{p}(t) = \mathbf{p}(x(t), y(t)), \tag{8.84}$$

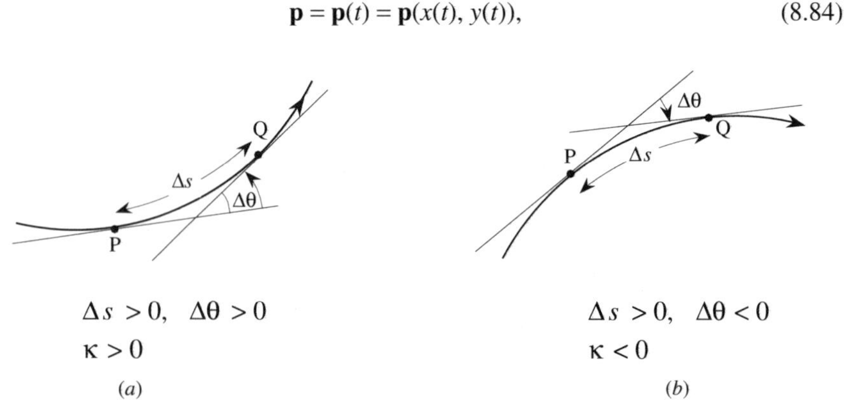

FIGURE 8.61 Direction of curvature k: (a) $\kappa > 0$ case; (b) $\kappa < 0$ case.

where $\mathbf{p}$ is a position vector on the curve γ and $(x(t), y(t))$ is a coordinate of the point. Then the length s is represented as a function of t as

$$s = \int_0^t |\dot{\mathbf{p}}(t)|dt = \int_0^t |\dot{\mathbf{p}}(u)|du, \tag{8.85}$$

where

$$\dot{\mathbf{p}}(t) = (\dot{x}(t), \dot{y}(t)).$$

Therefore

$$\dot{s}(t) = |\dot{\mathbf{p}}(t)|$$

holds. Naturally for all the t, if $\dot{\mathbf{p}}(t) \neq 0$, then $|\dot{\mathbf{p}}(t)| > 0$. Therefore s is a monotonic increasing function of t, which implies that t is represented as a function of s. That is,

$$\mathbf{p} = \mathbf{p}(s) = (x(s), y(s)), \tag{8.86}$$

and so

$$\mathbf{p}' = \mathbf{p}'(s) = (x'(s), y'(s)). \tag{8.87}$$

As stated before, it is easily proved that

$$|\mathbf{p}'(s)| = 1. \tag{8.88}$$

The $\mathbf{p}'(s)$ is newly represented as

$$\mathbf{e}_1 \equiv \mathbf{p}'. \tag{8.89}$$

The meaning of the vector $\mathbf{e}(s_0)$ is a tangent vector of a curve γ at $\mathbf{p}(s_0)$.

Now we take another vector $\mathbf{e}_2(s_0)$ which is perpendicular to $\mathbf{e}_1(s_0)$ and its length is equal to 1; see Fig. 8.62. However, we need to note that we have two choices in selecting $\mathbf{e}_2(s_0)$. In the figure the $\mathbf{e}_2(s_0)$ is chosen such that it lies to the left of $\mathbf{e}_1(s_0)$. We notice the following relations among the vectors defined so far:

$$\mathbf{p}' = \mathbf{e}_1, \quad \mathbf{e}_1 \cdot \mathbf{e}_1 = 1, \quad \mathbf{e}_2 \cdot \mathbf{e}_2 = 1, \quad \mathbf{e}_1 \cdot \mathbf{e}_2 = 0. \tag{8.90}$$

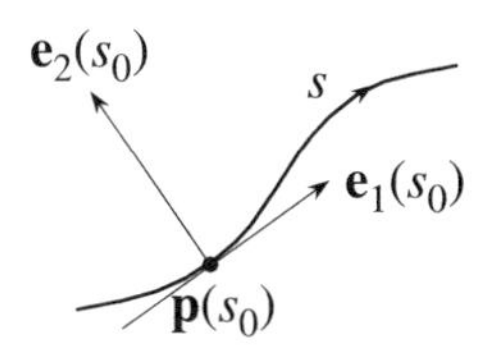

FIGURE 8.62 Definition of $\mathbf{e}_1(s_0)$ and $\mathbf{e}_1(s_0)$.

It is expected that the derivative of $\mathbf{e}_1$ is very close to the curvature. Therefore the relation $\mathbf{e}_1 \cdot \mathbf{e}_1 = 1$ is derivated about s, which results in $2\mathbf{e}_1' \cdot \mathbf{e}_1 = 0$. Therefore $\mathbf{e}_1'$ is orthogonal to $\mathbf{e}_1$, which is represented by using $\mathbf{e}_2(s)$.

Definition of Curvature-II *The curvature $\kappa(s)$ is defined as the relation between $\mathbf{e}_1'(s)$ and $\mathbf{e}_2(s)$ such that*

$$\mathbf{e}_1'(s) = \kappa(s)\mathbf{e}_2(s). \tag{8.91}$$

Thus curvature is a local feature and can be derived exactly when a curve is represented as a continuous function; this is called an analog approach. A corner is detected by examining the distribution of curvature values along a line on the curve. Likewise, by derivating the equation $\mathbf{e}_2 \cdot \mathbf{e}_2 = 1$, it is derived that $\mathbf{e}_2'$ is orthogonal to $\mathbf{e}_2$. Accordingly $\mathbf{e}_2'$ is represented using $\mathbf{e}_1$, (i.e., some multiple of $\mathbf{e}_1$). Then, by differentiating $\mathbf{e}_1 \cdot \mathbf{e}_2 = 0$, $\mathbf{e}_1' \cdot \mathbf{e}_2 + \mathbf{e}_1 \cdot \mathbf{e}_2' = 0$ is obtained. Substituting $\mathbf{e}_1' = \kappa\mathbf{e}_2$ into the equation, $\kappa + \mathbf{e}_1 \cdot \mathbf{e}_2' = 0$ is derived, which allows us to conclude that $\mathbf{e}_2'$ is multiple $-\kappa$ to $\mathbf{e}_1$. Thus we can write these interesting relations as

$$\begin{pmatrix} \mathbf{e}_1' \\ \mathbf{e}_2' \end{pmatrix} = \begin{pmatrix} 0 & \kappa \\ -\kappa & 0 \end{pmatrix} \begin{pmatrix} \mathbf{e}_1 \\ \mathbf{e}_2 \end{pmatrix}. \tag{8.92}$$

Yet another interpretation of curvature which we can base on the second definition is by way of the *Gaussian map*. For each point $\mathbf{p}(s)$ of a curve $\mathbf{p}$, let $\mathbf{p}(s)$ correspond to a vector from an origin 0 in Fig. 8.63 that is parallel to $\mathbf{e}_2(s)$. This mapped vector is also represented by $\mathbf{e}_2(s)$. Thus the mapping $\mathbf{p}(s)$ to $\mathbf{e}_2(s)$ is performed on a unit circle, which is denoted as g. When the point $\mathbf{p}(s)$ moves to $\mathbf{p}(s + \Delta s)$ along the $\mathbf{p}$, $\mathbf{e}_2$ moves from $\mathbf{e}_2(s)$ to $\mathbf{e}_2(s + \Delta s)$. Then, using Tayler expansion, we have

$$e_2(s + \Delta s) = e_2(s) + \mathbf{e}_2'(s)\Delta st + \cdots \tag{8.93}$$

$$= \mathbf{e}_2(s) - \kappa(s)\mathbf{e}_1(s)\Delta st + \cdots . \tag{8.94}$$

Thus $g(p)$ moves by a distance $\kappa(s)\Delta s$ toward the $-e_2(s)$ direction, which is counterclockwise on the unit circle. For a physical analogy we can write

$$\frac{\mathbf{e}_2(s + \Delta s) - \mathbf{e}_2(s)}{\Delta t} = -\kappa(s)\mathbf{e}_1(s)\frac{\Delta s}{\Delta t};$$

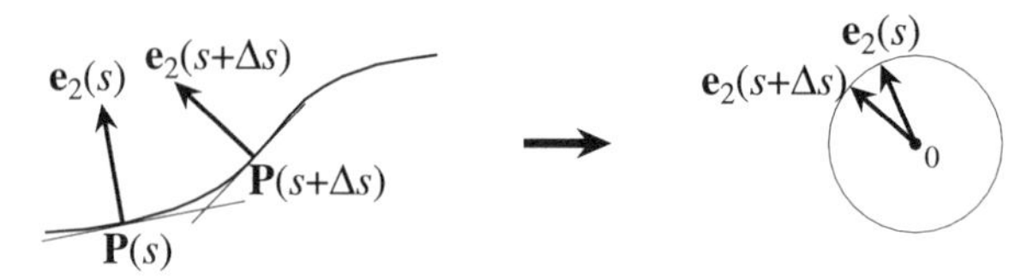

FIGURE 8.63 Gaussian map. From reference [33], © 1998, Shoukabou.

that is, we say that when $\mathbf{p}(s)$ moves on $\mathbf{p}$ by velocity 1, then $g(p)$ moves by velocity κ on a unit circle. When κ is negative, $g(p)$ moves toward the $\mathbf{e}_2(s)$ direction, which is clockwise.

On the other hand, we note that a curvature can be simply represented using symbols, coefficients of approximating polynominal functions. This is a key point of this approach. Let γ be a curve in parametric form,

$$x = y(t),$$
$$y = g(t), \tag{8.95}$$

where t is a parameter. The slope of a curve γ at a given point $A(t = t_1)$ is

$$\left[\frac{dy}{dx}\right]_{t=t_1} = \left[\frac{\frac{dg}{dt}}{\frac{df}{dt}}\right]_{t=t_1} \tag{8.96}$$

The curvature $C_v(t_1)$ at point A is the derivative of the slope with respect to the arc length. This was obtained using (8.82) and (8.96) as follows:

$$C_v(t_1) = \frac{((df/dt)(d^2g/dt^2) - (dg/dt)(d^2f/dt^2)}{((df/dt)^2 + (dg/dt)^2)^{3/2}} \tag{8.97}$$

The curve γ can be approximated by a cubic polynominal between $A(t = 0)$ and $B(t = 1)$, where $t \in [0, 1]$,

$$x = f(t) = a_1t^3 + b_1t^2 + c_1t + d_1,$$
$$y = g(t) = a_2t^3 + b_2t^2 + c_2t + d_2. \tag{8.98}$$

We assume that these coefficients can be calculated by a series of points consisting of γ, which will be described next in our discussion of B-splines. Then the curvature $C_v(t = 0)$ can be represented by coefficients as follows:

$$C_v(0) = 2\,\frac{c_1b_2 - c_2b_1}{(c_1^2 + c_2^2)^{3/2}}. \tag{8.99}$$

Review of B-Spline For reader's convenience, we review the B-spline functions. There are many books on spline approximation in graphics. Among these we recommend a very readable book *Geometric Modeling* by M. E. Morterson [34]. Here we borrow some figures and formulations presented in that book. For a more advanced discussion, we recommend the books by Farin [35] and by Hoschek and Lasser [36], both very readable.

Let us begin with a series of sampling points, say $\mathbf{p}_0$, $\mathbf{p}_1 \sim \mathbf{p}_5$, which are called control points in computer graphics. The objective is to approximate a series of discrete data points by continuous functions.[3] As a segment of the continuous functions, special kinds of polynominal functions are used called blending functions. That is, the overall approximating function is blended by these functions as

$$\mathbf{p}(u) = \sum_{i=0}^{n} \mathbf{p}_i N_{i,k}(u), \tag{8.100}$$

where $N_{i,k}(u)$ represent the blending functions and $n + 1$ is the number of sampling/control points. The parameter k in (8.100) is called order and a B-spline is made up of pieces of polynominals of degree $k - 1$ joined together with C^{k-2} continuity at the break points. The B-spline blending functions are defined recursively as follows for a given knot vector $T = (t_0, t_1, \ldots, t_{n-1}, t_n, t_{n+1}, \ldots, t_{t+k})$:

$$N_{i,1}(u) = \begin{cases} 1 & \text{if } t_i < u < t_{i+1}, \\ 0 & \text{otherwise}, \end{cases} \tag{8.101}$$

for $k = 1$, and

$$N_{i,k}(u) = \frac{(u - t_i)N_{i,k-1}(u)}{t_{i+k-1} - t_i} + \frac{(t_{i+k} - u)N_{i+1,k-1}(u)}{t_{i+k} - t_{i+1}}. \tag{8.102}$$

Thus k controls both the degree of blending functions and the continuity of the curve. Now let us examine the simplest case, $k = 1$. We see that no approximation is done, and the result is nothing but the original set of sampling points. However, we notice that the parametric value u is divided into intervals whose length are all equal to the unit interval for equally spaced knots, which is $t_i = i$. Each interval is assigned to a sampling point by the parameter t_i of (8.104) as

$$\mathbf{p}(u) = \mathbf{p}_i \qquad \text{for } i \le u < i + 1. \tag{8.103}$$

The parameters t_i are called *knot values*. They relate the parametric value u to the $\mathbf{p}_i$ control points. We define them as follows for an open nonuniform curve:

$$t_i = \begin{cases} 0 & \text{if } i < k, \\ i - k + 1 & \text{if } k \le i \le n, \\ n - k + 2 & \text{if } i > n, \end{cases} \tag{8.104}$$

where

$$0 \le i < n + k. \tag{8.105}$$

[3] Notice that we use a sampling point as in the next section, where sampling points are data points. However, we will treat the case where many data points are approximated by finding a small number of control points.

The range of the parametric value u is

$$0 \leq u \leq n - k + 2. \tag{8.106}$$

Note that the denominations of (8.102) can be zero, so 0/0 is defined to be equal to 1. These blending functions have the following properties:

1. $N_{i,k}(u) > 0$ for $t_i < u < t_{i+k}$.
2. $N_{i,k}(u) = 0$ for $t_0 < u < t_i$, $t_{i+k} < u < t_{n+k}$.
3. $\sum_{i=0}^{n} N_{i,k}(u) = 1$, $u \in [t_{k-1}, t_{n+1}]$.
4. $N_{i,k}(u)$ has continuity C^{k-2} at each of the knots t_l.

Property 4 is important and will be used in our approximating a curve by multiple segment B-spline curves later. For equally spaced knots, say $t_i = i$, the recurrence relation (8.102) becomes simpler and is referred to as *uniform B-spline.*

Next let us see the results of an approximation based on $N_{i,2}(u)$ blending functions with $n = 5$ and $k = 2$. According to (8.106), the range of the parameter u is [0, 5]. Furthermore the knot values are determined by (8.104) as follows:

$$\begin{aligned} t_0 &= 0 & t_4 &= 3 \\ t_1 &= 0 & t_5 &= 4 \\ t_2 &= 1 & t_6 &= 5 \\ t_3 &= 2 & t_7 &= 5 \end{aligned} \tag{8.107}$$

Notice here that we implicitly cannot use the $N_{0,1} \sim N_{5,1}(u)$ which were described above. They were pulse functions. This time, however, new $N_{i,1}(u)$ are all pulse functions that differ very little. They are found using the recursive expression (8.102) with the knot values listed above in (8.107). The results are very simple. The $N_{i,2}(u)$ are all linear piecewise functions, and the approximated functions are also piecewise linear functions:

$$\mathbf{p}(u) = (i + 1 - u)\mathbf{p}_i + (u - i)\mathbf{p}_{i+1}, \tag{8.108}$$

where $(i - 1) \leq u < i$. That is, each approximated function is determined by two sampling points and nothing but line segment connecting the two points, which constitutes a polygon. At any rate the approximated function has C^0 continuity at the connecting points, $u = 1, 2, \ldots, 4$.

Last we proceed to find the approximating functions based on the series of $N_{i,1}(u)$, $N_{i,2}(u)$, and $N_{i,3}(u)$ blending functions. Again we start by setting knot values with $n = 5$ and $k = 3$ as follows:

$$0 \le u \le 4, \tag{8.109}$$

$$\begin{aligned} t_0 &= 0 & t_5 &= 3 \\ t_1 &= 0 & t_6 &= 4 \\ t_2 &= 0 & t_7 &= 4 \\ t_3 &= 1 & t_8 &= 4 \\ t_4 &= 2 \end{aligned} \tag{8.110}$$

The resulting $N_{i,3}(u)$ are all piecewise quadratic functions on the range of [0, 4]. Each approximating function is determined by these consecutive points as follows:

$$\mathbf{p}_1(u) = (1-u)^2\mathbf{p}_0 + \frac{1}{2}u(4-3u)\mathbf{p}_1 + \frac{1}{2}u^2\mathbf{p}_2 \qquad \text{for } 0 \le u < 1,$$

$$\mathbf{p}_2(u) = \frac{1}{2}(2-u)^2\mathbf{p}_1 + \frac{1}{2}(-2u^2+6u-3)\mathbf{p}_2 + \frac{1}{2}(u-1)^2\mathbf{p}_3 \qquad \text{for } 1 \le u \le 2,$$

$$\mathbf{p}_3(u) = \frac{1}{2}(3-u)^2\mathbf{p}_2 + \frac{1}{2}(-2u^2+10u-11)\mathbf{p}_3 + \frac{1}{2}(u-2)^2\mathbf{p}_4 \qquad \text{for } 2 \le u < 3,$$

$$\mathbf{p}_4(u) = \frac{1}{2}(4-u)^2\mathbf{p}_3 + \frac{1}{2}(-3u^2+20u-32)\mathbf{p}_4 + (u-3)^2\mathbf{p}_5 \qquad \text{for } 3 \le u < 4. \tag{8.111}$$

An example is shown in Fig. 8.64 which approximates six points, $\mathbf{p}_0$, $\mathbf{p}_1$, ~ $\mathbf{p}_5$. The curve are smooth; however, four segments expressed in (8.111) are connected with C^1 continuity. Except for start and end points the curve does not pass through the sampling points; rather it is tangent to each successive side of the characteristic polygon. It is also tangent at $\mathbf{p}_1 - \mathbf{p}_0$ and $\mathbf{p}_5 - \mathbf{p}_4$ which are the start and end points, respectively. This property holds only for $k = 3$ curves. Therefore we can imagine the approximated curve by the drawing the characteristic polygon consisting of the sampling points. However, the curve depends on n. It is influenced by the two end points, which makes the derivation of these blending functions cumbersome, as seen so far. On the other hand, we can infer that the interior part of the curve is not influenced by the end points. Actually it is true. Based on this idea, we can construct a more convenient expression of the blending functions being independent of n. Now looking at the basic formula of (8.100) superficially, $\mathbf{p}_i(u)$ seems to be constructed of n blending functions. However, if we examine the $\mathbf{p}_i(u)$ of (8.111), we can find that the $\mathbf{p}_i(u)$ is formed by only three blending functions when $k = 3$. This is a nice property of the B-spline curves, which is locality. The three blending functions are designated as $N_{i,3}(u)$, $N_{i+1,3}(u)$, and $N_{i+2,3}(u)$. Furthermore we can assume that i is in the range of $k \le i \le n$ so that the t_i knot values are simple:

$$t_i = i - k + 1 = i - 2, \tag{8.112}$$

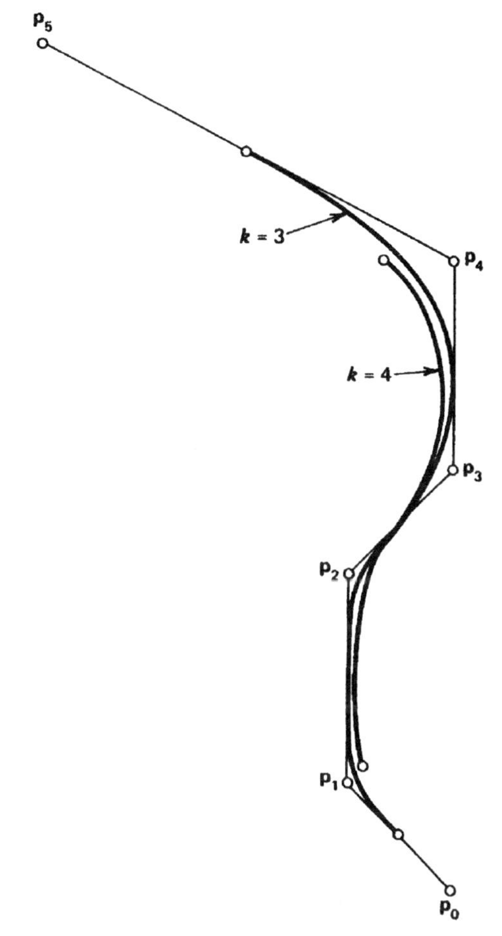

FIGURE 8.64 Nonperiodic *B*-spline curves: $n = 5$, $k = 3$; $n = 5$, $k = 4$. From reference [34], © 1998, Wiley.

and this simplifies the derivation of the blending functions accordingly. Thus, for the interval $i < u < i + 1$, the $\mathbf{p}(u)$ is constructed by the three points, $\mathbf{p}_i$, $\mathbf{p}_{i+1}$, and $\mathbf{p}_{i+2}$ and the blending functions $N_{i,3}(u)$, $N_{i+1,3}(u)$, and $N_{i+2,3}(u)$ as follows:

$$\mathbf{p}(u) = \frac{1}{2}(i + 1 - u)^2\mathbf{p}_i + \frac{1}{2}[(u - i + 1)(i + 1 - u) + (i + 2 - u)(u - 2)]\mathbf{p}_{i+1}$$

$$+ \frac{1}{2}(u - i)^2\mathbf{p}_{i+2}. \tag{8.113}$$

Since there is only one segment around $\mathbf{p}_i$, which is the interval of the $\mathbf{p}(u)$, we can rewrite the above expression more uniformly and conveniently reparametrizing

(8.113) as $u \to u + i$ so that $0 \leq u \leq 1$. The $\mathbf{p}(u)$ is designated as $\mathbf{p}_i(u)$ (the same notation as before) and expressed as

$$\mathbf{p}_i(u) = \frac{1}{2}\left[(1-u)^2\mathbf{p}_i + (-2u^2 + 2u + 1)\mathbf{p}_{i+1} + u^2\mathbf{p}_{i+2}\right] \tag{8.114}$$

This expression can be further represented more uniformly and conveniently replacing i by $i - 1$ (i now denotes the curve segment number), as follows:

$$\mathbf{p}_i(u) = \frac{1}{2}\left[u^2 \quad u \quad 1\right]\begin{bmatrix} 1 & -2 & 1 \\ -2 & 2 & 0 \\ 1 & 1 & 0 \end{bmatrix}\begin{bmatrix} \mathbf{p}_{i-1} \\ \mathbf{p}_i \\ \mathbf{p}_{i+1} \end{bmatrix} \tag{8.115}$$

for $i \in [1 : n - 1]$. Now we have the expression in canonical form in some sense, and without concern for the boundary. Equation (8.115) is called a *periodic* B-spline curve expression, so it can be directly applied to a closed curve. Its earlier incarnation (8.111) is called *nonperiodic.*

Analogously, the form for cubic B-spline ($k = 4$) is

$$\mathbf{p}_i(u) = \frac{1}{6}\left[u^3 \quad u^2 \quad u \quad 1\right]\begin{bmatrix} -1 & 3 & -3 & 1 \\ 3 & -6 & 3 & 0 \\ -3 & 0 & 3 & 0 \\ 1 & 4 & 1 & 0 \end{bmatrix}\begin{bmatrix} \mathbf{p}_{i-1} \\ \mathbf{p}_i \\ \mathbf{p}_{i+1} \\ \mathbf{p}_{i+2} \end{bmatrix} \tag{8.116}$$

for $i \in [1 : n - 2]$. This is usually represented simply as

$$\mathbf{p}_i(u) = \frac{1}{6} U_4 M_4 [\mathbf{p}_{i-1}, \mathbf{p}_i, \mathbf{p}_{i+1}, \mathbf{p}_{i+2}]^T. \tag{8.117}$$

8.8.2 Calculation of Curvature

Now we are ready to represent the curvature of curves directly by the coordinates of the sampling points based on cubic B-spline ($k = 4$). For convenience we change the notation back again to the one used by Medioni and Yasumoto. That is, $\mathbf{p}_i(u)$ is changed to $(x(t), y(t))^T$. Thus the expression (8.116) is separated into $x(t)$ and $y(t)$, each of which is expressed by $[x_{i-1}, x_i, x_{i+1}, x_{i+2}]^t$ and $[y_{i-1}, y_i, y_{i+1}, y_{i+2}]^t$, respectively. Either $x(t)$ or $y(t)$ is enough to obtain the final curvature expression. Accordingly the (8.116) is expressed for $x(t)$ as

$$x(t) = \frac{1}{6}\left[t^3 \quad t^2 \quad t \quad 1\right]\begin{bmatrix} -1 & 3 & -3 & 1 \\ 3 & -6 & 3 & 0 \\ -3 & 0 & 3 & 0 \\ 1 & 4 & 1 & 0 \end{bmatrix}\begin{bmatrix} x_{i-1} \\ x_i \\ x_{i+1} \\ x_{i+2} \end{bmatrix} \tag{8.118}$$

$$= \frac{1}{6}(-t^3 + 3t^2 - 3t + 1)x_{i-1} + \frac{1}{6}(3t^3 - 6t^2 + 4)x_i$$
$$+ \frac{1}{6}(-3t^3 + 3t^2 + 3t + 1)x_{i+1} + \frac{1}{6}x_{i+2}, \tag{8.119}$$

which tells how the x coordinates are blended. However, better for our purpose is another expression of the polygonal function form which corresponds to (8.98). That is,

$$x(t) = \frac{1}{6}(-x_{i-1} + 3x_i + 3x_{i+1} + x_{i+2})t^3 + \frac{1}{2}(x_{i-1} - 2x_i + x_{i+1})t^2 + \frac{1}{2}(-x_{i-1} + x_{i+1})t + \frac{1}{6}(x_{i-1} + 4x_i + x_{i+1}). \tag{8.120}$$

This allows us to obtain four coefficients:

$$a_1 = \frac{(-x_{i-1} + 3x_i - 3x_{i+1} + x_{i-2})}{6},$$

$$b_1 = \frac{(x_{i-1} - 2x_i + x_{i+1})}{2},$$

$$c_1 = \frac{(-x_{i-1} + x_{i+1})}{2},$$

$$d_1 = \frac{(x_{i-1} + 4x_i + x_{i+1})}{6}. \tag{8.121}$$

In the same manner we can obtain the coefficients of $y(t)$ as

$$a_2 = \frac{(-y_{i-1} + 3y_i - 3y_{i+1} + y_{i+2})}{6},$$

$$b_2 = \frac{(y_{i-1} - 2y_i + y_{i+1})}{2},$$

$$c_2 = \frac{(-y_{i-1} + y_{i+1})}{2},$$

$$d_2 = \frac{(y_{i-1} + 4y_i + y_{i+1})}{6}. \tag{8.122}$$

Using (8.99), and setting $t = 0$, we can calculate the curvature at a point corresponding to (x_i, y_i) from some point $\mathbf{p}_i$ and its neighboring points $\mathbf{p}_{i-1}$ and $\mathbf{p}_{i+1}$. That is,

$$C_{v,i}(t = 0) = 4\,\frac{(x_{i+1} - x_{i-1})(y_{i-1} - 2y_i + y_{i+1}) - (y_{i+1} - y_{i-1})(x_{i-1} - 2x_i + x_{i+1})}{((x_{i+1} - x_{i-1})^2 + (y_{i+1} - y_{i-1})^2)^{3/2}}. \tag{8.123}$$

For its relation to the cubic B-spline segments, see at Fig. 8.65. Note that the curve consists of four cubic B-spline segments $i - 1$, i, $i + 1$, and $i + 2$, in which i and $i + 1$

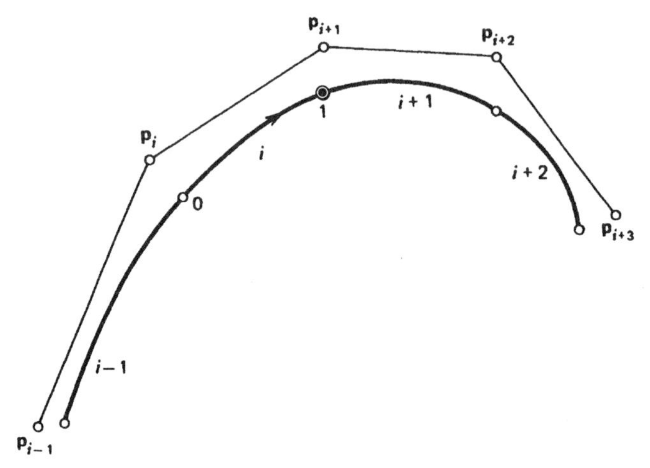

FIGURE 8.65 Four segments of a cubic B-spline curve and $k = 4$. From reference [34], © 1998, Wiley.

correspond to $\mathbf{p}_i(t)$ and $\mathbf{p}_{i+1}(t)$. The former is blended as $\mathbf{p}_i(t) = U_4M_4[\mathbf{p}_{i-1}, \mathbf{p}_i, \mathbf{p}_{i+1}, \mathbf{p}_{i+2}]^t$, and the latter as $U_4M_4[\mathbf{p}_i, \mathbf{p}_{i+1}, \mathbf{p}_{i+2}, \mathbf{p}_{i+3}]^t$. They connect very smoothly at $t = 1$ with C^2.

8.8.3 Corner Detection Based on Curvature Calculation

In order to detect a corner, a smoothing operation is essential as described before. In other words, to detect a corner at a given point, we need to take its neighboring points into account in such a way that their effects are smoothed. This can be done intuitively, for example, by something like $\mathbf{p}_i \leftarrow \mathbf{p}_{i-1}/4 + \mathbf{p}_i/2 + \mathbf{p}_{i+1}/4$. The weighting factors are $\frac{1}{4}, \frac{1}{2}, \frac{1}{4}$. In our case, however, this smoothing will be done alone based on cubic B-spline function. At each point $\mathbf{p}_i$, we want to calculate a displacement of a given point $\mathbf{p}_i$ from the cubic B-spline using (8.120) and setting t to zero. This is expressed as

$$\delta x = d_i - x_i = \frac{1}{6}x_{i-1} - \frac{1}{3}x_i + \frac{1}{6}x_{i+1},$$
$$\delta y = d_i - y_i = \frac{1}{6}y_{i-1} - \frac{1}{3}y_i + \frac{1}{6}x_{i+1}. \quad (8.124)$$

As shown in Fig. 8.66, if we calculate the displacement at each point, the series of the displaced points, $\mathbf{p}_i + \delta$, is a smoothed curve. Based on the smoothed curve, B-spline fitting is done through a displaced point. Since

$$\mathbf{p}_i + \delta = \frac{1}{6}\mathbf{p}_{i-1} + \frac{2}{3}\mathbf{p}_i + \frac{1}{6}\mathbf{p}_{i+1},$$

the smoothing operation takes effectively the following substitution operations

$$\mathbf{p}_i \rightarrow \frac{1}{6}\mathbf{p}_{i-1} + \frac{2}{3}\mathbf{p}_i + \frac{1}{6}\mathbf{p}_{i+1},$$
$$\mathbf{p}_{i-1} \rightarrow \frac{1}{6}\mathbf{p}_{i-2} + \frac{2}{3}\mathbf{p}_{i-1} + \frac{1}{6}\mathbf{p}_i,$$
$$\mathbf{p}_{i+1} \rightarrow \frac{1}{6}\mathbf{p}_i + \frac{2}{3}\mathbf{p}_{i+1} + \frac{1}{6}\mathbf{p}_{i+2}. \quad (8.125)$$

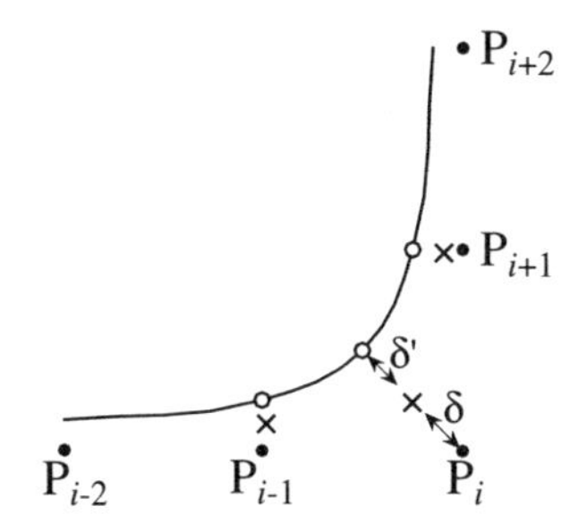

FIGURE 8.66 Smoothing process.

The new curvature C'_v based on the smoothed curve can be calculated easily. Substitute $\mathbf{p}_{i-1}$ through $\mathbf{p}_{i+1}$ into (8.121), (8.122), and (8.99):

$$C'_v = 2\,\frac{c'_1 b'_2 - c'_2 b'_1}{(c'^2_1 + c'^2_2)^{3/2}}, \tag{8.126}$$

where

$$\begin{aligned}
b'_1 &= \frac{1}{12}x_{i-2} + \frac{1}{6}x_{i-1} - \frac{1}{2}x_i + \frac{1}{6}x_{i+1} + \frac{1}{12}x_{i+2},\\
b'_2 &= \frac{1}{12}y_{i-2} + \frac{1}{6}y_{i-1} - \frac{1}{2}y_i + \frac{1}{6}y_{i+1} + \frac{1}{12}y_{i+2},\\
c'_1 &= \frac{(x_{i+1} - x_{i-1})}{3} + \frac{(x_{i+2} - x_{i-2})}{12},\\
c'_2 &= \frac{(y_{i+1} - y_{i-1})}{3} + \frac{(y_{i+2} - y_{i-2})}{12}.
\end{aligned} \tag{8.127}$$

One example is shown in Fig. 8.67, where y_i, y_{i-1}, and y_{i-2} take the value 0. Naturally the result is invariant to shift, $\mathbf{p}_i + \mathbf{n}$ for all the i, which is easily proved using (8.127). Thus

$$b'_1 = 0,\quad b'_2 = \frac{1}{3},\quad c'_1 = 1,\quad \text{and}\quad c'_2 = \frac{1}{2},$$

$$C'_v = \frac{2/3}{(\sqrt{1 + (0.5)^2})^3} \cong 0.477.$$

y_{i+2}	2					•
y_{i+1}	1				•	
y_{i-2}, y_{i-1}, y_i	0	•	•	•		
		x_{i-2}	x_{i-1}	x_i	x_{i+1}	x_{i+2}
		-2	-1	0	1	2

FIGURE 8.67 Data points used in the smoothing example.

As the measures of curvature, beside the curvature itself, Medioni and Yasumoto proposed to use displacement and curvature local maximum. The displacement for a recomputed B-spline curve is easily obtained in the same manner as C'_v. The total δ_t is $\delta' + \delta$, and δ' is given as

$$\delta'_x = \frac{1}{36}x_{i-2} + \frac{1}{18}x_{i-1} - \frac{1}{6}x_i + \frac{1}{18}x_{i+1} + \frac{1}{36}x_{i+2},$$

$$\delta'_y = \frac{1}{36}y_{i-2} + \frac{1}{18}y_{i-1} - \frac{1}{6}y_i + \frac{1}{18}y_{i+1} + \frac{1}{36}y_{i+2}. \tag{8.128}$$

Therefore for the total δ_t we have

$$\delta_{tx} = \frac{1}{36}x_{i-2} + \frac{2}{9}x_{i-1} - \frac{1}{2}x_i + \frac{2}{9}x_{i-1} + \frac{1}{36}x_{i-2},$$

$$\delta_{ty} = \frac{1}{36}y_{i-2} + \frac{2}{9}y_{i-1} - \frac{1}{2}y_i + \frac{2}{9}y_{i-1} + \frac{1}{36}y_{i-2}. \tag{8.129}$$

As a result the conditions for a given point i to be a corner are as follows:

1. The displacement δ_t must be larger than a given threshold d_c.
2. The curvature C'_v must be larger than a given threshold c_c.
3. The curvature C'_v must be a local maximum.

We examine these conditions next.

Experimental Results Medioni and Yasumoto presented detailed experimental results and a comparative study as well. Their results are very good compared with other methods such as k-curvature, iterative cubic polynominal approximation [37], and corner detection based on the facet model [44]. Fortunately, since detailed data and results are given, we trace the B-spline method using the same data, which are considerably complex but rich in concavity and convexity. Through the tracing experiment, use of the conditions above will be discussed.

The data used by Medioni and Yasumoto are shown in Fig. 8.68, together with corner points detected according to the threshold values specified in their paper, $d_c = 0.2$, $c_c = 0.4$. In this case we can regard the data as black boundary tracing points. If conditions 1 and 2 are used in logical OR, a few more corner points are detected. However, they are inclusive, so condition 3 is applied. More specifically, when the same curvature values that are greater than the threshold value continue consecutively, the first point is chosen. The direction of contour is defined clockwise. The double square marked points indicate that the points satisfy both conditions 1 and 2. The reader who is very careful will find that only one point is different from the result of their paper. However, this is obviously a mistake in their paper because the same shape is marked by a corner point that lies next to a questionable point in terms of corner points. The difference between them is that one is for the concave curve and the other for the convex curve. Notice that the curvature takes a signed value; the sign of concavity is plus and that of convexity is minus in this case. In the detection of the corner points shown above, absolute values were taken, since there was considerable fluctu-

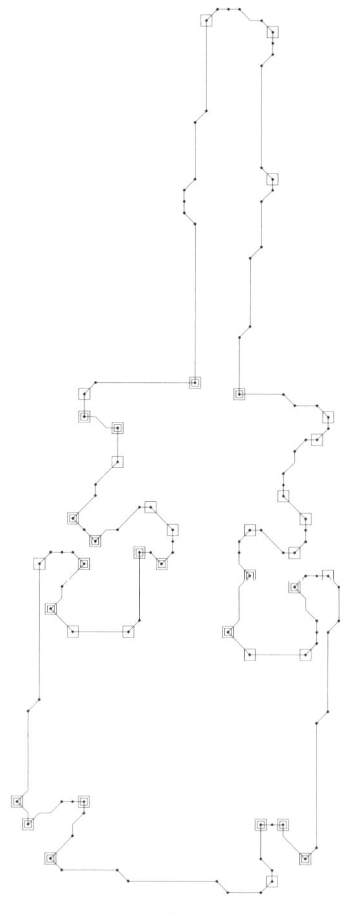

FIGURE 8.68 Data in the simulation by Medioni and Yasumoto. From reference [32], © 1998, Academic Press.

ation in the curvature values. This tells us that the method still reflects the digital nature of the sampled data.

For comparison, the same data were tested using Teh and Chin's algorithm; the result is shown in Fig. 8.69. While the results are similar, Teh and Chin's algorithm can be seen to be more sensitive to curvature differences than Medioni and Yasumoto's algorithm.

Detection of Curvature Difference So far we have used a mathematical concept of curvature in its formulation. The analog value of curvature was used only for corner detection. Now we want to detect curvature differences which is necessary for subtle changes in corners as shown in Fig. 8.70. The important thing is that we need to trace continuously any change of curvature so that we can describe any gray zones between two shapes belonging to different categories such as $2 \leftrightarrow Z$, and $4 \leftrightarrow 9$. This is where we depend on an analog feature [63].

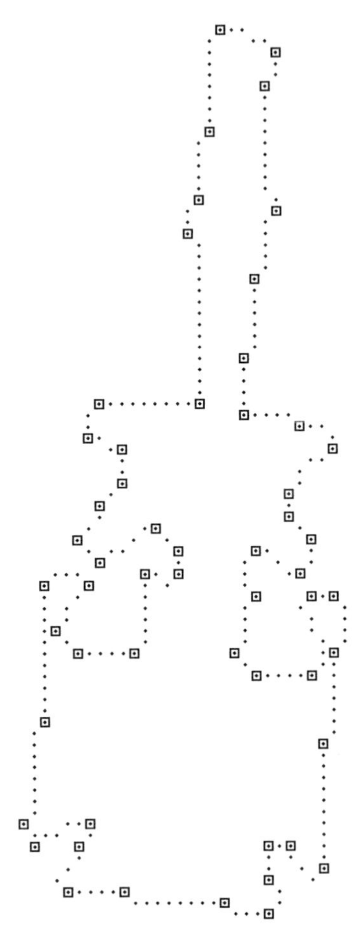

FIGURE 8.69 Result of Teh and Chin's algorithm when applied to the data shown in Fig. 8.68.

To make our purpose clear, let us consider an ideal data set such as the data set $2 \leftrightarrow Z$ shown in Fig. 8.70 (*a*). This was synthesized using a Bézier curve which, in general, is represented as

$$f(t, n) = \sum_{i=0}^{n} C_i^n (1-t)^{n-i} t^i \mathbf{p}_i,$$

where C_i^n is a binomial coefficient, t is a parameter, and $\mathbf{p}_i$ is a control point vector. Then we can calculate the curvature value directly using (8.82) or (8.97). The results are shown in Fig. 8.71 where we can see a continuous change of the curvature values according to the curvature change of the upper concave part of $2 \leftrightarrow Z$. The problem, however, is to detect curvature values continuously according to the continuous change of curvature values of a shape under the noise.

The method must be intrinsically global, which differs essentially from the methods introduced for corner detection so far. Certainly curvature itself is a local feature,

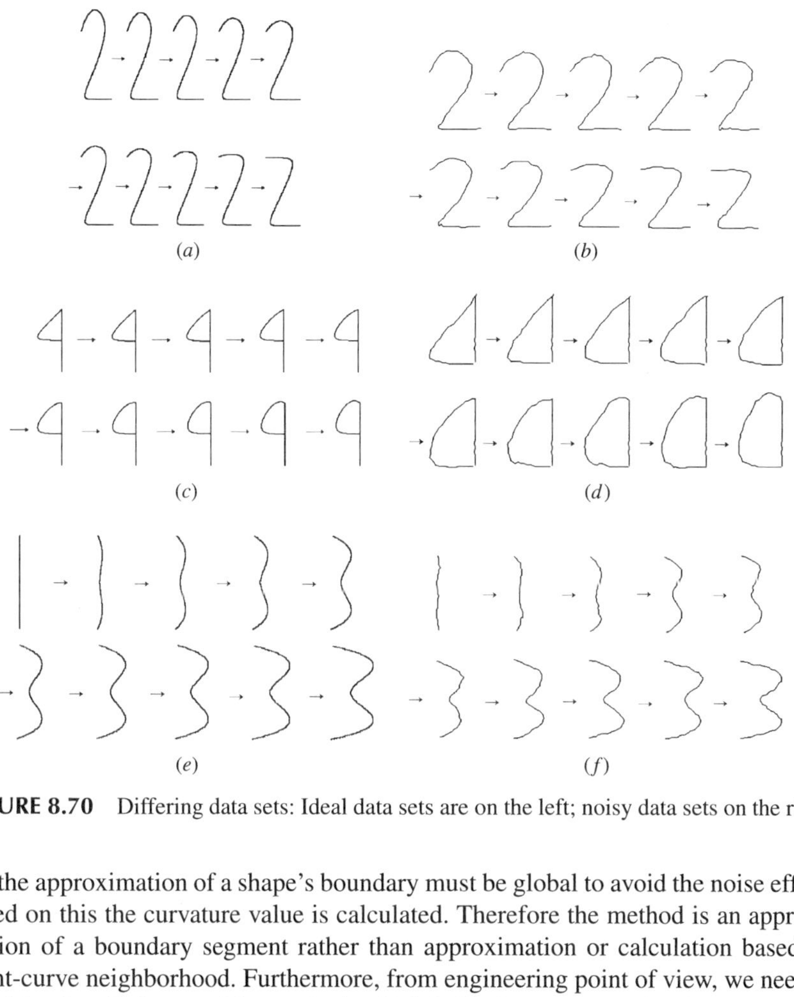

(*a*) (*b*)

(*c*) (*d*)

(*e*) (*f*)

FIGURE 8.70 Differing data sets: Ideal data sets are on the left; noisy data sets on the right.

but the approximation of a shape's boundary must be global to avoid the noise effect; based on this the curvature value is calculated. Therefore the method is an approximation of a boundary segment rather than approximation or calculation based on point-curve neighborhood. Furthermore, from engineering point of view, we need to consider the efficiency of the method, which is likely more complex than the corner detection methods introduced so far. For this reason we may consider a hierarchical approach [39]. Then obvious corner points such as end points can be detected using k-cosine method, for example. We further may consider segmenting the boundary of the shape appropriately. Actually the k-cosine method works quite well for this purpose, and a polygonal approximation can be used also.

We turn to the next problem which is what function class to use to approximate a line/boundary segment. The natural choice is a spline function because we need to connect continuous functions smoothly for a global approximation. Among the different functions that can be used, it is well-known that a cubic spline function is strongest in handling noise [40,41] and has minimum degree for C^2 continuity. Therefore we will use a cubic spline function.

Approximation by Multiple Bézier Curves and B-Spline Curves Here we explain our method of approximating a given line segment with connected multiple cubic

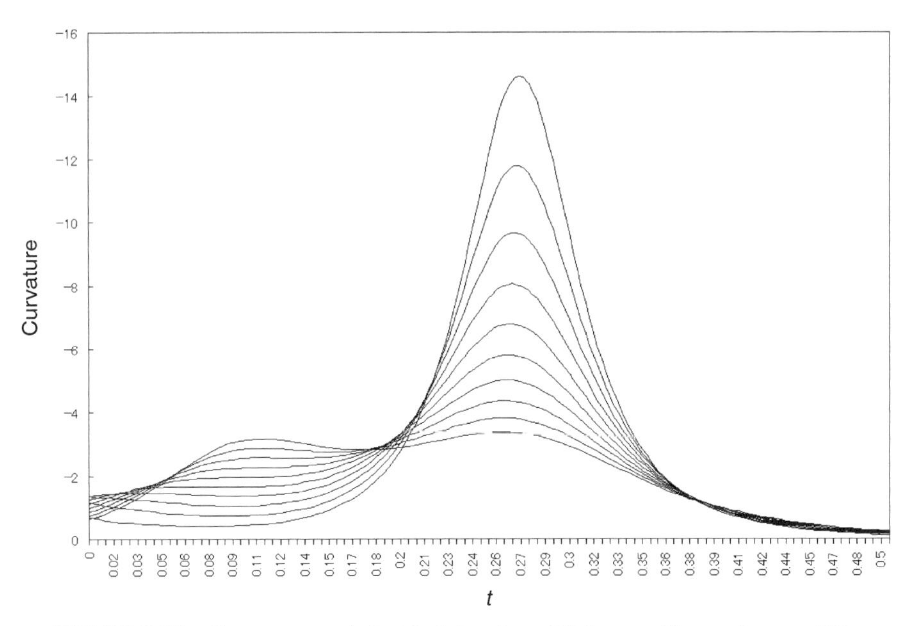

FIGURE 8.71 Curvature graph for ideal data from 2Z data set. From reference [63].

Bézier curves and B-spline curves [42]. While the two approaches are equivalent, the B-spline approach is simpler than the multiple Bézier one. However, the mathematical structure of the approximation is explicit in the multiple Bézier approach. Therefore first we explain it from tutorial point of view. The same scheme was mentioned by Saint-Marc et al. based on a quadratic B-spline [43].

Multiple Cubic Bézier Approach The Bézier method consists of the following two steps:

STEP 1: For each given data point, determine the corresponding parameter value of the approximation curve.

STEP 2: Specify the number of Bézier curves to be used. Determine the positions of the Bézier curves by the least squares method with the objective function **R** (explained later; in minimizing **R**, we will give the conditions to guarantee C and C^2 continuity at the connections of the generated Bézier curves). At the end points of the approximation curves, give the natural end condition, namely the second derivatives which are set equal to 0 to avoid undesirable undulations near the end points.
Now we will describe the details of the two steps.

DETAILS OF STEP 1: Given ordered data points $\mathbf{Q}_i (i = 0, 1, \ldots, n)$, let the distance between two points $\mathbf{Q}_i$ and $\mathbf{Q}_{i+1}$ be l_i. The total length L traced from the first point to the last one is given by

$$L = \sum_{i=0}^{n-1} l_i. \tag{8.130}$$

Let the approximation curve generated be $\mathbf{C}(t)$, $0 \le t \le 1$. In step 2, the least squares method will minimize the sum of the length between $\mathbf{Q}_j$ and $\mathbf{C}^j(t^j)$ where $t^j = \Sigma_{i=0}^{j} l_i/L$.

Here, we notice that the parameter t is used as a global variable and for each jth segment the parameter t^j is used.

DETAILS OF STEP 2: The cubic Bézier curve $\mathbf{C}^j(t^j)$ can be expressed by

$$\mathbf{C}^j(t^j) = \left(\frac{j+1}{N} - t^j\right)^3 \mathbf{P}_0^j + 3\left(\frac{j+1}{N} - t^j\right)^2 t^j\mathbf{P}_1^j + 3\left(\frac{j+1}{N} - t^j\right)(t^j)^2\, \mathbf{P}_2^j + (t^j)^3\, \mathbf{P}_3^j$$

$$\left(\frac{j}{N} \le t^j \le \frac{j+1}{N}, j = 0, 1, \ldots, N-1\right), \tag{8.131}$$

where $\mathbf{P}_i^j(i = 0, \ldots, 3 \quad j = 0, \ldots, N-1)$ are the control points and N is the specified number of Bézier curves. We use Lagrange's method of indeterminate coefficients to solve our minimization problem under several conditions. Let two connected Bézier curves be $\mathbf{C}^j$ and $\mathbf{C}^{j+1}$. Assume that their control points are $\mathbf{P}_i^j$ and $\mathbf{P}_i^{j+1}$ $(j = 0, \ldots, 3)$, respectively.

The conditions to guarantee $\mathbf{C}^0$, $\mathbf{C}^1$ and $\mathbf{C}^2$ continuity at their connection can be expressed by the following three equations:

$$\mathbf{C}^j(1) = \mathbf{C}^{j+1}(0), \tag{8.132}$$

$$\frac{d\mathbf{C}^j(1)}{dt} = \frac{d\mathbf{C}^{j+1}(0)}{dt}, \tag{8.133}$$

$$\frac{d^2\mathbf{C}^j(1)}{dt^2} = \frac{d^2\mathbf{C}^{j+1}(0)}{dt^2}, \tag{8.134}$$

The continuous expressions above are represented by the following place vectors using (8.131):

$$\mathbf{P}_3^j = \mathbf{P}_0^{j+1}, \tag{8.135}$$

$$\mathbf{P}_3^j - \mathbf{P}_2^j = \mathbf{P}_1^{j+1} - \mathbf{P}_0^{j+1}, \tag{8.136}$$

$$\mathbf{P}_3^j - 2\mathbf{P}_2^j + \mathbf{P}_1^j = \mathbf{P}_2^{j+1} - 2\mathbf{P}_1^{j+1} + \mathbf{P}_0^{j+1}. \tag{8.137}$$

Note that the number of the connections of the curve made by N Bézier curves are $N - 1$. This means we should have $3 \times (N - 1)$ conditions similar to the above. The natural end conditions can be expressed by

$$\frac{d^2\mathbf{C}(0)}{dt^2} = 0, \tag{8.138}$$

$$\frac{d^2\mathbf{C}(1)}{dt^2} = 0. \tag{8.139}$$

Hence we define the objective function **R** with the conditions as follows:

$$\mathbf{R} = \underbrace{\sum_{i=0}^{m-1} (\mathbf{C}^0(t_i) - \mathbf{Q}_i)^2 + \sum_{i=m}^{r-1} (\mathbf{C}^1(t_i) - \mathbf{Q}_i)^2 + \cdots + \sum_{i=l}^{n} (\mathbf{C}^{N-1}(t_i) - \mathbf{Q}_i)^2}_{N}$$

$$+ \underbrace{\lambda_0^0(\mathbf{P}_3^0 - \mathbf{P}_0^1) + \lambda_0^1(\mathbf{P}_3^1 - \mathbf{P}_0^2) + \cdots}_{N-1}$$

$$+ \underbrace{\lambda_1^0\{\mathbf{P}_1^1 - \mathbf{P}_0^1 - (\mathbf{P}_3^0 - \mathbf{P}_2^0)\} + \lambda_1^1\{\mathbf{P}_1^2 - \mathbf{P}_0^2 - (\mathbf{P}_3^1 - \mathbf{P}_2^1)\} + \cdots}_{N-1}$$

$$+ \underbrace{\lambda_2^0\{\mathbf{P}_3^0 - 2\mathbf{P}_2^0 + \mathbf{P}_1^0 - (\mathbf{P}_2^1 - 2\mathbf{P}_1^1 + \mathbf{P}_0^1)\} + \lambda_2^1\{\mathbf{P}_3^1 - 2\mathbf{P}_2^1 + \mathbf{P}_1^1 - (\mathbf{P}_2^2 - 2\mathbf{P}_1^2 + \mathbf{P}_0^2)\} + \cdots}_{N-1}$$

$$+ \lambda_3(\mathbf{P}_2^0 - 2\mathbf{P}_1^0 + \mathbf{P}_0^0) + \lambda_4(\mathbf{P}_3^{N-1} - 2\mathbf{P}_2^{N-1} + \mathbf{P}_1^{N-1}), \tag{8.140}$$

where $\lambda_1^i, \lambda_2^j, (i, j = 0, \ldots, N-2)$ are Lagrange's inderminate coefficients.

The integers $m, r, \ldots,$ are determined as follows: Let the parameter value of a data point determined in Step 1 be t_i'. If $t_i' \geq j/N$ and $t_i' < (j+1)/N$, the data point is compared with the jth Bézier curve and the distance of the data point and Bézier curve are added to **R.** The last data point is to be compared with the last Bézier curve. This process determines the integers $m, r, \ldots, n$. t_i $(i = 0, \ldots, n)$ in equation (8.140) is calculated using the parameter value t^j by

$$t_i = \frac{t^j - (j/N)}{1/N} \tag{8.141}$$

if the data point corresponding to t_i' is compared to the jth Bézier curve.

By differentiating **R** with each control point and inderminate coefficient, and equating them with 0, we get the following equations:

$$\frac{\partial \mathbf{R}}{\partial \mathbf{P}_i^j} = 0 \qquad (i = 0, \ldots, 3, j = 0, \ldots, N-1), \tag{8.142}$$

$$\frac{\partial \mathbf{R}}{\partial \lambda_0^j} = 0 \qquad (j = 0, 1, \ldots, N-2), \tag{8.143}$$

$$\frac{\partial \mathbf{R}}{\partial \lambda_1^j} = 0 \qquad (j = 0, 1, \ldots, N-2), \tag{8.144}$$

$$\frac{\partial \mathbf{R}}{\partial \lambda_2^j} = 0 \qquad (j = 0, 1, \ldots, N-2), \tag{8.145}$$

$$\frac{\partial \mathbf{R}}{\partial \lambda_3} = 0 \tag{8.146}$$

$$\frac{\partial \mathbf{R}}{\partial \lambda_4} = 0. \tag{8.147}$$

By solving the above system of linear equations, the positions of the control points of the Bézier curves are obtained. Note that the generated approximation curve can be converted to a cubic B-spline curve.

B-Spline Approach Here we explain our algorithm of the approximation by cubic B-spline curves. The B-spline curves use the concept of degree. When the one segment is defined, the number of the control points required is (degree + 1). The algorithm consists of the following two steps:

STEP 1: For each given data point, determine the corresponding parameter value of the approximation curve, as mentioned before.

STEP 2: Determine positions of the control points of the B-spline curves by the least squares method with the objective function **R** (explained below; note that we do not have to consider boundary conditions used by the Bézier curves, due to the property (d) of $N_{i,k}(t)$ mentioned before).

Determination of B-Spline Curves The B-spline curves are made on the basis of a defined sequence of points called control points too. Hence it is possible to obtain an approximate curve, by obtaining its control points.

The B-spline curves $\mathbf{C}(t)$ can be expressed by

$$\mathbf{C}(t) = \mathbf{P}_0 N_{0,4}(t) + \mathbf{P}_2 N_{2,4}(t) + \mathbf{P}_3 N_{3,4}(t) + \cdots + \mathbf{P}_j N_{j,4}(t)$$
$$(j = 0, \ldots, N+2), \tag{8.148}$$

where $\mathbf{P}_j$ ($j = 0, \ldots, N+2$) are control points and N is the specified number of segments of the B-spline curves. $N_{j,4}(t)$ ($j = 0, 1, \ldots, N+2$) is the blending function of B-spline curve. Note that we use uniform B-spline curves.

Hence we define the objective function $\mathbf{R}$ with the conditions as follows:

$$\mathbf{R} = \sum_{i=0}^{m} (\mathbf{C}(t_i) - (\mathbf{Q}_i))^2 + \lambda_1(\mathbf{P}_2 - 2\mathbf{P}_1 + \mathbf{P}_0)$$
$$+ \lambda_2(\mathbf{P}_{N+2} - 2\mathbf{P}_{N+1} + \mathbf{P}_n) \tag{8.149}$$

By differentiating $\mathbf{R}$ with the control points, and equating it to 0, we get the following equations:

$$\frac{\partial \mathbf{R}}{\partial \mathbf{P}_j} = 0 \qquad (j = 0, \ldots, N+2). \tag{8.150}$$

The equations generated by equation (8.150) are simple simultaneous equations of the control points. By solving the above system of linear equations, the positions of the control points of the B-spline curves are obtained. The size of the matrix necessary to solve the equation is determined depending on the number of segments. Hence the size of the matrix is $(N+3) + 2 = N + 5$. If the cubic B-spline curves are uniform, $d\mathbf{C}(t)/dt$ and $d^2\mathbf{C}(t)/dt^2$ at connecting point of each segment are continuous regardless of $\mathbf{P}_j$ ($j = 0, 1, \ldots, n$) vectors, as mentioned before. In general, the tangent vector and curvature vector are continuous for arbitrary $\mathbf{P}_j$. Hence we do not need to consider boundary conditions used in the Bézier curve case.

The size of the matrix required for B-spline curves is smaller than that for Bézier curves because we do not need to consider the condition of a connecting point.

Experimental Results We need to note that the usual data base is not enough to estimate the performance of the method. That is, a set of virtually continuously changing images in terms of convexity/concavity is necessary. For this purpose,

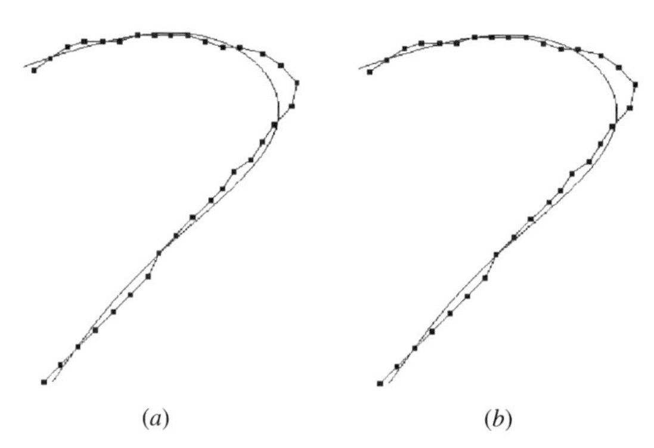

FIGURE 8.72 Approximation curves for data 2Z-N8 (*a*), and for data 2Z-N9 (*b*). From reference [42], © 1998, IEEE.

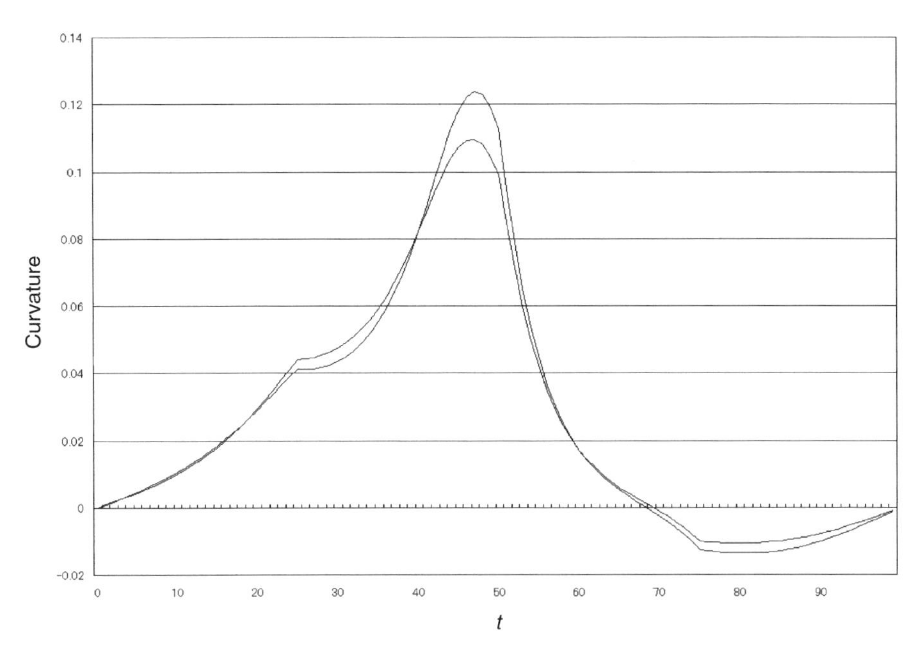

FIGURE 8.73 Curvature curves for the data 2Z-N8 (*bottom*) and 2Z-N9 (*top*); x-axis is parameter t and y-axis is curvature value.

some typical and notoriously close pairs of roman letters and arabic numbers are chosen; $2 \rightarrow Z$, $4 \rightarrow 9$, and $1 \rightarrow 3$. For each pair a set of data is generated using a tool that can be used to generate an almost arbitrary curve changing the degree of curvature of convex/concave segments of a curve interactively. Each segment is generated by a Bézier curve. Each date set consists of ten element character images as shown in Fig. 8.70 (left). Recall that in the figure, ideal character images are shown on the left side and noisy ones are shown on the right side. The sampled ideal data are numbered from leftmost to rightmost as 2Z–N9 and 49–N3.

The noisy data were generated using Ishii's method [64] which is simple but powerful enough to generate statistically distorted images using only two parameters for

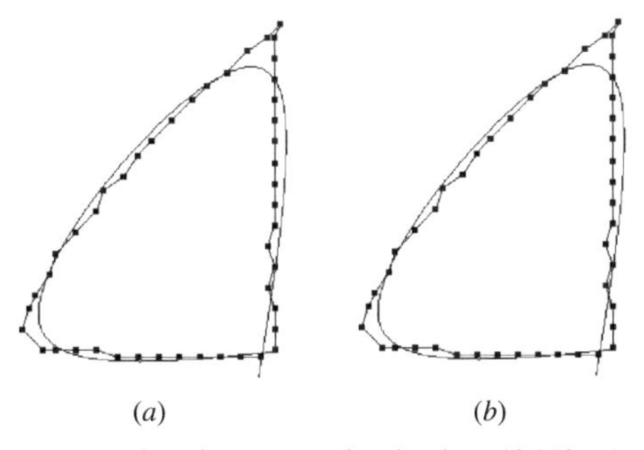

FIGURE 8.74 Approximation curves for the data 49-N2 (*a*) and 49-N3 (*b*).

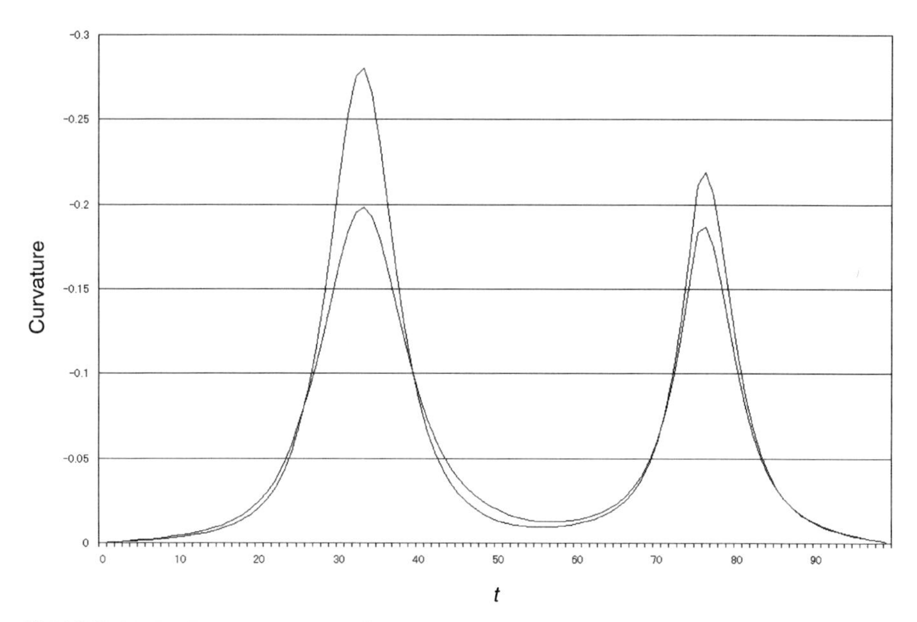

FIGURE 8.75 Curvature curves for the data 49-N2 (*top*) and 49-N3 (*bottom*); x-axis is parameter t and y-axis is parameter t and y-axis curvature value.

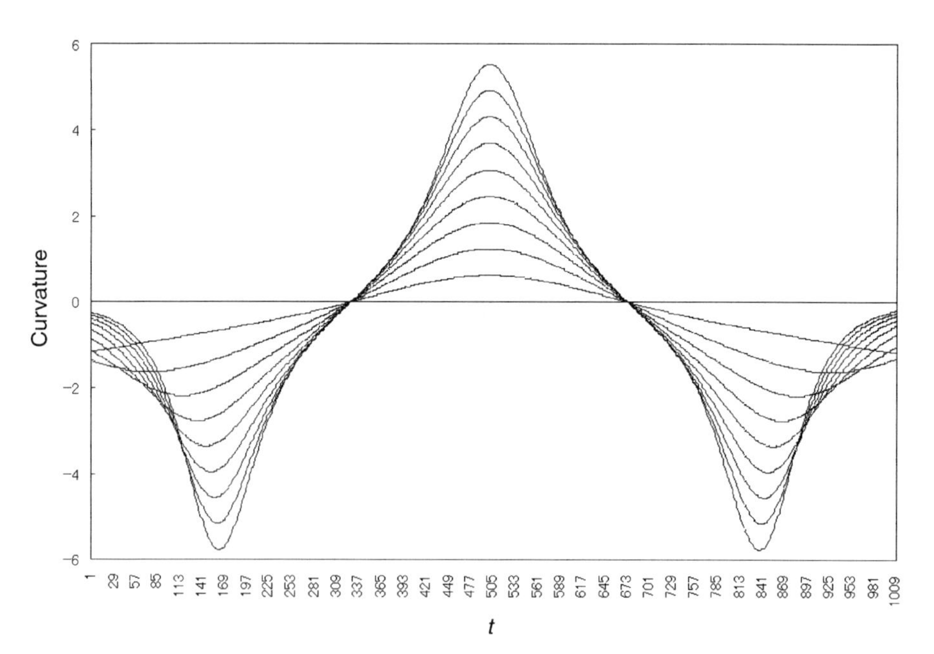

FIGURE 8.76 Curvature graph for ideal data from 13 data set. From reference [42], © 1998, IEEE.

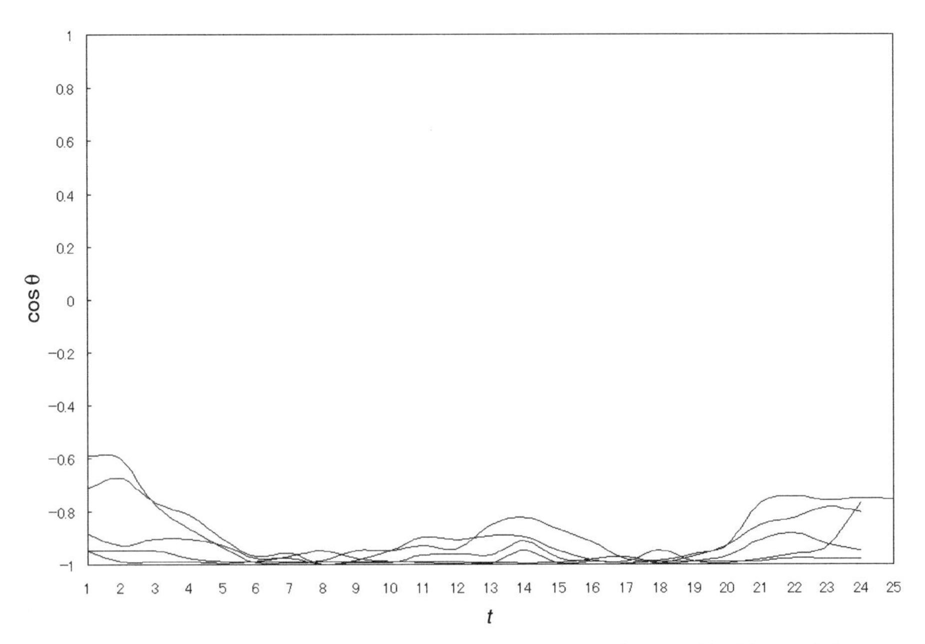

FIGURE 8.77 Curvature curves for the data set 13-N (13-N1 through 13-N5); x-axis is sample point number and y-axis is cosθ, where θ is angle. From reference [42], © 1998, IEEE.

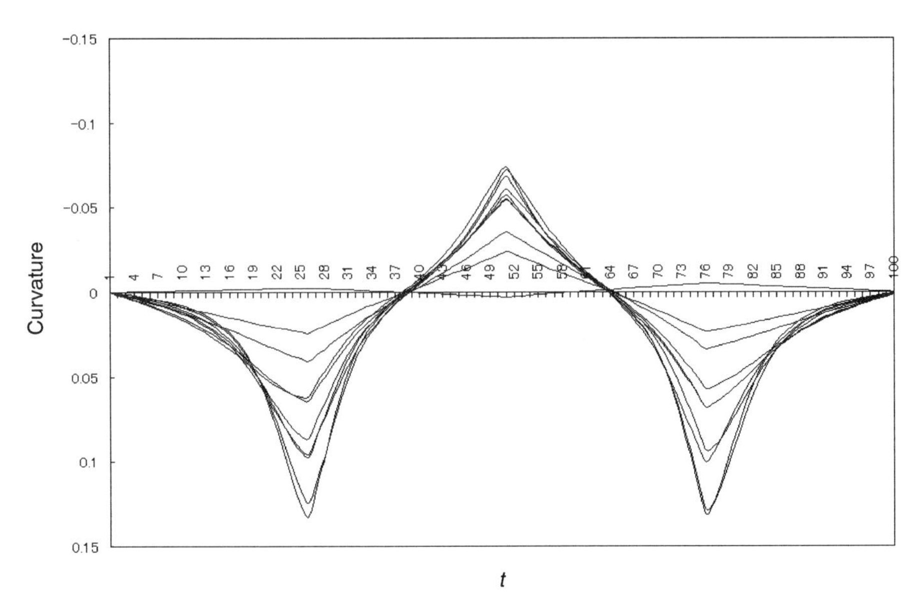

FIGURE 8.78 Curvature curves for the data set 13-N; x-axis is parameter t and y-axis is curvature value; n is 4. From reference [42], © 1998, IEEE.

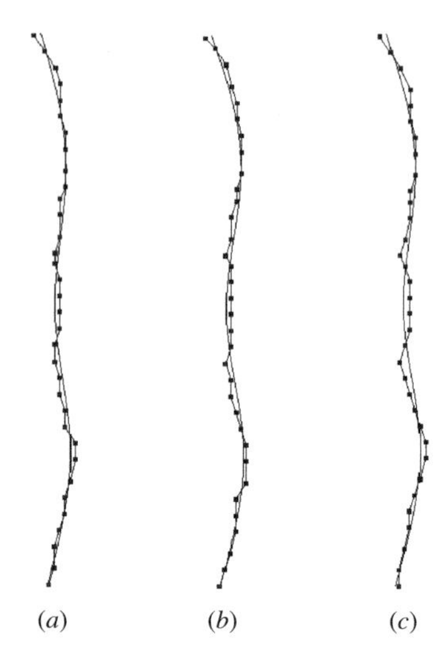

FIGURE 8.79 Approximation curves for the data 13-2 with noise parameter A, 0.04 (a); 0.05 (b); and 0.06 (c); n is 4. From reference [42], ©1998, IEEE.

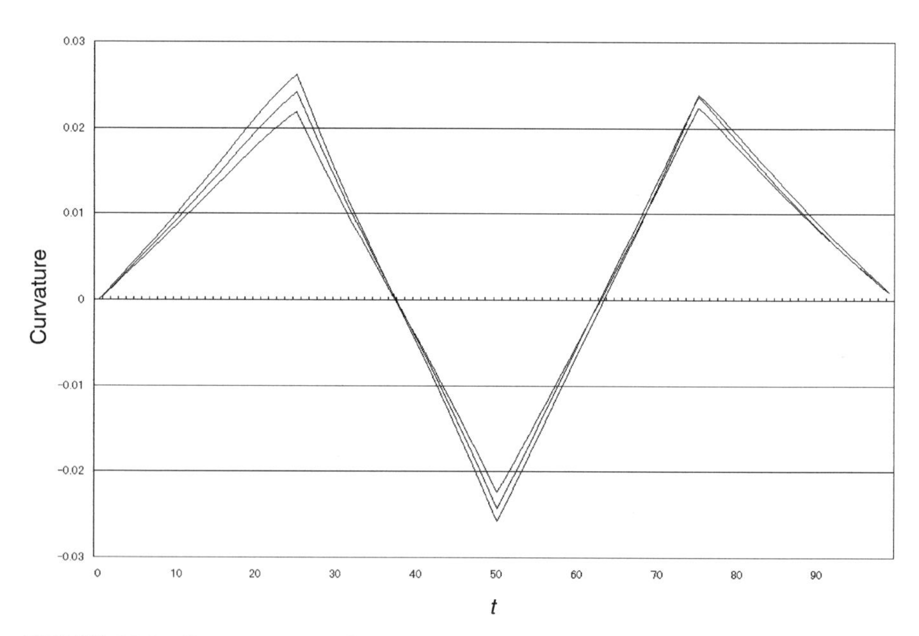

FIGURE 8.80 Curvature curves for the data 13-N-2 with noise parameter A, 0.04, 0.05, and 0.06; x-axis is parameter t and y-axis is curvature value. From reference [42], ©1998, IEEE.

the perturbation. The parameters used are denoted as A and W. A is an amplitude of the perturbation and W is a range of the constrain of continuity of a curve. Therefore, if A is large and W is small, then the curve becomes very ragged. If both A and W are large, then the curve is deformed globally. The sampled data sets being statistically distorted are denoted, 2Z-N, 49-N, and 13-N. They are shown in Fig. 8.70 (right) and numbered likewise 2Z-N1, 2Z-N2, . . . , 2Z-N9, 2Z-N10, and so on. For the noisy data, parameter values A and W are taken as 0.03 and 10, respectively. For the straight line segments, $A = 0.02$ is used compared with the curved line segments. In panel (d) only the interesting parts, the loops, are depicted.

In the change of shapes in 2Z-N, the most interesting pair 2N-N8 and 2N-N9 is shown in Fig. 8.72 with each approximated B-spline curve together with each original data curve. The corresponding curvature graph is shown in Fig. 8.73, in which the acute curvature parts are depicted with the very delicate plateau parts degraded for those noisy data. In the shape change in 49-N, the delicate pair 49-N2 and 49-N3 is shown, as explained, in Fig. 8.74. The corresponding curvature graph is also shown in Fig. 8.75, in which the expected results are obtained. In case of 49-N, the expected results are obtained for all of data set 49-N. Finally the curvature graph for the ideal data set of 13 is shown in Fig. 8.76. For the comparison, the Rosenfeld-Johnston method was tried on the perturbed data set for 13-N. The curvature graphs for a part of the data, 13-N-1 through 13-N-5, are shown in Fig. 8.77. The ordering of the curvature values obviously is not preserved, although the graph is somewhat busy. On the other hand, for the same noise data 13-N, the cubic B-spline method gave a very clear ordering in the curvature values as in Fig. 8.78 in which all the data results are shown.

In order to see the robustness of the cubic B-spline against noise, more noisy data were tested for the data set 13-2 and 13-3, taking the noise parameter A to 0.04, 0.05, and 0.06 with fixed $W = 10$. The approximation curves for 13-2 with the noise are shown in Fig. 8.79, and their curvature curves are shown in Fig. 8.80. We can see that the approximations are quite good for such heavy noise. Similar results were obtained for the data 13-3. However, it is shown that each curvature consists of four segments, each junction of which C^2 is preserved, but C^3 is not. This does not meet our intuition looking at the very smooth curves shown in Fig. 8.79. This is a limitation of the polygonal functions of third degree. For a more precise description of curvature higher order polygonal functions are necessary.

8.9 BIBLIOGRAPHICAL REMARKS

A general discussion on representation by primitives is provided by Marr in his famous book *Vision* [53]. His "primal sketch" gives a very useful conceptual framework. Alternatively, the so-called syntactic pattern recognition approach considers the representation of image by primitives based on a linguistics approach. This is fully discussed by Fu's book *Syntactic Pattern Recognition and Applications* [54]. Some formal definition of "representation" and other important terms are given by Simmon in his book *Patterns and Operators* [55], in which information is divided into representation and item interpretation. A representation is a string of elements from a finite alphabet representing an item of information. Identification is the action of giving a name to a representation, such as a stroke in a printed character, a phoneme in a spoken word, or an edge or texture in visual image.

In the text we have already described Pavlidis's idea for primitives, but we recommend his book *Structural Pattern Recognition* for the basic consideration of this essential approach and other principal topics [56].

There is considerable work in the field of psychology on the features of characters. A survey paper by R. N. Haber and L. R. Haber [57] points that difficulty and refers to some researchers' experiments and considerations. Their results seem to coincide with the engineer's viewpoint. Specifically, they investigated lowercase roman letters in which they introduced Dunn-Rankin's experiments [58]. They used a relatively similar judgment technique in which the subjects were presented with all possible triplets of letters. For each triplet the subject was asked whether the first letter was visually more similar to the second or to the third letter of the triplet. From these judgments over a large number of subjects, a scale of similarity was generated for each letter (j, g, v, x, and z omitted). It revealed that the letters b, p, and d are very similar to each other but quite different from others. Similarly $\{n, u, m, w\}$, $\{e, a, s, c, o, g\}$, and $\{h, f, l, t, k, i\}$ form clusters in terms of the similarity.

Convex Decomposition In this chapter only pure convex decomposition was described. Feng and Pavlidis extend this line to make their decomposition more flexible [59]. The basic idea of their decomposition is to allow a small degree of concavity in convex decomposition. However, their algorithm of decomposition is rather complex. Shapiro and Haralick [60] take a different approach to the decomposition problem, one that is even more flexible by using a clustering technique based on graph theory. However, it takes considerable computing time to execute, such as is usual with clustering algorithms. This method was improved by Guerra and Pieroni [61] more in an engineering sense. They considered their clustering in a graph matrix representation; this has made the algorithm much simpler. They show that the processing time is reduced by about 1/50 using the same data. Also, Bjorklund and Pavlidis have challenged, by generalizing, the Shapiro and Haralick's idea [62,60]. Their basic idea is as follows: An input shape is polygonally approximated. Then its vertices are set to nodes and its sides between vertices to edges, so a graph is constructed. When the nodes have some specific relations to other nodes, such as is visible, we can draw edges between the nodes that satisfy the visualization. However, this procedure requires computing time that is proportional to the second power of the number of nodes, which can be tremendous when the number is very large. Therefore only k relations per one node are allowed. The relations are limited basically to a local one. Further Bjorklund and Pavlidis noted the similarity between primitives that each is constructed by a local combination of nodes and edges, called LINK. As a result they produced a graph constructed by LINK related nodes, called a similarity graph. The algorithm is somewhat complex, but the LINK idea is very important.

Piecewise Linear Approximation On piecewise linear approximation many methods have been developed. They can be roughly classified according to norms in measuring errors. The norm L_2 is convenient for an analytical derivation to obtain an optimal approximation, such as by Cantoni's method introduced in this chapter, but it is not completely optimal. In this sense an optimal method was formalized by Bellman [45] based on his dynamic programming. For a norm L_∞, Ramer's method was introduced in this chapter. Recall that it was based on the iterative end point fit algorithm in which a straight line segments is first connected between end points of a curve (a sequence of data points). In the sense of the norm L_∞, the perpendicular distance from the segment is measured. If the distance is greater than some specified threshold value, the segment is replaced by two segments and the same procedure continues, as explained. Alternatively, there is another approach that constrains a straight line to pass within a radius around each data point. The line segment is projected from the first point until it falls outside the radius of a point. Then a new line segment is started. For this class of method, we can enumerate the following methods: Tomek's method [46], Williams's method [47,48], and J. Sklansky and V. Gonzalez's method [49]. The difference between Williams' two methods is that the former (1978) is somewhat not smooth but the latter (1981) is smooth. For L_1, there is Wall and Danielsson's method [50]. Intuitively the norm L_1 measures an area, so we can expect smoother approximation. Apart from norm classification, a minimax approach was taken by Kurozumi and Davis [51]. That is, line segment approximations are chosen to minimize the maximum distance between the data points and the approximating line segment. In their paper the computational complexities of the several methods are compared.

A comparative study was performed for the Ramer's method, split-and-merge method, and *curvature-guided polygonal approximation* by Ansari and Delp [52]. Curvature-guided polygonal approximation is a method devised by these authors to address some of the problems with the other two methods. The test images are set at different orientations and converted at different scales. Ramer's and the split-and-merge method were reported to be sensitive to both the orientation and the scale of the boundary in terms of the break points. The different starting sets of break points used in a polygonal approximation algorithm resulted in different approximated polygons. Thus a good starting set of break points is considered important to the polygonal approximation algorithm. Since extreme curvature points are likely to be break points, Ansari and Delp proposed to use these points as a starting set of break points for the polygonal approximation algorithm. The split-and-merge algorithm was used as the polygonal approximation algorithm to correct and modify the original starting set of break points. Such a combined approach was named a curvature-guided polygonal approximation. Here the curvature was approximated simply by the finite differences to approximate the differentiation operator

$$\dot{X} = X(t+1) - X(t-1),$$

$$\ddot{X} = X(t+1) + X(t-1) - 2X(t).$$

They concluded that although the curvature-guided polygonal approximation algorithm is sensitive to scaling, it is less sensitive to orientation.

Corner Detection Extensive work has been done to date on corner detection. Historically Freeman's work is earliest, so far as I know. He proposed smoothing method of local curvature code, P_i, defined as

$$P_i = \text{octmod}(u_i - u_{i-1} + 11) - 3,$$

where u_i is a Freeman code at a boundary point [65]. The smoothing is performed by a simple average as

$$f_i^N = \frac{1}{N} \sum_{k=0}^{N_1} P_{i+k},$$

where N is the number of pixels over the smoothing range. The f_i^N is called an *F-operation.* When Gallus and Neurath improved Freeman's method as

$$G_i^N = NP_i + \sum_{k=1}^{N_1} (N - k)(P_{i-k} + P_{i+k}),$$

they called it a *G-operation* [66]. There is a close relation between F- and G-operations. That is, the G-operation is equivalent to double operations of F. The F-operation can be represented in analytical form by using a δ function, and it can be neatly analyzed by Fourier analysis. Since Rosenfeld and Johnston's method was proposed, many methods of corner detections have been reported. First Rosenfeld and Weszka improved Rosenfeld and Johnston's method in terms of smoothing [29]. Freeman and Davis proposed another smoothing method that includes a moving straight line segment that connects the end points of a sequence of chain links [67]. The straight line segment, denoted as L_{is}, spans s chain links and terminates on the node to which link $\vec{c}_i$ is directed. That is, L_{is} is described as $\{\vec{c}_j, j = i - s + 1, \ldots, i\}$. As the line segment moves from one chain node to the next, the angular difference between successive segment positions is used as a smoothed measure of local curvature along the chain. Let x and y components of L_{is} be X_{is} and Y_{is}. Then the angle L_{is} that makes with the x-axis is given by

$$\theta_{is} = \begin{cases} \tan^{-1} \dfrac{Y_{is}}{X_{is}} & \text{if } |X_{is}| \geq |Y_{is}| \\ \tan^{-1} \dfrac{X_{is}}{Y_{is}} & \text{if } |X_{is}| < |Y_{is}| \end{cases}.$$

The *incremental curvature* δ_{is} is defined as twice the mean over two adjacent angular differences:

$$\delta_{is} = \theta_{i+i}, s - \theta_{i-1}, s.$$

As stated in the test, an angle is a global feature that is reflected by providing a measure of cornerity. That is, a corner is characterized by three incremental curvature regions. The center region is a region in which $\sum\delta_{is}$ takes a significant value. The other parts are two regions that lie at the two sides of the center region. Thus the *cornerity* at node i is defined by

$$K_i = \sqrt{t_{i1}} \times \sum_{j=i}^{i+s} \delta_{js} \times \sqrt{t_{i2}},$$

where

$$t_{i1} = \max\{t|\delta_{i-v,s} \in (-\Delta, \Delta), \forall\, a \leq v \leq t\},$$

$$t_{i2} = \max\{t|\delta_{i+s+v,s} \in (-\Delta, \Delta), \forall\, a \leq v \leq t\},$$

and

$$\Delta = \tan^{-1}\left(\frac{1}{s-m}\right)$$

with

$$m = 1 \text{ or } 2.$$

Finally only those points are retained where $|K_i| \leq |K_j|$ for all j such that

$$|i - j| \leq s$$

as the corner points.

O'Gorman [68] combined the corner detection methods mentioned above and Shirai's method [69] to come up with a *difference of slopes* (DOS) approach. In the DOS approach a curvature at a point is estimated as the angular difference between the slopes of two line segments fit to the data before and after the point. The plot of local curvature is called a θ plot. DOS methods in the literature differ in the separation distance M between the two segments about the point where the curvature is to be measured. In Freeman and Davis method, $M = 1$, and in Rosenfeld and Johnston/Weszka methods, and Shirai's method, $M = 0$. The case $M < 0$ can be also considered taking overlapping.

O'Gorman determined how the M affects the effectiveness of the DOS approach in terms of signal detection ability. The measure of signal detection ability was defined as the ratio of the peak on the θ-plot due to a corner versus that due to noise, called SNR. The noise was defined as the curvature due to one sample displacement from a straight horizontal or vertical line. He derived analytical form of SNR as a function of M and found a very interesting result that the best signal detection ability is obtained for values of M near zero but not equal to zero. The special case of the DOS approach where ($M \to +0$, $M \neq 0$) he named the DOS$^+$ method. He also compared the DOS approach with conventional Gaussian smoothing developed by Asada and Brady [70]. Specifically, the θ-plot is obtained first from the local second derivative estimate of the curvature, then a smoothing filter is applied to that plot. The smoothed curvature plot $\theta(s)$ is obtained from the local estimate $\theta'(s)$ by convolving it with a Gaussian window of chosen σ. Asada and Brady took this approach and obtained a multiple-scale curvature representation based on scale space filter invented by Witkin [71]. Their idea itself is very interesting and useful and so will be mentioned later. However, we will continue O'Gorman's work. In general, an advantage of Gaussian smoothing which is considered to improve the SNR is that noise is greatly attenuated by smoothing. However, the signal itself is also attenuated. In other words, Gaussian smoothing method treats a corner the same as it treats noise by attenuating its peak on the θ-plot. However, O'Gorman claims that the DOS^+ method is able to act as an adaptive matched filter, namely to retain the corner peak while attenuating the peak due to noise. He experimented with Gaussian smoothing and was able to obtain favorable results for DOS^+ compared with Gaussian smoothing.

Since so many kinds of methods have been developed for corner/curvature detection, some comparison studies on corner detection methods have been conducted. Rutkowski and Rosenfeld [72] compared five corner detection techniques for chain-coded curves and concluded that the weighted k-curvature result was the best. In that method the chain-coded curve is defined by a sequence of vectors, $\mathbf{v}_0, \mathbf{v}_1, \ldots, \mathbf{v}_{n-1}$, where $\mathbf{v}_i \in \{(1,0), (1,1), (0,1), (-1,1), (-1,0), (-1,-1), (0,-1), (1,-1)\}$. Taking a set of appropriate weights, $w_1, w_2, \ldots, w_k$, the weighted k-vectors at point i is defined as

$$\mathbf{V}_i^{(1)} = \sum_{j=1}^{k} w_j \mathbf{v}_{i-j},$$

$$\mathbf{V}_i^{(2)} = \sum_{j=1}^{k} w_j \mathbf{v}_{i+j-1}.$$

Then the weighted k-curvature at point i is determined as

$$\theta_i = \cos^{-1}\left\{ \frac{\mathbf{V}_i^{(1)} \cdot \mathbf{V}_i^{(2)}}{\|\mathbf{V}_i^{(1)}\| \, \|\mathbf{V}_i^{(2)}\|} \right\}.$$

Once the curvature has been estimated for all contour points, a corner is detected if the curvature is above a given threshold and is a local maximum within a range $[i-k, i+k]$.

Liu and Srinath compared six kinds of typical corner detection methods using five test images from simple to complex ones [73]. The six methods compared are the Medioni-Yasumoto corner detector, the Rosenfeld-Weszka corner detector, and the Cheng-Hsu corner detector [75]. The Beus-Tiu corner detector is an improvement of the Freeman-Davis corner detector mentioned before. Beus and Tiu found that Freeman and Davis's algorithm failed to detect some obvious corners and also detected spurious corners in some cases. Therefore they proposed two modifications to Freeman and Davis's algorithm. First, K_i is determined as the average of the K values. Second, the two sides are cut off by a specified threshold value. The Cheng and Hsu corner detector is more complex and defined by the degree of bending according to the direction changes of forward and backward arms. In Liu and Srinath's paper those six methods are introduced concisely.

The experiments conducted were fair to all methods. That is, the thresholds of all the methods were chosen so as to result in a maximum number of true corners and minimum number of spurious corners. Two criteria of good detection and good sensitivity were adopted. The Medioni-Yasumoto corner detector is very sensitive to the smoothness of boundary points. (This has been also my experience, as I stated in the text.) Liu and Srinath noted that it is understandable because this method detects the corners by using only five points in the boundary to compute the curvature. As mentioned in the text, Medioni-Yasumoto's method is close to local operation, although their idea is very good and the implementation is simple. Liu and Srinath evaluate Cheng-Hsu corner detector as also very sensitive to smoothness. In the weighted k-curvature corner detector only very few spurious corners were detected. The Rosenfeld-Johnston corner detector gave incorrect results when corners occurred too close together. The modified method, the Rosenfeld-Weszka corner detector, improved on the Rosenfeld-Johonston results in the difficult points mentioned above. Finally the Beus-Tiu corner detector showed the best results in all the examples tested. The methods were tested for noisy data set except for the smooth sensitive methods of Medioni-Yasumoto and Cheng-Hsu corner detectors. For the noisy test images the weighted k-curvature and Beus-Tiu corner detectors gave the best results. Thus Liu and Srinath concluded that it appears that the performance of the Beus-Tiu corner detector is closest to that of the human eye on both the original test image set and the noisy image set.

From these comparative studies mentioned above, and in the text, it appears that Rosenfeld-Johnston, Rosenfeld-Weszka, Beus-Tiu, and Teh-Chin achieved some of the best results. However, Legault and Suen criticized those studies because the number of images used in them was generally very small (from two to eight images). Therefore Legault and Suen conducted a practical and statistical comparative study [76]. The methods compared were those of Beus-Tiu, Rosenfeld-Johnston, Rosenfeld-Weszka, Teh-Chin, Lee-Lam-Srihari, and Legault-Suen. In the Lee-Lam-Srihari method [77] the so-called angle accumulation algorithm was used, in which differential chain code value were taken as a measure of local change in curvature. A concavity is defined as the longest sequence of perimeter points whose angular changes between consecutive elements are all nonnegative and their accumulated sum is greater than 1. Legault and Suen used this scheme, but also another polynominal approximation was employed, which will be explained later.

The method was tested on 100 binary images of handwritten digit selected from a subset of a 20,000-sample CENPARMI database. Ten samples were chosen for each numeral class, offering a variety of styles and sizes, which are shown in their paper. They introduced a measure of goodness (MG) for quantitative comparison. First of all, for each sample, regions of external contour corresponding to the global shape features to be detected were determined manually by them (Legault and Suen themselves). The MG was computed for each method. For sample i, let R_i be the set of the indices of all contour points belonging to the n_i regions selected by the humans. Furthermore let S_i^k be the set of the indices of all contour point belonging to the m_i^k regions selected for sample i by method k. Q_i^k and f are defined as $R_i \cap S_i^k$ and n_i/m_i^k or m_i^k/n_i whichever is less than (or equal to) 1.

Thus the MG of method k in determining the global feature regions of samples i is defined as

$$MG_i^k = f\left(\frac{\sum_{j \in Q_i^k} w_j}{\sum_{j \in R_i} w_j}\right),$$

where w_j is the weight associated with each point whose index belongs to R_i. Clearer convex points, namely end points, were weighted more than other convex/concavity points. That is, 5 for the former and 2 for the latter. This idea is reasonable because, if an obvious convex point like an end point is not detected, this is a fatal flaw in the recognition process, although detection of an end point is not necessarily easy for a thick stroke. The results are shown in Table 8.5.

As seen in the table, the Beus-Tiu, Legault-Suen, and Rosenfeld-Weszka methods give the best results in terms of their MG values. However, Legault and Suen claim that their method was found to be the most reliable in the sense that it detects some points within each and every feature region and only once is a feature region missed altogether. On the other hand, they stated that their method's major weakness is in the extraction of long smooth curves which are not detected as single global regions. The poor performance of Teh-Chin method is explained as its detection of a large number of tiny consecutive regions, thus preventing the method from extracting features at a higher, more global level.

Legault and Suen's method [78] consists of the following steps:

1. Smoothing and initial processing of the contour.
2. Selection of reference point.
3. Parametric approximation of each piece of contour between consecutive reference points.

Step 1 uses the usual simple averaging accumulation mentioned before. That is, an arc is defined as the longest sequence of contour points $[k, k + l - 1]$ satisfying the following conditions:

TABLE 8.5 Average Measure of Goodness for the Methods Tested

Method	0	1	2	3	4	5	6	7	8	9	Overall
Beus-Tiu	0.46	0.84	0.68	0.70	0.86	0.73	0.56	0.82	0.64	0.62	0.69
Lee et al.	0.46	0.57	0.67	0.68	0.73	0.64	0.72	0.71	0.67	0.55	0.64
Legault-Suen	0.39	0.84	0.73	0.74	0.82	0.72	0.56	0.80	0.61	0.64	0.68
Rosenfeld-Johnston	0.42	0.58	0.63	0.72	0.64	0.62	0.55	0.62	0.55	0.55	0.59
Rosenfeld-Weszka	0.44	0.75	0.76	0.80	0.73	0.71	0.59	0.70	0.60	0.63	0.67
Teh-Chin	0.14	0.21	0.20	0.22	0.22	0.21	0.16	0.23	0.22	0.17	0.20

From reference [78], © 1998, IEEE.

1. $|\theta_k| \geq T_1$.
2. θ_j has constant sign for $j \in [k+1, k+l-1]$.
3. $\theta_{j-1} \cdot \theta_j > T_1$ for $j \in [k+1, k+l-1]$.

Here θ_k is the direction change in an angle at point k, and T_1 is an appropriate threshold value. For the arc, $\theta_{sum} = \sum \theta_j$ and $\theta_{max} = \max[\theta_j | j \in [k, k+l-1]]$ are defined. Let j_{max} be the label of the point where θ_{max} is reached. Then reference points are selected. For example, if $|\theta_{sum}| \geq T_2$, a reference point is created within the arc. If $\theta_{max} > T_3$, the reference point will be j_{max}. Finally in step 3 a contour piece is selected, which is the sequence $[i, f]$, where points i and f are consecutive reference points. Each contour piece is approximated by a parametric cubic or quartic. The parameter value is normalized such that for each contour piece, $t = 0$ for the initial point and $t = 1$ for the final point. For the cubic method, a linear system of equations is fully determined by the requirement of going through two intermediate points in (i, f). The selection of the intermediate points is somewhat complex, and it is described in detail in their paper. Line fitness is estimated. If it is not satisfactory, then further improvement is performed. Thus their approximation is a heuristic optimization process.

T. Kadonaga and K. Abe conducted a systematic comparative study [79]. The 11 methods compared were as follows:

1. A method using n-code [80].
2. A method using slope information [81].
3. Freeman-Davis method.
4. Rosenfeld-Weszka method.
5. Koyama et al. method [82].
6. Arcelli et al. method [83].
7. Teh-Chin method.
8. DOS method.
9. Fischler-Bolles method [84].
10. Beus-Tiu method.
11. Held et al. method [85].

They noticed invariance of corner feature points under rotation, size change, and reflection. As for rotational invariance, good results were obtained for the n-code and DOS methods, and poor results for the Teh method. Concerning the size change, n-code and Arcelli methods yielded relating invariant results, while those of Freeman and Beus methods were unstable. On reflecting the input figures, perfect unchanged results were obtained with Rosenfeld, Koyama, and Fishler methods. Kadonaga and Abe used four different kinds of figures.

On the other hand, the evaluation by human subjects was conducted, using 10 various kinds of figures. The results were sarcastic, they said. That is, Rosenfeld-Weszka method was the best and slightly worse was the n-code method.

A comparison study from a different angle was performed theoretically by Worring and Smeulders [86]. A given curved line was digitized naturally so that in the digitization exact information on the continuous line was lost. Therefore the curvature could only be estimated. The methods presented were analyzed on the basis of their ability to estimate properly the predigitized curvature. They calculated the accuracy and precision of estimation, namely the bias and deviation under repeated placement of a continuous object in a random position with respect to the digitization grid.

Then the curvature estimation methods were classified by the orientation of the tangent, the second derivative of the curve considered as the path, and the local touching circle. Gaussian smoothing was employed to all the methods, so the truncation effect of the Gaussian kernel was examined. The orientation of the tangent method was classified further into three groups: Method I (chain code), Method II (resampling), and Method IV (linefit). The path-based method and circular fitting were methods that referred to Method IV and Method V, respectively.

The results are as follows:

1. Curvature estimates are improved by using a Gaussian differential kernel and a better arc length estimator.
2. Method I performs poorly, since the anisotropy of the grid is not accounted for.
3. Method II has the best overall performance.
4. Method V is poorest, since smoothing is hampered by the discontinuous derivatives of the coordinate functions.
5. Method V is only suited for curvature estimation when arcs are large and of constant radius.

The recommended method, Method II, was developed by Duncan, Lee, Smeulder, and Zaret [87], who naturally used it in the resampling they employed. Their resampling is mentioned in the Bibliographical Remarks of Chapter 2.

Anderson and Bezdek gave a theoretical consideration to the tangential deflection/angle and curvature of discrete curves [31]. They derived a neat form of the tangential deflection based on the eigenvalue and eigenvector structure of sample covariance matrices. Specifically, at a corner point, we can image two sets of sample points, say X and Y, distributed over both neighbor sides. From these sets, principal directions can be derived using principal component analysis constructing scatter matrixes, say A and B. The angle ($\cos \Delta\theta(X, Y)$) is given by the Euclidean inner product of these principal vectors/eigenvectors. However, Anderson and Bezdek used the property of commutator of matrix defined as

$$[A, B] = AB - BA.$$

This is further represented simply as

$$[A, B] = \begin{bmatrix} 0 & \delta(X, Y) \\ -\delta(X, Y) & 0 \end{bmatrix}.$$

The $\delta(x, y)$ is given as

$$\delta(X, Y) = \frac{1}{2} D_X D_Y \sin [2\Delta\theta(X, Y)],$$

where

$$D_X = \sqrt{1 - 4 \det A},$$
$$D_Y = \sqrt{1 - 4 \det B}.$$

On the other hand, $\delta(X, Y)$ is also represented as

$$\delta(X, Y) = \frac{(a_{22} - a_{11})(b_{22} - b_{11}) + 4a_{12}b_{12}}{2},$$

where

$$A = \begin{bmatrix} a_{11} & a_{12} \\ a_{21} & a_{22} \end{bmatrix}, \quad B = \begin{bmatrix} b_{11} & b_{12} \\ b_{22} & b_{22} \end{bmatrix}.$$

Thus the final result is given as

$$\cos 2\Delta\theta(X, Y) = \frac{(a_{22} - a_{11})(b_{22} - b_{11}) + 4a_{12}b_{12}}{\sqrt{(1 - 4 \det A)(1 - 4 \det B)}}.$$

Therefore their method is expected to be strong against noisy curves.

In the literature the terms, angle, and curvature at a corner point are interchangeably used. Anderson and Bezdek derived a clear relationship between them for discrete curves. They based this on a relation they derived as

$$\frac{\delta(X,Y)}{\Delta S(x)} = \frac{1}{2}\sqrt{(1 - \det A)(1 - \det B)}\,\frac{\sin 2\Delta\theta(X,Y)}{\Delta S(x)}$$

$$\approx \frac{\sin 2\Delta\theta(X,Y)}{2\Delta\theta(X,Y)} \cdot \frac{\Delta\theta(X,Y)}{\Delta S(x)}$$

$$\approx \frac{\Delta\theta(X,Y)}{\Delta S(x)} \approx k(t_0),$$

where $\mathbf{x} = (X(0), Y(0))$ and $\Delta S(\mathbf{x})$ is the Euclidean distance between $\mathbf{x}_{i_0}$ and $\mathbf{x}_{i_0+1}$. Specifically, let $\{a_k\}$ and $\{b_k\}$ be equally spaced sequences of numbers. Define

$$X(\varepsilon) = \{\mathbf{x}_k = \mathbf{r}(a_k\varepsilon) | k = 1, \ldots, m\},$$
$$Y(\varepsilon) = \{\mathbf{y}_k = \mathbf{r}(b_k\varepsilon) | k = 1, \ldots, m\}.$$

If $X(\varepsilon)$ is centered at $\mathbf{x} = \mathbf{r}(0)$ and $Y(\varepsilon)$ is centered at the point in the succession to $\mathbf{x}$ along the discrete curve, and if $\Delta S(\mathbf{x})$ is the distance from the center of $X(\varepsilon)$ to the center of $Y(\varepsilon)$, then

$$\lim_{\varepsilon \to 0} \frac{AB - BA}{\Delta S(\mathbf{x})} = \begin{bmatrix} 0 & k(0) \\ -k(0) & 0 \end{bmatrix}$$

is proved.

Comparative studies between the analytical result and discrete curvature, and their experimental method in relation to other methods (i.e., Freeman-Davis and Rosenfeld-Johnston methods), were given.

Asada and Brady defined a set of primitive curvature changes used in their representation and analyzed scale space behavior [70], as mentioned before. First they took two isolated curvature changes, the corner and smooth join which they defined as follows:

1. *Corner.* An isolated curvature change for which the tangent to the contour (and hence the curvature) is discontinuous.
2. *Smooth join.* An isolated curvature change for which the tangent is continuous but the curvature is discontinuous.

Curvature changes that are close complicate the analysis. The set of compound primitives that arise in practice are defined as follows:

1. *End.* A compound curvature changes consisting of two nearby corners of the same sign,
2. *Crank.* A compound curvature change consisting of two nearby corners of opposite signs,
3. *Bunk or dent.* A complex compound change that consists of two nearby crank changes.

Theoretically there are another combinations of nearly smooth joins and a smooth join near a corner. In practice, they are neglected because the former appears like a bump and the latter is typically perceived as an end or crank. For each primitive a simple geometrical model is constructed, and the analytic form for its convolution with the first and the second derivatives of Gaussian is neatly given. Thus we can see a guide line of the behavior of the scale space filter when it is applied as a function of the Gaussian of standard deviation σ.

Young, Walker, and Bowie developed a measure of the complexity of a simply connected plane object in terms of the curvature [88]. This is an extension of the well-known measure of the complexity of a figure, the (perimeter)^2/area, for example. They used an analogy of the bending energy of an elastic rod and derived an expression for the average bending energy per unit length, E, as

$$E = \frac{1}{P} \int_0^P |k(p)|^2 dp,$$

where P is the total length of the contour and $k(p)$ is a curvature at point p on the contour. Their definition of a curvature is based on the orientation of the tangent, and it is very simple and local using the Freeman chain code.

Taxt, Ólafsdóttir, and Dealen used a curvature function along a contour of a given character image, using B-splines approximation in the sense of the Fourier descriptor [89]. They conducted a handwritten character recognition experiment using a Bayesian classification rule with a multivariate normal probability density model. They achieved a very high correct recognition rate for numerals (independent test sets using B-splines approximation) of 99.8% which is comparable to the experimental results employing the elliptic Fourier descriptor. However, for the alphanumeric large sample set (all handwritten lowercase letters and handwritten digits), the parametric spline approximation method also gave good results, but clearly inferior to those of the elliptic Fourier descriptors. They analyzed it as follows: Part of the reason for this discrepancy between numerals and alphanumeric results was that some of the letters, but none of the digits, were only one pixel wide at the end of some strokes, where there exists a singular point of curvature. This can be removed.

Last we introduce corner detection methods operating directly on a gray tone image. Historically the earliest such corner detector is Beaudet's DET operator [90]. The DET is defined as follows along with its Laplacian:

$$\mathbf{DET} = \frac{1}{2}\sum_{ij}(\nabla_i\nabla_j I)(\nabla_i\nabla_j I) - \frac{1}{2}L^2,$$

$$\text{Laplacian} = L = \sum_i \nabla_i\nabla_i I,$$

where I is image intensity and ∇_i is a gradient operator. Thus $\nabla_i I$, $i = 1, 2$, is a gradient vector. $\nabla_i\nabla_j I$ is a tensor (second rank) which is contracted to be scalar as shown above. Beaudet constructed operators from zeroth to fourth order for array up to 8×8.

Following Beaudet's corner detector Kichen and Rosenfeld [91] and Dresehler and Nagel [92] developed their corner detectors, respectively. Kichen and Rosenfeld tried several methods of gray tone corner detectors among which the best one was obtained by measuring cornerness by the product of gradient magnitude and instantaneous rate of change in gradient direction evaluated from a quadratic polynomial gray tone surface fit. In case of Dreshler and Nagel the procedure is a somewhat sophisticated computation of Gaussian curvature, which is performed by fitting a local quadratic polynomial for each point and computing the Hessian matrix. Concerning such a differential geometric approach, we will discuss it in some detail later. As seen so far, this is closely related to edge detection. Haralick investigated a new edge detector based on zero crossing of second directional derivatives approximating a given image by a cubic polynomial [93]. Thus Zuniga and Haralick associated corners with two things: the occurrence of an edge and significant changes in edge direction [94]. Specifically, they applied Haralick's edge detector to the corner detector in three ways:

1. Incremental change along tangent line.
2. Incremental change along contour line.
3. Instantaneous rate of change.

According to their results, the best method is the second one and the next is the first, which is the simplest among the methods. They also conducted a comparison study with the Kitchine-Rosenfeld and Dreschler-Nagel corner detectors and demonstrated that their method is the best. The next best is the Kitchen-Rosenfeld corner detector, and the worst is the Dreschler-Nagel corner detector.

Wu and Rosenfeld developed a simpler corner detection method that can be applied directly to a gray tone image, using the x and y projections of the image [95]. This method aimed at avoiding heavy computation due to the need to compute higher-order difference operators in every position, as seen above.

Direct Gray-Scale Feature Extraction So far we have focused our attention on corner detection. However, once a gray-scale image is treated, we can image a general feature extraction scheme, namely the classification of the image's surface intensity into a complete set of topographic elements. This was done by Watson, Laffey, and Haralick [96]. This idea is not new, and considerable research has been done, such as introduced by the cited paper above. However, their scheme is very systematic and general based on differential geometry. Haralick, Watson, and Laffey noticed and proved that the major topographic peaks, pits, ridges, ravines, saddles, flats, and hill sides are invariant under monotonically increasing transformations of the gray tone values [97].

The surface properties are described by a Hessian matrix, namely for eigenvalues. Let $f(x, y)$ be the illumination intensity of a given image at a point (x, y); let the eigenvalues of the Hessian be λ_1 and λ_2. Then the topographic types are classified as follows:

Peak	$\|\nabla f\| = 0, \lambda_1 < 0$	$\lambda_2 < 0$
Pit	$\|\nabla f\| = 0, \lambda_1 > 0$	$\lambda_2 > 0$
Ridge	$\|\nabla f\| \neq \lambda_1 < 0$	$-f \cdot \omega^{(1)} = 0$ or
	$\|\nabla f\| \neq 0, \lambda_2 < 0$	$-f \cdot \omega^{(2)} = 0$
	$\|\nabla f\| = 0, \lambda_1 < 0$	$\lambda_2 = 0$
Saddle	$\|\nabla f\| = 0, \lambda_1 * \lambda_2 < 0$	
Flat	$\|\nabla f\| = 0, \lambda_1 = 0$	$\lambda_2 = 0$, etc.

Haralick and his associates made a detailed table of the mathematical properties mentioned above.

They took an image surface and approximated it by generalized quadratic B-splines and a discrete cosine basis. The experimental results were not so good as expected. For even artificial surfaces without noise, the scheme did not necessarily work well. Their interpretation is that the spline basis functions have a significantly smaller RMS error, yet the labeling based on them may be much poorer than DCT-based labeling. It still remains unclear what properties are desirable for basis functions.

Now we return to the subject of OCR. Pavlidis noticed the importance of direct gray-scale extraction of features in OCR. Actually now every engineer of OCR knows the effectiveness of using the gray level directly. In other words, uniform binarization preprocessing is dangerous because it sometimes produces broken lines. In particular, among the many kinds of envelopes passing through a mailing system are those whose backgrounds are not uniform or have different colors of ink used for the print. Therefore direct gray-scale processing is becoming a popular concept.

Wang and Pavlidis provide a systematic method of direct gray-scale extraction features for OCR [98]. They followed the labeling scheme of local topographic features developed by Haralick and associates mentioned above. In addition to those labels the global topographic features are defined as *ridge line, ravine line, flat region,* or *hillside region.* To investigate the ridge/ravine line extraction, an ideal bar is constructed with a Gaussian smoothing filter used as an approximation to the point spead function of the scanner. In this simulation experiment a spurious ridge line is found on the x-axis for the vertical bar model elongated along y-axis. To suppress such spurious ridge lines, the notion of directional curvature is employed. That is, the normal curvature of the surface S defined by an arbitrary parameterization $r = r(u, v)$, where $r = (x, y, z)^T$ is defined at point (u, v) in the direction of $(du : dv)$ as

$$K_0 = \frac{Ldu^2 + 2Mdudv + Ndv^2}{Edu^2 + 2Fdudv + Gdv^2},$$

where the denominator and numerator are *the first fundamental form* and *the second fundamental form,* respectively. The notation used is conventional. The direction (du, dv) on a surface is called the principal direction if the normal curvature of the surface at a given point attains an extreme value in this direction. The extreme values of K_0 are called the *principal curvature* of the surface. Therefore the next step is to find the relationship between (1) the principal directions of curvature and the eigenvectors of the Hessian and (2) the directional curvatures and the eigenvalues of the surface. Concerning these questions, the following theorem holds: The curvature of the surface defined by $Z = f(x, y)$ in the direction $\mathbf{w}$ is

$$\frac{1}{\sqrt{1 + \nabla f \cdot \nabla f}} \cdot \frac{\mathbf{w}^T H \mathbf{w}}{1 + (\nabla f \cdot \mathbf{w})^2}.$$

Furthermore the principal directions of curvature coincide with the eigenvectors of the Hessian, $\mathbf{w}_1$ and $\mathbf{w}_2$, if and only if one of the following conditions is satisfied:

1. $-f \cdot \mathbf{w}_1 = 0$.
2. $-f \cdot \mathbf{w}_2 = 0$.
3. $\det(H) = 0$.

Now the *strength of a ridge/ravine line* at point $(x, y)^T$ denoted by $S(x, y)$ is defined as the magnitude of the principal curvature in the direction orthogonal to the gradient at $(x, y)^T$:

$$S(x, y) = \frac{|\lambda|}{\sqrt{1 + \nabla f \cdot \nabla f}},$$

when λ is the eigenvalue corresponding to the eigenvector orthogonal to the gradient. When $\mathbf{w}$ is the unit eigenvector orthogonal to the gradient, the following relations hold: $\nabla f \cdot \mathbf{w} = 0$, $\mathbf{w}^T H \mathbf{w} = \lambda$. Naturally the strength of the ridge/ravine line gives a measure on whether or not the ridge/ravine line represents the skeleton of a character. Thus "weak" ridge/ravine lines such as the ridge line along x-axis in the vertical bar model can be discarded.

To study touching or merged characters, an ideal model consists of two closely placed bars that are perpendicular to each other. A saddle point between these two bar models is identified. The two ridge lines giving the skeletons of the objects are joined together at this saddle point. Therefore these two objects can be easily separated by locating the saddle point. Otherwise, the two objects are connected by the usual binarization preprocessing, and sophisticated techniques are required to separate them.

The extraction of the geometric features is performed as follows:

STEP 1: Topographic classification.

STEP 1a: Computation of first and second derivatives.

STEP 1b: Computation of eigenvalues and eigenvectors.

STEP 1c: Classification of elements. A label is assigned to each element using classification scheme defined before.

STEP 2: Extraction of basic structural information.

STEP 2a: Formation of characteristic regions. A connected characteristic region formed entirely by peak and ridge points, saddle points, or flat points is extracted. Each region is labeled by the name of its elements, such as ridge region, saddle region, and flat region.

STEP 2b: Formation of the TFG. The TFG is defined to be the adjacent graph of the characteristic regions obtained in Step 2a.

STEP 3: Extraction of geometric features.

STEP 3a: Transformation of characteristic regions into points or line segments. Each one of the saddle regions, flat regions, and compact peak/ridge regions in the TFG is replaced by a single point, and each elongated peak/ridge region is converted into one or more line segments, or more precisely, into a pseudopolyline.

STEP 3b: Formation of geometric feature graph (GFG). Some grouping and assembling of the points and line segments produced in Step 3a is performed to form a GFG.

Note on the Implementation Discrete Chebyshev polynomials are used up to the third degree as the bases spanning the vector space of these continuous functions. Wang and Pavlidis devised an effective way to avoid the expensive subpixel computation in the precise calculation of interesting points for the labeling as Haralick et al. did.

The systematic experiments were conducted and many examples were demonstrated depicting original character images. The studies were also compared using the dynamic thresholding method.

BIBLIOGRAPHY

[1] R. O. Duda and P. E. Hart, *Pattern Classification and Scene Analysis,* New York: Wiley, 1973.

[2] R. O. Duda and P. E. Hart, "Use of Hough transformation to detect lines and curves in pictures," *CMCM,* vol. 15, pp. 11–15, January 1972.

[3] T. Pavlidis, "Structural pattern recognition: Primitives and juxtaposition relations," S. Watanabe, ed., *Frontiers of Pattern Recognition,* Academic Press, 1972.

[4] T. Pavlidis, "Analysis of set patterns," *Pattern Recogn.,* vol. 1, pp. 165–178, 1968.

[5] M. Nadler, "Sequentially-local picture operators," *2nd IJCPR,* pp. 131–135, 1974.

[6] A. Cantoni, "Optimal curve fitting with piecewise linear function," *IEEE Trans. Computers,* vol. C-20, no. 1, pp. 59–67, 1971.

[7] R. Bellman, "On the approximation of curves by line segments using dynamic programming," *CACM,* vol. 4, no. 6, p. 284, 1961.

[8] Y. Nakajima and S. Mori, "A model-based classifier in a scheme of recognition filter," *Proc. Int. Conf. Document Anal. Recogn.,* pp. 68–71, October 20–22, 1993.

[9] G. C. Buck, "Advanced Calculus," 3rd ed., New York: McGraw-Hill, 1978.

[10] J. R. Rice, *The Approximation of Functions,* vols. 1 and 2, Reading, MA: Addison-Wesley, 1969.

[11] U. E. Ramer, "An iterative procedure for the polygonal approximation of plane curve," *CGIP,* vol. 1, pp. 244–256, 1972.

[12] T. Pavlidis and S. L. Horowitz, "Segmentation of plane curves," *IEEE Trans. Computers,* vol. C-23, no. 8, pp. 860–870, 1974.

[13] E. C. Greanias, P. E. Meagher, R. J. Norman, and P. Essinger, "The recognition of handwritten numerals by contour analysis," *IBM J. Res. Dev.,* vol. 7, pp. 2–13, 1963.

[14] J. H. Munson, "The recognition of hand-printed text," *Proc. IEEE Pattern Recogn. Workshop,* Puerto Rico, pp. 115, October 1966.

[15] M. D. Levine, "Feature extraction: A survey," *Proc. IEEE,* vol. 57, pp. 1391–1419, August 1969.

[16] F. P. Preparata and M. I. Shamos, "Computational Geometry," New York: Springer-Verlag, 1988.

[17] R. L. Graham, "An efficient algorithm for determining the convex hull of a finite planar set," *In for. Proc. Lett.,* vol. 1. pp. 132–133, 1972.

[18] K. Yamamoto and S. Mori, "Recognition of hand-printed character by outer most point method," *Proc. 4th IJCPR,* pp. 794–796, 1978.

[19] T. Saito, K. Yamamoto, and H. Yamada, "Polygonal approximation based on extreme point list method," *Proc. Annual Conf. IECE* in Japan. no. 1692, 1985.

[20] F. Attneave, "Some informational aspects of visual perception," *Psychol. Rev.,* vol. 61, pp. 183–293, 1954.

[21] H. Terasaka, *Basic Geometry,* Tokyo: Iwanami, 1973

[22] H. Freeman and L. S. Davis, "A corner finding algorithm for chain coded curves," *IEEE Trans. Compt.,* vol. C-26, pp. 297–303, 1977.

[23] A. Rosenfeld and E. Johnston, "Angle detection on digital curve," *IEEE Trans. Compt.,* vol. C-22, pp. 875–878, 1973.

[24] C. Teh and R. T. Chin, "On the detection of dominant points on digital curves," *IEEE Trans. Pattern Anal. Machine Intell.,* vol. 11, no. 8, pp. 859–872, August 1989.

[25] A. Rosenfeld and M. Thurston, "Edge and curve detection for visual scene analysis," *IEEE Trans. Compt.,* vol. C-20, no. 50, pp. 562–569, 1971.

[26] A. Witkin, "Scale-space filtering," in *Proc. 7th Int. Joint Conf. Artificial Intell.,* Kalsrühe, pp. 1019–1021, 1983.

[27] F. C. A. Groan and P. W. Verbeek, "Freeman-code probabilities of object boundary quantized contours," *Comp. Vision, Graphics Image Proc.,* vol. 7, pp. 391–402, 1978.

[28] L. S. Davis, "Understanding shape: Angles and sides," *IEEE Trans. Compt.,* vol. C-26, pp. 292–299, March 1977.

[29] A. Rosenfeld and J. S. Weska, "An improved method of angle detection on digital curve," *IEEE Trans. Comput.,* vol. C-24, pp. 940–941, September 1975.

[30] P. V. Sanker and C. V. Sharma, "A parallel procedure for the detection of dominant points on a digital curve," *Comp. Graphics Image Proc.,* vol. 7, pp. 403–412, 1978.

[31] I. M. Anderson and J. C. Bezdek, "Curvature and tangential deflection of discrete arcs: A theory based on the commutator of scatter matrix pairs and its application to vertex detection in planar shape data," *IEEE Trans. Pattern Anal. Machine Intell.,* vol. PAMI-6, pp. 27–40, January 1984.

[32] G. Medioni and Y. Yasumoto, "Corner detection and curve representation using cubic B-splines," *Comp. Vision, Graphics, Image Proc.,* vol. 39, pp. 267–278, 1987.

[33] S. Kobayashi, *Differential Geometry of Curve and Surface,* Tokyo: Shoukabou Publishing, 1969.

[34] M. E. Morterson, *Geometric Modeling,* New York: Wiley, 1985.

[35] G. Farin, *Curves and Surfaces for Computer Aided Geometric Design,* 3rd ed., New York: Academic Press, 1993.

[36] J. Hoschek and D. Lasser, *Computer Aided Geometric Design,* A. K. Peters, Wellesley, Massachusetts Ltd., 1993.

[37] Y. Yasumoto, "Corner detection using cubic polynomial with subpixel accuracy," Matsushita Electric Engineering Documentation, 1985.

[38] H. Yamada, "Continuous nonlinearity in character recognition," *IEICE Trans. Info. Syst.,* vol. E79-D, no. 5, pp. 423–428, May 1996.

[39] I. Sekita, K. Toraichi, R. Mori, K. Yamamoto, and H. Yamada, "Feature extraction of handwritten Japanese characters by spline functions for relaxation matching," *Pattern Recogn.,* vol. 21, no. 1, pp. 9–17, 1988.

[40] I. J. Schoenberg, "Spline functions and the problem of graduation," *Proc. National Academy of Science of the USA,* vol. 52, pp. 947–950, 1964.

[41] C. H. Reinsh, "Smoothing by spline functions" *Numerical Matematik,* vol. 10, pp. 177–183, 1967.

[42] K. T. Miura, R. Sato, and S. Mori, "A method of extracting curvature feature and its application to handwritten characters recognition," *Proc. ICDAR '97,* August 1997.

[43] P. Saing-Marc, H. Rom, and G. Medioni, "B-spline contour representation and symmetry detection," *IEEE Trans. Pattern Anal. Machine Intell.,* vol. 15, no. 11, pp. 1191–1197, November 1993.

[44] O. A. Zaniga and R. M. Haralick, "Corner detection using the facet model," *Proc. IEEE Conf. Comp. Vision Pattern Recogn.,* Washington, DC, pp. 30–37, July 1983.

[45] R. Bellman, "On the approximation of curves by line segments using dynamic programming," *Common ACM,* vol. 4, no. 6, p. 284, June 1961.

[46] I. Tomek, "Two algorithms for piecewise linear continuous approximation of functions of one variable," *IEEE Trans. Compt.,* vol. C-23, pp. 445–448, 1974.

[47] C. M. Williams, "An efficient algorithm for the piecewise linear approximation of planar curves." *CGIP,* vol. 8, pp. 286–293, 1978.

[48] C. M. Williams, "Banded straight-line approximation of digitized planar curve," *CGIP,* vol. 16, pp. 370–381, 1981.

[49] J. Sklansky and V. Gonzalez, "Fast polygonal approximation of digitized curves," *Pattern Recogn.,* vol. 12, pp. 327–331, 1980.

[50] K. Wall and P.-E. Danielsson, "A fast sequential method for polygonal approximation of digitized curves," *CGIP,* vol. 28, pp. 220–227, 1984.

[51] Y. Kurozumi and W. A. Davis, "Polygonal approximation by the minimax method," *CGIP,* vol. 19, pp. 248–264, 1982.

[52] N. Ansari and E. Delp, "On detecting dominant points," *Pattern Recogn.,* vol. 24, no. 5, pp. 441–451, 1991.

[53] D. Marr, *Vision,* New York: W.H. Freeman, 1982.

[54] K. S. Fu, "Syntactic pattern recognition and applications," Englewood Cliffs, NJ: Prentice-Hall, 1982.

[55] J. C. Simmon, "Pattern and Operators," McGraw Hill, NY, 1986.

[56] T. Pavlidis, *Structural Pattern Recognition,* New York: Springer-Verlag, 1977.

[57] R. N. Haver and L. R. Haber, "Visual components of the reading process," *Visible Lang.,* vol. XU2, pp. 147–172, 1981.

[58] P. Dunn-Rankin, "The similarity of lowercase letters of English alphabet," *J. Verbal Learning and Verbal Behavior,* vol. 7, pp. 990–995, 1968.

[59] H. F. Feng and T. Pavlidis, "Decomposition of polygons into simpler components: Feature generation for syntactic pattern recognition," *IEEE Trans. Comput.,* vol. C-24, no. 6, pp. 636–649, June 1975.

[60] L. G. Shapiro and R. M. Haralick, "Decomposition of two-dimensional shapes by graph-theoretic clustering," *IEEE Trans. Pattern Anal. Machine Intell.,* vol. 1, no. 1, pp. 10–20, January 1979.

[61] C. Guerra and G. C. Pieroni, "A graph-theoretic method for decomposing two-dimensional polygonal shapes into meaningful parts," *IEEE Trans. Pattern Anal. Machine Intell.,* vol. PAMI-4, no. 4, July, 1982.

[62] C. Bjorklund and T. Pavlidis, "Global shape analysis by κ-syntactic similarity," *IEEE Trans. Pattern Anal. Machine Intell.,* vol. PAMI-3, no. 2, March 1981, pp. 144–155.

[63] S. Mori, Y. Nakajima, and H. Nishida, "A grey zone between two classes—Case of smooth curvature change," *IECE Trans. Inf. Syst.,* vol. E79-D, no. 5, pp. 477–484, May 1996.

[64] K. Ishii, "Generation of distorted characters and its application," *IECE Trans.*, vol. J66-D, no. 11, pp. 1270–1277, 1983.

[65] H. Freeman, "On the digital computer classification of geometric line patterns," *Proc. Nat. Elect. Conf.*, vol. 18, pp. 312–324, 1962.

[66] G. Gallus and P. W. Neurath, "Improved computer chromosome analysis incorporating preprocessing and boundary analysis," *Phys. Med. Biol.*, vol. 15, pp. 435–445, 1970.

[67] H. Freeman and L. S. Davis, "A corner-finding algorithm for chain-coded curves," *IEEE Trans. Comp.*, vol. C-26, pp. 297–303, March 1997.

[68] L. O'Gorman, "An analysis of feature detect ability from curvature estimation," *Proc. Comp. Vision Pattern Recogn.*, pp. 235–240, June 1988.

[69] Y. Shirai, "Anlyzing intensity arrays using knowledge about scenes," in P. H. Winston, ed., *The Psychology of Computer Vision,* New York: McGraw-Hill, pp. 93–96, 1975.

[70] H. Asada and M. Brady, "The curvature primal sketch," *IEEE Trans. Pattern Anal. Machine Intell.*, vol. PAMI-8, no. 1, pp. 2–14, January 1986.

[71] A. Witkin, "Scale-space filtering," *Intell.*, Kalsrühe, West Germany, pp. 1019–1021, 1983.

[72] W. S. Rutkowski and A. Rosenfeld, "A comparison of corner detection techniques for chain-coded curves," Technical Report TR-623, Computer Science Center, University of Maryland, 1978.

[73] H. C. Liu and M. D. Srinath, "Corner detection from chain-code," *Pattern Recogn.*, vol. 23, no. 1–2, pp. 51–68, 1990.

[74] H. L. Beus and S. S. Tiu, "An improved corner detection curves," *Pattern Recogn.*, vol. 20, pp. 291–296, 1987.

[75] F. Cheng and W. Hsu, "Parallel algorithm for corner finding on digital curves," *Pattern Recogn. Lett.*, vol. 8, pp. 47–53, 1988.

[76] R. Legault and C. Y. Suen, "A comparison of methods of extracting curvature features," Proc. 11th IAPR, vol. 3, The Hague, August 30–September 3, pp. 134–138, 1992.

[77] D. S. Lee, S. W. Lam, and S. N. Srihari, "A structural approach to recognize hand-printed and degraded machine printed characters," Pre-Proc. IAPR Workshop on Syntactic and Structural Pattern Recognition, Murray Hill, NJ, pp. 256–272, June 1990.

[78] R. Legault and C. Y. Suen, "Contour tracing and parametric approximations for digitized patterns," in A. Krzyzak, T. Kasvand, and C. Y. Suen, eds; *Computer Vision and Shape Recognition,* pp. 225–240, Singapore: World Scientific., 1989.

[79] T. Kadonaga and K. Abe, "Comparison of methods for detecting corner points from digital curves," in R. Katuri and K. Tombre, eds., *Graphics Recognition Method and Applications,* pp. 23–34, 1995, First International Workshop, University Park, PA, August 10–11, 1995.

[80] G. Gallus and P. W. Neurath, "Improved computer chromosome analysis incorporating preprocessing and boundary analysis," *Phys. Med. Biol.*, vol. 15, pp. 435–445, 1970.

[81] T. Ibakaki, BS thesis, Dept. of Computer Science, Shizuoka University, 1991.

[82] T. Koyama, M. Shiono, H. Sanada, and Y. Tezuka "Corner detection on thinned pattern," IEICE Technical Report IE80–119, 1981.

[83] C. Arcelli, A. Held, and K. Abe, "A coase to fine corner-finding method," *Proc. IAPR Workshop on Machine Vision Applications,* pp. 427–430, 1990.

[84] M. A. Fishler and R. C. Bolles, "Perceptual organization and curve partioning," *IEEE Trans. PAMI,* vol. PAMI-8, pp. 100–105, 1986.

[85] A. Held, K. Abe, and C. Arcelli, "Towards a hierarchical contour description via dominant point detection," *IEEE Trans. SMC,* vol. 24, pp. 942–949, 1994.

[86] M. Worring and A. W. M. Smeulders, "Digital curvature estimation," *CVGIP: Image Understanding,* vol. 58, no. 3, pp. 362–382, November 1993.

[87] J. S. Duncan, F. Lee, A. W. M. Smeulder, and B. L. Zaret, "A bending energy model for measurement of cardiac shape deformity," *IEEE Trans. Med. Imaging,* vol. 10 (3), pp. 307–320, 1991.

[88] L. T. Young, J. E. Walker, and J. E. Bowie, "An analysis technique for biological shape I," *Info. Control,* vol. 25, pp. 257–370, 1974.

[89] T. Taxt, T. B. Ólatsdóttir, and M. Daohen, "Recognition of hand written symbols," *Pattern Recogn.,* vol. 23, no. 11, pp. 1155–1166, 1990.

[90] P. R. Beaudet, "Rotationally invariant image operators," 4th IJCPR, Kyoto, pp. 579–583, November 1978.

[91] L. Kichen and A. Rosenfeld, "Gray level corner detection," Technical Report 887, Computer Science Center, University of Maryland, College Park, MD20742, April 1980.

[92] L. Dreschler and H. Nagel, "Volumetric model and 3D—Trajectory of a moving car derived from monacular TV-frame sequence of a street scene," *IJCAI,* pp. 692–697, 1981.

[93] R. Haralick, "Digital step edges from zero crossing of second directional derivatives," *IEEE Trans. Pattern Anal. Machine Intell.,* vol. PAMI-5, no. 1, pp. 58–66, January 1982.

[94] O. A. Zuniga and R. M. Haralick, "Corner detection using the focet model," *Proc. Comp. Vision Pattern Recogn.,* pp. 30–37, 1983.

[95] Z. Q. Wu and A. Rosenfeld, "Filtered profections as an aid in corner detection," *Pattern Recogn.,* vol. 16, no. 1, pp. 31–38, 1983.

[96] L. Y. Watson, T. J. Laffey, and R. M. Haralick, "Topographic classification of digital image intensity surface using gemeralized splines and the discrete cosine transformation," *Comp. Vision, Graphics Image Proc.,* vol. 29, pp. 143–167, 1985.

[97] R. M. Haralick, L. T. Watson, and T. J. Laffey, "The topographic primal sketch," *Int. J. Robot. Res.,* vol. 2, pp. 50–72, 1983.

[98] L. Wang and T. Pavlidis, "Direct gray-scale extraction of features for character recognition," *IEEE Trans. Pattern Anal. Machine Intell.,* vol. 15, no. 10, pp. 1053–1067, October 1993.

[99] S. Mori and T. Sakakura, *Fundamentals of the Image Recognition I,* Tokyo: Ohm, 1986.

[100] S. Mori and T. Sakakura, *Fundamentals of the Image Recognition II,* Tokyo, Ohm, 1990.

[101] K. Yamamoto, "Research on handprinted character recognition based on structural approach," ETL, Research Report No. 831, February, 1983.

CHAPTER NINE

Algebraic Description

This chapter presents a systematic shape description technique, called algebraic description. We begin with Show's picture description language and some simpler toy models of the algebraic description scheme. After that we focus on a primitive that we use as a base of the description and illustrate some theoretical background. Based on the preparation, a complete algebraic description scheme will be presented, from both theoretical and practical points of view. Finally some experimental results will be given.

9.1 INTRODUCTORY CONSIDERATIONS

9.1.1 Operators of Primitives

For a while we direct our attention to operators between primitives in order to see how the shape is constructed in an algebraic scheme. An algebraic system is constructed by defining a set X and some operators on X so that the operation on X is closed. This is very simple description of an algebraic system and enough to understand the basic idea of the algebraic description. Such a scheme was introduced first by Show [1] and was known as *picture descriptive language* (PDL). He abstracted primitives to very simple line and curve segments. This is shown in Fig. 9.1 for a line segment primitive. That is, a primitive is abstracted by an arrowed line segment, extended from one end point, called the tail, to the other end point, called the head. These head and tail are the marks that work well in the operation, as seen below. Show defined four kinds of operators, $\{+, \times, -, *\}$. The definitions of these operators are described as follows:

$a + b$: connect head of a to tail of b and keep tail of a and head of b.

$a \times b$: connect tail of a to tail of b and keep head and tail of b.

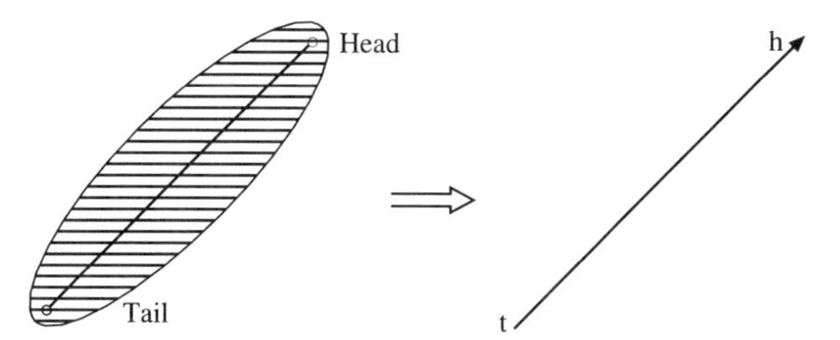

FIGURE 9.1 Show's simplification of a primitive and its attribute, head and tail.

$a - b$: connect head of a to head of b and keep tail of a and head of connected point.

$a * b$: connect tail of a to tail of b and connect head of a to head of b (keep head and tail of connected points).

These definitions are illustrated in Fig. 9.2. Thus, taking appropriate abstracted primitives, we can construct any line drawing. We don't pursue this topic further. Instead, we examine some toy models of algebraic structure of shape to understand the nature of the approach.

9.1.2 Algebraic Systems of Shape

Here we give two kinds of toy models and a very flexible model.

Toy Models Our basic assumption is that the alphabet is constructed by some implicit principles that are not primarily conscious. In other words, it reflects some human reasoning, algebraic or otherwise. This notion was originated by the famous anthropologist Claude Lévi-Strauss [2]. An obvious example of such an alphabet, known as Ogham characters, is shown in Fig. 9.3. It was used by the ancient Celts in writing their language and found in the British Isles only in the fifth century [3]. In contrast, arabic numerals are widely used throughout the world, so it is not easy to find a construction rule for the shape of each character. However, we can perceive some common property in arabic numerals in terms of their shape. What is it? This is our motivation.

To see this more clearly, two kinds of toy models are introduced. One is geometric and the other quasi-topological. For the former, only four kinds of generator are introduced, $\{|, -, /, \backslash\}$. The operation + is a simple superposition, assuming that each generator has a marked center point as shown in Fig. 9.4. In this figure all the elements that belong to the algebraic system and multiple table are shown. This algebraic system has no identity and is called a semigroup. Let us examine the automatically generated elements. Whether they are actually used depends on human nature. As is well-known, and a subject eloquently expounded by Herman Weyl, Nature favors symmetry [4]. This is true too of human nature. Actually, among all the elements, the following shapes are used: $\{|, -, \backslash, /, \times, +, *\}$. However, the last three

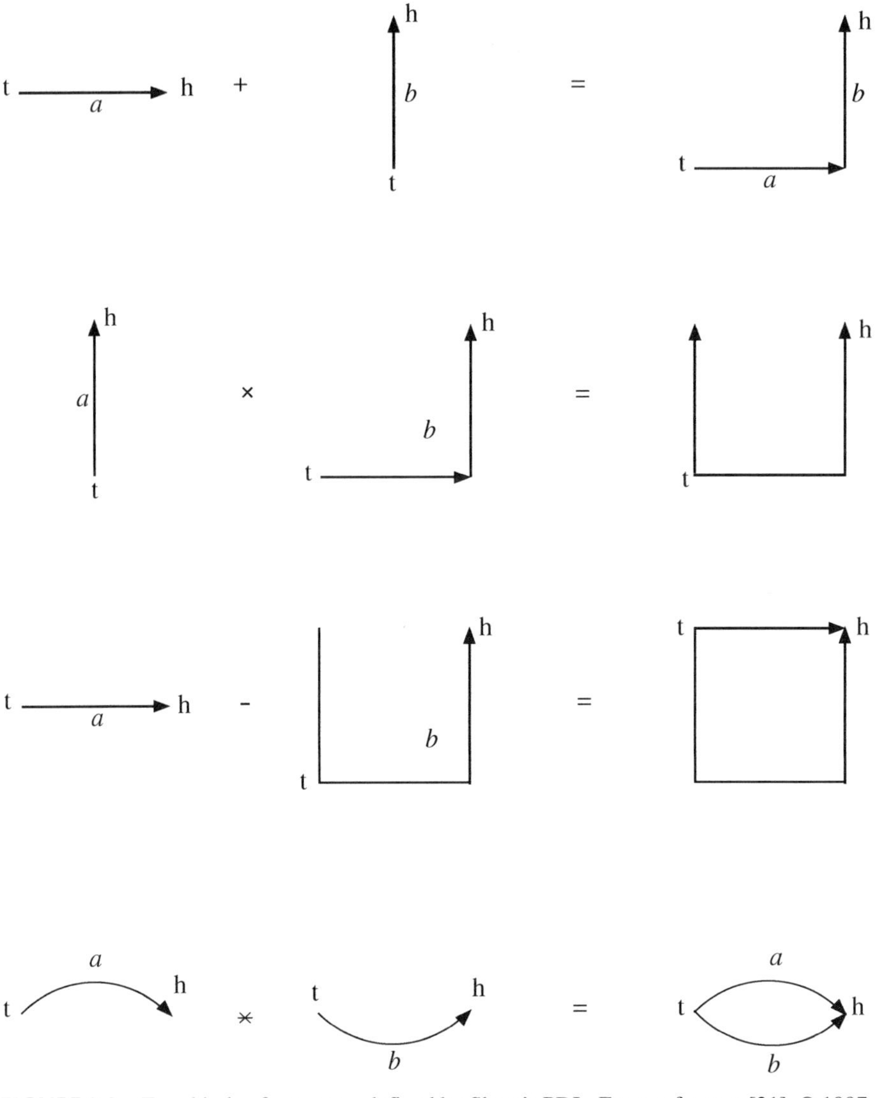

FIGURE 9.2 Four kinds of operators defined by Show's PDL. From reference [21], © 1997, Ohm.

shapes in the list of all the elements shown in Fig. 9.4 are used as "stars" in hand. Certainly they can be confusing to the human eye, and this was confirmed by the handprinted data base ETLI [5]. On the other hand, we notice that an extended algebraic model can be construed introducing head and tail devices as shown in Fig. 9.5, where a head is marked by a circle. In this algebraic system we can generate more shapes as shown. The semigroup described above is a proper partial set of the extended system.

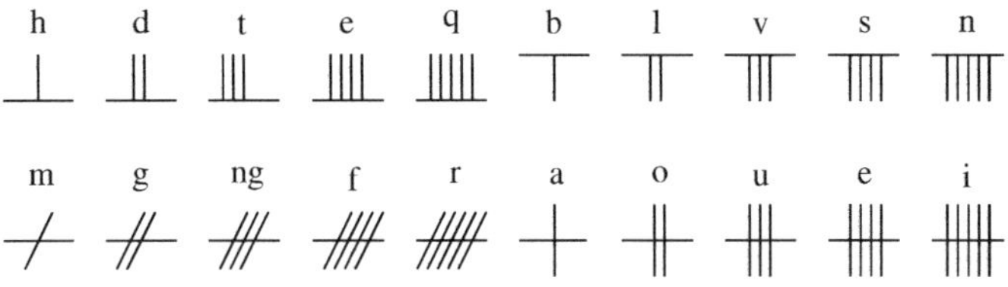

h d t c q b l v s n

m g ng f r a o u e i

FIGURE 9.3 Ogham character set. From reference [3].

Now let us move to the quasi-topological case. Among the elements generated by the extended system, we pick out so-called L type generators and define the other "star" operators such that two elements are superposed so that one concavity is generated as shown in Fig. 9.6. For generators of L type, all the elements generated by the algebraic system and multiple table are shown in the figure. The algebraic system is also a semigroup. In this case all the elements is used, although the number of the elements are small. Therefore the L-type generator and the quasi-topological system seem to agree with human reasoning patterns. At any rate this point will be elaborated more fully in the next section. After all, using both the geometrical and quasi-topological system, we can construct all the alphanumeric character shapes based on our primitive shape, as shown in Fig. 9.5.

An Extension of the Toy Model Here we focus on a natural extension of the quasi-topological toy model. Since the L-type generator was produced by the geometrical

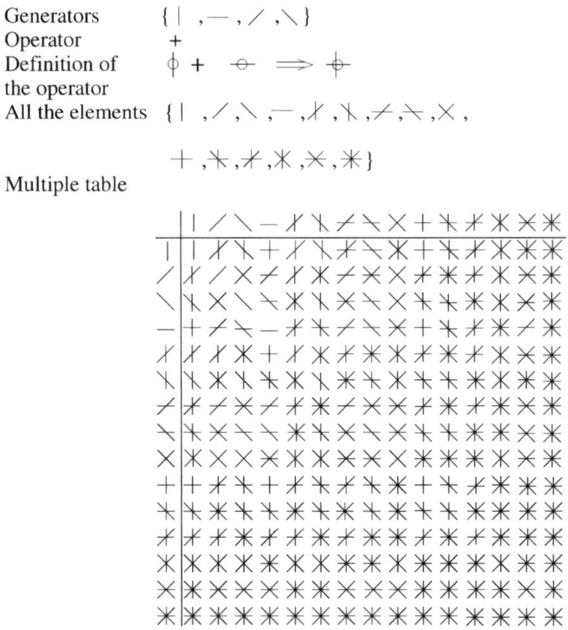

Generators

Operator +

Definition of the operator

All the elements

Multiple table

FIGURE 9.4 Basic geometrical algebraic system. From reference [21], © 1997, Ohm.

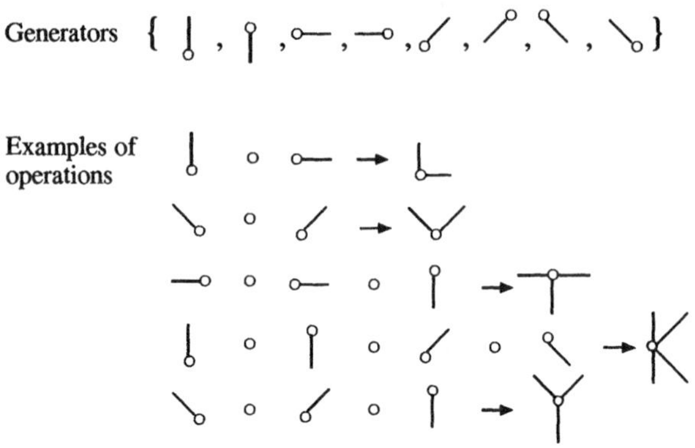

Generators

Examples of operations

FIGURE 9.5 Extension of the basic geometrical algebraic system. From reference [21], © 1997, Ohm.

algebraic system, we might take it to be the basis of further consideration. However, we have to take a leap from toy to real model in contemplating the primitives. In the case of Shaw, too much attention was paid to the operators and the primitive was far too abstracted so that nothing but line segments or curve segments remained. We proceed to investigate what should be primitive after the construction of an algebraic structure in order to find what type of primitive is promising. Let us recall the conditions of primitives given by Pavlidis. Primarily the primitive must be mathematically operational.

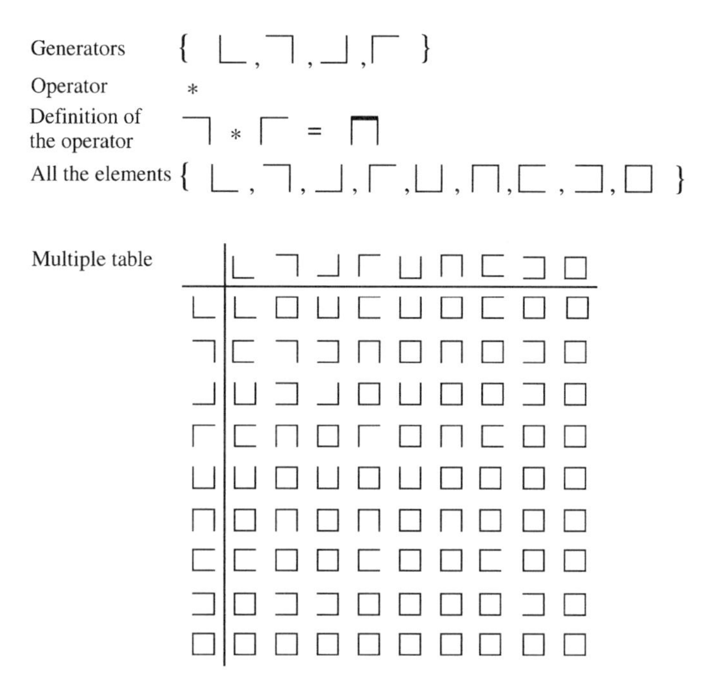

FIGURE 9.6 Basic quasi-topological algebraic system. From reference [21], © 1997, Ohm.

From this consideration, four kinds of monotone functions were considered as primitives [7]. The important point was to capture the infinite variation of shape. Each primitive had to include an infinite number of primitive shapes that satisfy the simple mathematical property of a monotone function. The monotone function has also the good mathematical property that neither maximum nor minimum points hold:

Theorem 9.1.1 *Let f be a real-valued continuous function defined on the interval $I = [a, b]$, and suppose that f is one-to-one on I. Then f is strictly monotonic on I.*

Proof For a proof, any book on advanced calculus can be consulted; for example, see [6]. ■

In other words, within the family of strictly monotonic functions each has its inversion function. This naturally allows segmentation of a continuous function by monotonic functions. Now we define four kinds of primitives corresponding to *L-type primitives* as follows:

Definition 9.1.1

$L0$: *monotonic descending contour from left to right.*

$L1$: *monotonic descending contour from right to left.*

$L2$: *monotonic ascending contour from right to left.*

$L3$: *monotonic ascending contour from left to right.*

These L-type primitives are schematically shown in Fig. 9.7. All the contours are segmented by these four primitives. We assume that a contour of the shape is traced along as if stroked by the right hand, namely by a right-hand system. We used a similar description for the grid plane but not for a Euclidean plane. A horizontal line segment,

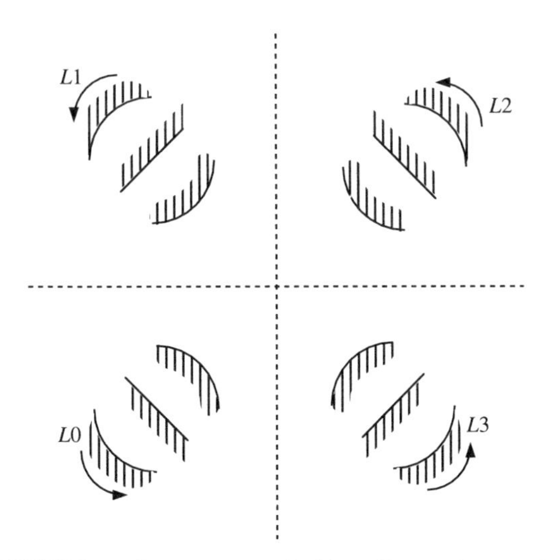

FIGURE 9.7 Some L-type primitives. From reference [7].

for example, is described by the four primitives as shown in Fig. 9.8. In this case there is considerable overlap between primitives, so the description is not effective for such a simple line segment. One idea is to introduce other primitives for horizontal and vertical line segments in order to apply the description scheme to abstracted lines on a Euclidean plane. The algebraic expression will be described in Section 9.2.

Construction of an Algebraic System Before proceeding further, we need to know more about the primitives. In the primitives the metric of length is disregarded. (Of course we can attach this attribute, and it actually helps in the description, since it is only disregarded at the first level.) Instead, for each primitive, the important measure is that of the angle. For example, a transition of $L2$ to $L1$ means rotation by 90° degree. Therefore a successive transition of $L2$ to $L1$, $L0$, $L3$, and $L2$, in tracing closed convex contour from a starting point to an end point that coincides with the starting point means one revolution, $4 \times 90°$ rotation (this meets our intuition if we imagine a convex simply connected shape). This interpretation suggest us that we should remove symbol L from L-type symbol notation Li. That is, we can treat L-type primitives as a kind of number system.

First we notice that the following property:

Property 9.1.1 *Let $\{i\}$ be a function family of type Li, then $\{i\} \cap \{j\} = \phi$, $i \neq j$, $i, j =$ $0 \sim 3$. That is, each family is disjoint to each other, which is a necessary condition of the families of primitives.*

Property 9.1.2 *The function family i can be connected to $i \pm 1 \pmod 4$ but cannot be connected to $i \pm 2$.*

This property is illustrated in Fig. 9.9. For example, if we want to connect primitive $L0$ to $L2$, then there exists always $L1$ between them. This restriction does not apply to the Euclidean plane. Further the relation between $\boldsymbol{i}$ and $i \pm 2$ is very important, as will be explained later, and this leads us to the following unary operation $\bar{i}$, called the *conjugate operation:*

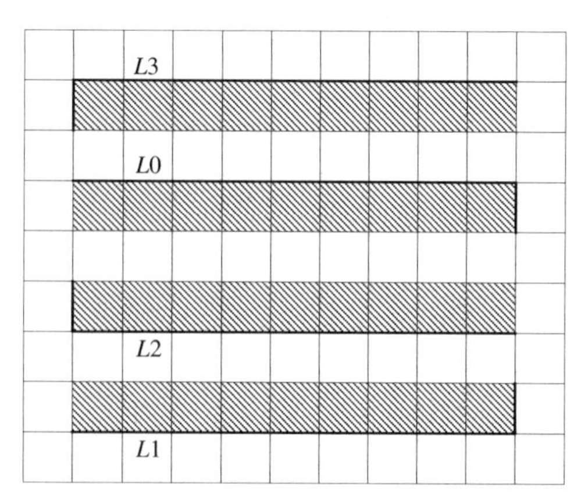

FIGURE 9.8 Four L-type primitives represented by a horizontal line. From reference [7].

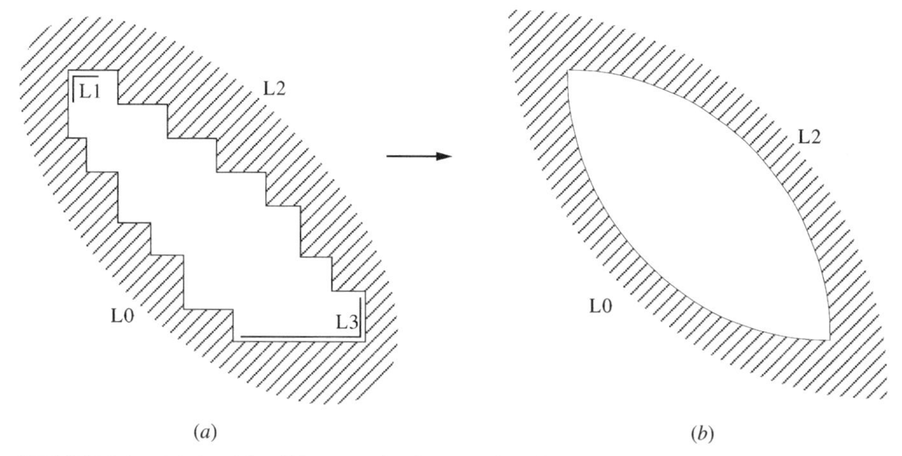

FIGURE 9.9 Noticeable differences in "connections." In both the grid and the Euclidean plane, for $L0$ and $L2$ to connect, for example, $L1$ must mediate between them. Shown is the case where $L0$ and $L2$ can be connected directly on both planes. Notice that in the grid plane, $L1$ exists between $L0$ and $L2$. This is the exact meaning of "cannot be connected." On the other hand, in the case of the Euclidean plane the length of $L1$ becomes infinitesimally small, and so can be neglected as an extreme case. From reference [7].

Definition 9.1.2

$$0 = \overline{2}, \quad \overline{0} = 2, \quad 1 = \overline{3}, \quad \overline{1} = 3.$$

As usual conjugate operation $\overline{\overline{i}} = i$ holds. Here it is appropriate to give some remarks on the binary operator +. The operator + maps an integer domain $\{0, 1, 2, 3\}$ onto the same domain based on mod 4. The + operator should not be considered as a "connection" operator but as a plus operator on the domain of quantized angles by 90°. For the "connection" another binary operator is introduced below. So far by these properties and definitions, the following property holds:

Property 9.1.3 *If $i = \overline{j}$, then for any integer k,*

$$i + k = \overline{j + k} \tag{9.1}$$

holds.

Now we are ready to describe a contour in an algebraic manner. To do so, we introduce two kinds of binary operators, denoted as $*$ and $\overline{*}$. They are defined as follows:

Definition 9.1.3

$*$: *$i * j$ denotes that i is connected to j with a right-hand system.*

$\overline{*}$: *$i \,\overline{*}\, j$ denote that i is connected to j with a left-hand system.*

Notice here the property 9.1.5. These examples are shown in Fig. 9.10. By all the cases are shown in the figure, and by the nature of monotone functions, the binary operators have the following properties:

Property 9.1.4 *The idempotence law holds,*

$$i * i = i, \tag{9.2}$$

$$i \mathbin{\bar{*}} i = i. \tag{9.3}$$

Property 9.1.5 *The commutative law does not hold, but the following modified commutative law holds:*

$$i * j = j \mathbin{\bar{*}} i. \tag{9.4}$$

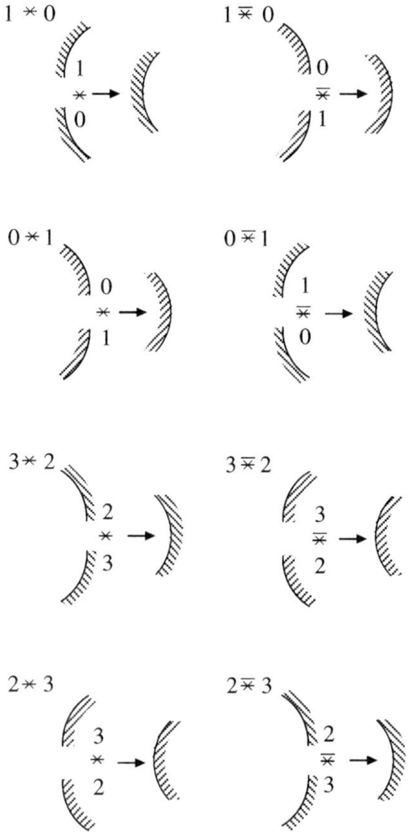

FIGURE 9.10 Two kinds of operators $*$ and $\bar{*}$. From reference [7].

Property 9.1.6 *The associative laws hold,*

$$(i * j) * k = i * (j * k), \tag{9.5}$$

$$(i \mathbin{\overline{*}} j) \mathbin{\overline{*}} k = i \mathbin{\overline{*}} (j \mathbin{\overline{*}} k). \tag{9.6}$$

Description of a Contour Based on the preceding description, we denote a closed contour as

$$(ijk \ldots)(*|\overline{*}).$$

If necessary, the binary operators are used explicitly as

$$(i * j * k * \ldots * n *).$$

That is, this is a circular expression.

Now we can describe some general properties of a contour. First, we show a role of conjugate operator. Imagine a simply connected shape F.

Property 9.1.7 *We assume that a simply connected shape F is expressed as*

$$F = (a * b * \ldots * p * g * r *), \tag{9.7}$$

Then a hole, which is generated in removing a blob from the grid plane on which it is placed, is represented using conjugate operators as

$$\overline{F} = (\overline{a} \mathbin{\overline{*}} \overline{b} \mathbin{\overline{*}} \overline{c} \mathbin{\overline{*}} \ldots \mathbin{\overline{*}} \overline{p} \mathbin{\overline{*}} \overline{q} \mathbin{\overline{*}} \overline{r} \mathbin{\overline{*}}). \tag{9.8}$$

$\overline{F}$ is also represented using $*$ as

$$\overline{F} = (\overline{r} * \overline{q} * \overline{p} * \ldots * \overline{c} * \overline{b} * \overline{a} *). \tag{9.9}$$

These characteristics of a simply connected shape are shown in Fig. 9.11 and computed as follows:

$$\begin{aligned} A &= (01232301)(*), \\ \overline{A} &= (\overline{0}\,\overline{1}\,\overline{2}\,\overline{3}\,\overline{2}\,\overline{3}\,\overline{0}\,\overline{1})(\overline{*}) \\ &= (23010123)(\overline{*}) \\ &= (32101032)(*). \end{aligned}$$

Next we apply this to an extension of Euclidean geometry. That is, in Euclidean geometry, the sum of the outer angles of a polygon is always $4 \times 90°$ regardless of number of sides. We can see a contour part of our algebraic description system. To see that, we define rotation number RN[1] as

[1] See the note given at the end of this section.

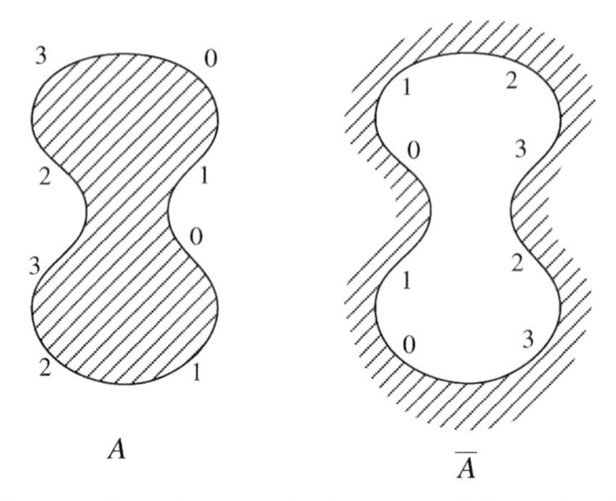

FIGURE 9.11 Correspondence between simply connected figure A and its conjugate one (hole) $\bar{A}$ with their primitive representations shown along the contours. From reference [7].

Definition 9.1.4

$$RN = \sum_{i=1}^{n} (a_{i+1} - a_i), \tag{9.10}$$

where $a_{n+1} = a_1$.

Here the binary operator "–" is different from the "+." In case of "+," it is a mapping from G to G, where $G = \{0, 1, 2, 3\}$. All the combination of the operations can be represented by $G \times G$, where X is direct sum. The $G \times G$ is mapped to $\{-2, -1, 0, 1, 2\}$ by the "–" operator. Therefore we need some premise that $3 - 0 = -1$ and $0 - 3 = 1$, between circularly neighboring numbers. Otherwise, an ordinal "–" operation is applied. Because of the property 9.1.2, the actual domain of the value $a_{i+1} - a_i$ is limited to $\{1, -1\}$, which has an important function as we will see later. Now we can prove the following theorem:

Theorem 9.1.1 *RN is always 4 for any contour of a simply connected shape.*

The exact proof of the theorem is lengthy, and thus it is omitted here. Mathematical induction is effectively used in the proof. See [7].

As suggested above, the series of $a_{i+1} - a_i$ gives useful features of a contour. For example, if we consider the series $a_{i+1} - a_i$ of Fig. 9.12,

$$1 \; -1 \; 1 \; 1 \; -1 \; 1 \; 1 \; -1 \; 1 \; 1 \; -1 \; 1.$$

we can count consecutive 1's and –1's, which is 2 and 1, respectively. The 1's and –1's just correspond to the convex and concave parts, respectively. Recall that the exact meanings of convexity/concavity is what we can detect the changes when scanning a shape horizontally and vertically, as shown in the figure. We described this sort of concavity in the stream-following method of Chapter 8. Some other examples

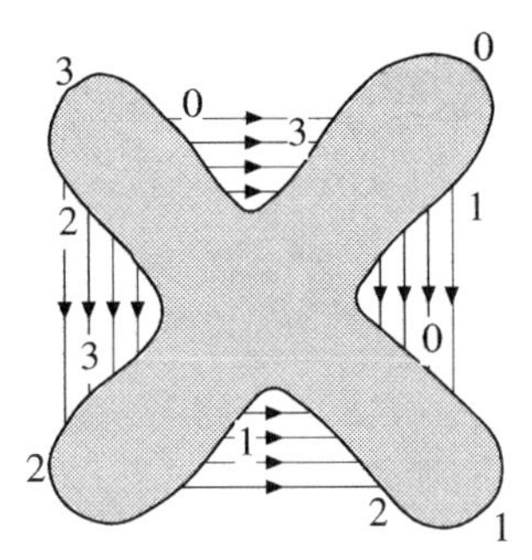

FIGURE 9.12 Shallow concavities and an illustration of what is meant by "concavity." From the reference [21], © 1997, Ohm.

are shown in Fig. 9.13. Both panels (*a*) and (*b*) have two convex and two concave parts. Panel (*c*) has one convex and one concave part. Here we notice that in these examples, the concave run, namely the consecutive number of −1's, is larger than that of Fig. 9.12 which has only 1's. In case of Fig. 9.13 (*a*), (*b*), it is 2, and in (*c*), it is 4. This number can be regarded as a measure of complexity of the concavity. In this sense we can call it the *degree of concavity*. The degree 1 is a weak concavity, so the concavity may be lost in a rotation. A degree greater than 1 is strong and retains its shape in any rotation. The above discussion is described as follows:

Definition 9.1.5 *The number (−1) of each consecutive −1 part in the series of* $a_{i+1} - a_i\,(i = 0 \sim n)$ *when the contour is represented as* $(a_0, a_1, \ldots, a_i, \ldots, a_n)$ *is defined as the degree of concavity of that part.*

Theorem 9.1.2 *The concavity whose degree of concavity complexity is larger than 1 keeps its concavity under rotation in such algebraic representation.*

The theorem is obtained easily by thinking about the strongest shape among *L*-type primitive, which is the exact *L* shape. The *L* shape can be described by two *L*-type primitives as a *V* shape at most by rotation. Therefore even the strongest *L*-type primitives can change their representation to two primitive concave representations at most. On the contrary, two primitive concavity representations can be changed to one primitive concave representation by rotation and never be lost. The concavity whose degree is larger than 1 spans three primitives at least, so it cannot have less than two concavity primitives in its representation. Therefore the theorem is proved. By the way, the strongest shape in terms of the concavity of two primitive representations is the exact ⊔ shape. The reader can obtain some theoretical insight to the roman letters *L, V,* and *U*.

As mentioned before, in general, an algebraic representation of a shape's contour changes its representation under rotation. However, for $n \times 90°$ quantized rotation, the change is described simply as

$$(a_0, a_1, \ldots, a_i, \ldots, a_n) \rightarrow (a_0 + n, a_1 + n, \ldots, a_i + n, \ldots, a_n + n), \quad (9.11)$$

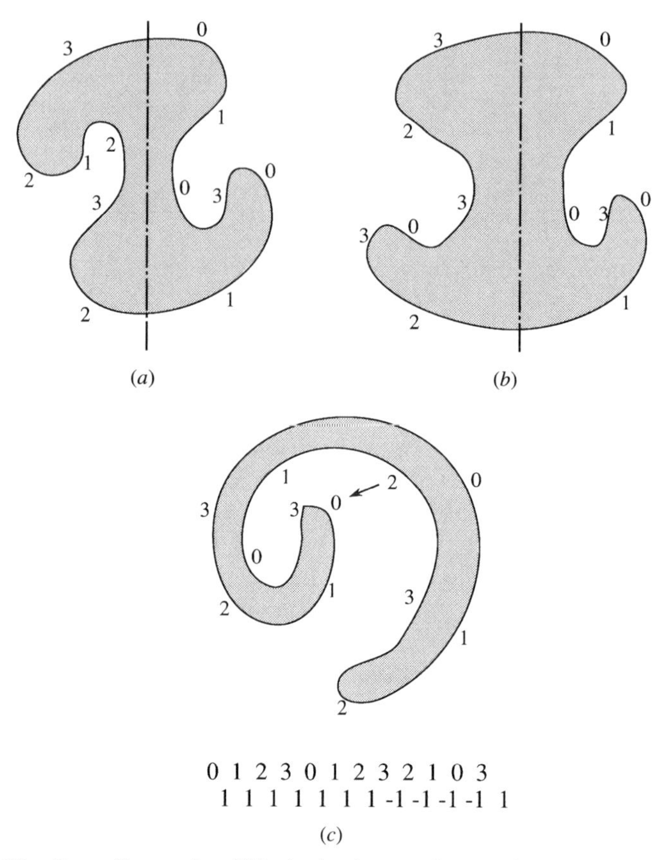

FIGURE 9.13 Some figures that differ in the degree of concavity: (*a*) Each concavity has degree of concavity of 3 and origin symmetry; (*b*) each concavity has degree of concavity of 3 and vertical symmetry; (*c*) one convexity and one concavity whose degree is 4. Calculation of the degree of concavity is shown in (*c*). From reference [21], © 1997, Ohm.

where the operator + means mod 4. Let us now consider symmetry. Let F be given as $F = (A, B)$, in which the number of primitives (degree) of A is equal to that of B. If B is equal to A's 180° rotation, then F is regarded as having symmetry about the origin. Specifically this is the same as the condition that $b_i = \bar{a}_i$ for every a_i and b_i, where a_i and b_i are corresponding primitives of A and B, respectively. An example is shown in Fig. 9.13 (*a*). Naturally this property must be invariant for any rotation in general. However, in our case it is proved that it holds for any $n \times 90°$ rotations. This is guaranteed by the property 9.1.3 that $i + l = \overline{j + k}$ for $i = \bar{j}$. We can give the conditions of symmetry about the vertical and horizontal center lines. As described above, a contour F is represented by (A, B). The degree of A is equal to that of B. We set B's primitive order in reverse. Then we consider the correspondences between A and the reversed B primitives. If the correspondences $0 \leftrightarrow 3$ and $1 \leftrightarrow 2$ hold about the vertical center line, then the shape is symmetric about the vertical center line. An example is shown in Fig. 9.13 (*b*). Likewise symmetry about the horizontal center line is defined. In this case the correspondences are $2 \leftrightarrow 3$ and $0 \leftrightarrow 1$. The simplest shape (0, 1, 2, 3) has both symmetry properties. These definitions of symmetry are clearly much more abstract than found in Euclidean geometry.

Last we consider an interesting case of a shape whose degree of concavity is 1 but the degree of concavity complexity is greater than 1. This shape is an eddy, which is shown in Fig. 9.14 (*a*). The degree of 2, the number of rotations in the eddy's shape, may seem intuitively possible. However, the shape in panel (*b*) is the same representation as the shape shown in panel (*a*). The difference comes from disregarding length in the representation. If we take length as an attribute of each primitive, then the eddy shape shows a monotonic decreasing sequence of lengths. The shape shown in panel (*b*) has no monotonic property in its length series.

Notice In differential geometry the rotation number m is defined as

$$m = \frac{1}{2\pi} \oint k(s)ds, \tag{9.12}$$

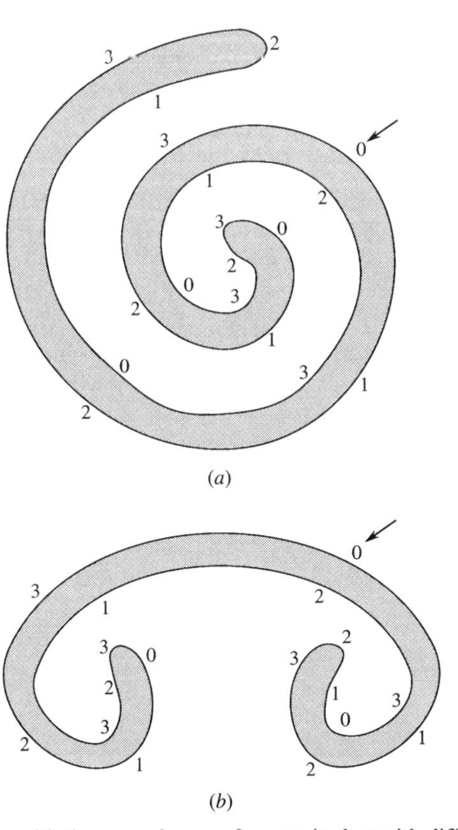

FIGURE 9.14 Figures with the same degree of concavity but with different appearances for humans because of an omission of the primitive's length. The algebraic representation of (*a*) and (*b*) is the same, which is (0123210321323012)(*).

where $k(s)$ is curvature of a given closed curve. The number m is always integral. In particular, if a given closed curve is smooth and has no crossing point, then its rotation number is 1. In this sense our rotation number can be regarded also as a quantized version of the rotation number defined above.

9.2 ALGEBRAIC APPROACH

9.2.1 Algebraic Approach to Shape Description

Now we proceed to the practical and more precise theoretical description. The prime difficulty in handwritten character recognition is the variety of shape deformations. Since machine-printed fonts are designed with some curve-fitting techniques using polynomials, quantitative shape description with straight lines, arcs, and corners is appropriate. However, throughout more than a quarter of a century of research, it is found that such *quantitative* shape description is not appropriate to handwritten characters, but some *qualitative* features are flexible and powerful such as

- Quasi-topological features (convexity, concavity, and loop).
- Directional features (for example, upward, downward, leftward, and rightward).
- Singularities (branch points and crossings).

On the basis of this observation, we describe an algebraic approach to representing character shapes in terms of a qualitative and global structure that is robust against shape deformation.

In handwriting, since different strokes are written independently, strokes have independent information from each other. The first step in analyzing handwritten character shapes is to decompose the line picture into independent parts (strokes). Each stroke is analyzed separately by integrating low-level structural features into high-level features in a systematic way. Then characters are described by few components with this rich and independent information.

Decomposition of Singular Points The output from thinning is a piecewise linear curve expressed in terms of the undirected graph $G = (V, E)$ which is a simple graph (in the sense that it has neither a self-loop nor a multiple edge). The coordinate function.

$$\text{coord} : V \rightarrow R^2$$

is defined for each element v in V, where coord (v) corresponds to the coordinates of v.

The vertex in V whose order (the number of edges incident to the vertex) is equal to or more than 3 is called a singular point of G. For instance, the vertices p, q, and r in Fig. 9.15 are singular points, whose orders are 4 for p and q, and 3 for r.

For each singular point v, we introduce n new vertices v_i such that coord (v_i) = coord (v), $i = 1, 2, \ldots, n$ (n is the order of v). Next for n vertices w_i $(i = 1, 2, \ldots, n)$

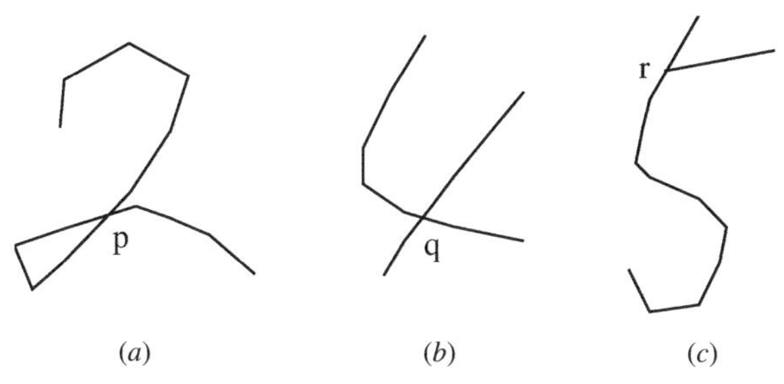

FIGURE 9.15 Thinned line pictures. p, q, and r are singular points of the orders 4, 4, and 3, respectively. From reference [11], © 1998, IEEE.

adjacent to v, we add virtual edges (v_i, w_i) $(i = 1, 2, \ldots, n)$ and remove the vertex v. Though the vertices v_i $(i = 1, 2, \ldots, n)$ have the same coordinates as v, they are regarded as different from each other. After applying this operation, the graph G is decomposed into connected components each of which is a simple arc. Figure 9.16 (*a*) depicts a singular point of the order 5, and Fig. 9.16 (*b*) illustrates the result of the operation mentioned above.

Let coord(v) be (x_0, y_0) and coord(w_i) be (x_i, y_i). Now compute $S(i, j)$ $(i < j,\ i, j = 1, 2, \ldots, n)$:

$$S(i, j) = \frac{\vec{q}_i \cdot \vec{q}_j}{|\vec{q}_i| \cdot |\vec{q}_j|}, \tag{9.13}$$

where

$$\vec{q}_k = (x_k - x_0, y_k - y_0), \qquad k = 1, 2, \ldots, n.$$

Next find the sequence

$$S(i_1, j_1) \leq S(i_2, j_2) \leq \cdots \leq S(i_m, j_m) \tag{9.14}$$

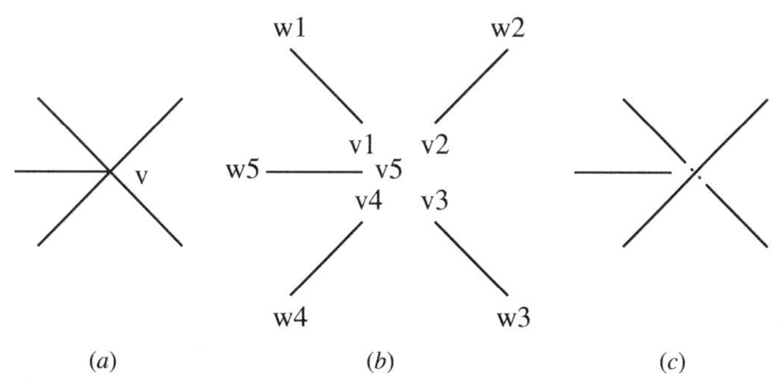

FIGURE 9.16 Decomposition of singular points whose order is five; from the reference [11], © 1998, IEEE.

such that

$$S(i_1, j_1) = \min\{S(i, j) \mid i, j \in I, i < j\},$$

$$S(i_k, j_k) = \min\{S(i, j) \mid i, j \in I - \bigcup_{\lambda=1}^{k-1} \left\{\{i_\lambda, j_\lambda\}, i < j\right\},$$

for $k = 2, 3, \ldots, m$, where $I = \{1, 2, \ldots, n\}$ and $m = \lfloor n/2 \rfloor$ ($\lfloor r \rfloor$ is the greatest integer that is equal to or smaller than r). Then connect a pair of edges (v_{i_k}, w_{i_k}) and (v_{j_k}, w_{j_k}) by regarding the two vertices v_{i_k} and v_{j_k} ($k = 1, 2, \ldots, m$) as identical.

A series of the above operations is called *singular point decomposition.* What decomposes is the vertex v into $\lceil n/2 \rceil$ ($\lceil r \rceil$ is the smallest integer that is equal to or greater than r) vertices, and this generates $\lceil n/2 \rceil$ pairs of edges incident to v. The graph obtained by applying singular point decomposition to the graph G is called the *stroke graph* of G. Each connected component of the stroke graph is called a *stroke,* and it is either a simple arc or a simple closed curve. Therefore the curve is decomposed into unicursal components in terms of the smoothness of directional changes at the singular points.

Figure 9.16 (*c*) illustrates the decomposition of the singular point shown in Fig. 9.16 (*a*). Since $n = 5$ in this case, three pairs of edges $\{1, 3\}$, $\{2, 4\}$, and $\{5, \phi\}$ are connected by the procedure mentioned above.

Definition of $S(\cdot, \cdot)$ depends on the application and complexity of a shape. We have given the simplest one suitable for simple shapes such as arabic numerals and the roman alphabet. When local features around the singular point are complex. For example, for some l,

$$S(i_l, j_l) \approx S(i_l, k), \qquad j_l \neq k,$$

or

$$S(i_l, j_l) \approx S(k, j_l), \qquad i_l \neq k,$$

the decision should be made in terms of global features of adjacent curves (sequences of edges) rather than the local rules for adjacent edges. The definition of $S(\cdot, \cdot)$ can be replaced by the absolute value of the curvature around the singular point estimated from curve fitting, for instance, cubic B-splines or Bernstein-Bézier curve fitting.

Example 9.2.1 The graphs shown in Fig. 9.17 are obtained by applying the singular point decomposition to the graphs in Fig. 9.15. The singular point p is decomposed into $a1$ and $a3$, q into $b1$ and $b4$, and r into $c1$ and $c3$. The graph in Fig. 9.15 (*a*) is transformed into a stroke graph that consists of a simple arc ($a0$, $a1$, $a2$, $a3$, $a4$), the one in Fig. 9.15 (*b*) into the stroke graph of two simple arcs ($b0$, $b1$, $b2$) and ($b3$, $b4$, $b5$), and the one in Fig. 9.15 (*c*) into the stroke graph of two simple arcs ($c0$, $c1$, $c2$) and ($c3$, $c4$).

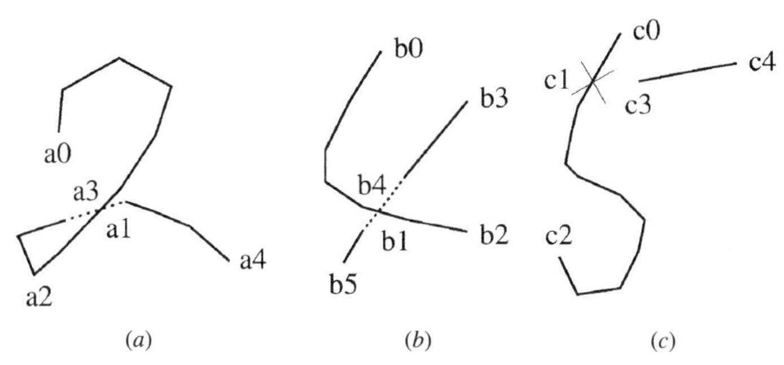

FIGURE 9.17 Singular point decomposition. The point p is decomposed into $a1$ and $a3$, and q into $b1$ and $b4$, and r into $c1$ and $c3$. From reference [11], © 1998, IEEE.

Quasi-topological Structure of Simple Curves By decomposing each singular point in parallel, a curve is decomposed into components (*strokes*) each of which is either a simple arc or a simple closed curve. Therefore, without loss of generality, we assume that a curve is a simple arc or a simple closed curve.

Directional features are essential for shape description of characters because characters are oriented and rotation invariance is of no importance. Therefore the description should depend on only directions of the coordinate axes but be independent of size and position. In this chapter we take the coordinate system as shown in Fig. 9.17: The x-axis goes from the left to the right horizontally, and the y-axis from the bottom to the top vertically.

Figure 9.18 gives a flowchart analysis and description of the quasi-topological structure of curves. In Fig. 9.18 the items in the rectangles are the objects on each representation level, and those in ellipses are the operations on the objects. The arrows in Fig. 9.18 mean that the higher-level representation is given by applying the operations to objects on the lower level. First, the four classes of curve primitives are defined explicitly. Each primitive has the two end points called *head* and *tail,* and a primitive can be concatenated to others only at its head or tail. The four types of binary operations on the primitives are defined by classifying the concatenations of primitives according to the direction of convexity. We show that the operators have algebraic properties explicitly. Next the primitive sequences are generated by linking the binary operations of primitives. A label is given to each primitive sequence according to the properties of the primitives and their concatenations forming the sequence. A primitive sequence can be connected with others by sharing its first or last primitive. In other words, two binary operations are introduced to primitive sequences. Moreover we show that the operations can be applied only if the labels of the two primitive sequences satisfy the specific relations. We describe the structure of the curve by the quasi-topological structure of each stroke and the adjacent structures of the primitive sequences on the singular point. Consequently a compact and concise description of the curve structure is obtained.

Primitives of Curves An xy-monotone curve is a curve whose x and y values are always either nonincreasing or nondecreasing as one traverses the curve. Let the two

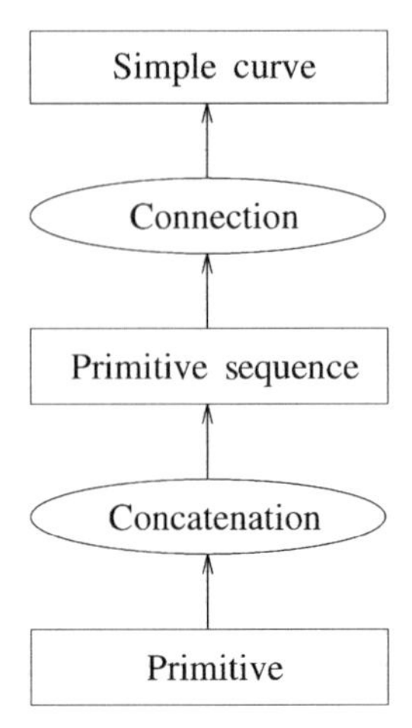

FIGURE 9.18 Analysis of quasi-topological structure of curves.

end points of an *xy*-monotone curve be $H(x_0, y_0)$ and $T(x_1, y_1)$, where H and T are determined in such a way that $x_0 < x_1$ if $x_0 \neq x_1$ and $y_0 < y_1$ if $x_0 = x_1$. The point H is called the *head* of the curve, and T is the *tail.* The *xy*-monotone curve is classified according to the order relation of the x and y coordinates of the head and the tail:

1. $x_0 < x_1$ and $y_0 = y_1$ (denoted by "–" below and illustrated in Fig. 9.19 (*a*)).
2. $x_0 < x_1$ and $y_0 < y_1$ ("/," in Fig. 9.19 (*b*)).
3. $x_0 = x_1$ and $y_0 < y_1$ ("|," in Fig. 9.19 (*c*)).
4. $x_0 < x_1$ and $y_0 > y_1$ ("\," in Fig. 9.19 (*d*)).

The primitives of curves are defined as these four classes of *xy*-monotone curves.

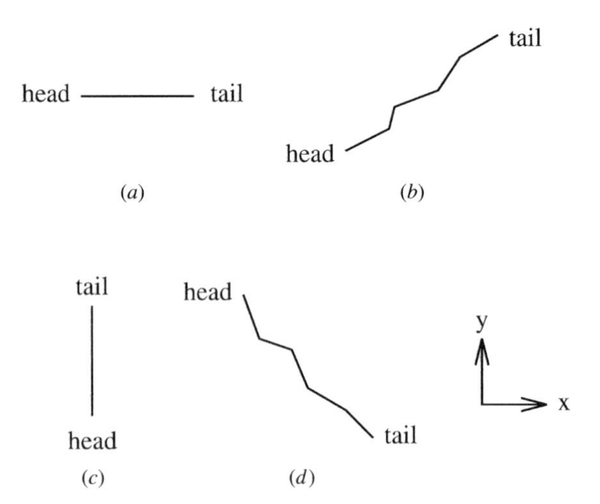

FIGURE 9.19 Four types of curve primitives. From reference [11], © 1998, IEEE.

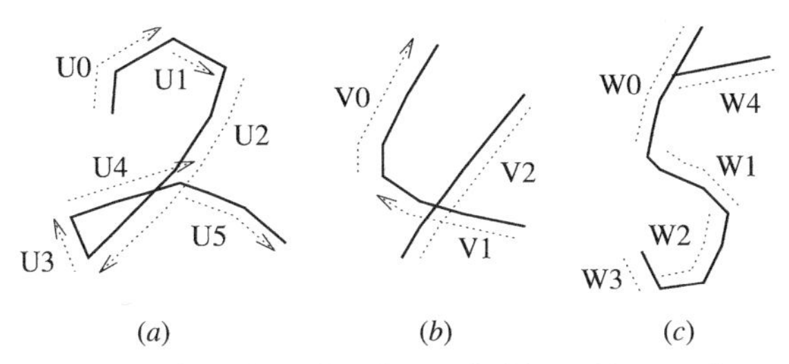

FIGURE 9.20 Decomposition of strokes into primitives. From reference [11], © 1998, IEEE.

Example 9.2.2 For curves in Fig. 9.17, each stroke is decomposed into primitives as shown in Fig. 9.20. *U*0, *U*2, *U*4, *V*0, *V*2, *W*0, *W*2, and *W*4 are primitives of type "/," and the others are "\."

Concatenation Operators of Primitives We define the binary operators of the primitives. Following Shaw's scheme, we assume that a primitive can be concatenated to others only at its head or tail. Furthermore we assume that a curve formed by a concatenation of two primitives is not a primitive (an *xy*-monotone curve) any more. Before introducing the binary operators, we define some notations and functions.

Definition 9.2.1 *$\Delta(a, b)$, where a and b are primitives. Suppose that two primitives a and b are concatenated. Let P be a joint of a and b, and let P_a and P_b be points sufficiently close to P such that P_a and P_b are contained in only a and only b, respectively. Then $\Delta(a, b)$ is defined as*

$$\Delta(a, b) = \text{sign} \begin{vmatrix} x_a - x_p & x_b - x_p \\ y_a - y_p & y_b - y_p \end{vmatrix}, \tag{9.15}$$

$$\text{sign}\, x = \begin{cases} -1 & \text{if } x < 0, \\ 0 & \text{if } x = 0, \\ 1 & \text{if } x > 0, \end{cases}$$

where (x_p, y_p), (x_a, y_a), and (x_b, y_b) are the coordinates of P, P_a, and P_b, respectively.

Definition 9.2.2 prm(*a*)*, where a is a primitive,* prm(*a*) *denotes the primitive type of a.* prm(*a*) ∈ { |, /, –, \}.

Definition 9.2.3 *[a, A, α ; b, B, β], where a and b are two primitives, A =* prm(*a*)*, B =* prm(*b*)*, and α, β ∈* {head, tail}. *[a, A, α ; b, B, β] means that primitives a* (prm(*a*) *= A) and b* (prm(*b*) *= B) are concatenated at α of a and β* of *b in such a way that $\Delta(a, b) = 1$ (see Fig. 9.21).*

Now, for the two primitives *a* and *b,* we define the binary operations

$$a \xrightarrow{j} b, \qquad j = 0, 1, 2, 3, \tag{9.16}$$

where j is the characteristic number of the operator and denotes the direction of convexity formed by the concatenation of primitives. Note that the operation (9.16) is not compatible with

$$b \xrightarrow{j} a$$

and that two primitives a and b satisfy $\Delta(a, b) = 1$, as shown in Fig. 9.21. The operation (9.16) is defined for each type of concatenation of primitives by the following rules:

Rule 9.2.1 (Downward convex)

$$a \xrightarrow{0} b$$

for [*a*, /, *head; b*, \, *tail*], [*a*, |, *head; b*, \, *tail*], [*a*, /, *head; b*, |, *head*], [a, /, *head;* b, /, *head*]. They are illustrated in Fig. 9.22.

Rule 9.2.2 (Leftward convex)

$$a \xrightarrow{1} b$$

for [*a*, \, *head; b*, /, *head*], [*a*, –, *head; b*, /, *head*], [*a*, \, *head; b*, –, *head*], [*a*, \, *head; b*, \, *head*]. They are illustrated in Fig. 9.23.

Rule 9.2.3 (Upward convex)

$$a \xrightarrow{2} b$$

for [*a*, /, *tail; b*, \, *head*], [*a*, |, *tail; b*, \, *head*], [*a*, /, *tail; b*, |, *tail*], [*a*, /, *tail; b*, /, *tail*]. They are illustrated in Fig. 9.24.

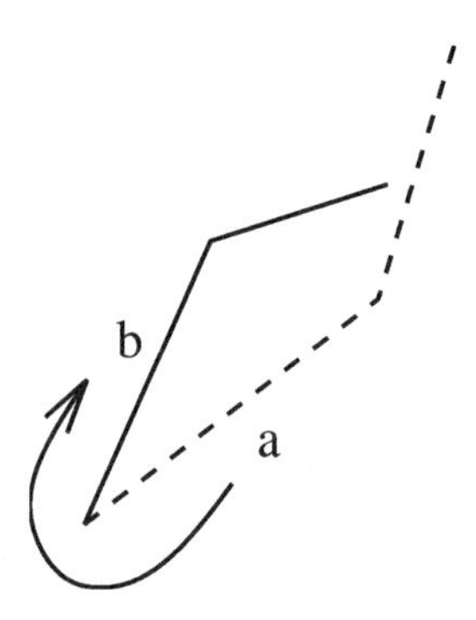

FIGURE 9.21 Concatenation of two primitives. Two primitives a and b are concatenated in such a way that we turn to the right around the joint of a and b when we traverse the curve from a to b.

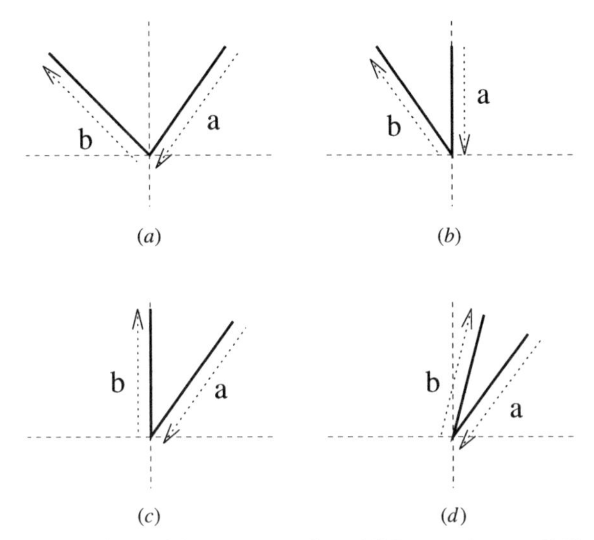

FIGURE 9.22 Illustrations of the concatenation "0." From reference [11], © 1998, IEEE.

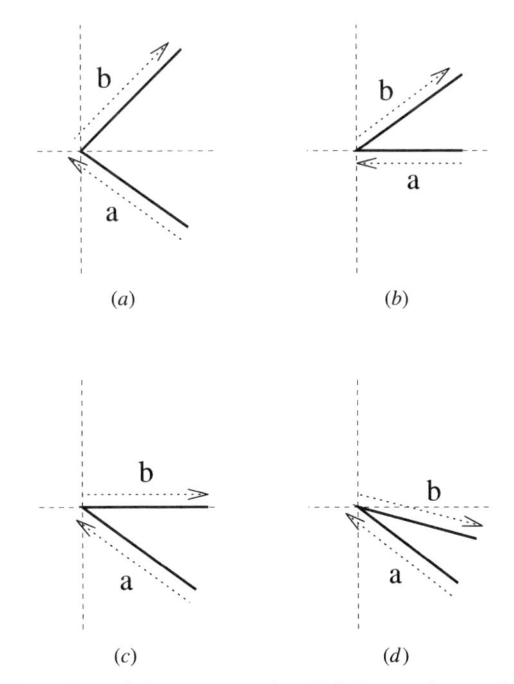

FIGURE 9.23 Illustrations of the concatenation "1." From reference [11], © 1998, IEEE.

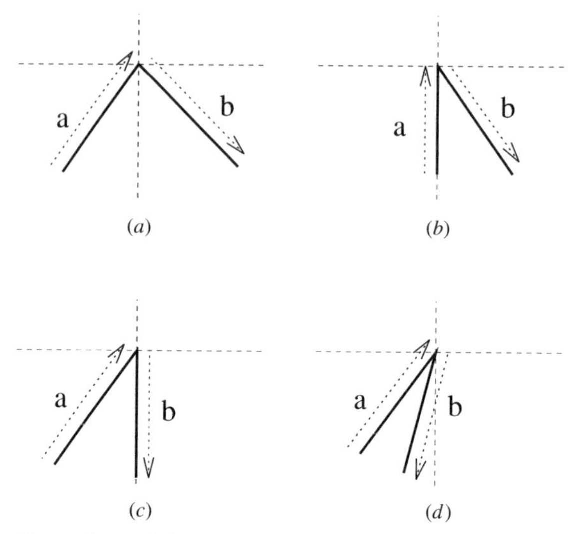

FIGURE 9.24 Illustrations of the concatenation "2." From reference [11], © 1998, IEEE.

Rule 9.2.4 (Rightward convex)

$$a \xrightarrow{3} b$$

for [*a*, \, *tail; b*, /, *tail*], [*a*, –, *tail; b*, /, *tail*], [*a*, \, *tail; b*, –, *tail*], [*a*, \, *tail; b*, \, *tail*]. They are illustrated in Fig. 9.25.

Rules 1, 2, 3, and 4 correspond to downward, leftward, upward, and rightward convexity, respectively. In other words, the directions of convexity have correspondence to elements of the set $\{0, 1, 2, 3\}$. In the following we will show that structure of curves with complex shape is represented in terms of simple arithmetic operations on the set $Z/(4)$, which is the residue-class system of integer with respect to 4. Table 9.1 summarizes these four rules.

Example 9.2.3 We apply the binary operations to the primitives in Fig. 9.20 as follows:

$$U0 \xrightarrow{2} U1, \quad U1 \xrightarrow{3} U2, \quad U2 \xrightarrow{0} U3,$$
$$U3 \xrightarrow{1} U4, \quad U4 \xrightarrow{2} U5,$$
$$V1 \xrightarrow{1} V0,$$
$$W1 \xrightarrow{1} W0, \quad W1 \xrightarrow{3} W2, \quad W2 \xrightarrow{0} W3.$$

Properties of Concatenation Operators We have defined the binary operations

$$a \xrightarrow{j} b, \qquad j = 0, 1, 2, 3,$$

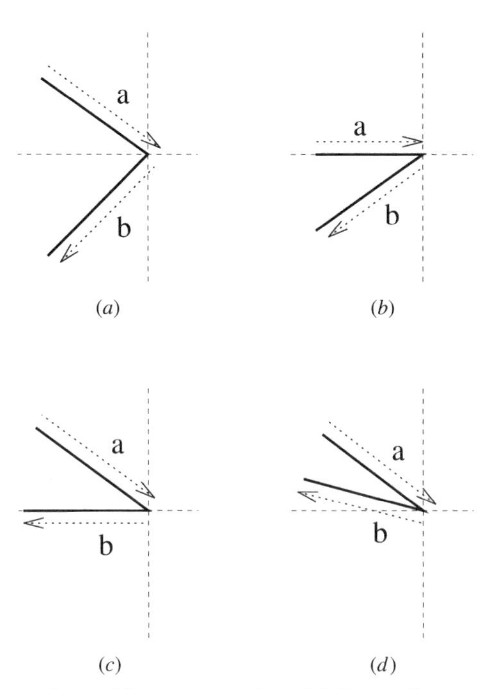

FIGURE 9.25 Illustrations of the concatenation "3." From reference [11], © 1998, IEEE.

for primitives a and b. Now we show the properties of the concatenation operators, from which we will derive algebraic properties of the higher-level representation of the curve structure.

Definition 9.2.4 corner$(a \xrightarrow{j} b)$, *where a and b are two primitives, and* $i \in \{0, 1, 2, 3\}$.

$$\text{corner}(a \xrightarrow{i} b) = \begin{cases} 1 & \text{if prm}(a) = \text{prm}(b), \\ 0 & \text{otherwise.} \end{cases} \tag{9.17}$$

Let a, b, and c be primitives, and let i and j $\in \{0, 1, 2, 3\}$ *be the characteristic numbers of the operators. The concatenation operator has algebraic properties as follows:*

TABLE 9.1 Concatenation Rules of Two Primitives

<table>
<tr><th>j</th><th colspan="2">a</th><th colspan="2">b</th><th colspan="2">a</th><th colspan="2">b</th><th colspan="2">a</th><th colspan="2">b</th><th colspan="2">a</th><th colspan="2">b</th></tr>
<tr><td>0</td><td>/</td><td>h</td><td>\</td><td>t</td><td>|</td><td>h</td><td>\</td><td>t</td><td>/</td><td>h</td><td>|</td><td>h</td><td>/</td><td>h</td><td>/</td><td>h</td></tr>
<tr><td>1</td><td>\</td><td>h</td><td>/</td><td>h</td><td>—</td><td>h</td><td>/</td><td>h</td><td>\</td><td>h</td><td>—</td><td>h</td><td>\</td><td>h</td><td>\</td><td>h</td></tr>
<tr><td>2</td><td>/</td><td>t</td><td>\</td><td>h</td><td>|</td><td>t</td><td>\</td><td>h</td><td>/</td><td>t</td><td>|</td><td>t</td><td>/</td><td>t</td><td>/</td><td>t</td></tr>
<tr><td>3</td><td>\</td><td>t</td><td>/</td><td>t</td><td>—</td><td>t</td><td>/</td><td>t</td><td>\</td><td>t</td><td>—</td><td>t</td><td>\</td><td>t</td><td>\</td><td>t</td></tr>
</table>

Property 9.2.1 *If*

$$a \xrightarrow{i} b, \quad a \xrightarrow{j} c,$$

then

$$i - j \equiv 2 \pmod 4.$$

Property 9.2.2 *If*

$$a \xrightarrow{i} b, \quad c \xrightarrow{j} b,$$

then

$$i + \text{corner}(a \xrightarrow{i} b) \equiv j + \text{corner}(c \xrightarrow{j} b) + 2 \pmod 4.$$

Property 9.2.3 *If*

$$a \xrightarrow{i} b, \quad b \xrightarrow{j} c \ (\text{prm}(b) \in \{\, /, \backslash \,\}),$$

then

$$j \equiv i + \text{corner}(a \xrightarrow{i} b) + 1 \pmod 4.$$

Property 9.2.4 *If*

$$a \xrightarrow{i} b, \quad b \xrightarrow{j} c \ (\text{prm}(b) \in \{\, |, - \,\}),$$

then

$$j - i \equiv 2 \pmod 4.$$

Figure 9.26 illustrates these properties:

(*a*) $a \xrightarrow{2} b,\ a \xrightarrow{0} c.$
(*b*) $a \xrightarrow{2} b,\ c \xrightarrow{0} b.$
(*c*) $a \xrightarrow{1} b,\ b \xrightarrow{2} c\ (\text{prm}(b) = /).$
(*d*) $a \xrightarrow{0} b,\ b \xrightarrow{2} c\ (\text{prm}(b) = |).$

Primitive Sequence of Curves For a simple arc or a simple closed curve, we can cover the curve with the minimum number of primitives and apply the concatenation operators to each pair of the primitives. By linking the binary operations on primitives, the following sequence is generated:

$$a_0 \xrightarrow{j_1} a_1 \xrightarrow{j_2} \cdots \xrightarrow{j_{n-1}} a_{n-1} \xrightarrow{j_n} a_n. \tag{9.18}$$

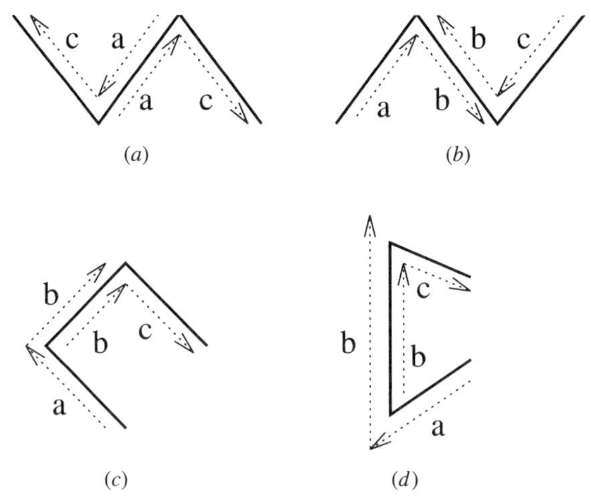

FIGURE 9.26 Properties of the concatenation of primitives. From reference [11], © 1998, IEEE.

The sequence (9.18) is called the *primitive sequence*. In the following we give a label to the primitive sequence according to the properties of the primitives and their concatenations.

When a curve consists of two primitives or more and the primitive sequence on the curve is not cyclic, it is decomposed into primitive sequences (9.18), where there exists neither b nor c such that

$$b \xrightarrow{j_0} a_0 \qquad (j_0 = 0, 1, 2, \text{ or } 3),$$

$$a_n \xrightarrow{j_{n+1}} c \qquad (j_{n+1} = 0, 1, 2, \text{ or } 3).$$

In addition the end points of the primitive sequence on the primitives a_0 and a_n are called the *h-point* and the *t-point,* respectively. Since any operation of two primitives

$$a \xrightarrow{*} b$$

satisfies $\Delta(a, b) = 1$, we always turn to the right at any joints of primitives when we traverse a curve composed of one primitive sequence from the h-point to the t-point (see Fig. 9.26(c), (d), and 9.27).

The label of the primitive sequence, *PS-label* for short, $\langle ps, idr \rangle$ is given to the sequence (9.18) composed of $n + 1$ primitives by the following formulas:

$$ps = (n + 1) + L + M, \tag{9.19}$$

$$idr = j_1, \tag{9.20}$$

where L is the number of concatenations

$$a_i \xrightarrow{j_{i+1}} a_{i+1} \qquad (i = 0, 1, \ldots, n - 1)$$

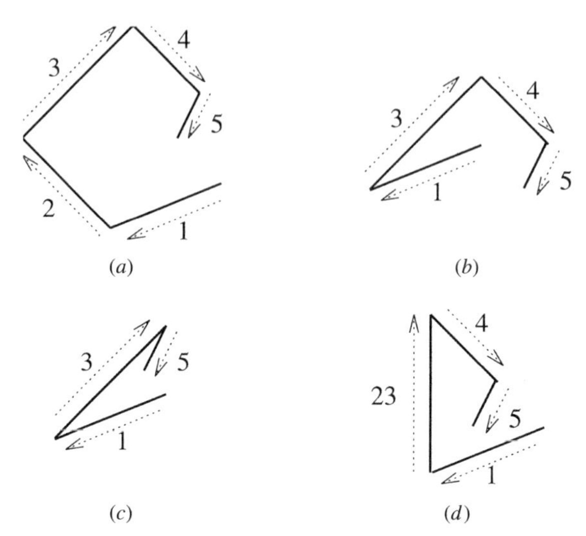

FIGURE 9.27 Instances of curves with the PS-label ⟨5, 0⟩ (*a*) $L = M = 0$; (*b*) $L = 1, M = 0$; (*c*) $L = 2, M = 0$; (*d*) $L = 0, M = 1$. From reference [11], © 1998, IEEE.

such that $\text{prm}(a_i) = \text{prm}(a_{i+1})$, and M is the number of a_i ($i = 1, 2, \ldots, n-1$) such that $\text{prm}(a_i) \in \{|, -\}$.

Figure 9.27 shows instances of curves with the PS-label ⟨5,0⟩. In the figure, *h* and *t* denote the h-point and the t-point of the primitive sequence, respectively. The curves shown in Fig. 9.27 (*a*), (*b*), (*c*) and (*d*) are examples of $L = M = 0$, $L = 1$, $L = 2$, and $M = 1$, respectively.

We use *idr* to denote the initial direction of the rotation. The meaning of *ps* can be interpreted in the following way: If $L = M = 0$ (Fig. 9.27 (*a*)), then "/" and "\" appear alternately in the primitive sequence:

$$1 \xrightarrow{0} 2 \xrightarrow{1} 3 \xrightarrow{2} 4 \xrightarrow{3} 5.$$

When $\text{prm}(a_i) = \text{prm}(a_{i+1})$ (Fig. 9.27(b) and (c)), one "\" ("/") is thought to be skipped between a_i and a_{i+1} because of an excessive change in the direction of the curve. For instance, if the primitive "2" on the curve in Fig. 9.27(*a*) becomes smaller, it is finally removed, and we get the curve in Fig. 9.27(*b*):

$$1 \xrightarrow{0} 3 \xrightarrow{2} 4 \xrightarrow{3} 5.$$

Then we get the curve in Fig. 9.27(*c*) by removing the primitive "4" from the curve in Fig. 9.27(*b*):

$$1 \xrightarrow{0} 3 \xrightarrow{2} 5.$$

Similarly, when $\text{prm}(a_i) \in \{|, -\}$ (Fig. 9.27(*d*)), "/" and "\" are thought to be merged into "|" or "–." For example, if we collapse the primitives "2" and "3" in Fig. 9.27(*a*), then they are finally merged into "23" in Fig. 9.27(*d*):

$$1 \xrightarrow{0} 23 \xrightarrow{2} 4 \xrightarrow{3} 5.$$

Thus it is reasonable to add L and M to the number of the primitives a_i so that ps is invariant in such variations. This argument is formalized by the following theorem:

Theorem 9.2.1 *For* $n \geq 2$,

$$ps - 2 = \sum_{i=1}^{n-1} \{(j_{i+1} - j_i) \pmod 4\} + \text{corner}(a_{n-1} \xrightarrow{j_n} a_n). \tag{9.21}$$

When the primitive sequence is cyclic on a simple closed curve as

$$\cdots \xrightarrow{j_{n-1}} a_{n-1} \xrightarrow{j_n} a_n \xrightarrow{j_0} a_0 \xrightarrow{j_1} a_1 \xrightarrow{j_2} \cdots \xrightarrow{j_n} a_n \xrightarrow{j_0} a_0 \xrightarrow{j_1} \cdots, \tag{9.22}$$

the PS-label $\langle 0, 0 \rangle$ *is given to the curve.*

When a curve is composed of one primitive, the PS-label is given to the curve in the following way: $\langle 1, 0 \rangle$ *for "–,"* $\langle 1, 1 \rangle$ *for "/,"* $\langle 1, 2 \rangle$ *for "|," and* $\langle 1, 3 \rangle$ *for "\."*

Example 9.2.4 By linking the binary operations in Example 9.2.3, the following primitive sequences are generated:

$$F_1 : U0 \xrightarrow{2} U1 \xrightarrow{3} U2 \xrightarrow{0} U3 \xrightarrow{1} U4 \xrightarrow{2} U5 \quad \langle 6, 2 \rangle,$$
$$F_2 : V1 \xrightarrow{1} V0 \quad \langle 2, 1 \rangle,$$
$$F_3 : V2 \quad \langle 1, 1 \rangle,$$
$$F_4 : W1 \xrightarrow{1} W0 \quad \langle 2, 1 \rangle,$$
$$F_5 : W1 \xrightarrow{3} W2 \xrightarrow{0} W3 \quad \langle 3, 3 \rangle,$$
$$F_6 : W4 \quad \langle 1, 1 \rangle.$$

Figure 9.28 depicts the two primitive sequences F_4 and F_5. $p1$ and $p2$ are the h-points, and $p0$ and $p3$ are the t-points of F_4 and F_5, respectively.

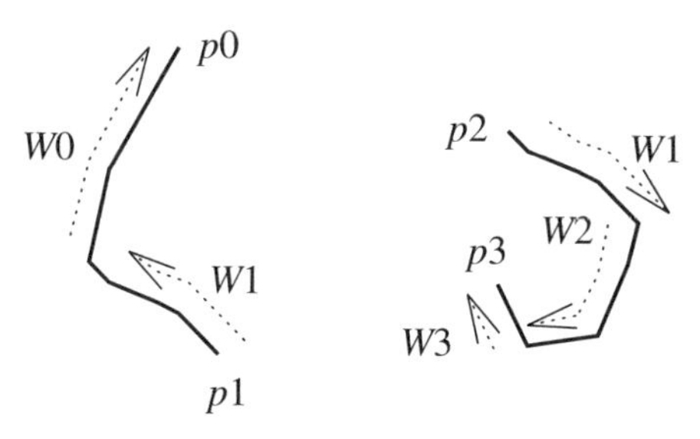

FIGURE 9.28 Decomposition of a stroke into primitive sequences. The primitive sequence on the left has the PS-label $\langle 2, 1 \rangle$, where $p1$ is h-point and $p0$ is t-point. The right one has the PS-label $\langle 3, 3 \rangle$, where $p2$ is h-point and $p3$ is t-point.

Connection of Primitive Sequences We have obtained primitive sequences on a simple arc or a simple closed curve. Next we consider the connection of primitive sequences. Let two primitive sequences e_0 and e_1 be

$$e_0 : a_0 \xrightarrow{i_1} a_1 \xrightarrow{i_2} \cdots \xrightarrow{i_m} a_m,$$
$$e_1 : b_0 \xrightarrow{j_1} b_1 \xrightarrow{j_2} \cdots \xrightarrow{j_n} b_n.$$

Let the PS-labels of e_0 and e_1 be $\langle ps_0, idr_0 \rangle$ and $\langle ps_1, idr_1 \rangle$, respectively. For any pair of primitives a and b that is concatenated to one another, either

$$a \xrightarrow{*} b$$

or

$$b \xrightarrow{*} a$$

is satisfied. Moreover there do not exist c_0, d_0, c_1, or d_1 such that

$$c_0 \xrightarrow{*} a_0, \quad a_m \xrightarrow{*} d_0, \quad c_1 \xrightarrow{*} b_0, \quad b_n \xrightarrow{*} d_1.$$

Therefore e_0 and e_1 are connected by sharing a_0 (b_0) or a_m (b_n). The connection is classified as follows:

h-connection. The primitive a_0 is identical to the primitive b_0. Two primitive sequences are connected to one another by sharing the first primitive of each one (see Fig. 9.26(a)).

t-connection. The primitive a_m is identical to the primitive b_n. Two primitive sequences are connected to one another by sharing the last primitive of each one (see Fig. 9.26(b)).

These are denoted by

$$e_0 \overset{h}{—} e_1 \qquad (\text{or } e_1 \overset{h}{—} e_0, \tag{9.23}$$

$$e_0 \overset{t}{—} e_1 \qquad (\text{or } e_1 \overset{t}{—} e_0), \tag{9.24}$$

respectively. Recall that we always turn to the right around the joint of primitives when we traverse a curve composed of one primitive sequence from the h-point to the t-point. Thus, when we traverse a curve composed of two primitive sequences e_0 and e_1, we turn to the right (left) around the joints of primitives on e_0, while we turn to the left (right) around the joints on e_1.

Example 9.2.5 We examine connection of the primitive sequences in Example 9.2.4 (Fig. 9.2). Since F_4 and F_5 share the primitive $W1$,

$$F_4 \overset{h}{—} F_5.$$

Next we show that the h-connection and the t-connection can be applied to two primitive sequences only if the two PS-labels satisfy the following specific relations:

Theorem 9.2.2 *If*

$$e_0 \stackrel{h}{—} e_1,$$

then

$$idr_0 - idr_1 \equiv 2\ (mod\ 4), \tag{9.25}$$

and if

$$e_0 \stackrel{t}{—} e_1,$$

then

$$ps_0 + idr_0 \equiv ps_1 + idr_1 + 2\ (mod\ 4). \tag{9.26}$$

Corollary 9.2.1

$$e_0 \stackrel{h}{—} e_1, \quad e_0 \stackrel{t}{—} e_1 \qquad (ps_0 \geq ps_1 \geq 2),$$

then

$$ps_0 \geq 3, \quad ps_0 \equiv ps_1\ (mod\ 4), \quad idr_0 - idr_1 \equiv 2\ (mod\ 4). \tag{9.27}$$

Description of Quasi-topological Structure of Curves We have constructed the primitive sequences and their connections for a simple arc or a simple closed curve. Now we can describe the quasi-topological structure of the curve by the labels and the connections of the primitive sequences.

1. *The curve is a simple arc.* Suppose that the curve consists of n primitive sequences e_i $(i = 1, 2, \ldots, n)$, and e_i is connected to e_{i+1} $(i = 1, 2, \ldots, n-1)$ in such a way that

$$e_i \stackrel{c_{i+1}}{—} e_{i+1}\ (c_i \in \{h, t\}),$$

where $c_{i+1} = t$ if $c_i = h$, and $c_{i+1} = h$ if $c_i = t$, $i = 2, 3, \ldots, n-1$. By linking the connections of the primitive sequences, the following string is generated:

$$e_1 \stackrel{c_2}{—} e_2 \stackrel{c_3}{—} e_3 \stackrel{c_4}{—} \cdots \stackrel{c_{n-1}}{—} e_{n-1} \stackrel{c_n}{—} e_n.$$

Let the PS-label of e_i be $\langle ps_i, idr_i \rangle$. By substituting e_i by $\langle ps_i, idr_i \rangle$, the quasi-topological structure of the curve is described as

$$\langle ps_1, idr_1 \rangle \stackrel{c_2}{—} \langle ps_2, idr_2 \rangle \stackrel{c_3}{—} \cdots \stackrel{c_n}{—} \langle ps_n, idr_n \rangle. \tag{9.28}$$

2. *The curve is a simple closed curve.* Similarly, a string of the primitive sequences is generated.

$$e_1 \stackrel{c_2}{—} e_2 \stackrel{c_3}{—} \cdots \stackrel{c_{n-1}}{—} e_{n-1} \stackrel{c_n}{—} e_n \stackrel{c_1}{—} e_1^*.$$

The asterisk (*) means that the string is cyclic and that the last element is identical to the first one. Since the above string is cyclic, and h and t appear alternately in the sequence $(c_1, c_2, \ldots, c_n)$, n must be an even number for a closed curve. Then the quasi-topological structure of the curve is described as

$$\langle ps_1, idr_1\rangle \stackrel{c_2}{—} \langle ps_2, idr_2\rangle \stackrel{c_3}{—} \cdots \stackrel{c_n}{—} \langle ps_n, idr_n\rangle \stackrel{c_1}{—} \langle ps_1, idr_1\rangle^*. \tag{9.29}$$

Structure of Singular Points We give description of singular points. As shown in Fig. 9.29, there are three types of adjacent structures of two primitive sequences on the singular point: X-type (crossing), K-type (touch), and T-type. Let two primitive sequences e_i and e_j be

$$e_i : a_0 \xrightarrow{i_1} a_1 \xrightarrow{i_2} \cdots \xrightarrow{i_m} a_m,$$
$$e_j : b_0 \xrightarrow{j_1} b_1 \xrightarrow{j_2} \cdots \xrightarrow{j_n} b_n.$$

The structure of the singular point v is described by the binary relation of two primitive sequences e_i and e_j containing v:

$$[e_i, \sigma] \stackrel{\chi}{—} [e_j, \tau], \tag{9.30}$$

where $\chi \in \{X, K, T\}$, and $\sigma, \tau \in \{hp, tp, h, t, \phi\}$. χ denotes the adjacent structure of e_i and e_j on v, and σ (τ) denotes the position of v on the primitive sequence e_i (e_j):

$$\sigma = \begin{cases} hp & \text{if } v \text{ is the h-point of } e_i, \\ tp & \text{if } v \text{ is the t-point of } e_i, \\ h & \text{if } v \text{ lies on } a_0 \text{ but is not the h-point of } e_i, \\ t & \text{if } v \text{ lies on } a_m \text{ but is not the t-point of } e_i, \\ \phi & \text{otherwise.} \end{cases}$$

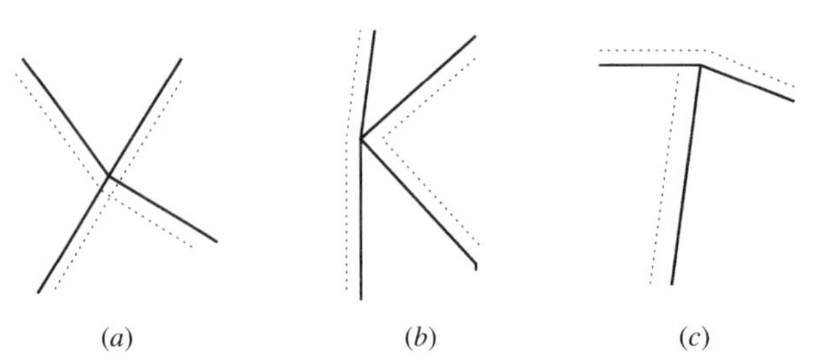

FIGURE 9.29 Adjacent structure of two primitive sequences on singular points. (*a*), (*b*), and (*c*) are the *X*-, *K*-, and *T*-types, respectively. From reference [11], © 1998, IEEE.

When e_i is composed of one primitive ($m = 0$),

$$\sigma = \begin{cases} hp & \text{if } v \text{ is the head of the primitive } a_0, \\ tp & \text{if } v \text{ is the tail of the primitive } a_0, \\ \phi & \text{otherwise.} \end{cases}$$

If the PS-label of e_i is $\langle 0, 0 \rangle$, σ is set to be ϕ. τ is determined in the same way as σ. Since $\lceil N/2 \rceil$ pairs of edges are generated on the singular point v of the order N, the structure of v is described by at least $\lceil N/2 \rceil \times (\lceil N/2 \rceil - 1)/2$ of the forms of (9.30).

Example 9.2.6 Along with the primitive sequences obtained in Example 9.2.4, we describe the structure of each of the singular points in Fig. 9.29 as follows (see also Figs. 9.17, 9.20, and 9.28). The singular point p in Fig. 9.15(a) lies on the primitives $U2$ and $U4$ in the primitive sequence F_1 (see Fig. 9.20(a)). The singular point q in Fig. 9.15(b) lies on the primitive $V1$ in the primitive sequence F_2 and $V2$ in F_3 (see Fig. 9.20(b)). The singular point r in Fig. 9.15(c) lies on the primitive $W0$ in the primitive sequence F_4 and the head $c3$ (Fig. 9.17(c)) of the primitive F_6 (see Fig. 9.20(c) and Fig. 9.28). Therefore we have

$$p : [F_1, \phi] \overset{X}{—} [F_1, \phi],$$

$$q : [F_2, h] \overset{X}{—} [F_3, \phi],$$

$$r : [F_4, t] \overset{T}{—} [F_6, hp].$$

Structural Description of Curves We have obtained the primitive sequences, their connections, and the adjacent structures of primitive sequences on the singular points. Now we can describe the structure of a curve by the quasi-topological structure of each stroke and the structure of each singular point. Since each stroke is either a simple arc or a simple closed curve, it is described in the form of (9.28) or (9.29). On the other hand, the structure of each singular point is described in the form of (9.30).

Example 9.2.7 For the curves represented as stroke graphs in Fig. 9.17, the structure of each curve is described as follows. For the curve shown in Fig. 9.17(a),

$$\text{stroke } (a0, a1, a2, a3, a4) : \langle 6, 2 \rangle (F_1),$$

$$\text{singular point } (a1, a3) : [F_1, \phi] \overset{X}{—} [F_1, \phi].$$

For the curve shown in Fig. 9.17(b),

$$\text{stroke } (b0, b1, b2) : \langle 2, 1 \rangle (F_2),$$

$$\text{stroke } (b3, b4, b5) : \langle 1, 1 \rangle (F_3),$$

$$\text{singular point } (b1, b4) : [F_2, h] \overset{X}{—} [F_3, \phi].$$

For the curve shown in Fig. 9.17(c),

$$\text{stroke } (c0, c1, c2) : \langle 2, 1\rangle (F_4 \overset{h}{-} \langle 3, 3\rangle (F_5),$$

$$\text{stroke } (c3, c4) : \langle 1, 1\rangle (F_6),$$

$$\text{singular point } (c1, c3) : [F_4, t] \overset{T}{-} [F_6, hp].$$

Structure of Closed Curves We analyze the structure of closed curves in terms of the properties of the PS-labels. The properties show the geometrical meanings of the PS-label. We assume that the closed curve C is composed of n primitive sequences e_1, $e_2, \ldots, e_n$ (n is an even number) such that

$$e_1 \overset{h}{-} e_2 \overset{t}{-} e_3 \overset{h}{-} \cdots \overset{h}{-} e_n \overset{t}{-} e_1^*. \tag{9.31}$$

Let the PS-label of e_i be $\langle ps_i, idr_i\rangle$. Now we give the total amount T of the directional change along the curve C in terms of the PS-labels.

Theorem 9.2.3 *The total amount T of the directional change along the curve C is given by*

$$|T| = \frac{\pi}{2} \left| \sum_{i=1}^{n} (-1)^i ps_i \right|. \tag{9.32}$$

Theorem 9.2.3 suggests that ps_i represents the rotation number along the primitive sequence e_i. Furthermore $T = \pm 2m\pi$ (m is an integer) for closed curves, and $T = \pm 2\pi$ for simple closed curves in particular. Thus we obtain the following corollary:

Corollary 9.2.2 *If the curve C is a closed curve, then*

$$\sum_{i=1}^{n} (-1)^i ps_i \equiv 0 \ (mod\ 4). \tag{9.33}$$

In particular, the curve C is a simple closed curve, then

$$\sum_{i=1}^{n} (-1)^i ps_i = \pm 4. \tag{9.34}$$

Finally, we mention the relationship between the PS-label set and singular points on a closed curve. The following corollary is derived from the theorem and corollary above:

Corollary 9.2.3 *If*

$$\sum_{i=1}^{n} (-1)^i ps_i \neq \pm 4,$$

then there is an X-type singular point on the closed curve C.

Application to Character Recognition We mention an application of our curve analysis and description technique to handwritten character recognition.

Since the description method ignores metric information such as length, curvature, angle, or position, sometimes objects representing different symbols have the identical structure. Some instance of "6" has the same structure as some of "0." Therefore parameterization of shapes is necessary to give more information to shape description.

Automatic construction of prototypes has been an open problem in structural pattern recognition, and therefore prototypes are manually created in structural approaches of character recognition. A strong advantage of our description scheme is that since the shape structure is represented in the well-organized form, character shapes can be classified automatically in the systematic way. Thus, clustering the thin line curves by the structural similarity in terms of our description scheme, we can construct the dictionary automatically from the training data. From the practical viewpoint, it is more important to reject too distorted or ambiguous data than to force them to be read. Since handwritten characters have various patterns of deformation, we have multiple models for one character. Each model has the quasi-topological structure of each stroke and the list of singular points, with each primitive sequence and singular point having the sets of eligible labels and structures, respectively. For instance, a model for the numeral "5" is described as follows:

- Symbol: 5.
- Stroke structure–Stroke 1: $PS_1 \overset{h}{-} PS_2$; Stroke 2: PS_3.
- Primitive sequences–PS_1: $\{ \langle 2, 0\rangle, \langle 2, 1\rangle, \langle 3, 0\rangle, \}$; PS_2: $\{ \langle 2, 3\rangle, \langle 3, 2\rangle, \langle 3, 3\rangle, \langle 4, 2\rangle, \langle 4, 3\rangle, \langle 5, 3\rangle \}$; PS_3: $\{ \langle 1, 0\rangle, \langle 1, 1\rangle, \langle 1, 3\rangle \}$.
- Singular points–SP_1 (optional): $\{[PS_1, t]\overset{T}{-}[PS_3, hp], [PS_1, o]\overset{T}{-}[PS_3, hp], [PS_1, tp]\overset{T}{-}[PS_3, o], [PS_1, t]\overset{X}{-}[PS_3, o], [PS_1, o]\overset{X}{-}[PS_3, o] \}$.

The primitive sequences P_1, P_2, and P_3 of the class correspond to F_4, F_5, and F_6 of Example 9.2.4. The singular point SP_1 of the class corresponds to the point r of Fig. 9.15(c). This model can cover all the shapes illustrated in Fig. 9.30, while techniques such as contour or background analysis require several models for those shapes. This example shows that our description method can cope with various deformation patterns with a smaller dictionary.

Structural matching of the object and models is based on the string matching of PS-labels on each stroke and the structural matching of singular points. Since the character shape is described in terms of global features, the description is not sensitive to local deformation. Therefore we employed neither cost functions nor editing operations for finding optimal correspondence between two strings, but we applied the *exact* string matching method, in which two string are matched against one another only if they are identical. If the object is matched with a model, Mahalanobis distance (Euclidean distance from the mean, scaled componentwise by standard deviation) is computed for parameters on structural components (primitive sequences, primitives, singular points), and the final decision is made on the basis of the distance.

FIGURE 9.30 Examples of the numeral "five." All instances can be represented by a single class.

Experiment on Loosely Constrained Data It should be noted that correct recognition ratio depends heavily on the quality of data given. In our case a set of handwritten data was obtained from 251 people. Shown the standard shapes of the handwritten numerals, they were asked to write numerals normally and neatly under no specification for a writing device. and many people ignored the standard shape. In this sense we may call the data *loosely constrained data* with a variety of writing styles and instruments. The data were digitized at 200 dpi with binary levels. Thus we acquired 13,400 characters in total for 10 numerals.

Training and test sets were disjoint, selected in the ratio 1:1 in a randomized manner from the data set. We used the thinning method by cross section sequence graph (see Chapter 5 [5], [7]) as a thinning algorithm. In postprocessing of thinning, small fluctuations nearly vertical or horizontal are smoothed, since the structural description is sensitive to them. The result on the test set is shown in Table 9.2, with the number of models created from the training set.

Experiment on Totally Unconstrained Data Another experiment was performed on totally unconstrained handwritten numerals. The data set consists of about 127,000 samples collected from 319 people (different from the writers of the loosely constrained data set) without any specification of writing styles or devices. The

TABLE 9.2 Experimental Result of Handwritten Numeral Recognition Using the New Curve Description Technique

	Recognition (%)	Rejection (%)	Substitution (%)	Model
0	99.3	0.4	0.3	3
1	99.7	0.2	0.1	3
2	98.9	0.5	0.6	5
3	99.3	0.1	0.6	4
4	97.7	2.0	0.3	5
5	99.4	0.6	0.0	5
6	99.1	0.9	0.0	2
7	98.7	0.8	0.5	5
8	95.7	3.8	0.5	10
9	98.9	0.9	0.2	4
	98.7	1.0	0.3	46

Note: The thinning method based on cross section sequence graph was used for thinning. The experiment was made on the loosely constrained data set of 13,400 samples.

loosely constrained data set was used as the training set, and the test was made on the whole of the totally unconstrained data set. Therefore the nature of the test set is quite different from the training set. Note that we neither added nor modified functionalities in the program or models/parameters in the dictionary to accommodate unconstrained data. The results are as follows:

- Recognition: 95.4%.
- Rejection: 2.9%.
- Substitution error: 1.7%.

Comparison with Contour Analysis Techniques To show flexibility of the curve description technique, we give experimental results on handwritten numerals recognition using contour analysis technique as well as our technique. Since a contour is a simple closed curve, the description technique can be applied to contours. A contour is smoothed by a Gaussian filter, and small fluctuations nearly vertical or horizontal are removed. Then the contour is described in the same way as thinned lines. The thinning method based on cell structure was used for thinning (see Chapter 5 [5]). For both techniques, structural matching and distance calculation were also done in the same way.

Since the result is influenced by image quality and preprocessing, we removed poor image quality samples from the loosely constrained data set mentioned above to give unbiased comparison of two methods. Thus, in the experiments, we used 11,500 characters in total written by 220 subjects. Training and test sets were disjoint, selected in the ratio 1:1 in a randomized manner from the samples.

We compare two methods from the viewpoint of recognition ratio and the number of models. The results on the test set are shown in Tables 9.3 and 9.4. From the tables it is found that our method gave a better result than contour analysis because both rejection and substitution error ratios are less than contour analysis technique, and the num-

TABLE 9.3 Experimental Result of Handwritten Numeral Recognition Using the New Curve Description Technique

	Recognition (%)	Rejection (%)	Substitution (%)	Model
0	99.2	0.8	0.0	3
1	99.6	0.0	0.4	2
2	99.5	0.5	0.0	5
3	98.0	2.0	0.0	4
4	98.9	1.1	0.0	2
5	98.5	1.5	0.0	4
6	99.8	0.0	0.2	2
7	99.2	0.6	0.2	2
8	98.5	1.5	0.0	9
9	99.4	0.4	0.2	4
	99.1	0.8	0.1	37

Note: The thinning method based on cell structure was used for thinning. The experiment was made on 11,500 samples, which is a subset of the loosely constrained data set.

TABLE 9.4 Experimental Result of Handwritten Numeral Recognition Using Contour Analysis Technique

	Recognition (%)	Rejection (%)	Substitution (%)	Model
0	98.6	1.0	0.4	6
1	98.9	1.1	0.0	3
2	98.6	1.4	0.0	4
3	98.0	1.9	0.1	4
4	94.4	4.6	1.0	9
5	97.9	2.1	0.0	5
6	97.9	2.1	0.0	5
7	98.5	1.1	0.4	4
8	96.6	3.2	0.3	12
9	94.0	4.8	1.2	6
	97.6	2.0	0.4	58

Note: The experiment was made on the same data as Table 9.3.

ber of models in our method is only two-thirds of that of contour analysis. Therefore our technique can cope with various patterns of deformation and agrees with human intuition. Flexibility and reliability of our method is verified by the experimental result.

Discussion Handwritten character recognition has been a main research theme in pattern recognition. The prime difficulty in research and development of the technology lies in the variety of deformation of the shape. Various methods such as contour analysis and background analysis have been proposed. Since most of the practical methods utilize some kinds of distance measures or matching methods on the feature space in which the features are position dependent, the normalization on the image is indispensable for coordinate transformation. In other words, image normalization has been a main method to cope with shape deformation. However, it is difficult to know the extent of displacement of the image with respect to the standard coordinate system before recognizing it. It implies that malfunction of normalization distorts the shape and results in rejection or substitution error. Furthermore, most curve representation techniques have been ad hoc, and dictionary creation relies on hand-tuning in structural methods.

We describe the shape of characters by two types of features, namely symbolic, qualitative, and discrete features (quasi-topological features and singular points) and statistical, quantitative, and continuous features (geometrical parameters). The former is regarded as dominant information, while the latter is secondary information attached to structural components. Since the structural description is size and translation invariant, normalization of the image is not necessary for global feature extraction. (Parameters of each shape are computed in the normalized scale, but the normalization has no side effect because it is applied to abstract geometrical parts such as curves and points.) Moreover statistical constraints for deformation can be put on the structural description in terms of the parameterization of shapes, and the categories with the same structure can be discriminated by statistical analysis of geometrical parameters. The structural description and parameterization are simple and agree with human intuition. Therefore we can construct hybrid recognition systems in cooperation with statistical methods, and key features to distinguish confusing characters can be analyzed quantitatively on the basis of the structural description and the parameterization of shapes.

Since different strokes were written independently, they should have independent information and their transformations should be independent from each other. We have shown that singular point decomposition is a clue to the compact and flexible description of curves, and that deformed characters can be recognized easily and effectively. Our recognition method is close to on-line character recognition techniques, while primitive sequences have nothing to do with time sequences of handwriting. The experimental result shows stability and effectiveness of our method.

Furthermore a systematic and rigorous description scheme of shapes is a basis for systematic clustering of character shapes. We have also developed a new method for clustering character shapes in terms of quasi-topological features and singular points, and applied it to dictionary creation.

9.3 BIBLIOGRAPHICAL REMARKS

In the text we have already described some background of the main theme of the Chapter 8, algebraic description. However, in addition to that we should mention the interesting work of French psychologists Jean Piaget and Barbel Inhelder. In particular, Piaget is famous for his work in developmental psychology [8]. Piaget and Inhelder classified infant development to three stages, as follows:

Stage 0 $\rightarrow$ 1 ~ 3 ages
Stage I $\rightarrow$ 3 ~ 4 ages
Stage II $\rightarrow$ 4 ~ 6 ages

In stage 0, children do not have ability to write any meaningful shape. However, in a stage I, they have usually remarkable ability to recognize "topological difference" in a simple figure. Three kinds of figures were provided to the children: a loop with a blob inside, a loop with a blob on the boundary, and a loop with a blob outside. The children at stage I were asked to copy these figures, and they were able to copy these figures keeping topological properties. On the other hand, they exhibited considerable difficulty in trying to copy such a figure as "X" or a rectangle. At stage II then, the children could have the ability to copy such geometrical figures as "X" and a rectangle. Therefore Piaget and Inhelder concluded that it is appropriate for children to start to learn characters from 6 years old. (My experience with my children has been completely the same.) Piaget and Inhelder's work is evidence that so-called quasi-topological features are very natural features in human perception.

Concerning algebraic description, some historical papers are introduced in Nishida and Mori [11]. A number of them were already described in this chapter. However, we need to mention that monotonic functions were described as primitives in the United States Patent by M. Oka, S. Yamamoto and S. Kadota [9]. The algebraic approach is related to a so-called syntactic approach that emphasizes "grammar." The historical aspects of this work are described fully by Fu and his associates in Fu's book [10]. On the other hand, recently Nishida has treated the algebraic approach more intensively and achieved remarkable results, which seems to prove that an algebraic description is essential in shape modeling. His results will be described here in three parts: shape analysis and description, shape transformation models, and model-based shape matching and recognition.

Shape Analysis and Description The algebraic description presented in this chapter was originally published by Nishida and Mori [11]. However, the number of directions is fixed to 4 in the analysis of directional features, and more directions such as 8 or 16 cannot be dealt with. For various and practical applications of the algebraic description, Nishida [12] extended the analysis to the case of 2^m-directional features ($m = 2, 3, 4, \ldots$), and Nishida [13] further generalized the analysis to the case of $2N$-directional feature ($N = 1, 2, 3, \ldots$).

Shape Transformation Models In general, structural shape transformations can be classified into two categories: continuous and discontinuous. The algebraic description is also a powerful tool for modeling these structural transformations clearly and rigorously.

A continuous transformation is a shape transformation that does not change the global structure of the shape. Furthermore a small amount of a continuous transformation should cause a small amount of change in a shape's features. In terms of the *PS-labels* characterizing the global and qualitative features of curve components, Nishida [14] presented a structural model for continuous transformation of shapes with simple, local operations that preserve the global structure. Another open problem in character recognition is in the analysis of *metamorphosis* whereby a character shape which is an instance of a class is transformed into an instance of another class via continuous transformations. An essential problem in handwriting recognition is how to cope with such complex shape deformation, and therefore the modeling of the metamorphosis is a key to breaking through the difficulties in handwriting recognition. Based on this structural deformation model, Nishida [15] also presented an experimental approach to analysis of *metamorphosis* of character shapes.

On the other hand, discontinuous transformations are catastrophic, since they change the global structure and features of a shape. Discontinuous transformations are also intractable, and there have been few systematic mathematical tools for practical use in computer vision and pattern recognition. Therefore practitioners in these fields have been obliged resort to heuristic methods for coping with discontinuous transformations. Among the various types of 2D patterns, handwritten characters are unique in their shape deformation. In particular, discontinuous transformations commonly occurring in handwriting can be classified as two types:

T1: Concatenating two end points of the curve's components by slight movement (stroke concatenation). If the end points of strokes coincide, the end points disappear.

T2: Connecting two end points of curve components with an additional curve.

These transformations are quite small, but they change global features and shape structures significantly. Since it is difficult to systematically analyze the a priori effects of these transformations, so far we are obliged to depend on heuristics to handle discontinuous transformations. Furthermore the extent of shape deformation is usually measured by a transformation cost determined heuristically or statistically. However, when we deal with complex patterns subject to complicated transformations, simple cost measures are inadequate for explaining the transformations, so some systematic, high-level model is necessary to guarantee the relevance of the matching process and class descriptions. Therefore it is now important to obtain systematic and complete knowledge of the effects of the transformations. In other words, the analysis must be carried out mathematically without depending on heuristics so that there can be obtained a small, tractable number of distinct cases. Once we have complete a priori knowledge of the possible cases, we can analyze other instances satisfying certain realistic conditions in a unified, systematic way without resorting to heuristics. To respond to these requirements, Nishida [16] carried out a complete and systematic analysis of the a priori effects of the discontinuous transformations $T1$ and $T2$. The results are his transformation laws which are based on a small and tractable number of distinct cases.

Automatic Construction of Structural Models and Learning Construction of class descriptions (prototypes) is an essential step in model-based machine vision systems. In particular, structural approaches have been widely adopted in recognition systems for complex objects or patterns. A substantive criticism against the structural approach is the difficulty of automatic inference of class descriptions. In many recognition systems based on structural methods, much time and human-power are usually spent in mannual construction of class descriptions. Another criticism is that a large number of class descriptions are required for taking account of shape variation systems, and the number of class descriptions constantly increases for dealing with a variety of possible cases. Maintenance of such recognition systems is difficult because of an intractably large number of class descriptions incorporated into the systems.

Many of early approaches to automatic construction of structural models are based on the idea of generalizing structural descriptions to class descriptions by applying some statistical or stochastic learning procedures to some similarity measures of attributes on top of structural descriptions. Such approaches are successful for simply structured patterns or objects. However, when we deal with complex patterns or objects subject to complicated transformations, simple similarity measures are inadequate as criteria for the generalization, and some systematic, high-level model is necessary to guarantee the relevance of the generalization process.

Nishida and Mori [17], Nishida [18] took different viewpoint for the problem of automatic, inductive construction of class descriptions from the data. This problem is considered as constructing *inductively, from the data set,* some shape descriptions that tolerate certain types of *shape transformations.* Shape transformations to be applied to patterns or objects are defined and analyzed explicitly in terms of some particular scheme of the algebraic description. In this way systematic, complete, a priori, high-level knowledge are obtained for the effects of all possible deformations caused by the transformations. The shape transformation models [14, 16] play essential roles in this respect. Then the problem of automatic construction of class descriptions can be stated as generalizing shape descriptions that can be transformed to each other by the particular types of shape transformations. The generalization process is controlled by high-level models of shape transformations, and is supported and guaranteed by the high-level knowledge.

Model-Based Shape Matching and Recognition A similar algebraic approach to structural shape matching is to describe the input pattern in a bottom-up manner and to apply some simple matching method to the class descriptions and the structural description of the input pattern. However, the algebraic shape description integrates low-level features into high-level features with some simple geometric conditions caused by stroke connection and breaking, and the global structure changed accordingly. Although the computational cost of this approach is quite low, we have to prepare a very large number of class descriptions taking account of all possible cases due to random variations as well as structural deformations. Therefore we need to incorporate some high-level knowledge and models into shape-matching algorithms in a *top-down* way. However, typical defects of *top-down* approaches are that strong assumptions are required for applying high-level knowledge (if the assumptions do not hold, then the method fails) and that knowledge-based systems tend to be ad hoc and difficult maintain.

Nishida [19] presented a clear and systematic approach to shape matching based on structural feature grouping. To cope with topological deformations caused by stroke connection and breaking, he incorporated some aspects of top-down approaches in a systematic way. The grouping of local structural features into high-level features is controlled by high-level knowledge, so the shape-matching algorithm requires a small number of prototypes (one class for one character). This shape-matching algorithm was originally developed for on-line handwriting recognition without any writing constraints; that is, the order of strokes is free, the number of strokes is free, and stroke connection and breaking are allowed. This algorithm has also been applied to off-line handwriting recognition [19] successfully.

On the other hand, the key to recognizing such complex objects as handwritten characters is through a shape description that is robust against shape deformation and quantitative estimation of the amount of deformation. In the quantitative analysis of deformation it is natural to assume that a point x' on the object is mapped to a point x on the *model* by a transformation T:

$$x = Tx'. \tag{9.35}$$

In general, the analysis of pattern deformations is difficult because deformation of such patterns as handwritten characters is nonrigid and elastic, and T is nonlinear. Furthermore the point correspondence between the model and the object must be known in advance in order to obtain T, which is also difficult to do for similar reasons.

An idea for finding the point correspondence and handling the nonlinearity of T is as follows:

1. Describe the shape in terms of a qualitative and global structure that is robust against deformation, and match the object against the built-in models. Then the correspondence of components (units in the structural description) is found between the model and the object. The point correspondence can be found by choosing some reference points on the components.
2. Regard the transformation T as the composition of a simple geometrical transformation (coordinate transformation) T_g and a statistical transformation T_s representing complex transformation that cannot be described by the geometrical transformation.

Based on the algebraic description, Nishida [20] proposed a shape-matching algorithm and a method of analysis and description of shape transformation for handwritten characters; he also presented a systematic approach to the problem of coping with pattern deformations of handwritten characters by integrating structural descriptions with geometrical/statistical transforms. Global, qualitative, and discrete features (quasi-topological features, directional features, and singular point) are regarded as the dominant information, whereas quantitative and continuous features (geometrical parameters, e.g., size and position) are used as secondary information attached to structural components. The object is described in terms of a qualitative and global structure that is robust against deformation, and the description is matched against built-in models. Then the correspondence of components (units in the structural description) is found between the model and the object. The point correspondence can be found by choosing some reference points on the components. T_g is obtained on the basis of the correspondence of components between the object and the model. By transforming each point on the object by T_g, the object is normalized with respect to the model. Then T_s is estimated on the normalized shape, and the decision of recognition or rejection is made on the basis of T_g and T_s.

BIBLIOGRAPHY

[1] A. C. Shaw, "The formal picture description scheme as a basis for picture processing system," *Infor. Control,* vol. 14, pp. 9–52, 1969.

[2] C. Lévi-Strauss, *Les Structures élementaires de la parente,* Paris: Plon, 1949.

[3] A. C. Moorhouse, *Writing and the Alphabet,* London: Gobett Press, 1946.

[4] H. Weyl, *Symmetry,* Princeton: Princeton University Press, 1952.

[5] H. Yamada and S. Mori, "Analysis of handprinted data base I," *Bull. Electrotechn. Lab.,* vol. 33, no. 8, pp. 580–589, 1975.

[6] G. C. Buck, *Advanced Calculus,* 3rd ed., New York: McGraw-Hill, 1978.

[7] S. Mori, "An algebraic structural representation of shape," *Trans. IECE Japan,* vol. J64-D, no. 8, pp. 705–712, August 1981. (in Japanese)

[8] J. Piaget and B. Inhelder, *The Child's Conception of Space,* trans. F. J. Langdon and J. L. Lunzer, London: Routledge, 1956.

[9] M. Oka, S. Yamamoto, and S. Kadota, "Pattern feature detection system," United States Patent, no. 3, 863, 218, January 28, 1975.

[10] K. S. Fu, *Syntactic Pattern Recognition and Applications,* Englewood Cliffs, NJ: Prentice-Hall, 1982.

[11] H. Nishida and S. Mori, "Algebraic description of curve structure," *IEEE Trans. Pattern Anal. Machine Intell.,* vol. 14, no. 5, pp. 516–533, May 1992.

[12] H. Nishida, "Structural feature extraction using multiple bases," *Comp. Vision Image Understanding,* vol. 62, no. 1, pp. 78–89, July 1995.

[13] H. Nishida, "Curve description based on directional features and quasi-convexity/concavity," *Pattern Recogn.,* vol. 28, no. 7, pp. 1045–1051, July 1995.

[14] H. Nishida, "A structural model of shape deformation," *Pattern Recogn.,* vol. 28, no. 10, pp. 1611–1620, October 1995.

[15] H. Nishida, "Model-based shape matching with structural feature grouping," *IEEE Trans. Pattern Anal. Machine Intell.,* vol. 17, no. 3, pp. 315–320, March 1995.

[16] H. Nishida, "Automatic construction of structural models incorporating discontinuous transformations," *IEEE Trans. Pattern Anal. Machine Intell.,* vol. 18, no. 4, pp. 400–411, April 1996.

[17] H. Nishida and S. Mori, "An algebraic approach to automatic construction of structural models," *IEEE Trans. Pattern Anal. Machine Intell.,* vol. 15, no. 12, pp. 1298–1311, December 1993.

[18] H. Nishida, "A structural model of curve deformation by discontinuous transformations," *Graph. Models Image Processing,* vol. 58, no. 2, pp. 164–179, March 1996.

[19] H. Nishida, "An approach to integration of off-line and on-line recognition of handwriting," *Pattern Recogn. Lett.,* vol. 16, no. 11, pp. 1213–1219, November 1995.

[20] H. Nishida, "Shape recognition by integrating structural descriptions and geometrical/statistical transforms," *Comp. Vision Image Understanding,* vol. 64, no. 2, pp. 248–262, September 1996.

CHAPTER TEN

Background Analysis

So far we noticed black portions, namely strokes that constitute a character. However, white portions of a character are also important ingredients of a character, since they are related to the black portions complementarily. Here we introduce another angle of feature extraction, called background analysis. General characteristics of the features obtained by background analysis lie in its globality. To extract meaningful feature, the neighbor of a white point must necessarily be large so that black portions can be captured. In this sense the background analysis approach is complementary to the usual feature extraction approach that we have described so far. We explain this approach from one-dimensional, digital one- to two-dimensional, and analog systematically.

10.1 CHARACTERISTIC LOCI

Characteristic loci was devised by Glucksman [1], who is a precursor of background analysis. *Glucksman's method* was highly developed in Japan, as described later, and has contributed greatly to OCR technology. Glucksman named his features *characteristic loci.*

Glucksman's method is very simple and a natural extension of the sonde method. As shown in Fig. 10.1, at every white point, four stroke detecting lines are spanned, namely to the top, bottom, right, and left. Then the crossing number of each line is counted. In this way at every white point we obtain a four-dimensional feature vector (1, 2, 3, 0), where the counting begins from the top detecting line and arranges the crossing numbers clockwise as in Fig. 10.1. Actually Glucksman contracted the expression above such that the counting number being greater than 2 was set to 2. He tried his method on the roman alphabet whose complexity of characters is not high, and this contraction contributed to the compact representation. Obviously the originality here lies in the expansion of focal points in the sonde method to all the white

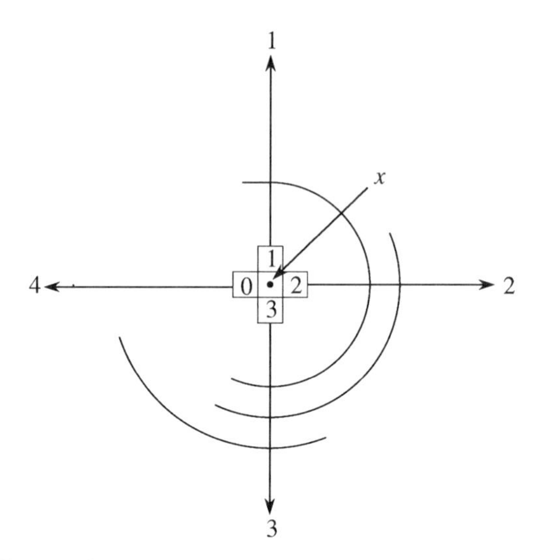

FIGURE 10.1 Glucksman's method for extracting features at point x.

points. Therefore Glucksman's method is uniform avoiding the ad hoc setting of focal points in the sonde method. Finally we will illustrate how this method was used by Glucksman himself.

Figure 10.2 shows both sonde and Glucksman's method. As described before, the sonde method is occasionally weak against a small change in a character. In panel (*b*) we see such a small change, and the image (*b*) is identified as an "8" as shown in panel (*b*) and (*c*). Of course, this point is improved easily by adding another detecting line, as shown by the dotted lines in the figure. Actually such improvement was tried and some complicated detecting lines were used to handle the variation in handprinted characters [2]. However, such improvement is ad hoc and cannot be uniformly applied.

On the other hand, the distribution of the frequencies of the four-dimensional feature vectors of each character image is stable as shown in Fig. 10.3. Here we notice that for simplicity the small feature vectors that appear at the places marked by an asterisk in Fig. 10.2 are neglected. The detailed illustration is shown in Fig. 10.4. These feature vectors are not important and not stable. Such simplification was done by Glucksman himself. The amount of the frequency change is proportional to the increase of area which has a corresponding four-dimensional feature vector such as (1, 1, 2, 1). There is no sudden jump in the frequency of a feature vector for a small continuous change of an image. This is a very important point in the recognition based on features. Formally, a small continuous change in the image plane is reflected in the feature space as a small change also. This reflects also the property that the feature is global. Therefore we can expect reasonable clustering in the feature space of the four-dimensional feature vectors, and this makes the discrimination easy in the feature space. Thus Glucksman constructed four-dimensional feature space to which he applied linear discrimination functions, and he obtained good results in reading a printed roman alphabet character set. His way of recognition is

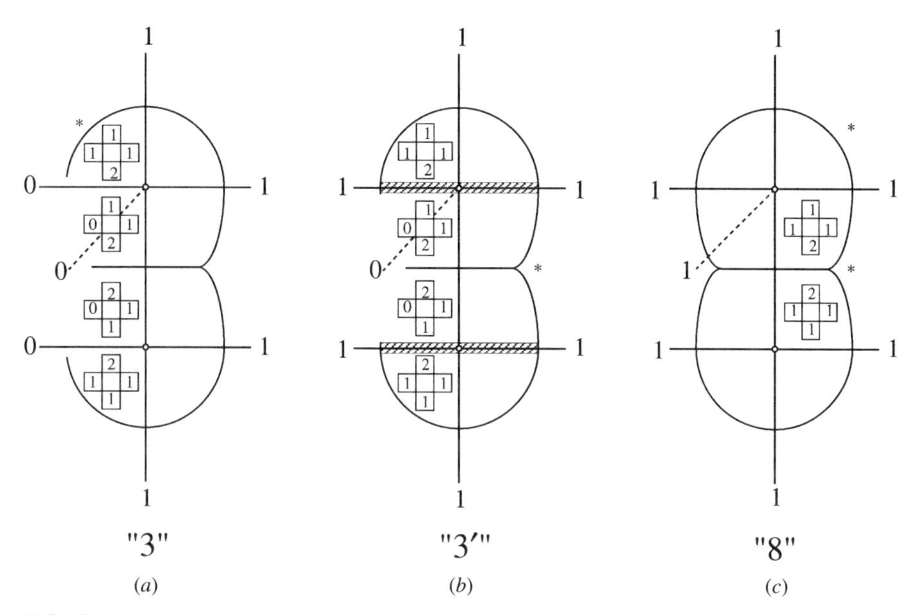

FIGURE 10.2 Effectiveness of Glucksman's method over the sonde method: (*a*) Character image "3"; (*b*) slight change of the image (*a*); (*c*) character "8" whose considerable part constitutes "3."

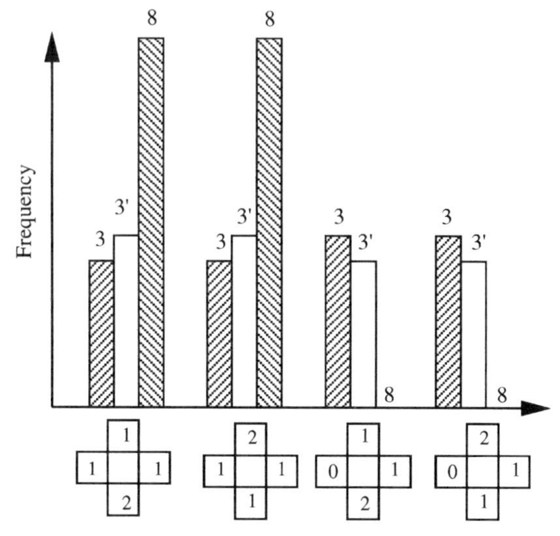

FIGURE 10.3 Histogram of the 4D features in the examples shown in Fig. 10.2, where 3 and 3′ denote the histogram change according to the little deformation of "3" to "3′" in Fig. 10.2.

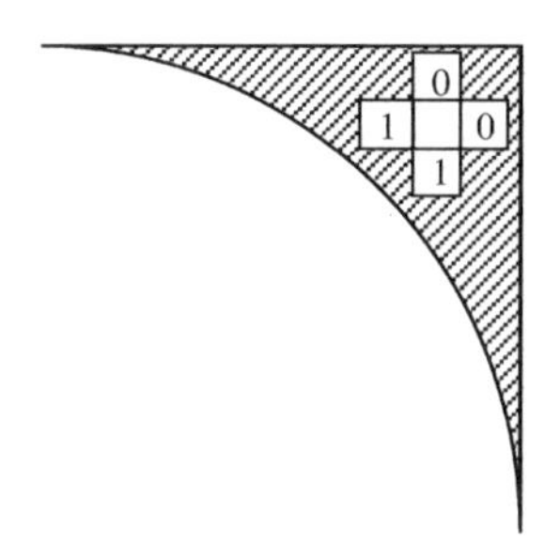

FIGURE 10.4 Example of a nonessential feature vector neglected in the Glucksman's experiment.

simple and has certain strong points as mentioned above, but it sacrifices the structural potentiality of the method. Next we will mention that structural point.

10.2 SEGMENTATION TO CONCAVITY REGIONS

The histogram method using four-dimensional feature vector is one-dimensional in the sense that it does not take the concept of neighborhood into account after the feature extraction. We can merge the same fragmentary feature vector regions so that it constructs a meaningful concavity region as shown in Fig. 10.5. Notice here that a hole is described as its extreme case of concavity. That is, the image of "6" can be regarded as consisting of two concavity regions, namely the concavity region open to the right and the hole. Therefore the image is described by the two regions very simply and very globally something like ⊂ /○. The next problem is how to merge four-dimensional feature vector regions.

For this purpose we classify the four-dimensional feature vectors so that it meets our intuition. This is shown in Fig. 10.6. Based on the classification the four-

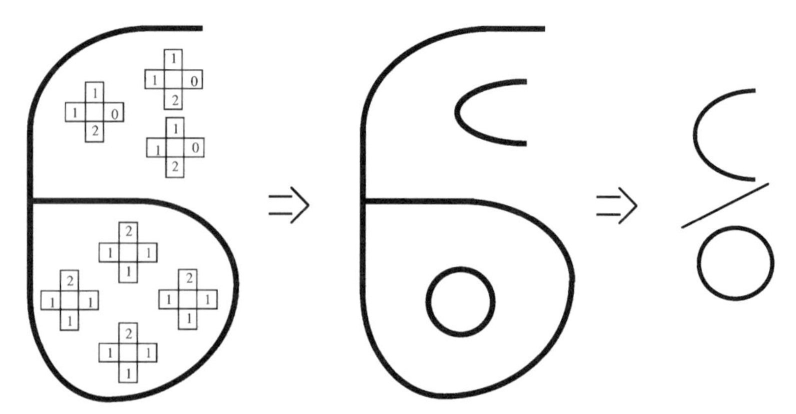

FIGURE 10.5 Schematic of the integration of 4D feature distribution to the structural description.

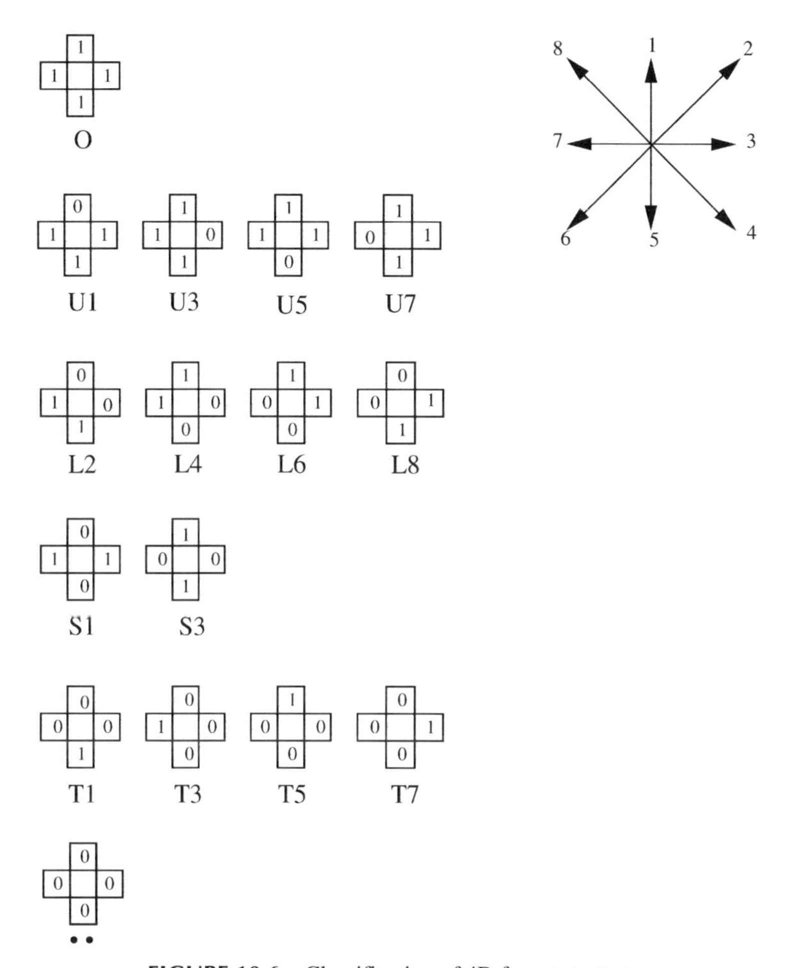

FIGURE 10.6 Classification of 4D feature vectors.

dimensional feature vectors are labeled as shown, called *white region primitives* (*WR primitives*). This merging process is performed by finding the connected region of each label, which is done by a standard labeling method [3]. However, one problem arises as shown in Fig. 10.7. That is, a superficial hole appears. Next we describe how to improve this point. The first is an approach from linguistics, and the second is a functional approach.

10.2.1 Linguistics Approach

Linguistics analyses have broad applications, so they have been used by many researchers. A standard book by Fu [4] and papers by Narashimhan [5]–[7] are proponents of this approach. So we will give as much an introduction as necessary for the linguistics approach to be understood.

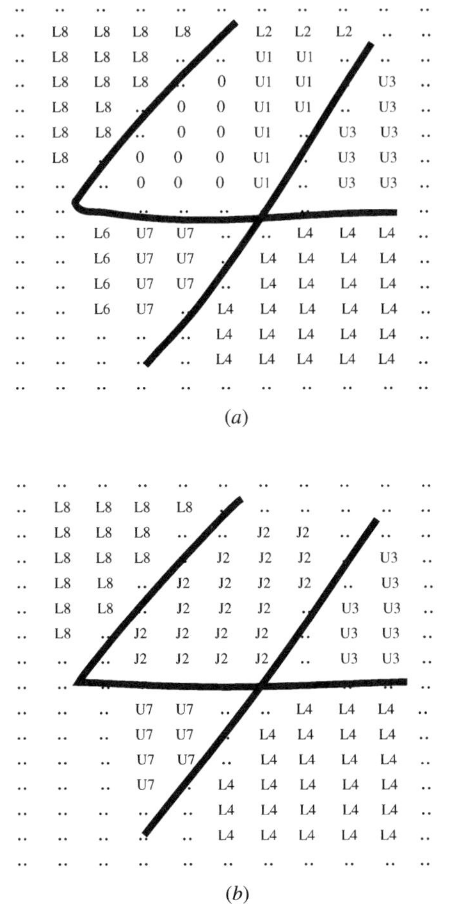

FIGURE 10.7 Example of region integration based on the linguistics approach: (*a*) An apparent hole region is constructed; (*b*) the apparent hole region is assessed by the linguistics rules.

Narashimhan gave a general framework for pictorial language which is represented by

$$G = G(P, A, R, C, T), \tag{10.1}$$

where G is a set of rules constructing a class description; P represents primitives whose attributes are denoted by A; R, C, and T are relational set, connection set, and transformation set, respectively. For example, the connection rule

$$f \leftarrow r(P_1, P_2) \tag{10.2}$$

denotes that a new primitives f is generated when P_1 and P_2 satisfy the relation r. In general, a connection rule is expressed as

$$f_3(\cdot) \leftarrow r(f_1(\cdot), f_2(\cdot)), \tag{10.3}$$

where $(\cdot)$ denotes some kind of attribute. The above expression can be described recursively as

$$f_1(\cdot) \leftarrow r(f_1(\cdot), f_2(\cdot)). \tag{10.4}$$

Finally transformation includes deleting and rotation as typical examples and is expressed as

$$f \leftarrow T(P). \tag{10.5}$$

Now we illustrate the linguistics approach based on the framework mentioned and more concretely referring Fig. 10.7. First, the following connection rules are applied to panel (*a*),

$$J2 = r(U1,\ O^7), \tag{10.6}$$

$$J2 = r(O,\ U1^3), \tag{10.7}$$

$$J2 = r(U1,\ J2^7|J2^5), \tag{10.8}$$

$$J2 = r(O,\ J2^3|J2^1). \tag{10.9}$$

For example, (10.6) means that the primitive $U1$ is replaced by $J2$ if $U1$ has O in the direction 7 at the point labeled $U1$. In general, such a connection rule is expressed as

$$f \leftarrow \gamma(P_1, P_2^i) \qquad i = 1 \sim 8. \tag{10.10}$$

That is, the primitive P_1 is replaced by f if P_1 has the primitive P_2 in its i direction. The symbol "|" means OR. Thus the superficial hole region of the image "4" in panel (*a*) is altered to the region labeled $J2$ as shown in panel (*b*). The simplification process mentioned in Glucksman's method is also done such that the $L2$ region adjacent to the $J2$ and $L6$ region adjacent to $U7$ are deleted as less significant regions. We don't mention this approach further because this is somewhat complex and the functional approach is compact and systematic. Our aim here is to give the reader a flavor of the linguistics approach.

10.2.2 Functional Approach

Although a linguistics approach is flexible, it results in a great number of rules in general. On the contrary, a functional approach [8] in which transformation rules are expressed by a small number of formulas is compact. We will show the case here.

Let each element of a *WR* primitive at point $\mathbf{x}$ be $D^v(\mathbf{x})$, and let the unit vector be $\mathbf{r}_i$. Notice that the direction of the four-dimensional feature vector element differs from that in Fig. 10.6, as shown in Fig. 10.8. Each *WR* primitive is denoted as Q_i, and its schematic shape is shown in Fig. 10.9. Furthermore Q_i is coded for its compact

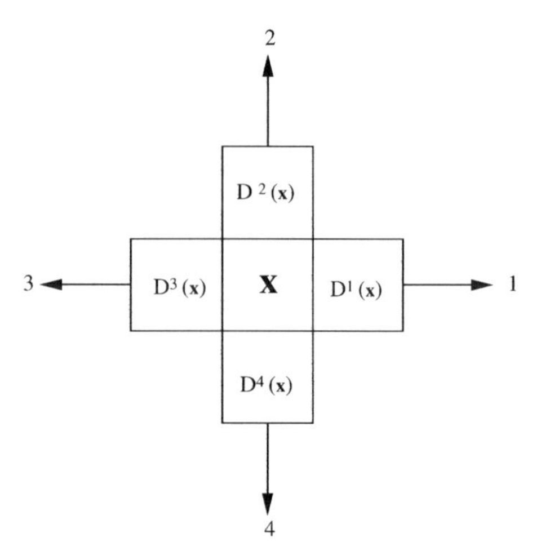

FIGURE 10.8 Notation used in the *WR* primitives.

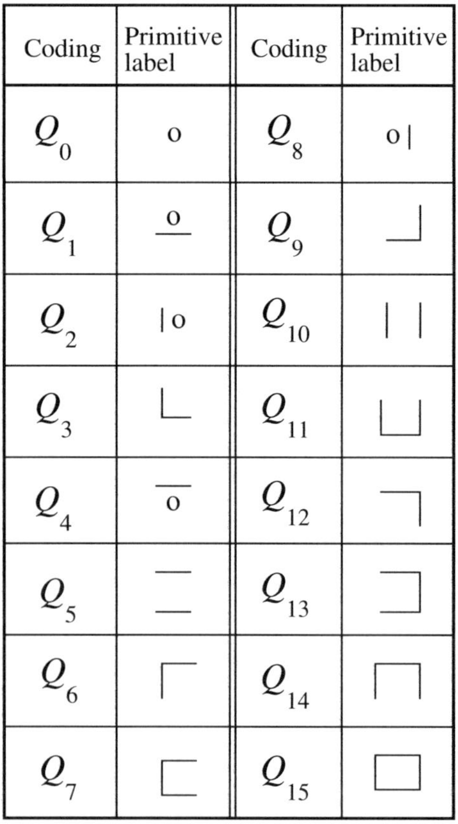

Coding	Primitive label	Coding	Primitive label
Q_0	o	Q_8	o |
Q_1	o̲	Q_9	┘
Q_2	| o	Q_{10}	| |
Q_3	└	Q_{11}	└┘
Q_4	o̅	Q_{12}	┐
Q_5	=	Q_{13}	]
Q_6	┌	Q_{14}	┌┐
Q_7	[	Q_{15}	□

FIGURE 10.9 Coding and labels of the *WR* primitives. From reference [8], (c) 1998, IEEE.

representation in its hardware implementation; namely 16 kinds of Q_i are implemented economically in a half byte.

Specifically, $D^{\nu}(\mathbf{x})$ is defined as

$$D^{\nu}(\mathbf{x}) = \begin{cases} 1 & \text{if there is black portion in the direction } \nu, \\ 0 & \text{otherwise.} \end{cases}$$

$D^0(\mathbf{x})$ denotes the value at point $\mathbf{x}$ as 1 or 0, corresponding to black or white, respectively. Then the rewriting rules mentioned before can be represented compactly. That is, if

$$\sum_{\nu=1}^{4} D^{\nu}(\mathbf{x})D^{\nu}(\mathbf{x}+\mathbf{r}_i)D^{\nu+2}(\mathbf{x})D^{\nu+2}(\mathbf{x}+\mathbf{r}_i) = 1, \qquad i = 1 \sim 4, \tag{10.11}$$

holds for some i, then the j of the code Q_j at point $\mathbf{x}$ is expressed by the following formula for the i:

$$j = \sum_{\nu=1}^{4} 2^{4-\nu} D^{\nu}(\mathbf{x})D^{\nu}(\mathbf{x}+\mathbf{r}_i). \tag{10.12}$$

Notice that in (10.11) the first i is selected in its increasing order from 1 to 4. This causes a problem as will be illustrated below. The result is too compact and not intuitive. Therefore the schematic representation of the rules is shown in Fig. 10.10. Looking at the figure, we can deduce some simple intuitive rules. That is, if a center has label □ and its neighbor has a weak label in the sense of degree of concavity, then the center label changes to the weak label. For U-type labels we can take AND between schematic shape at a center and its neighbor's one, such as ⊓ AND ⊔ = ||. We note here that the result of the rules application depends on the order of the applied rules. A typical example is shown in Fig. 10.11. The result is reasonable because the direction is quantized only in four directions, and the diagonal direction is just middle of = and ||, so it cannot be represented in this limited scheme appropriately. This point can be improved taking eight quantized directions instead of the four quantized directions, and we will describe the procedure next. A simple improved example adopting the eight quantized direction scheme is shown in Fig. 10.12.

10.3 FAN TYPE OPEN-DIRECTION MAP

As stated before, increasing of the directional resolution from 4 to 8 is a natural expansion of the *WR* primitive. In addition further improvements are considered [10]. One is to homogenize the directions. As is well known, on a grid plane, 8 directions are not homogeneous. The other is to change the view from line to plane. That is, in the old *WR* primitive, four detection lines are ejected from each point to obtain information in its neighborhood. However, the information is only limited to the four detecting lines, and the other neighborhood information is neglected. Therefore, to be exact, it is favorable to get planar information.

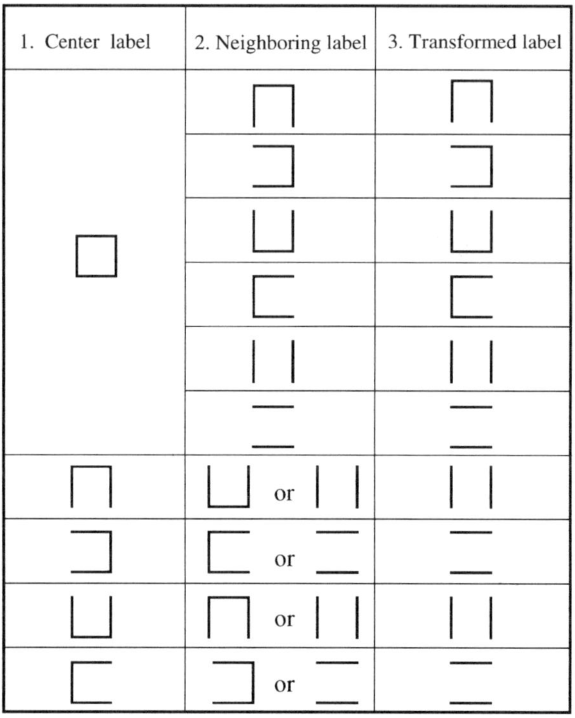

1. Center label	2. Neighboring label	3. Transformed label
	or	
	or	
	or	
	or	

FIGURE 10.10 Schematic of the functional rule. From reference [8]; (c) 1998, IEEE.

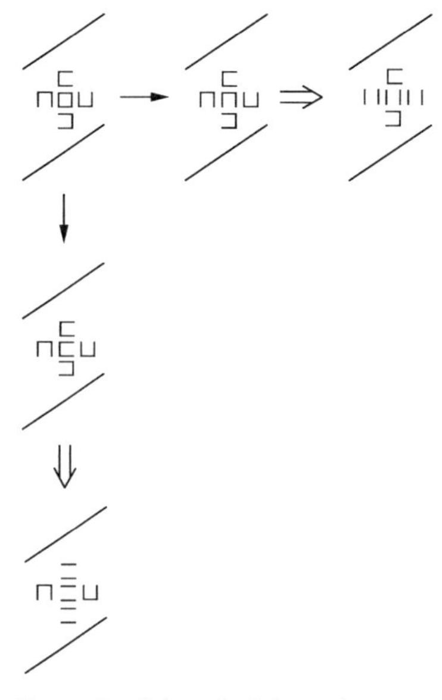

FIGURE 10.11 Example of the rules' dependency on application order.

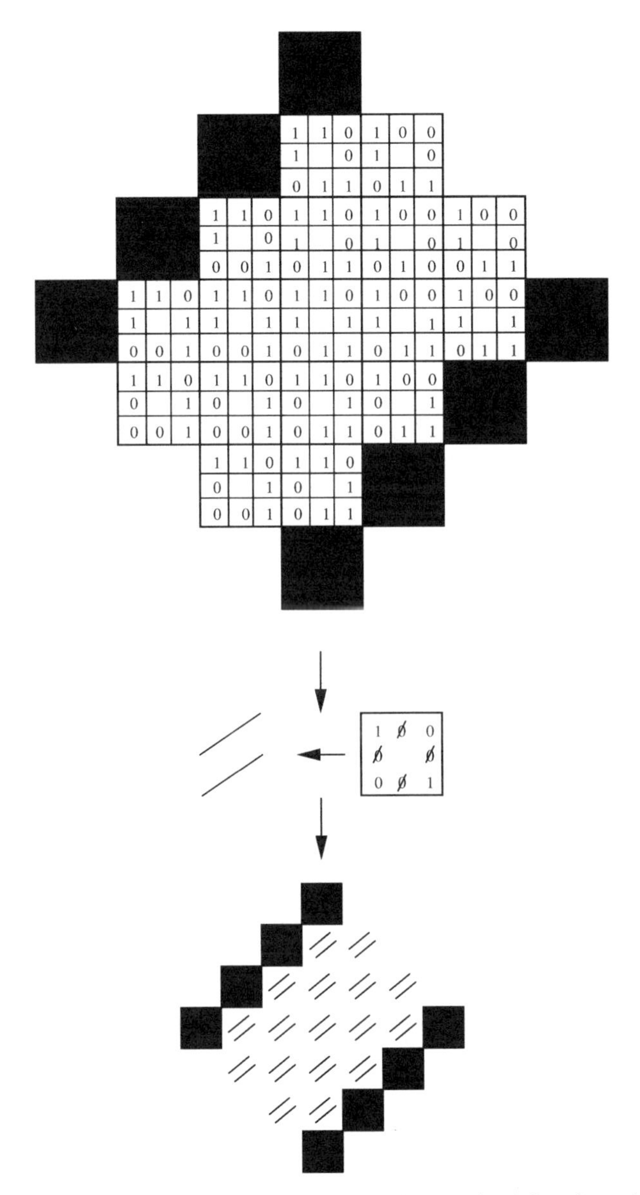

FIGURE 10.12 Example of the rule for the eight quantized direction scheme.

Specifically, this is illustrated in Fig. 10.13 (*a*) for the eight direction lines and (*b*) for the new planar directions each of which takes the shape of a fan. That is, centering on the (x, y) point, eight directional elements M^{ν} are shown where ν takes the values, $1\frac{1}{2}$, $2\frac{1}{2}, \ldots, 8\frac{1}{2}$. In this sense, M^{ν} indicates the closed state from the direction $\nu - \frac{1}{2}$ to $\nu + \frac{1}{2}$. For example, setting $\nu = 1\frac{1}{2}$, $M^{1\frac{1}{2}}$ gives the planar fan type direction from direction 1 to direction 2. In this sense the method treats eight directions on the plane homogeneously. That is the reason why this method was called a *fan type open-direction map.*

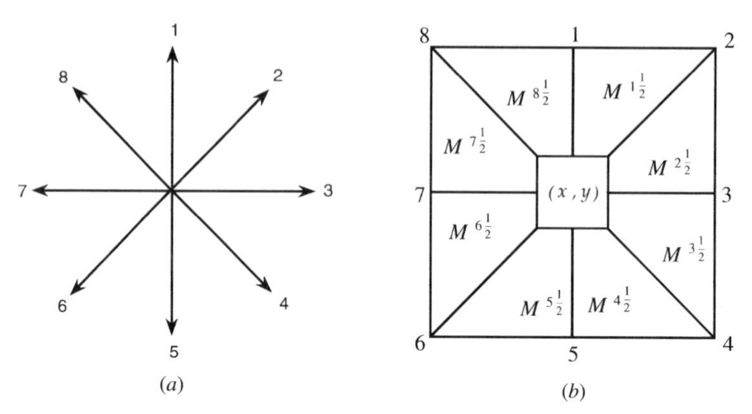

FIGURE 10.13 Notation for fan type open-direction map: (*a*) Eight direction lines; (*b*) planar directions. From reference [10].

Now let us see what the advantage of this method is. This is shown in Fig. 10.14. The way information propagation on a subplane can be considered is from two angles. In one way, called the MIN type, $M^{\nu} = 1$, means that we cannot rearch the frame segment cut by the fan open to the ν direction from the center (x, y) no matter what paths are taken, panel (*e*). In contrast, $M^{\nu} = 0$ if we can choose a path from the center to the frame segment, panels (*a*) through (*d*). Therefore in only the case shown in panel (*e*), $M^{\nu} = 1$ holds (MIN type). The other way to divide the subplane is by a MAX type method, in which $M^{\nu} = 1$ if there exists a black portion in the area of the M^{ν} fan type subplane. Therefore in only the case shown in panel (*a*), $M^{\nu} = 0$ holds (MAX type). Intuitively speaking the MIN type is strict and the MAX type is loose in terms of closeness.

Next we explain how to get the value M^{ν}. This is done simply by a propagation. Specifically, for the MIN type, we begin with an initial value

$$M^{\nu} = P(\mathbf{x}). \tag{10.13}$$

That is, all the $M^{\nu}(\mathbf{x})$ are set to their center value $P(\mathbf{x})$. Then the following propagation formula is applied uniformly:

$$M^{\nu}(\mathbf{x}) = P(\mathbf{x}) \cup [M^{\nu}(\mathbf{x} + \mathbf{r}^{\nu-(1/2)} \cap M^{\nu}(\mathbf{x} + \mathbf{r}^{\nu+(1/2)})] \qquad \nu = 1\tfrac{1}{2}, 2\tfrac{1}{2}, \ldots, 8\tfrac{1}{2}, \tag{10.14}$$

where $\mathbf{x} = (x, y)$, $\mathbf{r}^{\mu}$ is unit vector on the feature plane related to ν planar direction. For example, when $\nu = 1\frac{1}{2}$, $(\mathbf{x} + \mathbf{r}^1) = (x, y + 1)$, $(\mathbf{x} + \mathbf{r}^2) = (x + 1, y + 1)$; that is, the fan is spanned by $x + \mathbf{r}^1$ and $x + \mathbf{r}^2$. These two rules

$$(x, y) \to (x, y + 1),$$
$$(x, y) \to (x + 1, y + 1),$$

are applied repeadedly such that

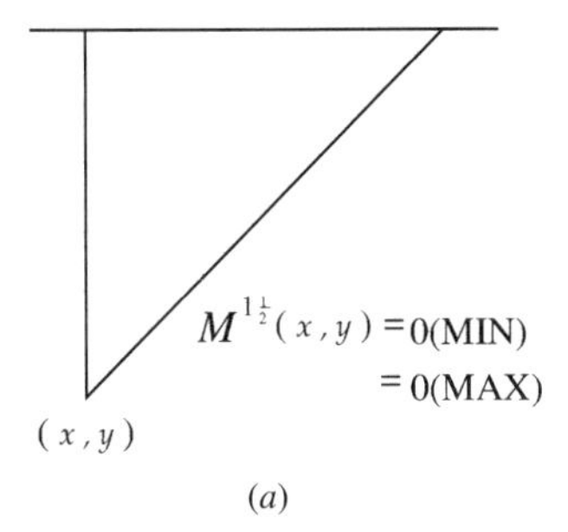

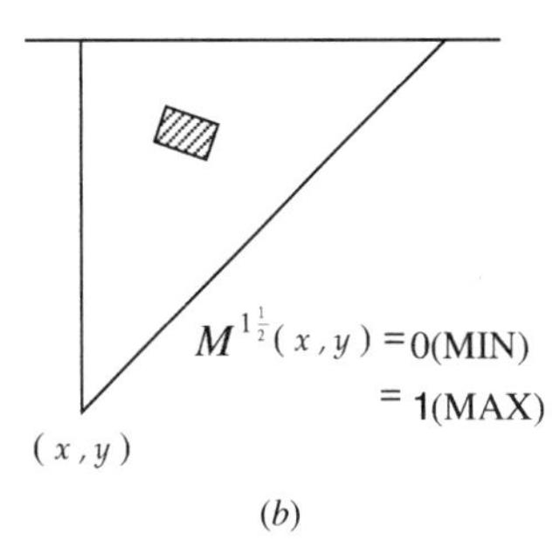

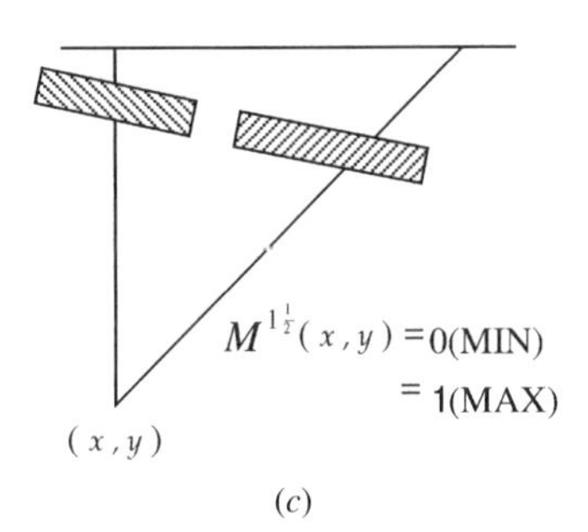

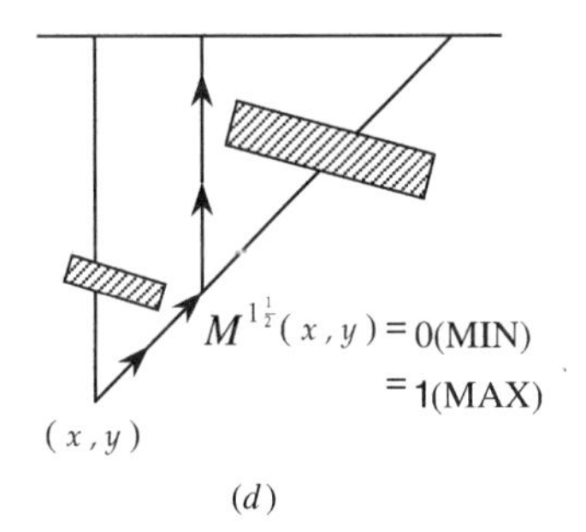

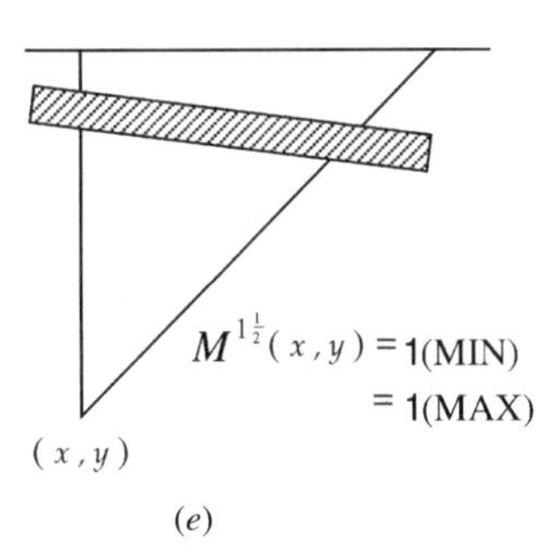

FIGURE 10.14 Two types of propagation methods, MIN and MAX and their conceivable combinations. From reference [10].

$$(x, y+1) \rightarrow (x, y+2),$$

$$(x, y+1) \rightarrow (x+1, y+2),$$

$$(x+1, y+1) \rightarrow (x+2, y+2) \ldots,$$

until a set of points that constitutes the fan is obtained.

For the MAX type, the propagation formula is expressed as

$$M^{\nu} = P(\mathbf{x}) [\cup M^{\nu}(\mathbf{x} + \mathbf{r}^{\nu-(1/2)}) \cup M^{\nu}(\mathbf{x} + \mathbf{r}^{\nu+(1/2)})] \qquad \nu = 1\tfrac{1}{2}, 2\tfrac{1}{2}, \ldots, 8\tfrac{1}{2}. \tag{10.15}$$

Thus we can obtain the feature plane based on MIN or MAX type propagation. That is, at each point on the plane an eight-dimensional vector is placed that reflects a global feature looking at that point. Representations and labels of these new *WR* primitives are given in Table 10.1.

Examples of region segmentation based on the MIN type and the MAX type are shown in Fig. 10.15. As seen, the MIN type is strict, so no complete closed label *Q* exists. On the contrary, in the MAX type large regions of label *Q* appear because of the looseness. In this sense the MIN type describes the concave regions exactly. The next problem is how to integrate these fragments of labeled regions. The orthodox approach is to construct a graph in which the nodes and edges correspond respectively to the labeled regions and boundaries in between them. Each edge value then takes the boundary length of the two adjacent subregions. This is shown in Fig. 10.16 in the MIN-type case, where *L* and *I* labeled regions have been neglected. We can see two clusters of the labeled regions, the upper half and the bottom half. In particular, the upper half cluster constitutes a complete graph which means that the region is compact. Thus we can describe the inner structure of the large concave region. On the other hand, the MAX type is strong against broken noise. The problem is how to integrate the MIN and MAX types, which are open.

From the engineering viewpoint, we can introduce a very simple integration. That is, for the MIN type, we integrate the labeled regions that connect to the *I*-labeled

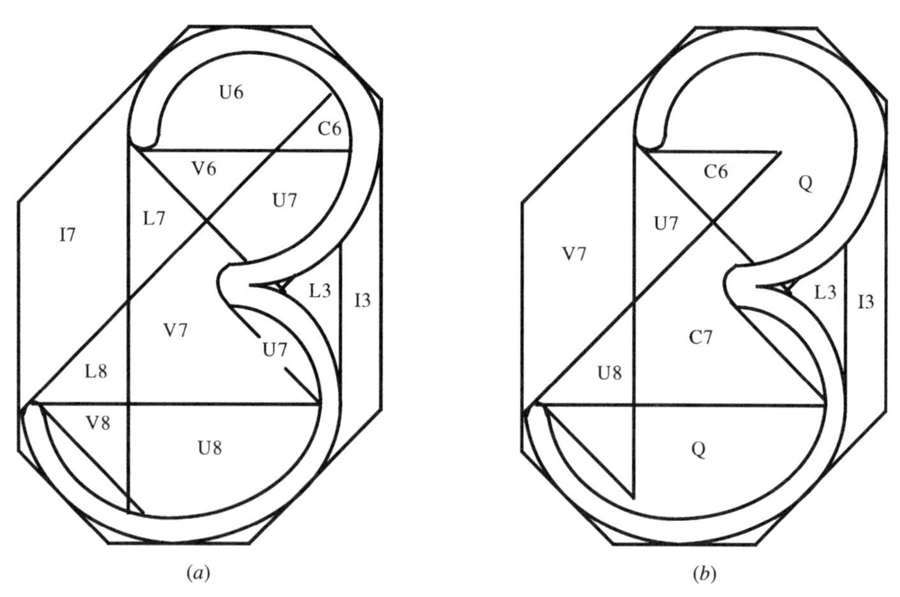

FIGURE 10.15 Examples of region segmentation by MIN type (*a*) and MAX type (*b*). From reference [10].

TABLE 10.1 Correspondence Between the Fan Type Open-Direction Map and its Labels

Typical Pattern	Closure Rate	Possible Open Direction	Label	Label with Open Direction
1 1 1 1 1 1 1 1 1	Undefined	None	*	*
1 1 1 1 0 1 1 1 1	8	None	Q	Q
1 0 1 1 0 1 1 1 1	7	$1\frac{1}{2}, 2\frac{1}{2}, \ldots, 8\frac{1}{2}$	C	$C1$
1 0 1 0 0 1 1 1 1	6	$1, 2, \ldots, 8$	U	$U2$
1 0 1 0 0 1 0 1 1	5	$1\frac{1}{2}, 2\frac{1}{2}, \ldots, 8\frac{1}{2}$	V	$V2$
1 0 1 0 0 1 0 1 0	4	$1, 2, \ldots, 8$	L	$L3$
1 0 1 0 0 1 0 0 0	3	$1\frac{1}{2}, 2\frac{1}{2}, \ldots, 8\frac{1}{2}$	I	$I3$
1 0 1 0 0 0 0 0 0	2	$1, 2, \ldots, 8$	T	$T4$
1 0 0 0 0 0 0 0 0	1	$1\frac{1}{2}, 2\frac{1}{2}, \ldots, 8\frac{1}{2}$	·	· 4
0 0 0 0 0 0 0 0 0	0	None	␣	
1 0 1 1 0 1 1 0 0 1 0 0 1 0 0 0 1 1		Instead of direction, number "1" is used	S	$S5$ $S4$

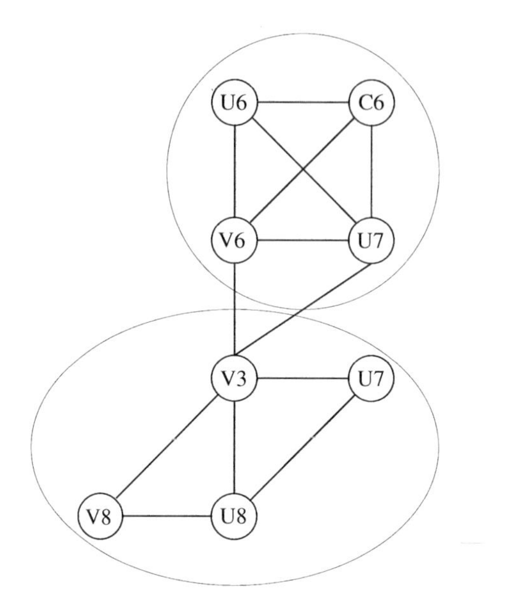

FIGURE 10.16 Graphic representation of the subregions shown in Fig. 10.15 (*a*).

region. Therefore the inner structure is disregarded. Instead, the state of concavity is calculated as an analog attribute. The most primitive attribute then is the area of a concavity.

This is found by dividing the area into two cases as follows:

Area

1. If there is no *IV* point (crossing point from the *I*-labeled region to the *V* one), then

$$A_G = A_I + A_L + \cdots - \frac{1}{2} l_1 l_2 \sin \theta. \tag{10.16}$$

2. If there is an *IV* point, then

$$A_G = A_L + A_V + \cdots + \frac{1}{2} l_1' l_2' \sin \theta, \tag{10.17}$$

where A_G is the total area of integrated region and A_I, A_L, . . . are areas of the labeled regions. An illustration of area calculation is provided in Fig. 10.17. The notation l_1, l_2, l_1', l_2', and θ shown there can be understood as the direction l_1 being 1's direction at +90° clockwise and l_2 being 2's direction at +90° clockwise. These directed lines are tangent to the end point strokes, by which we can obtain the crossing point of the l_1 and l_2 directed lines. The length l_1 is that between one end point tangent to the l_1 directed line and the crossing point. The length l_2 can be obtained in the same manner. Thus in the case of no *IV* point, the area found is the effective area of the concave region which is shaded in panel (*a*). Such concavity is called a shallow. The enlarged

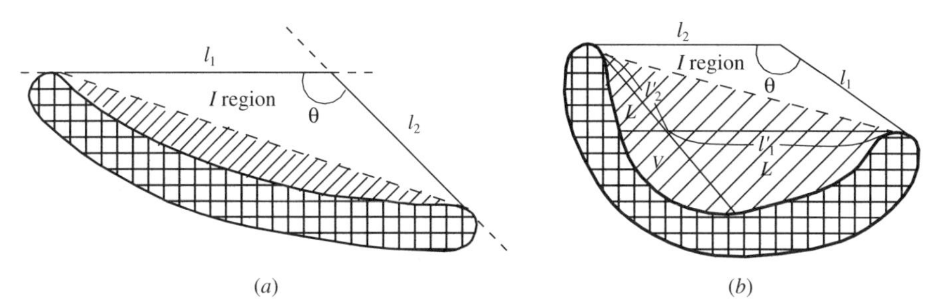

FIGURE 10.17 Area calculation of the integrated region: (a) No IV point; (*b*) IV point exists. From reference [10].

area of the *I*-labeled region is reduced by $\frac{1}{2}l_1l_2 \sin\theta$. For a deeper concavity region, the *IV* point and l_1' and l_2' are obtained by starting from the *IV* point in the same way but omitting the *I*-labeled region. Instead, the triangular area of $\frac{1}{2}l_1'l_2' \sin\theta$ is added so that a compact and effective area is obtained.

Center Coordinate of an Integrated Region The next natural attribute is a center of the coordinate of an integrated region. Let it be *X* and *Y*, then *X* is the center value between the maximum and minimum *x*-coordinates of the region. *Y* is obtained by the same manner. We notice here that (*X*, *Y*) does not necessarily exist within the integrated region.

Closure Rate An attribute that is an effective measure of the closeness of a concavity is the closure rate, which was described in Table 10.1 for the *WR* primitive. This is defined by treating two cases as follows, where C_G designates the closure rate of an integrated region.

1. If there is no *IV* point, then

$$C_G = \frac{A_I + A_L + \cdots - \frac{1}{2}l_1l_2 \sin\theta}{(l_1 + l_2)^2}. \tag{10.18}$$

2. If there is an *IV* point, then

$$C_G = \frac{A_L + A_V + \cdots + \frac{1}{2}l_1'l_2' \sin\theta}{(l_1' + l_2')^2}. \tag{10.19}$$

We notice here that the above closure is not normalized.

Open Direction The final attribute is the open direction normal to the effective concavity boundary being connected to the external background region. This is calculated approximately. For example, the open direction D_G is approximated on average by two cases:

1. If there is no *IV* point, then

$$D_G = \frac{\mathbf{1} \cdot l_1 + \mathbf{2} \cdot l_2}{l_1 + l_2}. \tag{10.20}$$

2. If there is an *IV* point, then

$$D_G = \frac{\mathbf{1} \cdot l_1' + \mathbf{2} \cdot l_2'}{l_1 + l_2}. \tag{10.21}$$

Notice here that l_1 and l_2 are scalar lengths but that **1** and **2** are vectors in directions **1** and **2**. We will give an example of how these attributes are located in Fig. 10.18.

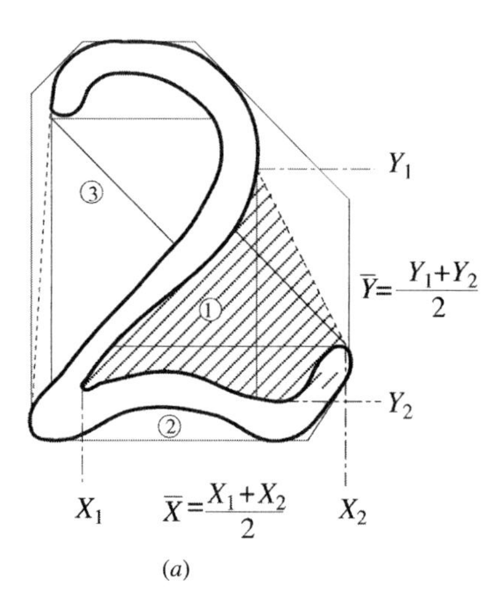

(*a*)

Integrated region	Closure rate C_G	Open direction D_G	Area A_G (%)	Center coordinate	
				$\overline{X}$	$\overline{Y}$
①	1.8	2.55	36	21	24
②	0.2	5.00	8	12	32
③	1.1	7.25	32	11	15

(*b*)

FIGURE 10.18 Integrated regions (*a*) and their attributes (*b*), where *Y* 's direction is taken from top to bottom. From reference [10].

Panel (*a*) gives the region segmentation and integration, and panel (*b*) the total results of the attributes shown; these results meet with our intuition.

The Experiment A comparative study is described that shows the effectiveness of region integration experimentally. It is matched with discrimination methods of cross-correlation, a linear discrimination function on multi-dimensional feature space, feature matching, and simple matching based on the integrated regions. The correct recognition rates were 88.4%, 94.0%, 95.1%, and 98%, respectively.

The data collected were 200 sample characters of ten numerals showing the standard numeral shapes. Therefore the quality of the handprinted character images was relatively good compared with a loosely constrained case. The observed character images had 72×76 sampling points with 16 gray levels, which were reduced to 36×36 with 16 levels, denoted as f_{ij} at (i, j) points. Then the images were binarized according to the simple average method. That is, the threshold value θ was decided at $\theta = \bar{f} + \alpha$, where $\alpha = 2$ and

$$\bar{f} = \frac{1}{36 \times 36} \sum f_{ij}. \tag{10.22}$$

The binarized images were then preprocessed by the half-cell expansion method [11], which is a noise removing process. The final binarized images were denoted as $\{g_{ij}\}$. Now let us turn to the details of each method.

Cross-correlation Taking a cross-correlation is a simple and standard procedure; it is used sometimes as a measure of variation in the quality of handprinted characters. The $\{g_{ij}\}$ was normalized in size and re-sampled to 8×10 sampling points, denoted as $\{h_{ij}\}$. A mask for each category was made by calculating the probability of black points as

$$P_{ij}^{(l)} = \frac{1}{N} \sum_{k=1}^{N} h_{ij}^{(k)}, \tag{10.23}$$

where N is the number of characters in each category and l is the name of each category. Each mask was made by setting a threshold value λ such that

$$\mu_{ij}^{(l)} = \begin{cases} 0 & \text{if } P_{ij}^{(l)} < \lambda, \\ 1 & \text{if } P_{ij}^{(l)} \geq \lambda. \end{cases} \tag{10.24}$$

Thus the cross-correlation was done by

$$S^{(l)} = \frac{1}{IJ} \sum_{ij} h_{ij} \cdot \mu_{ij}^{(l)}, \tag{10.25}$$

where I and J are lengths of each row and column, respectively. The matches are shown in the confusion matrix of Table 10.2. The correct recognition rate was 88.4% and the values of parameters used are $N = 100$ and $\lambda = 0.7$. We note here that half of the data were used for making the masks, so these data were also used for the test.

TABLE 10.2 Confusion Matrix for the Cross-Correlation Recognition Experiment

In \ Out	0	1	2	3	4	5	6	7	8	9
0	99						1			
1		81	8		2			3	8	4
2		1	96						3	
3	2		1	95		1			1	
4		4			76	1	7	2	3	7
5				6		83	5		6	
6	6			2			90		2	
7	2	3	5			1		83		6
8	1			2			1	1	94	1
9		3	3	1		1		5		87

Source: From reference [10].

Discrimination on Feature Space The discrimination method was the same as that used by Glucksman [1]. The features used were those that had closure rates greater than 2, so the dimension of the feature space was 8 (directions) × 5 (labels C through I) + 1(Q) = 41. The recognition rate was 94.0%. The data used and the test were the same as above.

Feature Matching The size-normalized data $\{h_{ij}\}$ were used. Each mask for each category was generated by using the probability calculations

$$PB_{ij}^{\nu(l)} = \frac{1}{N} \sum_{k=1}^{N} [h_{ij} = 0 \wedge M_{ij}^{\nu} = 1], \tag{10.26}$$

$$PW_{ij}^{\nu(l)} = \frac{1}{N} \sum_{k=1}^{N} [M_{ij}^{\nu} = 0], \tag{10.27}$$

where the notation N and l are the same as before and $[\cdot]$ denotes 1 or 0 depending on whether the expression inside was true or false. Accordingly the probability of a point (i, j) being black was set equal to $1 - PB_{ij}^{\nu(l)} - PW_{ij}^{\nu(l)}$. Thus each mask was given threshold values λ_B and λ_W such that

$$\mu_{ij}^{\nu(l)} = \begin{cases} 1 & \text{if } PB_{ij}^{\nu(l)} \geq \lambda_B, \\ -1 & \text{if } PW_{ij}^{\nu(l)} \geq \lambda_W, \\ 0 & \text{otherwise.} \end{cases} \tag{10.28}$$

The input character images were transformed after the feature extraction such that

$$h_{ij}^{\nu(l)} = \begin{cases} 1 & \text{if } M_{ij}^{\nu} = 1 \wedge P_{ij} = 0, \\ -1 & \text{if } M_{ij}^{\nu} = 0, \\ 0 & P_{ij} = 1. \end{cases} \tag{10.29}$$

The cross-correlation on the feature plane, $S^{(l)}$, was defined as

$$S^{(l)} = \frac{1}{IJ} \sum_{i,j} \sum_{\nu} h_{ij}^{\nu} \cdot \mu_{ij}^{\nu(l)}, \tag{10.30}$$

where $i = 1 \sim I$ and $j = 1 \sim J$.

The correct recognition rate was 95.1%. The values of parameters λ_B and λ_W were 0.7 for both parameters. The data used and the test were the same as above.

Matching Based on the Region Integration This matching process is very simple because the features extracted are global and the structure of numerical characters is simple enough for corresponding matching. That is, the matching is done by simple calculating penalties according to feature existence and feature nonexistence. More specifically, the feature existence is done by checking attributes one by one. The attributes have general ranges. Let min_1 and max_1 be the minimum value and maximum value of each attribute range, then the penalty is zero if the corresponding value of each attribute of input falls into this range. However, otherwise analog matching was taken. That is, let min_2 and max_2 be the second minimum value and the second maximum value that satisfy the conditions min_2 < min_1 and max_1 < max_2. If the value of each attribute of the corresponding value of each attribute of input falls within either [min_2, min_1] or [max_1, max_2], then analog penalty value is imposed. This is for a smooth matching, and it is frequently used in practice. The results are shown in Table 10.3 which is a confusion matrix table. The correct recognition rate, rejection rate, and substitution error rate were 98.0%, 1.7%, and 0.3%. In this case half of the data (100 samples) were used to make the masks, and the test was done on the rest of the data (remaining 100 samples). Thus the test conditions were stricter than the cases mentioned before. Despite that the result was significantly superior to those of other methods. Therefore the power of region integration was confirmed.

TABLE 10.3 Confusion Matrix of the Recognition Experiment Based on Region Integration Matching

In \ Out	0	1	2	3	4	5	6	7	8	9	Rej
0	99										1
1		95		1							4
2			99								1
3				99							1
4					97						3
5						96					4
6							99				1
7				2				97			1
8									99		1
9										100	

Source: From reference [10].

10.4 FIELD EFFECT METHOD

So far we have discussed the digital approach to background analysis. Here we will introduce an analog approach, which is a generalization of the digital approach. The analog concept is derived from the field of physics [9]. Steinbuch imagined a background field for feature extraction in which the black portion was assumed to have a constant electric potential value. According to the electric potential theory, a ϕ "electric field" is formed on the white region. Thus he considered that the static electric potential's form could be calculated in principle by knowing any derivative of ϕ. In the analogy of physics the important point here is that an electrostatic field can be considered to be generated by an "action through a medium" rather than by "action at a distance," as Faraday had conceived it. More specifically, we let the electrostatic field at point $\mathbf{r}$ be denoted by $E(\mathbf{r})$; then we have the following differential equation according to Gauss's law:

$$\text{div } E(\mathbf{r}) = \frac{1}{\varepsilon_0} \rho(\mathbf{r}), \quad (10.31)$$

where $\rho(\mathbf{r})$ is the charge density at point $\mathbf{r}$ and ε_0 is a dielectric constant in vacuum. This is written componentwise as

$$\frac{\partial E_r(\mathbf{r})}{\partial x} + \frac{\partial E_y(\mathbf{r})}{\partial y} + \frac{\partial E_z(\mathbf{r})}{\partial z} = \frac{1}{\varepsilon_0} \rho(\mathbf{r}). \quad (10.32)$$

If a one-dimensional case is considered, then the equation above becomes

$$E_x(x + \Delta x) = E_x(x) + \frac{\rho(x)\Delta x}{\varepsilon_0}. \quad (10.33)$$

This tells us the standpoint of the action through the medium, since the field at $x + \Delta x$, $E_x(x + \Delta x)$ is determined by its neighboring field $E_x(x)$ and by the charge distribution $\rho(x)$ at the point x. We note here more precisely that the other condition rot $E(x) = 0$ is necessary to determine the field. This is an important point, and we follow this model.

In other words, the model concerns the propagation of information. We will show that the mechanism of propagation extracts a global feature eventually. Before the strict formulation of the propagation mechanism, it is instructive to determine how such a mechanism was considered. That is, we want to learn what intermediate product bridged the symbol to the analog method.

10.4.1 Labeled Field and Macro-processing

Our objective in the field effect method is schematically shown in Fig. 10.19. A character image "6" is drawn, and the gray level of the black portion is regarded as a scalar potential ϕ. The field according to the potential ϕ is generated on the white regions. The field generated is integrated so that feature fields such as a hole and a concavity

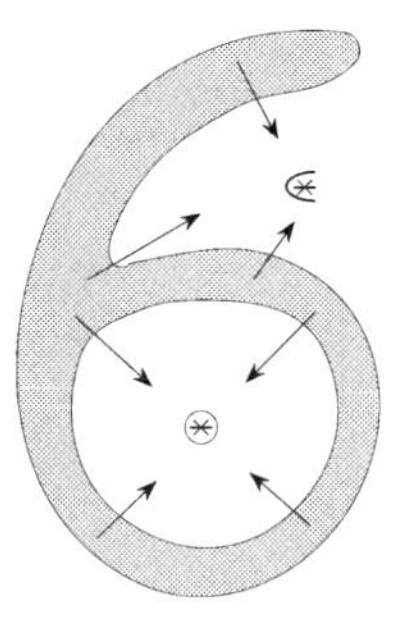

FIGURE 10.19 Schematic diagram showing the objective of the field effect method.

are constructed at appropriate points marked by an asterisk. The feature fields—the hole and the concavity—are schematically represented by ○ and ⊂, respectively.

Now let us look at a simple way to perform under limited conditions the objective mentioned above. The first step is a coarse sampling of an image, which is shown in Fig. 10.20 (*a*) where the image "8" is sampled by a 10 × 10 matrix. We need to distinguish the feature field from the potential, and this in shown schematically in Fig. 10.21.

(*a*)

0	0	260	0	0	260	0	270	0	0
290	0	320	260	300	0	0	0	0	0
310	280	350	360	360	350	0	0	0	0
0	290	360	360	360	360	270	0	0	0
0	0	360	360	360	350	0	0	0	0
0	0	360	360	360	360	0	0	0	0
0	250	360	320	320	360	340	0	0	0
0	0	360	360	360	360	0	0	0	0
0	0	0	310	330	250	0	0	0	0
0	0	0	0	0	0	0	0	0	0

(*b*)

.	*	P	L2	F8	P	*	A	.	.
P	U1	P	U1	P	V3	*	.	.	.
P	B	P	B	B	P	F2	.	.	.
F6	B	B	P	P	B	C3	.	.	.
.	L6	B	P	P	B	L4	.	.	.
.	L8	P	B	B	P	L2	.	.	.
.	C7	P	U3	U7	B	P	.	.	.
.	F6	P	P	P	P	L4	.	.	.
.	.	F6	B	B	F4	*	.	.	.
.	.	.	*	*	*	.	.	.	.

(*c*)

..	..	..	..	..	..	..	..	..	..
..	U1	PI	U1	..	U3	..	..	..	..
..	P-	P+	P-	P+	..	F2	..	..	..
F6	..	P+	P+	P+	P+	..	..	..	..
..	U7	PI	P+	P+	PI	U3	..	..	..
..	..	PI	P-	P-	PI	L2	..	..	..
..	..	PI	O.	..	PI	..	..	..	..
..	F6	..	P-	P-	..	L4	..	..	..
..	..	F6	..	..	F4	..	..	..	..
..	..	..	..	..	..	..	..	..	..

(*d*)

FIGURE 10.20 Primitive field effect method: (*a*) Original image data; (*b*) analog sampling result of the data; (*c*) the labeling; (*d*) the macro processed result.

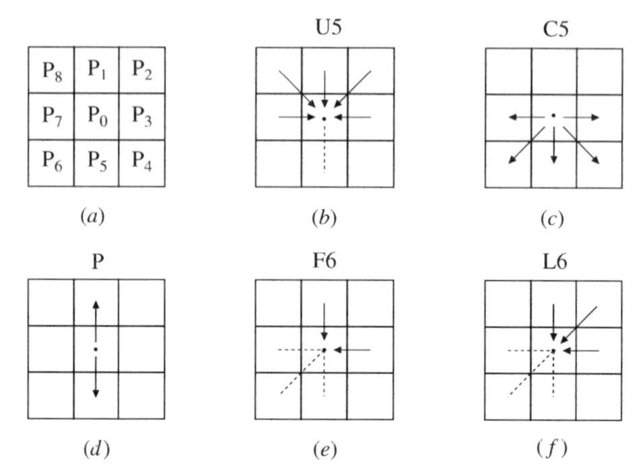

FIGURE 10.21 The notation of 3 × 3 local field frame (*a*) and illustration of some labeling (*b*) through (*f*).

A 3 × 3 neighborhood is considered in which the center potential value is denoted as P_0, the top value is denoted as P_1, and so on. The feature field is labeled according to the pattern of the differences between P_0 and P_i, $i = 1 \sim 8$. Some typical examples are shown in panels (*b*), (*c*), and (*d*). The labels *U5* and *F6* indicate concavity examples and *P* indicates a convex type. To be more precise, say a given point is labeled as *F6* if

$$(P_1 - P_0 > \lambda) \wedge (P_3 - P_0 > \lambda) \wedge (P_5 - P_0 \simeq 0) \wedge (P_6 - P_0 \simeq 0) \wedge (P_7 - P_0 \simeq 0) = \text{True}$$

holds. If $(P_0 - P_i > \lambda') \wedge (P_0 - P_{i+4} > \lambda') = \text{True}$ holds, then that point is labeled as *P*.

The labeled field shown in Fig. 10.20 (*b*) gives the results in cases where these rules were applied. We can see that the local feature fields, L_4, L_2, for example, are formed at the middle right part of the image. However, the desired feature fields, the concavity and the hole, are not formed at the desired points. The reason is that the integration of feature fields was not done; this is called a *macro-process.*

A macro-process based on symbols can be done by making rules that take two-dimensional configurations into account, as shown in Fig. 10.22. We can obtain the desired feature fields of concavity and a hole in Fig. 10.20 (*d*) by applying rules 1 through 3, respectively, of Fig. 10.22, although the top hole is filled up. This is reflected by a macro-processed field *P*+. However, a considerable number of other rules are necessary, 33 rules in all, although black portions are also labeled in finding a crossing point. The desired macro-process in fact used tri-level labels for smoothing the macro-process. The initial tri-levels, 0, 1, and 2 were defined by the potential difference between center value and its neighboring potential value. The intermediate labeling was done by converting a tri-level label to a symbol label and further converting a symbol label to tri-level label, which makes a loop of the macro-process. If there is no change in the field, then the final field is set, which is the result of macro-processing. This was considerably awkward in the back-and-forth movement between quasi-analog labeling and symbol labeling processing. However, the important thing is that this cut and try led to the following smart analog macro-process.

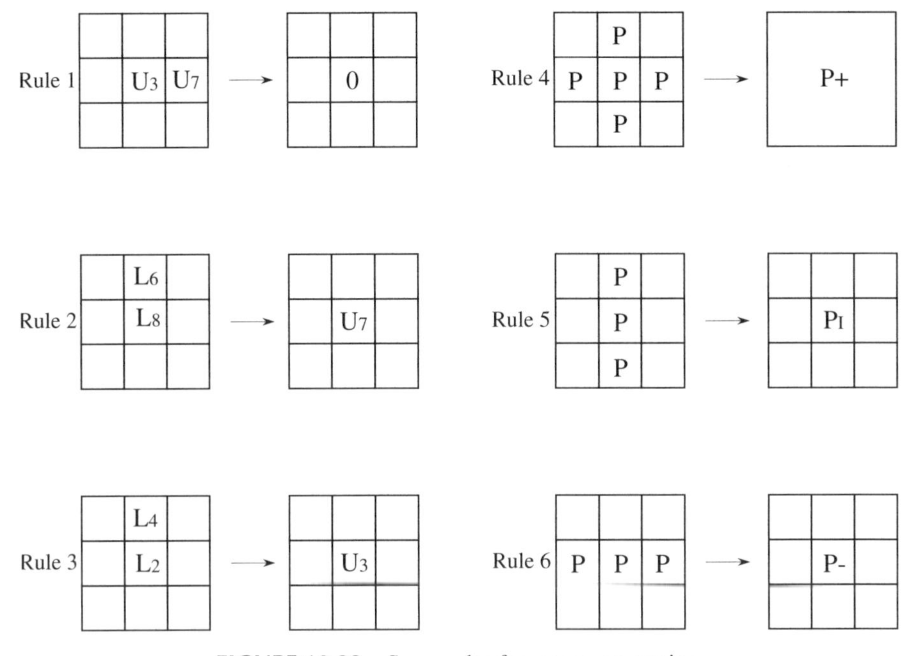

FIGURE 10.22 Some rules for macro-processing.

Another important thing was its high performance against heavy noisy in the character image such that no noise removing and binary preprocessing were needed. However, considerably coarse sampling was done, as seen in panel (*b*), where values less than 199 were set to 0 for the display. Success here provided strong motivation to continue the research.

10.4.2 Field Effect Formalization

Now we formalize the analog approach by first setting up its mathematical framework [12].

Framework

Definition 10.4.1 *If some value (feature) $\psi(\mathbf{x})$ is defined at every point $\mathbf{x}$ in some region(s) in a plane, S is called the "field" and denoted as $\psi(S)$. In particular, if more features are defined at every point $\mathbf{x}$, that is $\psi_1(\mathbf{x})$, $\psi_2(\mathbf{x})$, . . . , then S is called a multiple field. Then $\psi(\mathbf{x})$ is represented as $\psi_1(\mathbf{x}) \times \psi_2(\mathbf{x}) \times \psi_3(\mathbf{x}) \times \ldots$.*

The simplest case is the density potential field where the feature $\psi(\mathbf{x})$ is represented as $P(x)$ and $\psi(S)$ is nothing but an original image (monochromatic). A typical field is a vector field in which $\psi(\mathbf{x})$ is defined as $(\partial P/\partial x, \partial P/\partial y)$. We use "field" collectively and "feature field" at every point $\psi(\mathbf{x})$. Next we define an "effect" in the field effect method which is also called a macro-operation in the sense that we have already used it so far.

Definition 10.4.2 *Provided some field* ${}^{t}\psi(S)$, *the operation f to generate an infinite series* $\{{}^{t}\psi(S) | t = 0, 1, 2, \ldots\}$, *according to the formula*

$$ {}^{t+1}\psi(S) = f[{}^{t}\psi(S)] \tag{10.34} $$

is called the "field effect" when the suffix t denotes a discrete time sequence.

Specifically, we impose a restriction on the f for simplification of the operation such that ${}^{t+1}\psi(\mathbf{x})$ is determined by only the neighboring $u(\mathbf{x})$'s feature fields at time t:

$$ {}^{t+1}\psi(\mathbf{x}) = f[{}^{t}\psi[u(\mathbf{x})]]. \tag{10.35} $$

Naturally this sequence must converge to some feature in some finite time T:

$$ {}^{T}\psi(S) = {}^{T+1}\psi(S) \tag{10.36} $$

must hold. This field is called a *final field* or *macro-field.*

Finally we describe a representation of a feature field; namely as $\psi(\mathbf{x})$ we consider a 3×3 neighborhood of a point $\mathbf{x}$ in eight-connection. Furthermore the feature field is given by double fields as

$$ \psi(\mathbf{x}) = \psi_1(\mathbf{x}) \times \psi_2(\mathbf{x}), $$

when $\psi_1(\mathbf{x})$ and $\psi_2(\mathbf{x})$ are density and length fields, respectively. The notation for each feature field element is shown in Fig. 10.23 (*a*) and (*b*). The general representations are $D^{\nu}(\mathbf{x})$ and $\nu^{\nu}(\mathbf{x})$, $\nu = 0, 1, \ldots, 7$, where ν indicates direction, which is shown in panel (*c*).

Macro-operation To formalize the field effect f, let us recall the discussion given in the previous section. That is, propagation is a basic mechanism of the macro-operation. We consider only propagation on a white region in a black configuration/region. A schematic diagram of the propagation is shown in Fig. 10.24. The reader can begin to form a mental image of the propagation, which will become clearer as we proceed. The driving force of the propagation is a density difference between a given point and its neighboring point in each direction. That is,

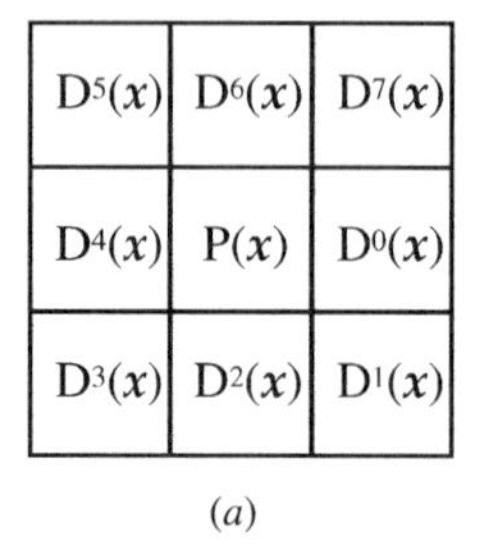

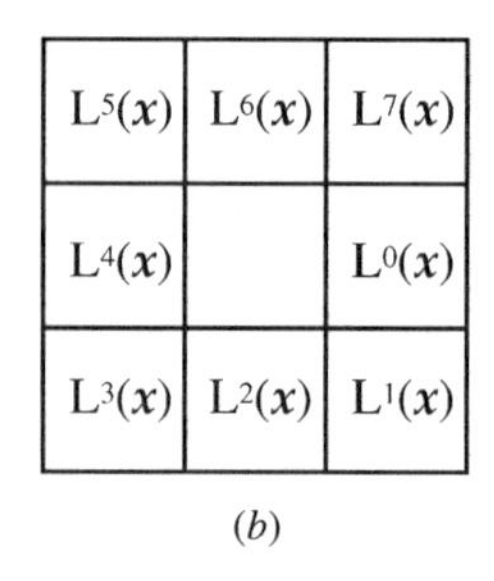

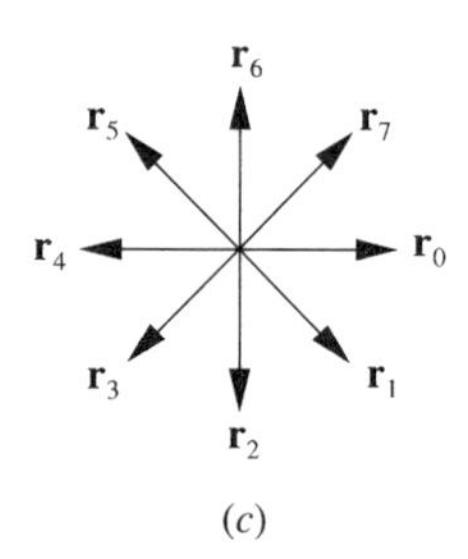

(*a*) (*b*) (*c*)

FIGURE 10.23 Notation for (*a*) density feature field $\Psi_1(x)$, (*b*) distance feature field $\Psi_2(x)$, and (*c*) unit directional vector $\mathbf{r}_i$.

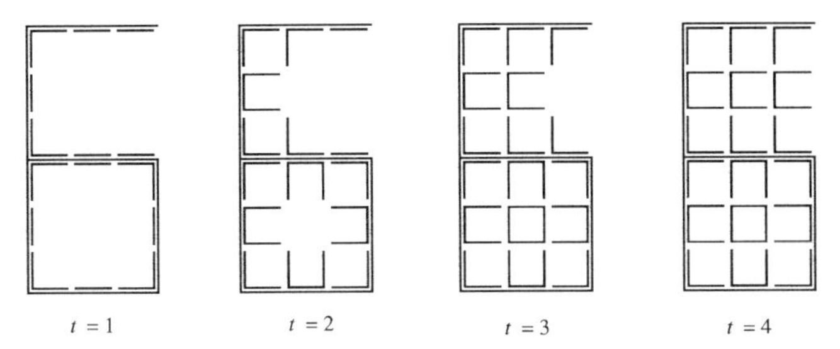

FIGURE 10.24 Schematic diagram of the propagation based on the field effect method. From reference [12].

$$P(x + \mathbf{r}_\nu) - P(x), \qquad \nu = 0, 1, \ldots, 7,$$

where $\mathbf{r}_\nu$ is a unit vector for each direction ν, as shown in Fig. 10.23 (c). These values are stored at $D^\nu(\mathbf{x})$ as the representation of the feature field at point $\mathbf{x}$, $\psi_1(\mathbf{x})$.

Here the initial value of $D^\nu(\mathbf{x})$ is zero, that is, at $t = 0$. Thus the first propagation ($t = 1$) of the black region to the white region is normally performed up to the boundary between the black and white regions. Refer to Fig. 10.25 ($t = 1$). In general, as is natural, the boundary is a gray zone but this does not matter due to the analog property of the propagation.

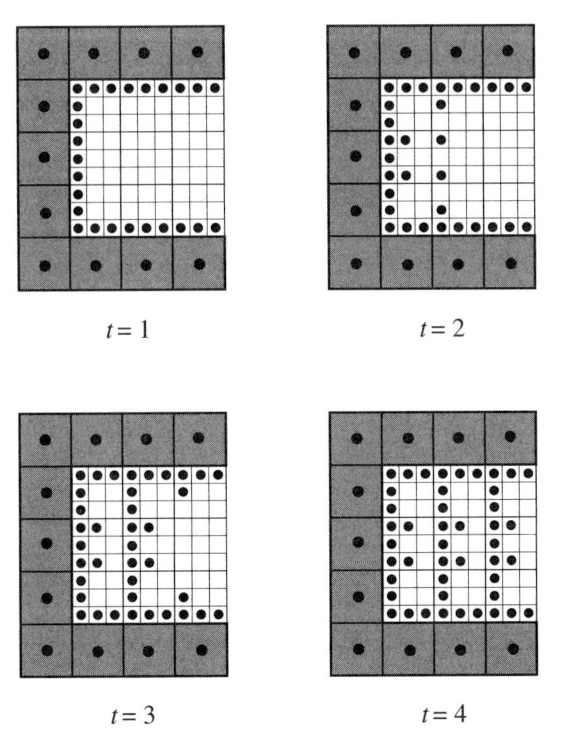

FIGURE 10.25 Mechanism of propagation seen schematically. From reference [15], © 1996, Ohm.

Next we need to take $D^{\nu}(\mathbf{x} + \mathbf{r}_{\nu})$ into account in the propagation to the feature field at $\mathbf{x}$, $\psi(\mathbf{x})$. The propagation is divided into two directions because of the heterogeneous character of a grid plane. One is where $\nu = 0, 2, 4, 6$, so the directions are right and left, and top and down. The other is where $\nu = 1, 3, 5, 7$, which takes two diagonal directions. These two cases are illustrated in Fig. 10.26 (*a*) and (*b*). First propagation in the diagonal direction (*b*) will be explained. That is, we consider propagation from the top-left feature field to D^5 which is an element at $\nu = 5$ of the feature vector at point $\mathbf{x}$ located at the bottom right. Here we can assume that only $\psi(\mathbf{x} + \mathbf{r}_5)$ can affect the value of D^5 at $\mathbf{x}$. Now the question is how and which of $D^{\nu}(\mathbf{x} + \mathbf{r}_5)$ of $\psi(\mathbf{x} + \mathbf{r}_5)$, $\nu = 0, 1 \sim 7$, is stored at $D^5(\mathbf{x})$. Notice that in the figure D_5^{ν} denotes $D^{\nu}(\mathbf{x} + \mathbf{r}_5)$. What we want to propagate is some kind of concavity. So, for example, if there is a fragment of concavity (*L* type) in the diagonal direction, then it will appear at $D^4(\mathbf{x} + \mathbf{r}_5)$, $D^5(\mathbf{x} + \mathbf{r}_5)$, and $D^6(\mathbf{x} + \mathbf{r}_5)$. Since we assume an eight-connection, the *L*-type concavity can be represented by $D^4(\mathbf{x} + \mathbf{r}_5)$ and $D^6(\mathbf{x} + \mathbf{r}_5)$. Specifically, min $(D^4(\mathbf{x} + \mathbf{r}_5), D^6(\mathbf{x} + \mathbf{r}_5))$ is stored at $D^5(x)$. This means that if either D_5^4 or D_5^6 is very small, then $D^5(\mathbf{x})$ is also very small, even if one of these values is very large. The operator min takes the role of AND in the logical world. Thus analog propagation is performed as we intended. What we have described can easily be formalized for the diagonal direction. The general propagation in the diagonal direction is written as

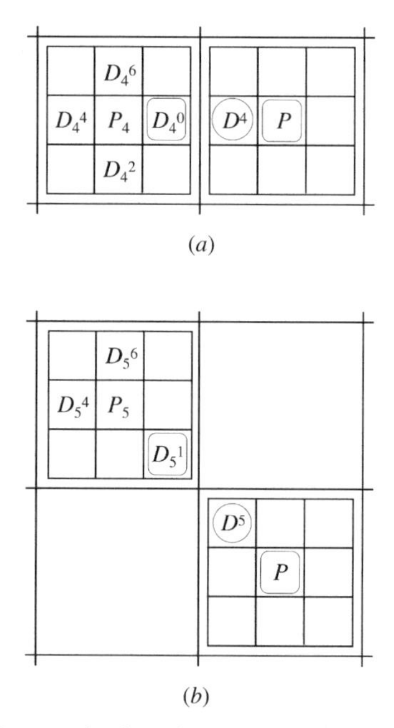

FIGURE 10.26 Propagation mechanism that causes a representation feature field on the field. The values marked by a square act to suppress the values marked by a circle, and other values act to promote it.

$$
\begin{aligned}
{}^{t}D^{\nu}(\mathbf{x}) &= P(\mathbf{x}+\mathbf{r}_{\nu}) - P(\mathbf{x}) \\
&\quad + \min\left({}^{t-1}D^{\nu+1}(\mathbf{x}+\mathbf{r}_{\nu}),\ {}^{t-1}D^{\nu-1}(\mathbf{x}+\mathbf{r}_{\nu})\right) \qquad (\nu = 1, 3, 5, 7). \qquad (10.37)
\end{aligned}
$$

Now we proceed to vertical and horizontal propagation. This process is more complicated, although the same idea is applied. Let us consider propagation from $\psi(\mathbf{x}+\mathbf{r}_4)$ to $D^4(\mathbf{x})$ which is marked by a circle in panel (*a*). Here we have to devise a mechanism such that both *U* and *L* types propagate, but not as a one-directional line. Specifically, since these can be two *L* types, two min terms are necessary, and they are connected by a max operator. The general form of propagation to the horizontal or vertical direction is expressed as

$$
\begin{aligned}
{}^{t}D^{\nu}(\mathbf{x}) &= P(\mathbf{x}+\mathbf{r}_{\nu}) + P(\mathbf{x}) \\
&\quad + \max\left[\min\left({}^{t-1}D^{\nu}(\mathbf{x}+\mathbf{r}_{\nu}),\ {}^{t-1}D^{\nu+2}(\mathbf{x}+\mathbf{r}_{\nu})\right),\right. \\
&\quad \left.\min\left({}^{t-1}D^{\nu}(\mathbf{x}+\mathbf{r}_{\nu}),\ {}^{t-1}D^{\nu-2}(\mathbf{x}+\mathbf{r}_{\nu})\right)\right] \qquad (\nu = 0, 2, 4, 6). \qquad (10.38)
\end{aligned}
$$

So far we have considered only propagation. However, our purpose is to construct a representative feature field at an appropriate position. Such a representative feature field can be expected to be generated where the propagations collide. Specifically, notice that the element $D^1(\mathbf{x}+\mathbf{r}_5)$ denoted as D_5^1 is marked by a square in panel (*b*). This $D^1(\mathbf{x}+\mathbf{r}_5)$ stores the propagation from the opposite direction $\nu = 1$ against $\nu = 5$ which we can see by looking at $\mathbf{x}+\mathbf{r}_5$ point. This means that this term can be used to control propagation in the direction $\nu = 5$. If there is enough propagation from the direction $\nu = 1$, then the value of $D^1(\mathbf{x}+\mathbf{r}_5)$ becomes large, so the propagation to $\nu = 5$ is suppressed. This idea is formalized by setting formulas (10.37) and (10.38) to ${}^{t}C^{\nu}(\mathbf{x})$. The new ${}^{t}D^{\nu}(\mathbf{x})$ is expressed as

$$
{}^{t}D^{\nu}(\mathbf{x}) = [{}^{t-1}D^{\nu+4}(\mathbf{x}+\mathbf{r}_{\nu}) < \lambda] \cdot {}^{t}C^{\nu}(\mathbf{x}), \qquad (10.39)
$$

where $[\cdot]$ means that if the contents are true, it takes 1 and otherwise it takes 0. Accordingly, if the suppression term $D^{\nu+4}(x+\mathbf{r}_{\nu})$ is greater than or equal to λ, then propagation to ν direction is prohibited. Otherwise, propagation is performed. In this sense the term $[\cdot]$ acts as a gate to propagation. Finally we need to formalize the condition to obtain a global field as stated in (10.36). That is, we redefine ${}^{t}D^{\nu}(x)$ as

$$
{}^{t}D^{\nu}(\mathbf{x}) = \max\left[{}^{t-1}D^{\nu}(\mathbf{x}),\ [{}^{t-1}D^{\nu+4}(\mathbf{x}+\mathbf{r}_{\nu}) < \lambda] \cdot {}^{t}C^{\nu}(\mathbf{x})\right]. \qquad (10.40)
$$

The meaning now is obvious: So long as ${}^{t-1}D^{\nu}(x) < {}^{t}C^{\nu}(x)$, the higher value ${}^{t}C^{\nu}(x)$ is selected, but once a collision happens, the term $[\cdot]$ is set 0 so that ${}^{t}D^{\nu}(x) = {}^{t-1}D^{\nu}(x)$. This satisfies the condition of the final field described in (10.36). On the other hand, propagation of the distance feature field is simply expressed because it is accompanied by density propagation. That is, the distance feature field ${}^{t+1}L^{\nu}(\mathbf{x})$ at point $\mathbf{x}$ is determined by the expression

$$
{}^{t+1}L^{\nu}(\mathbf{x}) = [{}^{t}D^{\nu}(\mathbf{x}) \geq \lambda] \cdot [{}^{t}L^{\nu}(\mathbf{x}+\mathbf{r}_{\nu}) + \text{unit length}]. \qquad (10.41)
$$

Figure 10.27 shows the results of the density feature field (*a*) and distant feature field (*b*) for real data on the character image "6." As intended, the results show clearly the analog and two-dimensional properties of field effect method.

Uniform Propagation Mode Equation (10.39) covers a general form of propagation. In a special case of the formula, the term $[{}^{t-1}D^{\nu+4}(\mathbf{x}+\mathbf{r}_\nu) < \lambda]$ can be set to 1. Then the potential differences can propagate freely into white regions. In this case we get a lake region, where the region has the same feature filed at every point. For this reason this approach is called the *uniform propagation mode.* It gives almost the same results as region segmentation described earlier. Actually both are a special case of (10.39).

Beside uniform propagation mentioned above, there are several modes in which the closure rate is the key value [13]. Although the closure rate was defined before, we define it again:

Definition 10.4.3 *The closure rate, w is defined as follows:*

$$w \equiv \sum_{\nu=0}^{7} W_\nu \cdot [{}^{t}D^{\nu}(\mathbf{x}) > \lambda'], \qquad (10.42)$$

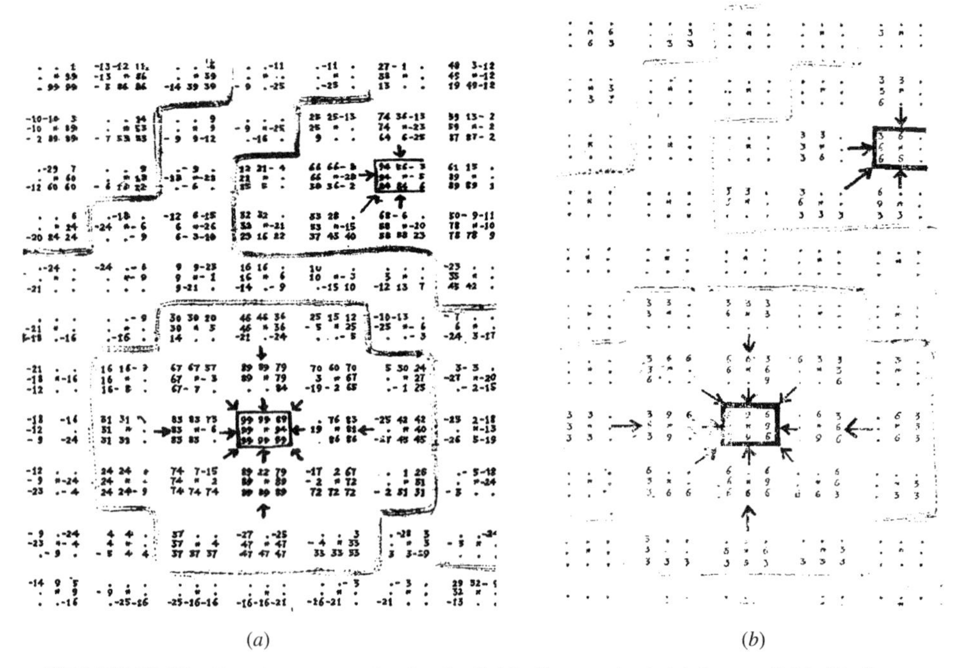

(*a*) (*b*)

FIGURE 10.27 Density propagation by the field effect method: (*a*) feature field; (*b*) distant feature field.

where

$$W_\nu = \begin{cases} 2 & \text{if } \nu = 0, 2, 4, \text{ and } 6, \\ 1 & \text{if } \nu = 1, 3, 5, \text{ and } 7, \end{cases}$$

and λ' is an appropriate threshold value.

Therefore $w \le 12$ holds. The propagation is controlled by the closure rate and so we introduce a parameter δ as a controlling closure rate. That is, if $w \ge \delta$ or $P(\mathbf{x}) > \lambda''$ hold at point $\mathbf{x}$, then to the position $(\mathbf{x} + \mathbf{r}_{\nu+4})$ in each $\nu + 4$ direction the propagation is performed from point $\mathbf{x}$. The elements of the feature field ${}^tD(\mathbf{x} + \mathbf{r}_{\nu+4})$ are given as follows:

$$
\begin{aligned}
{}^tD(\mathbf{x} + \mathbf{r}_{\nu+4}) &= P(\mathbf{x}) - P(\mathbf{x} + \mathbf{r}_{\nu+4}) \\
&\quad + \max\,[0, \min\,({}^{t-1}D^\nu(\mathbf{x}), {}^{t-1}D^{\nu+1}(\mathbf{x})), \min\,({}^{t-1}D^\nu(\mathbf{x}), {}^{t-1}D^{\nu-1}(\mathbf{x}))].
\end{aligned}
$$

The differences in these propagation modes appear in the propagation outward from the body of a character image. This is shown in Fig. 10.28, in which the case $\delta = 0$ is no regulation and cases $\delta \ge 7$ are too constrained. Therefore meaningful closure rates are $2 \le \delta \le 6$. Just as in the $\delta = 5$ case, the propagation is confined to the limits of a box. The case $\delta = 6$ is compact, and the closure more reasonable.

Let us look at how these differences occur. In Fig. 10.29, at point $\mathbf{x}_1$, propagation is possible only from one diagonal direction. Therefore $D^3(\mathbf{x}_1) > \lambda'$, and all the other $D^\nu(\mathbf{x}_1) = 0$ except for $\nu = 3$. Since $W_3 = 1$, the closure rate $w(\delta)$ is 1. At point $\mathbf{x}_2$ propagation takes place only from the bottom up. Therefore $D^2(\mathbf{x}_2) > \lambda'$, and all the other $D^\nu(\mathbf{x}_2) = 0$ except for $\nu = 2$. Since $W_2 = 2$, the closure rate $w(\delta)$ is 2. The same is true for point $\mathbf{x}_3$. At point $\mathbf{x}_4$ there are propagations from the right and from the bottom right. That is, both $D^0(\mathbf{x}_4)$ and $D^1(\mathbf{x}_4) > \lambda'$, and the rest $D^\nu(\mathbf{x}_4) = 0$. Since $W_0 = 2$ and $W_1 = 1$, the closure rate is 3. The same thing occurs at point $\mathbf{x}_5$ where $\delta = 4$.

Edge Propagation So far we have described density propagation that has an intrinsic weakness. That is, it is weak in detecting geometrical features such as the straightness of a line segment. However, we are not limited to density propagation. Edge propagation can also be considered. A line consisting of consecutive edge fragments can be integrated in the process of propagation to form a representative feature field. Actually this idea has been systematically investigated and implemented on a commercial OCR device already. The work was done by Oka [14].

He discarded the field analogy and came up with a two-dimensional array of automata. The two-dimensional array consists of uniform cells that have the same functions and structure and communicate only with neighboring cells. In this sense the idea is the same as that of the "action through media" mechanism we considered earlier. Each cell has its intra-cells, which are placed on a circle S^1. In a continuous representation, each intra-cell has a geometrical attribute, angle θ. That is, a two-dimensional distribution of edges is projected on S^1. Therefore, if a cell is placed at the center of a circle, then the strength distribution at the circles' center cell becomes uni-

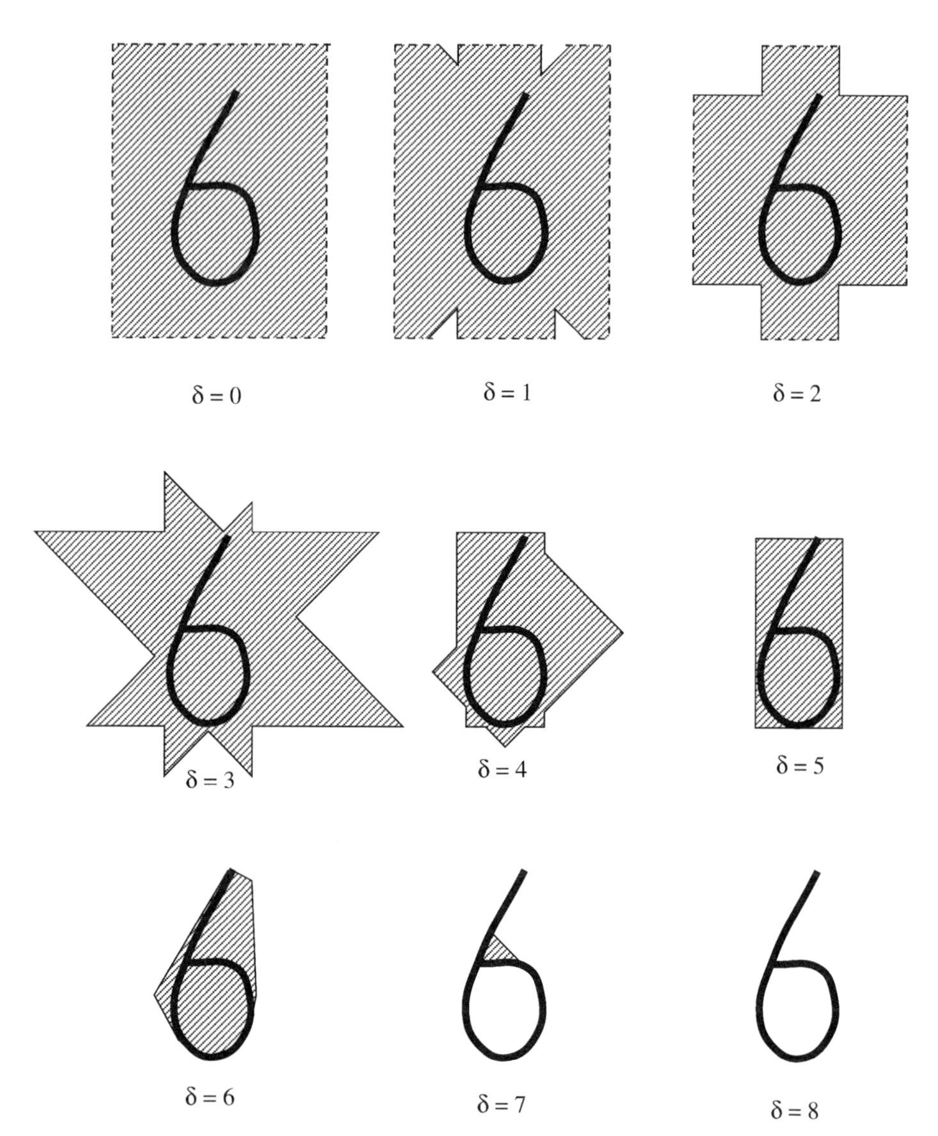

FIGURE 10.28 Several different propagation modes caused by changing the closure rate δ. The thin line segments mark boundaries where the propagations stop. From reference [13].

form after the edge propagation. In the case of a rectangle, after edge propagation a cell placed at the center of the rectangle will have four peak values corresponding to the four normal directions of the four sides of the rectangle. This case is illustrated for discrete values of θ, $\theta = 2\pi n/8$, $n = 0, 1, \sim 7$, in Fig. 10.30. In principle, the propagation mechanism is the same as that of density propagation. Some experimental results are given in Fig. 10.31 which shows the effectiveness of edge propagation as well.

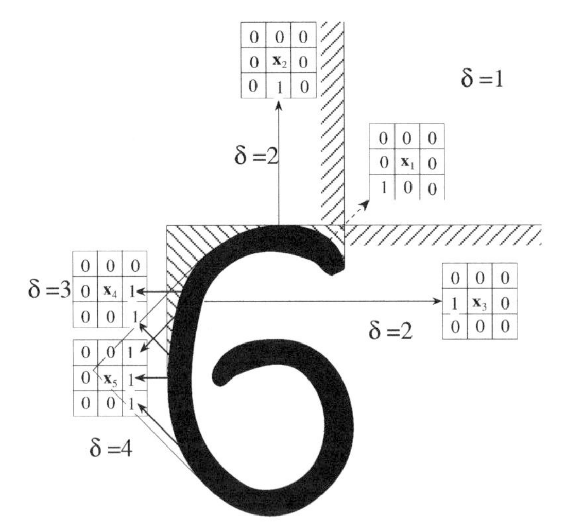

FIGURE 10.29 Determination of the propagation region based on closure rate δ. The propagation regions are partially represented; the propagation regions corresponding to δ = 3 and δ > 4 are not shown.

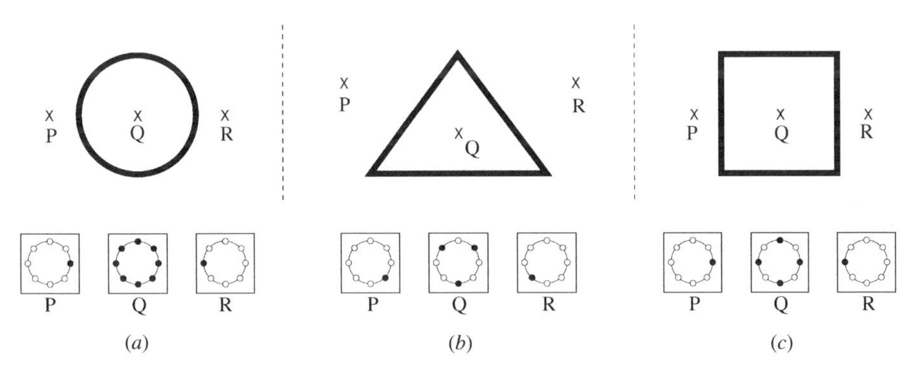

FIGURE 10.30 Examples of edge propagation and their cell representations. From reference [14].

Further Integration of Feature Field After integration of each concavity or hole region, further integration is done over strokes. A typical example is shown in Fig. 10.32. A character image of "V" is integrated into a 3 × 3 framework, each square represents an integrated region feature field. This field is global, so the "V" is represented by only one 3 × 3 representation framework. Other examples are shown in Fig. 10.33. Integration of the black region is shown in panel (*a*). *P*+ represents a crossing point. A more flexible example is that shown in Fig. 10.34. After integrating each region, a graph is made of regions that show corresponding nodes so that a total rep-

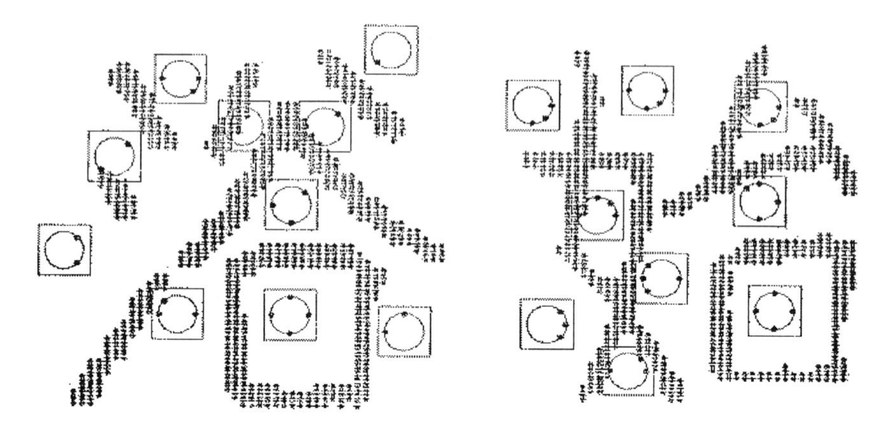

FIGURE 10.31 Some experimental results for edge propagation. From reference [14].

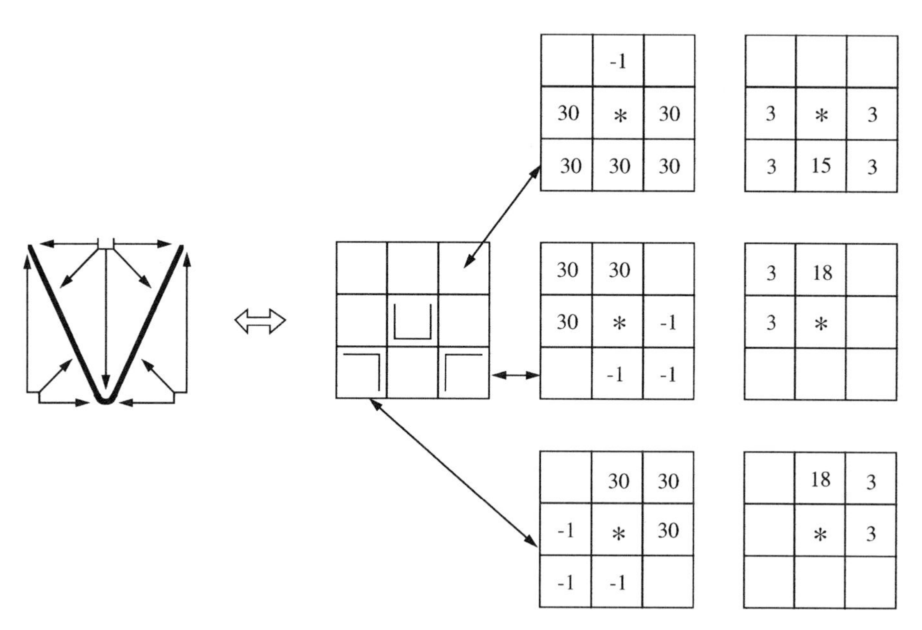

FIGURE 10.32 Example of further integration of feature fields.

resentation is obtained, and each boundary between the regions is represented by the corresponding edges connecting them. The edge value is obtained by the propagation of white regions into their adjacent black regions. When the propagated regions collide, they stop, and the length of this collision section is the edge value between the regions. As the other edge value the boundary direction of each pair of the adjacent regions can be added. The example of Fig. 10.34 is shown in Tables 10.4, 10.5, and

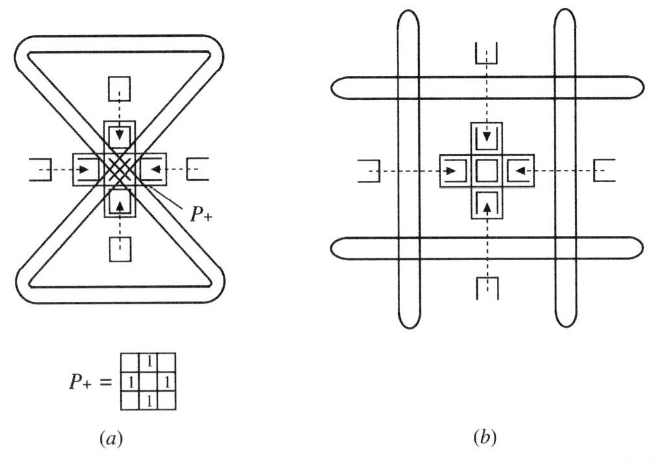

FIGURE 10.33 Examples of higher-level integration of feature fields.

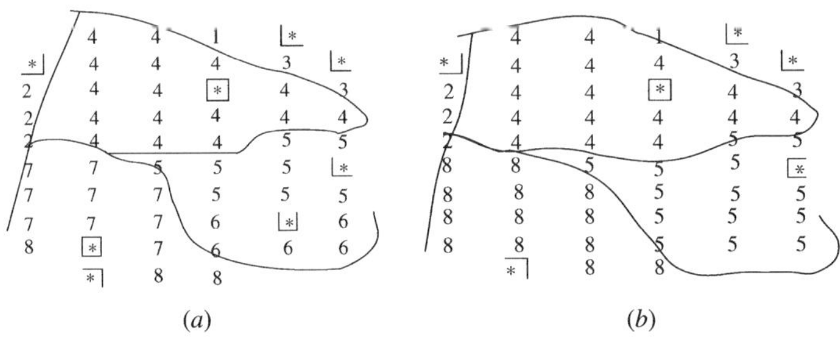

FIGURE 10.34 Higher-level integration of feature fields showing (*a*) first-level integration and (*b*) second-level integration in which each concavity and each hole are almost completely integrated.

10.6. The edge values are represented by a relational matrix. In this way graph matching is effectively performed.

Experiment We will describe an experimental result for the field effect method applied to loosely constrained handprinted numeric characters. The data were provided by 21 subjects who wrote 100 numerals (10 sets). In the experimental results shown in Table 10.7, we see that "4" often was incorrectly recognized as "9." The reason for the error becomes clear in looking at the actual data shown in Fig. 10.35. Take the data placed at the fourth row and tenth column as an example. Also note the first column of that row, which was substituted to "0." The templates were made manually and were similar to that shown in Fig. 10.32. From the table we can see both strong and weak points of the numeral. That is, it is strong in quasi-topological variation, so a high correct recognition rate of 98% is achieved. However, it is weak in geometrical features such as curvature, in other words, weak in corner detection.

TABLE 10.4 Relational Matrix of the Segmented Regions in Image "R" Shown in Fig. 10.34

Position	Region Number	(1)	(2)	(3)	(4)	(5)	(8)
(6, 12)	(1)	*		1	4		
(7, 8)	(2)		*		8	1	2
(7, 13)	(3)	1		*	6	1	
(8, 11)	(4)	4	8	6	*	9	5
(11, 13)	(5)		1	1	9	*	13
(15, 9)	(8)		2		5	13	*
	Total	5	11	8	32	24	20

TABLE 10.5 Directions of the Boundaries of the Segmented Regions in Image "R" Shown in Fig. 10.34

Region Number	(1)	(2)	(3)	(4)	(5)	(8)
(1)	*		3.0	2.2		
(2)		*		5.1	4.0	3.3
(3)	7.0		*	1.9	3.0	
(4)	6.2	1.1	5.9	*	3.5	2.6
(5)		8.0	7.0	7.5	*	1.5
(8)		7.3		6.6	5.5	*

TABLE 10.6 Some Representative Feature Fields of Image "R" Shown in Fig. 10.34

	Ψ^4			Ψ^5			Ψ^8		
Label		4			5			8	
	96	92	72	99	87	0	72	72	65
Ψ^1	99		76	99		0	6		33
	99	99	99	99	81	0	0	0	0
	4	5	3	3	7	0	4	11	6
Ψ^2	6		5	7		0	0		8
	6	8	6	6	8	0	0	0	0
Area		17			15			14	

Another typical example is "1" being substituted for "2." The reason is that the row dimension used, which was for 60 × 60 observed images, was reduced to 20 × 20 because it took too long to calculate at that time, the 1970s, at 60 s/ch. Also even "8" was incorrectly recognized as "9" when the under loops of "8" were small and smeared. The important point is that we can easily see how the substitution errors occurred. These experimental results also show the effectiveness of field effect method based on edge propagation.

TABLE 10.7 Confusion Matrix of the Recognition Experiment, Correct Recognition, Substitution Error, and Rejection Rates are 98%, 1.8%, and 0.2% Respectively

Input \ Output	1	2	3	4	5	6	7	8	9	0	Substitution Error	Rejection
1	207	1									1	2
2	1	207					1				2	1
3		1	207	1			1		1		3	0
4				203					5	1	6	0
5					206	1		3			4	0
6						209					0	1
7							207		2		3	0
8	1				2			205	5	2	10	0
9	2				2				203	1	5	1
0						3				207	3	0

10.5 BIBLIOGRAPHICAL REMARKS

A number of researchers contributed original work for the background analysis approach. Unger used image cellular automata which led him to propose propagation of black points in eight directions on a white region, which is effectively a special case of Glucksman's features at white points [16]. Glucksman then considered propagation on the background from the four sides of a frame in which a character image is placed [17]. While Unger's paper is earlier than Glucksman's, Unger cited Glucksman's research report at Air Force Cambridge Laboratories. In 1969 Knoll at Honeywell conducted systematic experiments on handprinted alphanumeric characters based on Glucksman's method [18]. The data used were written by 12 subjects; the number of the characters included in the data was 1222. The correct recognition rate and substitution error rate were reported as 98.9% and 0.2%, respectively. In his paper, Knoll reported that (D, O), (U, V, Y), and (H, N, M) were hard to differentiate from others in their combination sets so that other special processing was necessary. On the field effect method, Kazmierczak developed Steinbuch's idea in more realistic terms of implementation [19].

The field effect method was developed as a national project at ETL (Electro Technical Laboratory) by the Pattern Information Processing System. At the same time Fujitsu developed a handprinted character recognition system based on the background analysis approach [20]. They called their system a reflection method, since a line constructed in a concavity/loop region was a result of propagation from the stroke(s). The line is a representative of the concavity/loop which is generated only by raster scanning. This is a simple version of field effect method. On the other hand, Hagita and Masuda proposed a kind of stroke distribution method which they called a local directional contribution method [21]. It can be regarded as a special case of the field effect method, since they apply distance representation to the black portion of a character image. A four-dimensional distance vector was defined to represent the directional feature in a global sense. This method was applied to handprinted Kanji

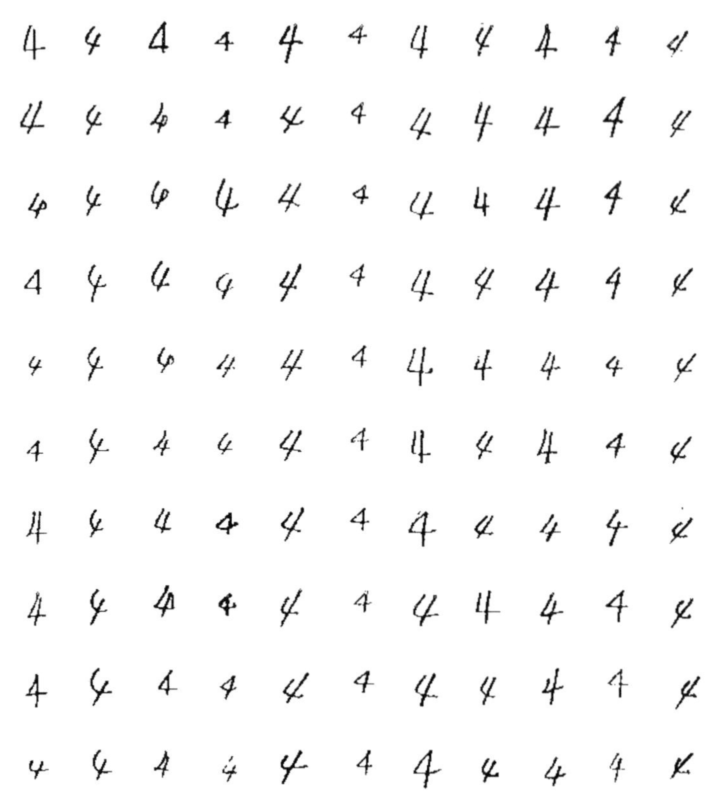

FIGURE 10.35 Data of "4."

recognition. The paper cited above is written in Japanese, so we recommend a survey paper by Mori et al. in which this method is concisely illustrated [22]. Readers can also find in this survey paper another variation of Glucksman's method and applications of the background analysis approach to Kanji recognition.

BIBLIOGRAPHY

[1] H. A. Glucksman, "Classification of mixed font alphabetics by characteristic loci," Digest 1st Ann. IEEE Comp. Conf., pp. 139–141, September 1967.

[2] L. A. Kamentsky and C. M. Liu, "Computer-automated design of multifont print recognition logic," *IBM J.*, vol. 7, pp. 2–13, 1963.

[3] A. Rosenfeld and J. L. Pfaltz, "Sequential operations in digital picture processing," *J. ACM*, vol. 13, no. 4, pp. 471–494, October 1966.

[4] K. Fu, "Syntactic Pattern Recognition and Applications," Englewood Cliffs, NJ: Prentice-Hall, 1982.

[5] R. Narashimhan, "Labeling schematic and syntactic descriptions of pictures," *Inf. Control*, vol. 7, 1964.

[6] R. Narashimhan, "Syntax-directed interpretation of classes of pictures," *CACM,* no. 3, March, 1966.

[7] R. Narashimhan, "A syntax-aided recognition scheme for handprinted English letters," *Pattern Recogn.,* vol. 3, 1971.

[8] K. Komori, T. Kawatani, K. Ishii, and Y. Iida," A feature concentration method for character recognition," *Proc. IFIP Congress 77,* pp. 29–34, 1977.

[9] K. Steinbuch, "Automatische Zeichnerkenung," SEL Nachrichten, Heft 3, p. 127, 1958.

[10] H. Yamada, S. Mori, and Suge, "Recognition of handprinted numeral characters based on fan type open-direction map," Technical Report of IECE Japan, PRL 75-4, April, 1975.

[11] J. Skalansky, "Measuring concavity on a rectangular mosaic," *IEEE Trans.,* vol. C-21, no. 12, pp. 1355-1364, December 1972.

[12] T. Mori, S. Mori, and K. Yamamoto, "Feature extraction based on field effect method-extraction of closure state," *Trans. IECE Japan,* vol. J57-D, no. 5, pp. 308–315, 1974.

[13] K. Yamamoto, "Research on handprinted character recognition based on structural analysis approach," Research of the Electrotechnical Laboratory, no. 762, June 1976.

[14] R. Oka, "Cell type feature extraction method," *Trans. IECE Japan,* vol. J65-D, no. 6, pp. 1219–1226, 1982.

[15] S. Mori and T. Sakakura, *Fundamentals of Image Recognition I,* Tokyo: Ohm, 1986.

[16] S. H. Unger, "Pattern recognition using two-dimensional, bilateral, iterative, combinational switching circuits," *Proc. Symp. Math. Theory of Automata,* Polytechnic Press, pp. 577–588, 1963.

[17] H. A. Glucksman, "A propagation pattern classifier," *IEEE Trans. Electronic Computer,* vol. EC-14, no. 3, pp. 434–443, 1965.

[18] A. L. Knoll, "Experiments with "characteristic Loci" for recognition of handprinted characters," *IEEE Trans. Computers,* pp. 366–372, April 1969.

[19] H. Kazmierczak, "The potential field as an aid to character recognition," *Proc. Int. Conf. Info. Processing,* p. 244, June 1959.

[20] Y. Tokunaga, M. Yoshida, O. Kato, and H. Akimoto, "Development of common use handprinted character recognition system," *Proc. Symp. PIPS Project,* pp. 45–57, July 7–8, 1977.

[21] N. Hagita and I. Masuda, "Handprinted Chinese character recognition," Technical Report of IECE Japan, vol. PRL81-13, 1981.

[22] S. Mori, K. Yamamoto, and M. Yasuda, "Research on machine recognition of handprinted characters," *IEEE Trans. Pattern Anal. Machine Intell.,* vol. PAMI-6, no. 4, pp. 286–405, 1984.

CHAPTER ELEVEN

Linear Matching

Template matching is the simplest and the oldest matching method used in OCR. It is represented by an inner product which is mathematically a simple procedure. Template matching can be extended naturally in terms of its mathematical structure to a subspace method and also feature matching. The subspace method is an extension of one template to a template space. That is, a much greater number of standard patterns are matched against an input pattern. Feature matching, on the other hand, is an extension of point matching to a local neighborhood matching. Mathematically it is an increase of dimension of a standard pattern, $N \times M$ of a given frame on which a character is observed. In both extensions the inner product is basically used. For this reason we name these methods *linear matching.*

11.1 TEMPLATE MATCHING

We covered template matching in Chapter 1 where we discussed the experimental results and gave a mathematical description. So already we have some notion of linear matching.

It is nevertheless instructive to note a well-known relationship between the template matching and the so-called MDD (*minimum distance decision*) rule. The MDD rule is intuitive. When an input character vector $\mathbf{x}$ and a standard pattern $\mathbf{f}_i$ are given for each class, a distance d_i between $\mathbf{x}$ and $\mathbf{f}_i$ is measured as

$$d_i = \|\mathbf{x} - \mathbf{f}_i\| \qquad (i = 1, \cdots, K). \tag{11.1}$$

Therefore a minimum d_i is found, say $d_m = \min_{i \in K}[d_i]$, where the input character is closest to the standard pattern. The $\mathbf{x}$ is identified as belonging to the class m. However, if the difference between $d_m - d_{m'}$ less than T, then the input character $\mathbf{x}$ is rejected, and $d_{m'}$ is the next minimum distance in the definition (11.1) with T an appropriate threshold value. Because of its importance in practice, we assume this point implicitly in the following formulations.

Now the relation between the MDD rule and the template matching is obtained by expanding the right side of the definition (11.1):

$$d_i^2 = \|\mathbf{x}\|^2 - 2\mathbf{f}_i^t\mathbf{x} + \|\mathbf{f}_i\|^2. \tag{11.2}$$

Usually each standard pattern $\mathbf{f}_i$ is normalized in such a way that its norm is 1 and $\|\mathbf{x}\|^2$ is common to all classes. Therefore (11.2) tells us that to seek the minimum distance d_i is equivalent to finding the maximum inner product $\mathbf{f}_i^t\mathbf{x}$, $(i = 1, \ldots, K)$. Thus matching based on the inner product is a very basic discrimination method. For the formulation of matching based on the inner product we follow [1]. An input character vector $\mathbf{x}$ is assumed to belong to R^N, N-dimensional Hilbert space. Each standard pattern $\mathbf{f}_i$ is also assumed to belong the R^N. Naturally the inner space is defined in the inner product, and a matching coefficient s_i is defined as

$$s_i = \mathbf{f}_i^t\mathbf{x} \qquad (i = 1, \ldots, K). \tag{11.3}$$

Thus the matching is defined so as to find a maximum coefficient, say,

$$s_m = \max_{1 \le i \le K} \{s_i\} \tag{11.4}$$

and to identify the $\mathbf{x}$ class m.

We can treat all the coefficient s_i $(i = 1, \ldots, K)$ as a whole as

$$\mathbf{s} = F^t\mathbf{x}, \qquad F = [\mathbf{f}_1, \ldots, \mathbf{f}_K] \tag{11.5}$$

where $\mathbf{s}$ is a K-dimensional vector in R^K and F is a $N \times K$ matrix. Thus the first step of the matching processes can be regarded as linear mapping from R^N to R^K, and the mapping process is represented by the matrix F. The $\mathbf{s}$ is called a *matching output vector*.

The problem is how to select F. Usually F is taken as

$$F \to M \equiv [\boldsymbol{\mu}_1, \boldsymbol{\mu}_2, \ldots, \boldsymbol{\mu}_K] \qquad (N \times K), \tag{11.6}$$

$$\boldsymbol{\mu}_i = E_{c_j}\mathbf{x},$$

where c_j is a subset consisting of image data belonging to category j. That is, M is constructed by arranging averaging vectors of the classes. Such matching is called *simple matching,* abbreviated to *SM*. Thus the matching output vector $\mathbf{s}_{SM}$ is given as

$$\mathbf{s}_{SM} = M^t\mathbf{x}. \tag{11.7}$$

If each input character is selected to be an averaging vector corresponding to each class, then the output is represented by a matrix as a whole as

$$\mathbf{s}_{SM} = M^t M. \tag{11.8}$$

The $M^t M$ is called *correlation matrix* of M denoted as R according to the convention. That is,

$$R = M^t M = [\boldsymbol{\mu}_i^t \boldsymbol{\mu}_i] \qquad (K \times K). \tag{11.9}$$

In the matching above, a standard pattern whose value of norm is higher than those of others has the advantage of being identifiable. Therefore the normalization is performed as

$$F \to MD^{-1}, \tag{11.10}$$

$$D = \mathrm{diag}(\|\boldsymbol{\mu}_1\|, \cdots, \|\boldsymbol{\mu}_K\|). \tag{11.11}$$

Then the matching output vector $\mathbf{s}_{SM}$ is obtained as

$$\mathbf{s}_{SM} = D^{-1} M^t \mathbf{x}, \tag{11.12}$$

and the corresponding matrix S_{SM} is given as

$$S_{SM} = D^{-1} M^t M = D^{-1} R. \tag{11.13}$$

To further improve the mapping matrix F, let us consider an optimum problem. We assume that $\boldsymbol{\mu}_i$, $i = 1 \sim K$, are linear independent. The input character is approximated as a linear combination of $\boldsymbol{\mu}_i$, $i = 1 \sim K$. Actually, if an input character is $\mathbf{x}$, then it is formulated as

$$\|\hat{\mathbf{x}} - \mathbf{x}\|^2 \to \min, \ \hat{\mathbf{x}} = \sum_{i=1}^{K} a_i \boldsymbol{\mu}_i = M\mathbf{a}, \tag{11.14}$$

where $\mathbf{a}$ is found so that $\|\hat{\mathbf{x}} - \mathbf{x}\|$ is a minimum. This approximation scheme is the same as given before when linear feature extraction was mentioned. That is, the minimum is given as a projection to the space spanned by a set $\{\boldsymbol{\mu}_i\}_{i=1}^{K}$ as

$$\hat{\mathbf{x}} = P_M \mathbf{x}, \tag{11.15}$$

$$P_M = M(M^t M)^{-1} M^t = M R^{-1} M^t. \tag{11.16}$$

According to the (11.14), $\mathbf{a}$ is given as

$$\mathbf{a} = R^{-1} M^t \mathbf{x}. \tag{11.17}$$

This is called *canonical matching,* abbreviated as *CM.* When an input character set is $\{\mu_i\}_{i=1}^{K}$, S_{CM} is given accordingly as

$$S_{CM} = R^{-1}M^tM = R^{-1}R = I_k. \tag{11.18}$$

As intended, for average input characters this matching scheme gives an ideal result. We note here that if the set $\{\mu_i\}_{i=1}^{K}$ is orthogonal, R^{-1} is reduced to D^{-1}, and if it is normalized, then R^{-1} is reduced to I_k. In other words, the above scheme *CM* can be interpreted such that an input character is mapped to the space in which $\{\mu_i\}_{i=1}^{K}$ is normally orthogonal, and a simple matching is performed there. Mathematically the set of standard patterns $\{\mathbf{f}_j\}_{j=1}^{K}$ is just a dual or orthogonal space of the set $\{\mu_i\}_{i=1}^{K}$, namely $\mathbf{f}_i^t\mu_j = \delta_{ij}$.

Here we have to consider the meaning of "average." The key question is whether or not the average of patterns that belong to a class represents the shape of the class. Unfortunately, it is not true. There are too many variations in shape, even if only one class is taken. Therefore we need further improvement, which will be mentioned next.

11.2 SUBSPACE MATCHING

Through the experiments on character recognition people learned that they have to be modest. In other words, it is very hard to cope with the variations and noise of characters relying on only one average standard pattern in the matching. Therefore, a natural direction is to use plural number of standard patterns for one class. Of course, such practice is very common in the handprinted character recognition in which the variation of slope is tremendous. However, here we consider this problem in the theoretical framework based on functional analysis. Roughly there are two approaches to this problem. One is the so-called subspace method, which was proposed by Watanabe [2]. The other is the so-called multiple similarity method, which was proposed by Iijima [3]. They are quite similar, but we provide another section on the multiple similarity method because the formulations are very different.

The basic idea of the subspace method is simply that the vector components in F are changed. Specifically, F is replaced by F_i corresponding to class i. The vector components in each F_i are appropriately chosen as standard patterns belonging to class i. Naturally they are all linear independent. Therefore we can consider the subspace spanned by them. When an input character $\mathbf{x}$ is given, the minimum distance to the set of subspaces spanned by F_i, $i = 1 \sim L$, is found. Fortunately, thanks to Hilbert space, the same argument mentioned at the end of the previous section can be applied directly. The distance δ_i to each subspace denoted simply as F_i is given as

$$\delta_i = inf_{x' \in F_i}\|\mathbf{x} - \mathbf{x}'\| = \|\mathbf{x} - \hat{\mathbf{x}}_i\|$$

Theoretically $\hat{\mathbf{x}}_i$ exists and is unique. Furthermore $\hat{\mathbf{x}}_i$ is given by the projection operator

$$P_i \equiv F_i(F_i^tF_i)^{-1} F_i^t,\ P_i^t = P_i,\ P_i^2 = P_i, \tag{11.19}$$

$$\hat{\mathbf{x}}_i = P_i\mathbf{x}, \tag{11.20}$$

$$\|\hat{\mathbf{x}}\|^2 = \mathbf{x}^tP_i^tP_i\mathbf{x} = \mathrm{x}^tP_i\mathbf{x}. \tag{11.21}$$

Therefore,

$$\|\mathbf{x} - \hat{\mathbf{x}}_i\|^2 = \|\mathbf{x}\|^2 - 2\mathbf{x}^t\hat{\mathbf{x}}_i + \|\hat{\mathbf{x}}_i\|^2 = \|\mathbf{x}\|^2 - \|\hat{\mathbf{x}}_i\|^2 \tag{11.22}$$

to find the minimum distance δ_i is equivalent to finding the maximum of $\|\mathbf{x}_i\|$, $i = 1 \sim L$.

11.2.1 Dual Orthogonal Subspace Method

In contrast to subspace construction of the preceding discussion, an elegant method is to treat the set of all subspaces F_i, $i = 1 \sim L$, as a whole [1]. That is, the matrix F is constructed by F_i, $i = 1 \sim L$, as

$$F = [F_1, F_2, \ldots, F_L], \quad (N \times d), \quad d = \sum_{i=1}^{L} m_i. \tag{11.23}$$

The objective now is to find the projection operator that gives good discrimination power among the classes. To do so, first a projection operator to the whole space F is constructed. We use the Moor-Penrose general inverse matrix [4]. If $|F^tF| \neq 0$ and $N > d$, then the general inverse matrix of F denoted as F^+ is given as

$$F^+ = (F^tF)^{-1}F^t. \tag{11.24}$$

See the note at the end of this section.

On the other hand, if we set

$$P \equiv FF^+, \tag{11.25}$$

then

$$P^2 = FF^+\,FF^+ = FI_dF^+ = FF^+ = P, \tag{11.26}$$

since rank $(F) = d$ and so $F^+F = I_d$. Thus we can construct the projection operator to F. Notice that we do not have necessarily $P^t = P$ as an orthogonal projection operator. That is, we work with a so-called projection operator. The structure of the F^+ is written as

$$F^+ \equiv G^t = \begin{bmatrix} G_1^t \\ \vdots \\ G_L^t \end{bmatrix}, \tag{11.27}$$

where G_i is $N \times m_i$.

Thus P is neatly represented as

$$P = FF^{+} = \sum_{i=1}^{L} F_i G_i^t = \sum_{i=1}^{L} \hat{P}_i, \tag{11.28}$$

where $\hat{P}_i \equiv F_i G_i^t$. This equation shows a decomposition of P corresponding to each space. Because of the relations $\text{rank}(F) = d$, $F^{+}F = I_d$, we have

$$G_j^t F_i = \delta_{ji} I_{m_i}, \tag{11.29}$$

which is a *dual orthogonal system.*

Now

$$\hat{P}_j \hat{P}_i = F_j G_j^t F_i G_i^t = F_j \delta_{ji} G_i^t = \delta_{ji} F_j G_j^t = \delta_{ji} \hat{P}_j, \tag{11.30}$$

and using the equation (11.28), we have

$$P\hat{P}_i = \hat{P}_i P = \hat{P}_i. \tag{11.31}$$

Therefore

$$\hat{P}_i F_j = F_i G_i^t F_j = \delta_{ji} F_i \tag{11.32}$$

hold, which means that projection $\hat{P}_i$ leaves F_i as it is but maps the standard patterns of other classes to zero.

Note: For $A(m \times n)$, A^t is defined as follows:

$$A^t = \begin{cases} A^t(AA^t) & \text{if } |AA^t| \neq 0, \text{ rank } (A) = m, \\ (A^tA)^{-1}A^t & \text{if } |A^tA| \neq 0, \text{ rank } (A) = n, \\ A^{-1} & \text{if } |A| \neq 0, \text{ rank } (A) = m = n. \end{cases}$$

11.3 MULTIPLE SIMILARITY METHOD

As stated before, the basic idea behind the multiple similarity method is similar to that of the subspace method. However, in practice, it has had a monumental presence in the history of OCR, and it is said that this theory proved itself first in the field. In this sense it is instructive to introduce the original contribution by Iijima [5]. However, before we do so, we give a general scheme of multiple similarity method.

Consider a set of all normalized characters that belong to a class, denoted as $\{g_\alpha(\mathbf{r}); w_i(\alpha)\}$, where α is an index of data and i is that of class. The average of the data on a domain D_i is taken as

$$\overline{g(\mathbf{r})} = \int_{D_i} w_i(\alpha) g_\alpha(\mathbf{r}) d\alpha. \tag{11.33}$$

Now we define a function φ_0 as

$$\varphi_0(\mathbf{r}) = \frac{\overline{g(\mathbf{r})}}{\|\overline{g(\mathbf{r})}\|}. \tag{11.34}$$

Using the $\varphi_0(\mathbf{r})$, $g_\alpha(\mathbf{r})$ is canonicalized as

$$h_\alpha(\mathbf{r}) = g_\alpha(\mathbf{r}) - (g_\alpha, \varphi_0)\varphi_0(\mathbf{r}). \tag{11.35}$$

As mentioned in our discussion of Karhumem-Loeve expansion in Chapter 7, a characteristic kernel based on the set $\{h_\alpha(\mathbf{r})\}$ is constructed as

$$K(\mathbf{r}, \mathbf{r}') = \int_{D_i} w(\alpha)h_\alpha(\mathbf{r})h_\alpha(\mathbf{r}')d\alpha, \tag{11.36}$$

and the operator defined as

$$(Kg)(\mathbf{r}) = \int_S K(\mathbf{r}, \mathbf{r}')g(\mathbf{r}')d\mathbf{r}', \tag{11.37}$$

where S is an appropriate domain in general. According to (11.35) and (11.36), it is easily proved that

$$(K\varphi_0)(\mathbf{r}) = 0. \tag{11.38}$$

That is, the space spanned by $K(\mathbf{r}, \mathbf{r}')$ is orthogonal to $\varphi_0(\mathbf{r})$. We can regard $\varphi_0(\mathbf{r})$ and the space spanned by $K(\mathbf{r}, \mathbf{r}')$ to represent the normalized character set $\{g_\alpha(\mathbf{r})\}$. Therefore they are adopted as the standard patterns belonging to the class i.

In general, a norm of an operator K is defined as

$$\|K\| = \sup_\varphi (K\varphi, \varphi)/\|\varphi\|^2. \tag{11.39}$$

If the maximum of the eigenvalue K is λ_1, then $\|K\|$ is given as

$$\|K\| = \lambda_1$$

because $K\varphi = \lambda\varphi$ and λ_1 is the maximum among the φ's. Now multiple similarity is defined for any input character $g(\mathbf{r})$ as

$$S_m[g(\mathbf{r})] = \frac{(g, \varphi_0)^2 + \mu(Kg, g)}{\|g\|^2}, \tag{11.40}$$

where $0 \leq \mu < 1/\|K\|$.

To evaluate the value of $S_m[g(\mathbf{r})]$, we rewrite $g(\mathbf{r})$ as

$$g(\mathbf{r}) = (g, \varphi_0)\varphi_0(\mathbf{r}) + h(\mathbf{r}), \tag{11.41}$$

where $h(\mathbf{r})$ is a canonicalized pattern of $g(\mathbf{r})$. Then

$$\|g\|^2 = (g, \varphi_0)^2 + \|h\|^2 \tag{11.42}$$

holds.

Notice that the preceding definition includes the kernel K whose representation is implicit. Therefore it is convenient for us to represent K explicitly. We do so writing the following useful formula:

$$K(\mathbf{r}, \mathbf{r}') = \sum_{n=1}^{\infty} \lambda_n \varphi_n(\mathbf{r}) \varphi_n(\mathbf{r}'), \tag{11.43}$$

where $\{\varphi_n\}$ is orthogonal bases of the space spanned by $K(\mathbf{r}, \mathbf{r}')$. By this formula, (Kg, g) is written as

$$(Kg, g) = \sum_{n=1}^{\infty} \lambda_n (g, \varphi_n)^2. \tag{11.44}$$

Substituting it into the definition of multiple similarity (11.40), we obtain

$$S_m[g(\mathbf{r})] = \frac{1}{\|g\|^2} \{(g, \varphi_0)^2 + \mu \sum_{n=1}^{\infty} \lambda_n (g, \varphi_n)^2\}. \tag{11.45}$$

Since the eigenvalue λ_n converges to zero, increasing n to ∞, an appropriate N is chosen for the sum in (11.45) in practice. In particular, if $\mu\lambda_n = 1$ $(n = 1, 2, \ldots, N)$ is taken, then

$$S_m[g(\mathbf{r})] = \frac{1}{\|g\|^2} \sum_{n=0}^{N} (g, \varphi_n)^2 \tag{11.46}$$

is obtained, which is just the same as in the subspace method, in a narrow sense.

11.3.1 Multiple Similarity in Invariant Position Displacement

We now turn to some original work on the multiple similarity method so that the reader can see how this idea was borne. As in all creative work at an early stage, the procedure is very accessible and specific. After that the abstraction sets in. In developing high-performance OCR, a joint team of ETL and Toshiba encountered a big problem. The simple and usual similarity method proved weak in position displacement, although the position normalization could be taken. Because of the noise, however, a complete normalization was very difficult to accomplish. See the Bibliographical Remarks at the end of this chapter.

For a given pattern $h_0(\xi_1, \xi_2)$, its position displacement pattern $h(\xi_1, \xi_2)$ is written as

$$h(\xi_1, \xi_2) = h_0(\xi_1 + \delta_1, \xi_2 + \delta_2). \tag{11.47}$$

Assuming that δ_1 and δ_2 are small values, the above h_0 is expanded by Taylor series formula as

$$h = h_0 + h_1\delta_1 + h_2\delta_2 + \frac{1}{2} h_{11}\delta_1^2 + h_{12}\delta_1\delta_2 + \frac{1}{2} h_{22}\delta_2^2, \tag{11.48}$$

where

$$h_0 = h_0(\xi_1, \xi_2), \quad h_1 = \frac{\partial}{\partial \xi_1} h_0(\xi_1, \xi_2), \quad h_2 = \frac{\partial}{\partial \xi_2} h_0(\xi_1, \xi_2),$$

$$h_{11} = \frac{\partial^2}{\partial \xi_1^2} h_0(\xi_1, \xi_2^2), \quad h_{12} = \frac{\partial^2}{\partial \xi_1 \partial \xi_2} h_0(\xi_1, \xi_2), \quad h_{22} = \frac{\partial^2}{\partial \xi_2^2} h_0(\xi_1, \xi_2). \tag{11.49}$$

Partial integration yields the following relations:

$$\begin{aligned}
&(h_0, h_1) = -(h_1, h_0) = 0, \quad (h_0, h_2) = 0, \\
&(h_0, h_{11}) - -(h_1, h_1) = \ \|h_1\|^2, \\
&(h_0, h_{22}) = -(h_2, h_2) = -\|h_2\|^2, \\
&(h_0, h_{12}) = -(h_1, h_2).
\end{aligned} \tag{11.50}$$

We note here that $h_0(\xi_1, \xi_2)$ is 0 along its boundary.

At first the inner product between $h(\xi_1, \xi_2)$ and $h_0(\xi_1, \xi_2)$ is calculated, namely the $h_0(\xi_1, \xi_2)$ component in $h(\xi_1, \xi_2)$ is obtained:

$$\begin{aligned}
(h, h_0) &= \|h_0\|^2 + \frac{1}{2}(h_{11}, h_0)\delta_1^2 + (h_{12}, h_0)\delta_1\delta_2 + \frac{1}{2}(h_{22}, h_0)\delta_2^2 \\
&= \|h_0\|^2 - \frac{1}{2}\{\|h_1\|^2\delta_1^2 + 2(h_1, h_2)\delta_1\delta_2 + \|h_2\|^2\delta_2^2\}.
\end{aligned} \tag{11.51}$$

The above results are neatly represented by rewriting the parameters as follows:

$$\begin{aligned}
&\Delta_1^2 \equiv \frac{\|h_0\|^2}{\|h_1\|^2}, \quad \Delta_2^2 \equiv \frac{\|h_0\|^2}{\|h_2\|^2} \\
&\theta \equiv \frac{(h_1, h_2)}{\|h_1\|\,\|h_2\|}.
\end{aligned} \tag{11.52}$$

That is,

$$\begin{aligned}
\left(h, \frac{h_0}{\|h_0\|}\right) &= \|h_0\|\left[1 - \frac{1}{2}\left\{\frac{\delta_1^2}{\Delta_1^2} + 2\theta\frac{\delta_1}{\Delta_1}\frac{\delta_2}{\Delta_2} + \frac{\delta_2^2}{\Delta_2^2}\right\}\right] \\
&\approx \|h_0\|\sqrt{1 - \left\{\left(\frac{\delta_1}{\Delta_1}\right)^2 + 2\theta\left(\frac{\delta_1}{\Delta_1}\right)\left(\frac{\delta_2}{\Delta_2}\right) + \left(\frac{\delta_2}{\Delta_2}\right)^2\right\}}.
\end{aligned} \tag{11.53}$$

In the same manner the components of $h_1(\xi_1, \xi_2)$ and $h_2(\xi_1, \xi_2)$ in $h(\xi_1, \xi_2)$ are calculated using the following relations:

$$(h_1, h_0) = 0, \quad (h_1, h_{11}) = 0, \quad (h_1, h_{12}) = 0, \quad (h_1, h_{22}) = 0,$$
$$(h_2, h_0) = 0, \quad (h_2, h_{11}) = 0, \quad (h_2, h_{12}) = 0, \quad (h_2, h_{22}) = 0. \tag{11.54}$$

That is,

$$\left(h, \frac{h_1}{\|h_1\|}\right) = \|h_1\|\delta_1 + \frac{(h_2, h_1)}{\|h_1\|}\delta_2$$
$$= \|h_0\| \left\{\left(\frac{\delta_1}{\Delta_1}\right) + \theta\left(\frac{\delta_1}{\Delta_2}\right)\right\},$$
$$\left(h, \frac{h_2}{\|h_2\|}\right) = \|h_0\| \left\{\theta\left(\frac{\delta_1}{\Delta_2}\right) + \left(\frac{\delta_2}{\Delta_2}\right)\right\}. \tag{11.55}$$

Now the following functions are defined by h_0, h_1, and h_2. We expect them the orthogonal bases to $h(\xi_1, \xi_2)$.

$$\varphi_0(\xi_1, \xi_2) \equiv \frac{h_0}{\|h_0\|}, \quad \varphi_1(\xi_1, \xi_2) \equiv \frac{\dfrac{h_1}{\|h_1\|} + \dfrac{h_2}{\|h_2\|}}{\sqrt{2(1+\theta)}},$$
$$\varphi_2(\xi_1, \xi_2) \equiv \frac{\dfrac{h_1}{\|h_1\|} - \dfrac{h_2}{\|h_2\|}}{\sqrt{2(1+\theta)}}. \tag{11.56}$$

From the definition, we can confirm that they are orthogonal and normalized as expected.

That is,

$$\|\varphi_0\|^2 = 1,$$
$$\|\varphi_1\|^2 = \frac{1 + 2\theta + 1}{2(1+\theta)} = 1,$$
$$\|\varphi_2\|^2 = \frac{1 - 2\theta + 1}{2(1-\theta)} = 1,$$
$$(\varphi_0, \varphi_1) = 0, \quad (\varphi_0, \varphi_2) = 0,$$
$$(\varphi_1, \varphi_2) = \frac{1 - 1}{2(1-\theta^2)} = 0. \tag{11.57}$$

Furthermore this normalized orthogonal system has the desired property of completeness, that is, the Perseval equation $\|h\|^2 = (h, \varphi_0)^2 + (h_1, \varphi_1)^2 + (h, \varphi_2)^2$ holds. To show this property (h, φ_m), $m = 0, 1, 2$, is calculated as

$$(h, \varphi_0) = \|h_0\|\sqrt{1 - \left\{\left(\frac{\delta_1}{\Delta_1}\right)^2 + 2\vartheta\left(\frac{\delta_1}{\Delta_1}\right)\left(\frac{\delta_2}{\Delta_2}\right) + \left(\frac{\delta_2}{\Delta_2}\right)\right\}},$$

$$(h, \varphi_1) = \frac{\|h_0\|}{\sqrt{2(1+\theta)}}(1+\theta)\left\{\left(\frac{\delta_1}{\Delta_1}\right) + \left(\frac{\delta_2}{\Delta_2}\right)\right\}$$

$$= \|h_0\|\sqrt{\frac{(1+\theta)}{2}}\left\{\left(\frac{\delta_1}{\Delta_1}\right) + \left(\frac{\delta_2}{\Delta_2}\right)\right\},$$

$$(h, \varphi_2) = \|h_0\|\sqrt{\frac{(1-\theta)}{2}}\left\{\left(\frac{\delta_1}{\Delta_1}\right) - \left(\frac{\delta_2}{\Delta_2}\right)\right\}. \tag{11.58}$$

Using these results, we calculate

$$(h, \varphi_0)^2 + (h, \varphi_1)^2 + (h, \varphi_2)^2$$

$$= \|h_0\|^2\left[1 - \left\{\left(\frac{\delta_1}{\Delta_1}\right)^2 + 2\theta\left(\frac{\delta_1}{\Delta_1}\right)\left(\frac{\delta_2}{\Delta_2}\right) + \left(\frac{\delta_1}{\Delta_1}\right)^2\right\}\right.$$

$$+ \frac{(1+\theta)}{2}\left\{\left(\frac{\delta_1}{\Delta_1}\right)^2 + 2\left(\frac{\delta_1}{\Delta_1}\right)\left(\frac{\delta_2}{\Delta_2}\right) + \left(\frac{\delta_2}{\Delta_2}\right)^2\right\} \tag{11.59}$$

$$\left. + \frac{(1-\theta)}{2}\left\{\left(\frac{\delta_1}{\Delta_1}\right)^2 - 2\left(\frac{\delta_1}{\Delta_1}\right)\left(\frac{\delta_2}{\Delta_2}\right) + \left(\frac{\delta_2}{\Delta_2}\right)^2\right\}\right].$$

From equation (11.47), we have

$$\|h\|^2 = (h_0(\xi_1 + \delta_1, \xi_2 + \delta_2), h_0(\xi_1 + \delta_1, \xi_2 + \delta_2))$$

$$= (h_0(\xi_1, \xi_2), h_0(\xi_1, \xi_2)) = \|h_0\|^2. \tag{11.60}$$

Notice that the norm is preserved in the positional displacement.

That is,

$$\int f(x+\delta)f(x+\delta)dx = \int f(X)f(X)dX;\ X = x + \delta.$$

Thus the Perseval equation is obtained as

$$\|h\|^2 = (h, \varphi_0)^2 + (h, \varphi_1)^2 + (h, \varphi_2)^2 \tag{11.61}$$

The simple similarity is defined as

$$S = \frac{(h, h_0)}{\|h\|\|h_0\|} \tag{11.62}$$

and is represented by the position displacement parameters according to (11.56) and (11.58) as

$$S(f_1, f_2) = \sqrt{1 - \left\{\left(\frac{\delta_1}{\Delta_1}\right)^2 + 2\theta\left(\frac{\delta_1}{\Delta_1}\right)\left(\frac{\delta_2}{\Delta_2}\right) + \left(\frac{\delta_2}{\Delta_2}\right)^2\right\}},$$

which means, the similarity changes as the squares of δ_1/Δ_1 and δ_2/Δ_2. On the other hand, the new similarity $S^*(f_1, f_2)$ is defined as

$$S^*(f_1, f_2) \equiv \frac{1}{\|h\|}\sqrt{(h, \varphi_0)^2 + (h, \varphi_1)^2 + (h, \varphi_2)^2} \tag{11.63}$$

and is approximately equal to 1 in the order of $(f_1/\Delta_1)^3$ and $(f_2/\Delta_2)^3$. An example is shown in Fig. 11.1, in which the advantage of the multiple similarity is shown clearly.

11.3.2 Compound Similarity

The multiple similarity method treats each class independently as well as the primitive subspace method. However, sometimes it happens that other classes must be taken at the same time when one class is considered This is because some character shapes that belong to different classes are considerably similar, such as "綱" and "網" in Chinese characters. Therefore some improvement of the multiple similarity is necessary to raise its discrimination power. Such a consideration was presented in the newer version of subspace method in which whole classes are considered together in order to maintain distinction among classes. In this subsection this improvement in multiple similarity is given.

Let us denote a set of normalized characters belonging to a class as $\{g_\alpha(\mathbf{r}); w(\alpha)\}$ and another set of normalized characters that are close to $\{g_\alpha(\mathbf{r}); w(\alpha)\}$ as $\{g_\beta(\mathbf{r}); v(\beta)\}$. Here, for simlicity, we omit the index i for denoting a specific class i. Note that

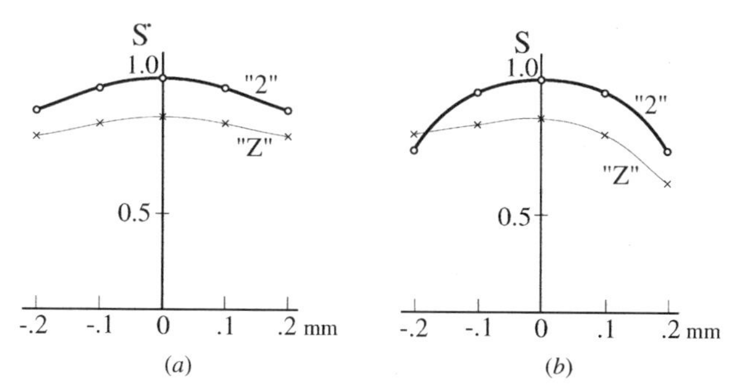

FIGURE 11.1 Comparison of multiple similarity (a) with simple similarity (b) in terms of displacement noise. The x-axis is scaled with displacement in millimeters. Note that in the case of simple similarity a reversal occurs at the displacement of 0.2 mm; the standard width of a stroke of printed Roman is 0.35 mm. From the reference [14].

in general, the set $\{g_\beta(\mathbf{r}); v(\beta)\}$ might consist of any number of classes and not necessarily only one class.

As mentioned before, first the average of $\{g_\alpha(\mathbf{r})\}$ is taken as

$$\overline{g(\mathbf{r})} = \int_D w(\alpha) g_\alpha(\mathbf{r}) d\alpha, \tag{11.64}$$

and it is normalized; then $\varphi_0(r)$ is defined as

$$\varphi_0(\mathbf{r}) = \frac{\overline{g(\mathbf{r})}}{\|\bar{g}\|}. \tag{11.65}$$

Using this $\varphi_0(r)$, the normalized character images $g_\alpha(\mathbf{r})$ and $g_\beta(\mathbf{r})$ are canonicalized as

$$\begin{aligned} h_\alpha(\mathbf{r}) &= g_\alpha(\mathbf{r}) - (g_\alpha, \varphi_0)\varphi_0(\mathbf{r}), \qquad \alpha \in D, \\ h_\beta(\mathbf{r}) &= g_\beta(\mathbf{r}) - (g_\beta, \varphi_0)\varphi_0(\mathbf{r}), \qquad \beta \in E, \end{aligned} \tag{11.66}$$

where E is a range to which β belongs.

For the two sets $\{h_\alpha(\mathbf{r}); w(\alpha)\}$ and $\{h_\beta(\mathbf{r}); v(\beta)\}$, the following functional is considered to be maximum under the condition $\|\varphi\|^2 = 1$:

$$J_0[\varphi(\mathbf{r})] = \int_D w(\alpha)(h_\alpha, \varphi)^2 d\alpha - \nu \int_E v(\beta)(h_\beta, \varphi)^2 d\beta, \qquad \nu > \mathbf{0}. \tag{11.67}$$

We cannot consider $(h_\alpha, \varphi)^2$ independent of $(h_\beta, \varphi)^2$. So maximum $(h_\alpha, \varphi)^2$ cannot be done independently. So "as much as *possible*" is used for $(h_\alpha, \varphi)^2$ and "as little as *possible*" is used for $(h_\beta, \varphi)^2$. Here, "possible" means that we attain the maximum value of $\mathbf{J}_0[\varphi\,(r)]$ under the condition of φ's norm is 1. The parameter ν is the degree of separation from other classes. If ν is too large, recognition of the set of characters $\{h_\alpha(\mathbf{r})\}$ is unlikely.

Such a variation problem was introduced in Chapter 7 in our discussion of Karhunen-Loève expansion, so only the result is shown. First the characteristic kernel is

$$K(\mathbf{r}, \mathbf{r}') = \int_D w(\alpha) h_\alpha(\mathbf{r}) h_\alpha(\mathbf{r}') d\alpha - \nu \int v(\beta) h_\beta(\mathbf{r}) h_\beta(\mathbf{r}') d\beta. \tag{11.68}$$

Next an integral operator K is defined as

$$(Kg)(\mathbf{r}) = \int_s K(\mathbf{r}, \mathbf{r}') g(\mathbf{r}') d\mathbf{r}'. \tag{11.69}$$

Then

$$(K\varphi_0)(\mathbf{r}) = 0 \tag{11.70}$$

is easily verified. For any character the following equality holds:

$$(Kg, g) = (g, Kg); \tag{11.71}$$

that is, K is symmetric. However, K is neither positive definite nor negative definite. Therefore the eigenvalues of K can be positive and negative; they are denoted as $\{\lambda_n\}$ and $\{-\mu_n\}$, respectively and ordered as

$$\lambda_1 \geq \lambda_2 \geq \cdots > 0 > \cdots \geq -\mu_2 \geq -\mu_1. \tag{11.72}$$

The corresponding eigenfunctions are denoted as $\varphi_n(\mathbf{r})$ for positive eigenvalues and $\psi_m(\mathbf{r})$ for negative values, respectively. As a result the following equations hold:

$$\begin{aligned} (K\varphi_n)(\mathbf{r}) &= \lambda_n \varphi_n(\mathbf{r}) \qquad (n = 1, 2, \ldots), \\ (K\psi_m)(\mathbf{r}) &= -\mu_m \psi_m(\mathbf{r}) \qquad (m = 1, 2, \ldots). \end{aligned} \tag{11.73}$$

Thus, for any input character $g(\mathbf{r})$,

$$S_c[g(\mathbf{r})] = \frac{(g, \varphi_0)^2 + \mu(Kg, g)}{\|g\|^2} \tag{11.74}$$

is defined, which defines *compound similarity.* Notice that the form of the definition is the same as that of multiple similarity, but only the K differs.

It is easily proved that the value of the compound similarity functional is bounded, since

$$-\mu\mu_1 \leq S_c[g(\mathbf{r})] \leq 1. \tag{11.75}$$

As before, this definition of compound similarity is not convenient in practice, and it is altered using the following formula:

$$(K\mathbf{r}, \mathbf{r}') = \sum_{n=1}^{\infty} \lambda_n \varphi_n(\mathbf{r}) \varphi_n(\mathbf{r})' \equiv \sum_{m=1}^{\infty} \mu_m \psi_m(\mathbf{r}) \psi_m(\mathbf{r})'. \tag{11.76}$$

Therefore

$$(Kg, g) = \sum_{n=1}^{\infty} \lambda_n (g, \varphi_n)^2 - \sum_{m=1}^{\infty} \mu_m (g, \psi_m)^2. \tag{11.77}$$

Furthermore, in practice, the two ∞ are replaced by some appropriate numbers N and M, respectively, so the compound similarity can be redefined as

$$S_c^*[g(\mathbf{r})] = \frac{1}{\|g\|^2}\left\{(g, \varphi_0)^2 + n\sum_{\mu=1}^{N}\lambda_n(g, \varphi_m)^2 - \mu\sum_{m=1}^{M}\mu_m(g, \psi_m)^2\right\}. \tag{11.78}$$

11.4 FEATURE MATCHING

Yet another expansion of template matching is to increase its dimensions in terms of mathematical formulation. This expansion is interpreted as feature matching. The gray-scale value at each pixel is the most primitive feature. Some other important features can be extracted at each point. The most effective feature is the direction of a given stroke or gradient when a given pattern is treated as scalar potential distribution. That is, the matching is performed among vector fields. This is a physical aspect of feature matching. Of course we can consider another feature. The example of the "field" obtained by the field effect method is a good one. This is what is done in actual practice, as mentioned before.

At any rate, the mathematical formulation of feature matching is very simple. The definition is given as

$$S_f = \frac{(F, G)}{\|F\| \cdot \|G\|}, \tag{11.79}$$

where

$$(F, G) = \sum_i \sum_j [F, G]_{i,j},$$

$$[F, G]_{i,j} = \sum_k F(i, j, k) \cdot G(i, j, k),$$

$$\|F\| = \left(\sum\sum [F, F]_{i,j}\right)^{1/2}, \quad \|G\| = \left(\sum\sum [G, G]_{i,j}\right)^{1/2}. \tag{11.80}$$

That is, we have again a simple similarity in form. To include the feature at point (i, j), the term $[F, G]_{i,j}$ is embedded, which is an inner product of k-dimensional vectors. If we treat the $N \times N$ frame of a pattern, then the total dimension is $k \times N \times N$. By convention, we image k planes to be processed independently. This has important implications in pattern recognition. First, we notice that feature components are treated as analog values, so they are real numbers. On the other hand, each point includes its neighbor which is an essential idea in pattern recognition. In practice, this concept of the neighbor is effectively a blurring process. In feature matching, the matching must be done by "neighbor" matching not by "point" matching as is the case discussed before in inner product pointwise matching. The blurring process' resemblance in "neighbor" matching has been confirmed by many experiments. We proceed to discuss in detail feature matching based on Yasuda's work [6,7].

11.4.1 Experiment of Feature Matching

Yasuda conducted a systematic experiment on the feature matching method. He divided the data into two sets. One was a numeral set and the other the roman alphabet and some symbols. Both had handprinted characters, whose quality was loosely unconstrained. However, we focus on the set of numeral characters. The number of subjects was 25 and the total number of the data was 2000. The human subjects were asked to write a character within a specified box, whose dimensions were 7 mm high and 5 mm wide.

The sampling pitch was 0.12 mm. The observed patterns were binarized and stored onto magnetic tape, which were represented in 48 × 30 pixels. All the data were thinned to one pixel width and are shown in Appendix C.

The feature used was the direction along a line of a given character in which the thinned line was coded by a Freeman code. However, there was no need to describe a contour/following direction. Therefore the eight quantized directions were reduced to four as shown in Fig. 11.2, in which *V, L, H,* and *R* or 1, 2, 3, and 4 have nothing but indexes to the directions and were taken for only convenience in further processing. The important thing is that there are four feature planes, the *V, L, H,* and *R* planes, in each of which the feature value at each point is an analog value. The four planes were processed independently but in the same manner.

After the feature planes had been constructed, a series of preprocesses were carried out. The first preprocessing was the normalization of position. The center of a given character image was placed at the center of the frame. The next one was the normalization of size where the original size of 48 × 30 was reduced to 9 × 12. Looking through the data shown in the Appendix, we can see considerable variation in sizes, among which the largest is 48 × 30. Actually this largest size determined the size of the planes.

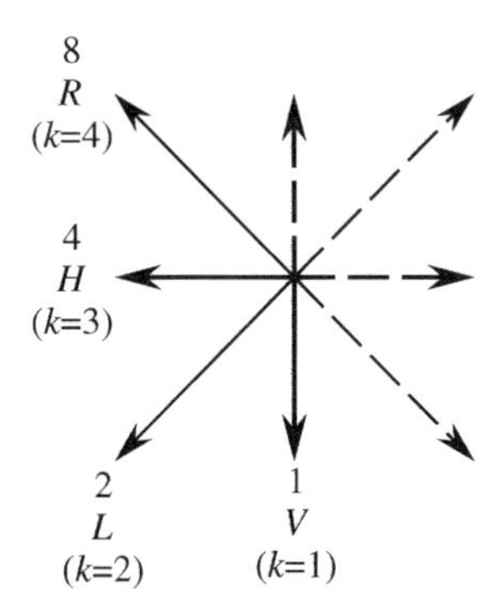

FIGURE 11.2 Definition of quantized directions and their notation, where plus and minus are regarded as one direction.

In effect the size normalization was only a size reduction. Therefore, if the size of a given character was less than 9×12, then it was treated as it was. In other words, no expansion processing was performed, since this requires comparatively complex processing.

The next preprocessing was blurring, which was based on the Gaussian function $A \exp(-(i^2 + j^2)/\sigma^2)$, where σ takes 2. The mathematical formulation of the blurring is expressed as

$$F^{(m)}(i, j, k) = \sum_{0 \le |r| \le 2,} \sum_{0 \le |s| \le 2} B(r, s) \times F^{(m-1)}(i - r, j - s, k). \qquad (11.81)$$

The coefficient $B(r, s)$ is shown in Table 11.1 where each number entry has a related meaning. Appropriate integers were chosen for the real values of the Gaussian function mentioned above. In addition to these illustrations, m in the formula denotes the number of iterations of the blurring process. This blurring process was done at each point on the reduced 9×12 plane, so the resultant plane changed in size from 9×12 to 13×16. This was because of the blurring formula in which r and s take 0, 1, 2. At the four corners of the 9×12 plane, two pixels were added to both the vertical and the horizontal directions. Therefore, as m increased, the size of the plane increased proportionately, as is shown in the upper row of Fig. 11.3. The lower row shows the proportionately reduced plane sizes which avoided heavy computations of the inner product. In the end five test sample variations, 0 ~ 4 in changing blurring, were obtained, all of the same size.

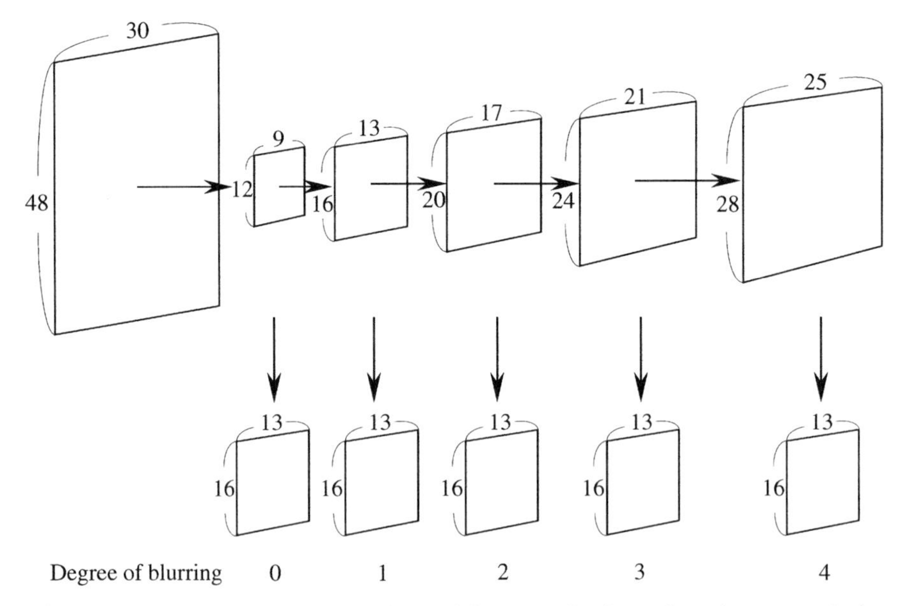

FIGURE 11.3 Relation between blurring and size normalizations where the top row depicts the degree of the blurring. Each is normalized so that the frame sizes are equal to 13×16. Input images sized to 9×12 included smaller images. From reference [6].

TABLE 1.1 Discrete Blurring Coefficients, B(*r*, *s*)

$\|s\|$ \ $\|r\|$	0	1	2
0	15	12	6
1	12	9	4
2	6	4	2

The standard patterns were constructed taking an average of all the samples at the final stage of preprocessing. Only one standard pattern for each category was made, which is unusual in practice, particularly for handprinted characters. However, the aim of this experiment lies in demonstrating its performance even if the data are handprinted. For this reason no rejection region was set. Some examples of the standard patterns are shown in Fig. 11.4, in which character "0" is taken, The left side (*a*) shows the unreduced planes, and the right side (*b*) shows the reduced planes. From the left column, *V, L, H,* and *R* planes are shown, and from the top row, the degrees of blurring 0, 1, 2, 3, and 4 are shown.

The experimental recognition results are shown in Fig. 11.5. The substituted error rates for the changing conditions are imposed on the y-axis namely for scalar (no-feature similarity) two-direction features, and four-direction features are shown taking on the x-axis the degrees of blurring, $0 \sim 4$. The positive effects of both directional features and blurring are shown clearly. The best score was the 97.1% recognition rate when four directions and the degree of the blurring 1 were taken. On the other hand, as stated before, position normalization is usually not so correct, so fur-

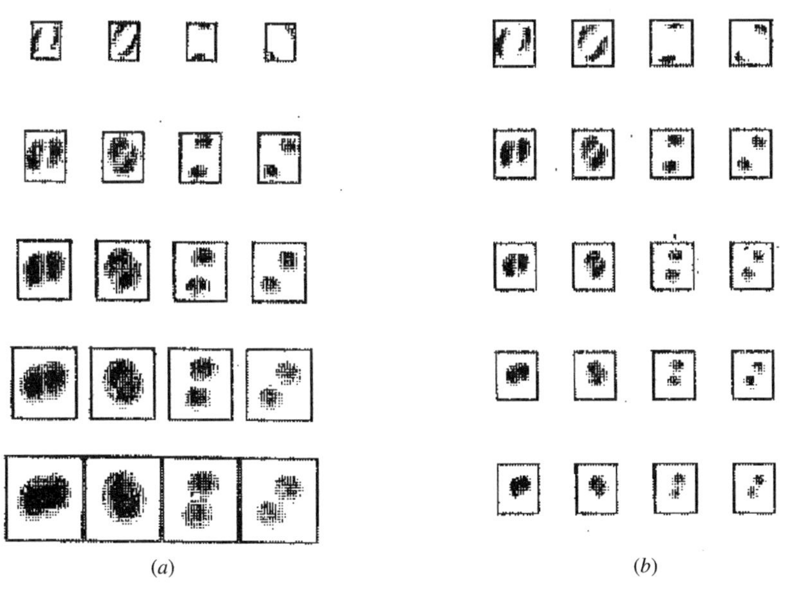

FIGURE 11.4 Blurred standard templates for "0" in four directions: (*a*) Nonnormalized templates; (*b*) normalized templates. From reference [6].

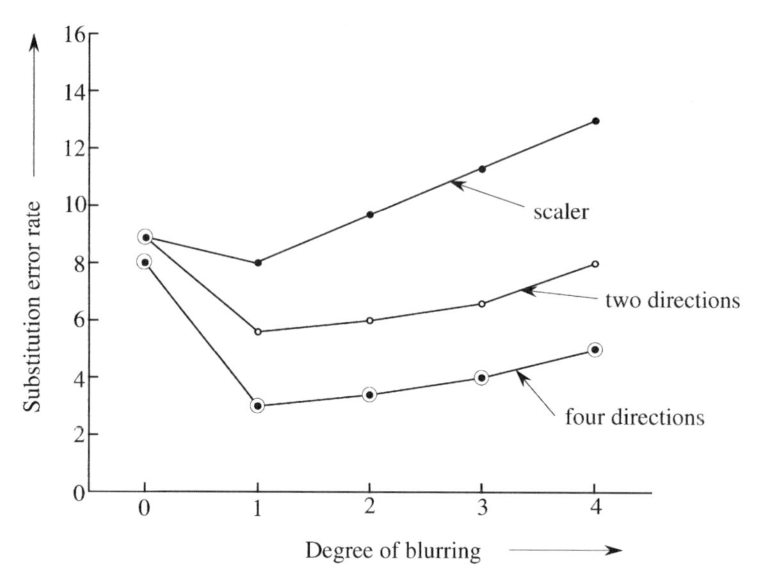

FIGURE 11.5 Effects of introducing direction and blurring. From reference [6].

ther processing of the displacement compensation is performed; that is, a given character or standard pattern is shifted toward eight directions by one pixel, and the maximum output of the similarity value is taken. This is a time-consuming processing but with considerably good effects. Overall the tests conducted to all the data, 2000 samples, were used to make the standard pattern. In other words, learning data and test data are the same, but in the case of the inner product matching scheme, the recognition rate is usually stable to the data variations so long as they have the same degree of variation. Improvement were seen on the experiment. The best score of the correct recognition rate of 98.3 was obtained at the same conditions.

11.5 BIBLIOGRAPHICAL REMARKS

Kurita et al. devised an effective directional feature extraction method, called the *weighted directional histogram* [8]. Their method was applied to handwritten Kanji character recognition. First of all, a binarized input image is normalized to the frame of size 68×68. Then each stroke is followed counterclockwise along its contour. Each point on the counter direction is quantized into 16. The frame is divided into 7×7 subregions, in each of which directional histogram is constructed as

$$H_{ij}(k), \quad [i, j \in \{1, 2, \ldots, 7\}, k = 1, 2, \ldots, 16].$$

For the histogram distribution on the 7×7 domain, blurring is performed at 9 positions: center, right, and left position on the middle horizontal line of the domain, top-center, top-right, and top-left positions on the top horizontal line, bottom-center, bottom-right, and bottom-left positions on the bottom horizontal line. That is, the frame is reduced to 3×3 subregions. The weights of the blurring process are shown in the matrix form as

$$\begin{bmatrix} 0 & 0.024 & 0.03 & 0.024 & 0 \\ 0.024 & 0.06 & 0.08 & 0.06 & 0.024 \\ 0.03 & 0.08 & 0.11 & 0.08 & 0.03 \\ 0.024 & 0.06 & 0.08 & 0.06 & 0.024 \\ 0 & 0.024 & 0.03 & 0.024 & 0 \end{bmatrix}.$$

The direction histogram is also blurred and reduced in its directions from 16 to 8. That is, at each odd numbered direction, its two neighbor directions are weighted by 0.5 and summed to the odd direction where its weight is set to 1. As a result $72 = 3 \times 3 \times 8$ dimensional feature vectors are obtained. Then Kurita and his colleagues used the so-called *quasi-Mahalanobis* distance in the descrimitation stage. The related story will be mentioned later.

On the subspace method, Oja's book [9] is the standard. It begins with a comprehensive introduction, gives a detailed description, and refers to the related field of each as learning to which Oja contributed. On the other hand, concerning the multiple similarity method, Iijima, Genchi, and K. Mori [10] gave a detailed description with some experimental result. However, the formulation is based on Iijima's blurring theory and it is somewhat hard to read without his basic theory on pattern recognition. Iijima wrote a book on that, but it is written in Japanese unfortunately [11].

There is interesting research by Kurosawa that attempts a theoretical relationship between the subspace method and MBDF (*modified Bayes's discriminant function*) [12]. Kurosawa obtained a new discrimination function assuming a kind of Gaussian distribution on a hyperspherical surface. The formula of the function is described as a weighted subspace method, which is similar to MBDF. The subspace method is obtained from the function with certain approximation. Concerning MBDF, Kimura et al. gives a systematic description with detailed experiments on Kanji characters using ETL8 database [22].

Both research results have achieved very high recognition performance. However, historically the so-called *pseudo-Bayesian Mahalanobis* distance denoted as $d^*(\mathbf{x}, k)$ was proposed by Katsuragi in 1970 [13]. That is, $d^*(\mathbf{x}, k)$ was given as

$$d^*(\mathbf{x}, k) = \frac{1}{\lambda^{(k)}}(\mathbf{x} - \overline{\mathbf{x}}^k)^2 - \sum_{j=1}^{M_k} \left(\frac{1}{\lambda^{(k)}} - \frac{1}{\lambda_j^{(k)}} \right) (\mathbf{x} - \overline{\mathbf{x}}^k, \mathbf{e}_j^k)^2,$$

where $\mathbf{x}(x_i, i = 1, 2, \ldots, N)$ is an input vector, $\overline{\mathbf{x}}^k$ is an average vector that belongs to the kth class, and $\mathbf{e}_j^k$ and $\lambda_j^{(k)}$ are eigenvectors and eigenvalues of the covariance matrix of the data belonging to the kth class. The basic idea of the $d^*(\mathbf{x}, k)$ is that the distance is the the combination of Euclidean distance in the $(M - M_k)$-dimensional subspace, where M is the total dimension. That is, the lower contribution subspace is replaced by a Euclidean distance weighted by $1/\lambda^{(k)}$, which is a representative eigenvalue for the remaining subspace of the $(M - M_k)$-dimension.

On the other hand, MQDF2, $g_2(x)$ proposed by Kimura et al. is described as

$$g_2(\mathbf{x}) = \frac{1}{h^2}\left[\|\mathbf{x} - \boldsymbol{\mu}_M\|^2 - \sum_{i=1}^{k}\left(1 - \frac{h^2}{\lambda_i}\right)\{\boldsymbol{\varphi}_i^t(\mathbf{x} - \boldsymbol{\mu}_M)\}^2\right] + \log\,(h^{2(n-k)}\prod_{i=1}^{k}\lambda_i).$$

Now we are ready to describe briefly Kurosawa's research. In the subspace method, the similarity is given as

$$S = \sum_{i=0}^{k-1}(\mathbf{x}, \boldsymbol{\varphi}_i)^2,$$

where $\mathbf{x}$ is normalized and $\boldsymbol{\varphi}_i$ is ith orthogonal vector, which is calculated from the covariance matrix as

$$K = \sum_{\alpha}\mathbf{x}_\alpha\mathbf{x}_\alpha^T,$$

where $\mathbf{x}_\alpha$ are learning data. Naturally λ_i is defined as ith eigenvalue of the covariance matrix K. On the other hand, the pseudo-Bayes method is defined as

$$D = \|\mathbf{x} - \mathbf{m}\|^2 - \sum_{i=0}^{k-1}\left(1 - \frac{\delta}{\lambda_i}\right)(\mathbf{x} - \mathbf{m}, \boldsymbol{\varphi}_i)^2 + \delta\sum_{i=0}^{k-1}\ln\lambda_i + \delta\sum_{i=k}^{M-1}\ln\delta,$$

where $\mathbf{x}$ is not normalized and $\mathbf{m}$ is a mean vector. However, λ_i and $\boldsymbol{\varphi}_i$ here are calculated from the following covariance matrix:

$$K = \sum_{\alpha}(\mathbf{x}_\alpha - \mathbf{m})(\mathbf{x}_\alpha - \mathbf{m})^T.$$

The D above is theoretically calculated by assuming a normal distribution of the input data. Assuming the Gaussian distribution on a pseudospherical surface, the Baysian similarity is given as

$$S = 2\delta\ln(\mathbf{x}, \varphi_0) + (\mathbf{x}, \varphi_0)^2 + \sum_{i=1}^{k-1}\left(1 - \frac{\delta}{\lambda_i}\right)(\mathbf{x}, \varphi)^2 - \delta\sum_{i=1}^{k-1}\ln\lambda_i - \delta\sum_{i=k}^{M-1}\ln\delta.$$

To compare the above expression with the D, the D is converted to the similarity using the conditions of $\mathbf{m} = (\mathbf{x}, \boldsymbol{\varphi})\boldsymbol{\varphi}_0$ and $\|\mathbf{x}\| = 1$ as

$$S = 1 - D = (\mathbf{x}, \boldsymbol{\varphi}_0)^2 + \sum_{i=1}^{k-1}\left(1 - \frac{\delta}{\lambda_i}\right)(\mathbf{x}, \boldsymbol{\varphi}_i)^2 - \delta\sum_{i=0}^{k-1}\ln\lambda_i - \delta\sum_{i=k}^{M-1}\ln\delta.$$

The form in which the term ln λ is neglected in the D is called the pseudo-Maharanobis distance as shown below:

$$D = \|\mathbf{x} - \mathbf{m}\|^2 - \sum_{i=0}^{k-1} \left(1 - \frac{\delta}{\lambda_i}\right) (\mathbf{x} - \mathbf{m}, \boldsymbol{\varphi}_i).$$

On the other hand, the *pseudospherical surface Mahalanobis similarity* is defined as

$$S = 2\delta \ln (\mathbf{x}, \boldsymbol{\varphi}_0) + (\mathbf{x}, \boldsymbol{\varphi}_0)^2 + \sum_{i=1}^{k-1} \left(1 - \frac{\delta}{\lambda_i}\right) (\mathbf{x}, \boldsymbol{\varphi}_i)^2.$$

If the first term is neglected and each coefficient of each inner product is set to 1, namely $\lambda_i \gg \delta$ is assumed, then we obtain the similarity in the subspace method formally. That is,

$$S = (\mathbf{x}, \boldsymbol{\varphi}_0)^2 + \sum_{i=1}^{k-1} (\mathbf{x}, \boldsymbol{\varphi}_i)^2 = \sum_{i=0}^{k-1} (\mathbf{x}, \boldsymbol{\varphi}_i)^2.$$

Likewise, in the pseudo-Mahalanobis distance, if $\lambda_i >> \delta$ is assumed, the following formula is obtained:

$$D = \|\mathbf{x} - \mathbf{m}\|^2 + \sum_{i=0}^{k-1} (\mathbf{x} - \mathbf{m}, \boldsymbol{\varphi}_i)^2,$$

which corresponds to the subspace method. Therefore we can see a relationship between the subspace method and the modified Bayes discriminant function.

Iijima et al.'s paper presented little application. Therefore a major project on OCR was conducted based on Iijima's theory and a very high performance OCR was made, called ASPET/71, which is an abbreviation for Analog Special Processing Electro technical laboratory and Toshiba 1971. This is due to Katsuragi who was a member of the joint development team and an excellent engineer. Unfortunately, he died soon after this success at a very young age. As a member of the team, I want to describe it in his memory and to express my respect. His original work has come in fashion as mentioned. Anyway, ASPET/71 was intended for reading alphanumeric characters and symbols of OCR-B font with a speed of 2000 characters per second. For the details see the paper written by Iijima, Genchi, and Mori [3].

Later the multiple similarity method was applied to printed Kanji recognition in which a two-stage recognition was employed because of the large number of categories in Kanji—2000 in this case. This work was done by Sakai, Hirai, Kawada, Amano, and Mori [15]. The original Chinese characters amounted to 50,000 characters. In Japan 2000 characters are enough in daily life. That is, in the first stage, an input character is classified into a group, and then in the second stage a multiple similarity method is applied to the classified input character. However, Kanji characters includes some pairs with very similar shapes such as 微-徴, 憶-億, 未-末. About 3% of the Chinese characters have such similar pattern pairing. To cope with this difficulty, the so-called compound similarity method was applied. The results for the multiple similarity method and compound similarity method were 96.43% (correct), 2.72% (reject), and 0.85% (error), and 98.94% (correct), 0.76% (reject), and 0.30% (error), respectively. Finally, to improve the total recognition performance, this step was rerun based on optimum thresholding and the correct recognition rate and rejection rate were improved to 99.20% and 0.49%, respectively.

Furthermore the multiple similarity method was applied to handprinted Hiragana characters by Maeda, Murashige, Hirai, and Sakai [16]. However, because of the variation of handprinted characters, some supplementary structural features were also used; this helped reduce the error rate. The correct recognition and error rate were 98.4% and 0.4%, respectively. For the multiple similarity method only, the results were 98.7% (correct) and 0.6 (error). In this case the data were collected under the condition that 200 subjects were asked to write characters neatly looking a standard character shape table of Hiragana. Kurosawa, Maeda, Asada, and Sakai [17] applied the multiple similarity method further to handprinted Kanji characters, a somewhat restricted set of 106 kinds of Kanji characters. The number of subjects was 250. The results for the learning data were 99.86% (correct) and 0.14% (error). As test data an ETL-8 database was used in which 76 kinds of Kanji characters and 160 characters/category were used. The results were 94.6% (correct) and 5.4% (error). On the other hand, it is expected that the multiple similarity method is effective for the multiple font Kanji character recognition. A systematic experiment was done by Ariyoshi and Asada [18]. Seven kinds of fonts were used in the experiment, but only three kinds of fonts were selected as learning sets. The number of Kanji characters was 656 kinds. The recognition results were very good, almost 100% recognition rate for all the fonts tested.

The multiple similarity method was applied from a new angle to handprinted Kanji character recognition including Hirakana and Katakana characters by Maeda, Kurosawa, Asada, Sakai, and Watanabe [19]. To cope with the variation of handprinted characters, the multiple similarity method was extended to gray-level and local features, and renamed the structural multiple method. To extract local features, Hermite's polynomials with Gaussian weight were used. This approach can be considered in the class of feature matching. Using this approach, more detailed and unified research was done by Maeda and Watanabe [20]. From the engineering viewpoint, this work was improved to 2×2 local feature distribution by Irie and Yaguchi [21]. A character frame was divided into 7×7 blocks, from each of which these local features were extracted. Thus 49×5 high-dimensional feature vectors were constructed, on which multiple similarity method was applied. The experiment was done in using ET-9B (handprinted characters, 3036 categories, and 200 subjects). For the learning data there was achieved a 98.4% correct recognition rate, and for the test data the correct recognition rate was 93.4%.

BIBLIOGRAPHY

[1] N. Otsu, "Mathematical studies on feature extraction in pattern recognition," Researches of The Electrotechnical Laboratory, no. 818, July 1981.

[2] S. Watanabe, P. F. Lambert, C. A. Kulikowski, J. L. Buxton, and R. Walker, "Evaluation and selection of variables in pattern recognition," in J. Tou, ed., *Computer and Information Science,* II, pp. 91–122, New York: Academic Press, 1967.

[3] T. Iijima, H. Genche, and Kenichi Mori, "A theory of character recognition by pattern matching method," *Proc. 1st IJCPR,* pp. 50–56, 1973.

[4] A. Ben-Israel and T. N. E. Greville, *Generalized Inverses,* New York: Wiley, 1974.

[5] T. Iijima, "Similarity being invariant to positional displacement," Research memo of National Project on OCR, July 1968.

[6] M. Yasuda, "Research on character recognition based on cross-correlation," Doctoral thesis, University of Tokyo, 1981.

[7] M. Yasuda and H. Fujisawa, "An improved correlation method for character recognition," *Scripta Electronica Japonica* IV, 10, pp. 29–38, 1979.

[8] M. Kurita, S. Tsuruoka, S. Yokoi, and Y. Miyake, "Handprinted Kanji and Hirakana character recognition using weight direction index histograms and quasi-Mahalanobis distance," IECE Japan, Technical report, vol. PRL82-79, January 1983.

[9] E. Oja, "Subspace Methods of Pattern Recognition," Research Studies Press, 1983.

[10] T. Iijima, H. Genchi, and K. Mori, "A theory of character recognition by pattern matching method," *Proc. 1st IJCPR,* pp. 50–56, 1973.

[11] T. Iijima, *Pattern Recognition,* Tokyo: Korona Press, 1973.

[12] Y. Kurosawa, "New pattern recognition method based on relationship between subspace method and modified bayes discriminant function," Technical Report of IECE Japan, PRMU96-104, 1996.

[13] S. Katsuragi, "Factor analysis of character pattern distribution and its application to discrimination," *Proc. Ann. Conf. IECE* Japan, p. 2816, 1970.

[14] T. Iijima, S. Mori, H. Genchi, and K. Mori, "Simulation for theory, design and tests of the OCR ASPET/71," *Proc. Simulation of Complex Systems,* Society of Analog Technique of Japan, F9/1–9/6, September 1971.

[15] K. Sakai, S. Hirai, T. Kawada, S. Amano, and K. Mori, "An optical Chinese character reader," *Proc. 3rd IJCPR,* pp. 122–126, 1976.

[16] K. Maeda, K. Murashige, S. Hirai, and K. Sakai, "Handprinted Hiragana recognition," *Proc. Ann. Conf. IECE in Japan,* no. 1315, 1980.

[17] Y. Kurosawa, K. Maeda, H. Asada, and K. Sakai, "Handprinted Kanji character recognition experiment based on multiple similarity method," *Proc. Ann. Conf. IECE* in Japan, no. 79, 1981.

[18] S. Ariyoshi and H. Asada, "Multi-font Chinese character recognition by generating distorted patterns," *Proc. Ann. Conf. IECE in* Japan, no. 1465, 1987.

[19] K. Maeda, Y. Kurosawa, H. Asada, K. Sakai, and S. Watanabe, "Handprinted Kanji recognition by pattern matching method," *Proc. 6th IJCPR,* pp. 789–792, October, 1982.

[20] K. Maeda and S. Watanabe, “A pattern matching method with local structure,” *Trans. IECE Japan,* vol. J68-D, no. 3, 1085.

[21] B. Irie and K. Yaguchi, “Handprinted Kanji character recognition using the 2×2 neighbor pattern distribution,” *Proc. Ann. Conf. in Japan,* no. D-453, 1988.

[22] F. Kimura, K. Takashina, S. Tsuruoka, and Y. Miyake, “Modified quadratic discriminant functions and the application to Chinese character recognition,” *IEEE Trans. Pattern Anal. Machine Intell.,* vol. PAMI-9, pp. 149–153, 1987.

CHAPTER TWELVE

Graph Matching

As we have shown in Chapter 9, shape is often represented by *parts* and *relations among the parts.* A mathematical structure composed of parts and relations among the parts is called a *relational structure.* If shape prototypes (models) are also given by a relational structure of parts and relations among the parts, *recognition* is transformed into a mathematical problem of matching two relational structures. In this chapter we introduce some mathematical theories and algorithms for relational matching problems.

12.1 RELATIONAL HOMOMORPHISM

Let O_A be an object or entity and A be the set of its parts. An N-ary relation R over A is a subset of the Cartesian product

$$R \subseteq A^N. \tag{12.1}$$

When $N = 2$, the relational structure is represented as a *graph* (A,R) by mapping elements in A onto vertices (nodes) and elements in R to edges (links) of the graph (A,R).

Given two objects O_A and O_B, let A be the set of parts of object O_A, and B be the set of parts of object O_B. Let $R \subseteq A^N$ be an N-ary relation over the part set. Let $f: A \rightarrow B$ be a function that maps elements of A to B. We define the *composition* of $R \circ f$ of R with f by

$$\begin{aligned} R \circ f = \{(b_1, \ldots, b_N) \in B \mid & \text{ there exists} \\ & (a_1, \ldots, a_N) \in R \text{ with } f(a_i) = b_i, i = 1, \ldots, N\}. \end{aligned} \tag{12.2}$$

Figure 12.1 illustrates the composition of a binary relation with a mapping.

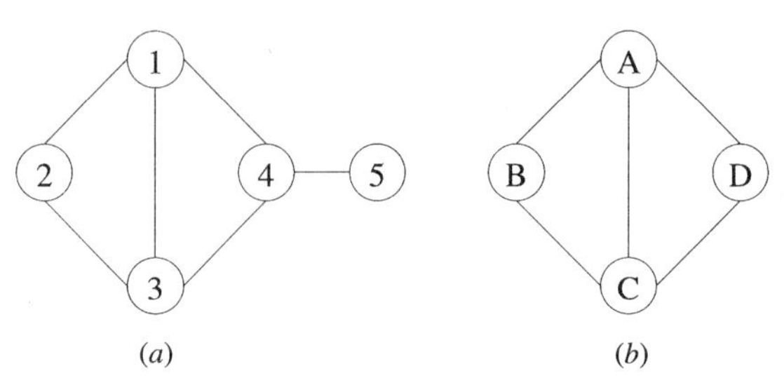

FIGURE 12.1 Composition of binary relation R with mapping h, where $h(1) = A$, $h(2) = B$, $h(3) = C$, $h(4) = D$, $h(5) = D$: (*a*) Binary relation R; (*b*) composition $R \circ h$.

Let $T \subseteq A^N$ and $S \subseteq B^N$ be two N-ary relations over A and B. A function $f : A \rightarrow B$ that satisfies $T \circ f \subset S$ is called a *relational homomorphism.* Given two arbitrary N-ary relations, the *relational homomorphism problem* is the problem of determining all relational homomorphism between them. Figure 12.2 illustrates a relational homomorphism.

Let $A = \{w, v, 0, 1\}$ and $R \subseteq A^3$ be a ternary relation defined on the set A. Suppose that R is given by

$$R = \{(v, v, 0), (w, v, 0), (v, w, 1), (w, w, 1)\} \subseteq A^3. \tag{12.3}$$

Let $B = \{a, b, c, d, 0, 1\}$ and $h : A \rightarrow B$ be a mapping from A to B defined by $h(w) = a$, $h(v) = d$, $h(0) = 1$, and $h(1) = 0$. Then $R \circ h$ is a set of triplets:

$$R \circ h = \{(d, d, 1), (a, d, 1), (d, a, 0), (a, a, 0)\}. \tag{12.4}$$

Let $S \subseteq B^3$ be a ternary relation defined on the set B. Suppose that S is given by

$$\begin{aligned} S = \{&(a, a, 0), (b, a, 0), (c, a, 0), (d, a, 0), \\ &(a, d, 1), (b, d, 1), (c, d, 1), (d, d, 1)\}. \end{aligned} \tag{12.5}$$

Since $R \circ h \subseteq S$, h is a relational homomorphism from R to S.

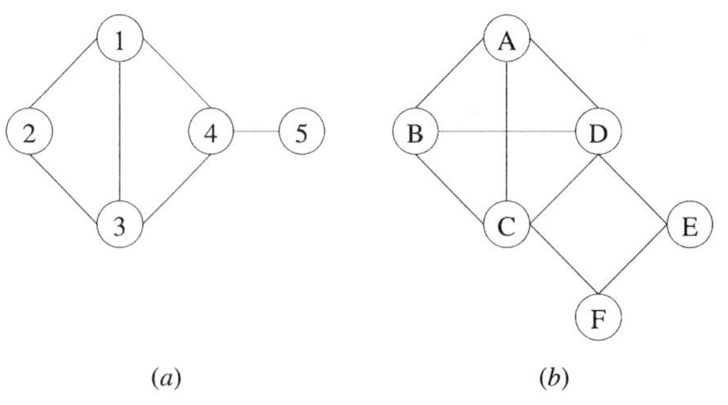

FIGURE 12.2 Relational homomorphism h from R to S, where $h(1) = A$, $h(2) = B$, $h(3) = C$, $h(4) = D$, $h(5) = B$: (*a*) Binary relation R; (*b*) binary relation S.

Let $f : B \rightarrow A$ be a mapping from B to A defined by $f(a) = v$, $f(b) = v$, $f(c) = w$, $f(d) = w$, $f(0) = 0$, and $f(1) = 1$. Then $S \circ f$ is a set of triplets:

$$\{(v, v, 0), (w, v, 0), (v, w, 1), (w, w, 1)\}. \tag{12.6}$$

Notice that $S \circ f \subseteq$ R, making f a homomorphism of S into R. Since $S \circ f = R$, R is a homomorphic image of S and f is said to be onto R.

Let $C = \{a, b, 0, 1\}$ and $T \subseteq C^3$ be the ternary relation defined on the set C. Suppose that T is given by the triplets:

$$T = \{(a, a, 1), (b, a, 1), (a, b, 0), (b, b, 0)\}. \tag{12.7}$$

Let $g : A \rightarrow C$ be a mapping from A to C defined by $g(w) = b$, $g(v) = a$, $g(0) = 1$, and $g(1) = 0$. Then $R \circ g = T$. This makes g a one-to-one onto homomorphism. One-to-one onto homomorphisms are called *isomorphisms.* Two relations that are isomorphic are exactly the same except for the name of the symbols used. To convert one relation to the other, we need just to translate all symbols in the first relation through the isomorphism, and we will obtain the second relation.

A relational homomorphism maps the primitives of A to a subset of the primitives of B having the same interrelationships that the original primitives or A had. If A is a set much smaller than B, then finding a one-to-one relational homomorphism is equivalent to finding a copy of a small object as part of a larger object. If A and B are about the same size, then finding a relational homomorphism is equivalent to determining that the two objects are similar. A relational monomorphism is a relational homomorphism that is one-to-one. Such a function maps each primitive in A to a unique primitive in B. A monomorphism indicates a stronger match than a homomorphism.

A *relational isomorphism* f from an N-ary relation to an N-ary relation S is a one-to-one homomorphism from R to S, and f^{-1} is a relational homomorphism from S to R. In this case A and B have the same number of elements; each primitive in A maps to a unique primitive in B, and every primitive in A is mapped to some primitive in B. Also every tuple in R has a corresponding tuple in S, and vice versa. An isomorphism is the strongest kind of match. Figure 12.3 illustrates a relational isomorphism, and Figure 12.4 shows the difference between a relational isomorphism and a relational monomorphism.

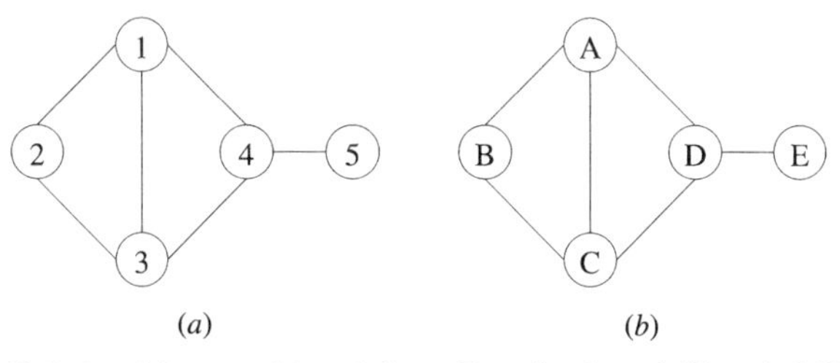

FIGURE 12.3 Relational isomorphism h from R to S, where $h(1) = A$, $h(2) = B$, $h(3) = C$, $h(4) = D$, $h(5) = E$: (*a*) Binary relation R; (*b*) binary relation S. $R \circ h = S$ and h is one-to-one, or equivalently, $R \circ h \subseteq S$, $S \circ h^{-1} \subseteq R$, and h is one-to-one.

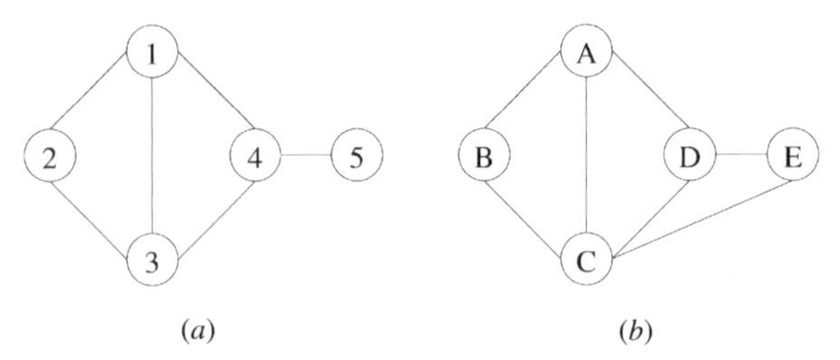

FIGURE 12.4 Relation monomorphism h from binary relation R onto binary relation S, where $h(1) = A$, $h(2) = B$, $h(3) = C$, $h(4) = D$, $h(5) = E$: (a) Binary relation R; (b) binary relation S. This mapping h is not a relational isomorphism because h^{-1} is not a relational monomorphism from S to R.

12.2 HOMOMORPHISM FOR SCENE LABELING

Suppose that a scene has been divided into segments $S = \{S_1, \ldots, S_K\}$. A low-level feature extractor with decision rules using gray tone, color shape, and texture of each segment assigns some possible description from a set of descriptions to each segment. Let D be the set of descriptions and $F \subseteq S \times D$ be a segment-description relation defined by this operation. Then $(s, d) \in F$ if and only if the segment s has a description d.

Each segment may be associated with multiple descriptions. In order to reduce the ambiguity, we use some a priori information such as constraints from a higher-level world model. Suppose that L is the set of relationship labels. Then the set of spatially related segments could be specified by the relation $A \subseteq S \times S \times L$, where $(s, t, i) \in A$ if and only if label i describes the way the segment s relates the segment t. In general, the relationship in L can describe the way N segments are related so that the relation $A \subseteq S^N \times L$ is a labeled N-ary relation.

The world model contains constraining information. For example, pairs of segments whose relationship label is i can be constrained by the world model to associate with them only certain allowable description pairs. In this case the world model is specified as a relation $C \subseteq D \times D \times L$ where $(d_1, d_2, i) \in C$ if and only if it is legal for a pair of segments s_1 and s_2 having relation i to have respective descriptions d_1 and d_2. In general, the relation $C \subseteq D^N \times L$ is a labeled N-ary relation that includes all labeled N-tuples of compatible descriptions for an ordered set of N related segments.

To summarize the information we have available

1. $F \subseteq S \times D$: the assignments of descriptions given by a low-level operations;
2. $A \subseteq S^N \times L$: the labeled set of related N-tuples of segments;
3. $C \subseteq D^N \times L$: the N-ary relational labeling constraints specified by the world model.

The scene labeling problem is expressed in terms of relational homomorphisms in the following way:

STEP 1: We extend the relation F to the relation $F' \subseteq (S \cup L) \times (D \cup L)$ by

$$F' = \{(s, d) \mid (s, d) \in F \text{ or } s = d \in L\}. \quad (12.8)$$

STEP 2: We consider the relation $A \subseteq S^N \times L$ as an $(N + 1)$-ary relation on $S \cup L$, namely $A \subseteq (S \cup L)^{N+1}$, and the relation $C \subseteq D^N \times L$ as an $(N + 1)$-ary relation on $D \cup L$, namely $C \subseteq (D \cup L)^{N+1}$.

STEP 3: The problem is to find all functions

$$G : (S \cup L) \rightarrow (D \cup L) \quad (12.9)$$

satisfying (1) $G \subseteq F'$ and (2) $A \circ G \subseteq C$.

12.3 ALGORITHMS FOR RELATIONAL MATCHING

The only known algorithms that can solve relational matching problems employ a tree search. We describe the standard backtracking tree search and one of its variants.

12.3.1 Backtracking Tree Search

Let R be an N-ary relation over part set A, and let S be an N-ary relation over part set B. We will refer to the elements of set A as *units* and the elements of set B as *labels*. We wish to find the set of all mappings $f : A \rightarrow B$ that satisfies $R \circ f \subseteq S$. Of course the set may be empty, where the algorithm should fail. The backtracking tree search begins with the first unit of A. This unit can potentially match each label in set B. Each of these potential assignments is a node at level 1 of the tree. The algorithm selects one of these nodes, makes the assignment, selects the second unit of A, and begins to construct the children of the first node, which are nodes that map the second unit of A to each possible label of B. At this level some of the nodes may be ruled out because they violate the constraint that $(u_1, \ldots, u_N) \in R$ implies $(f(u_1), \ldots, F(u_N)) \in S$. The process continues to the level of $|A|$ of the tree. The paths from the root node to any successful nodes at level $|A|$ are the relational homomorphism. Figure 12.5 shows a portion of the backtracking tree search for a simple directed graph matching problem. The algorithm for a backtracking tree search is as follows:

```
procedure treesearch(A, B, f, R, S)
a := first(A)
for each b ∈ B
  {
  f' = f ∪ {(a, b)};
  OK := true;
```

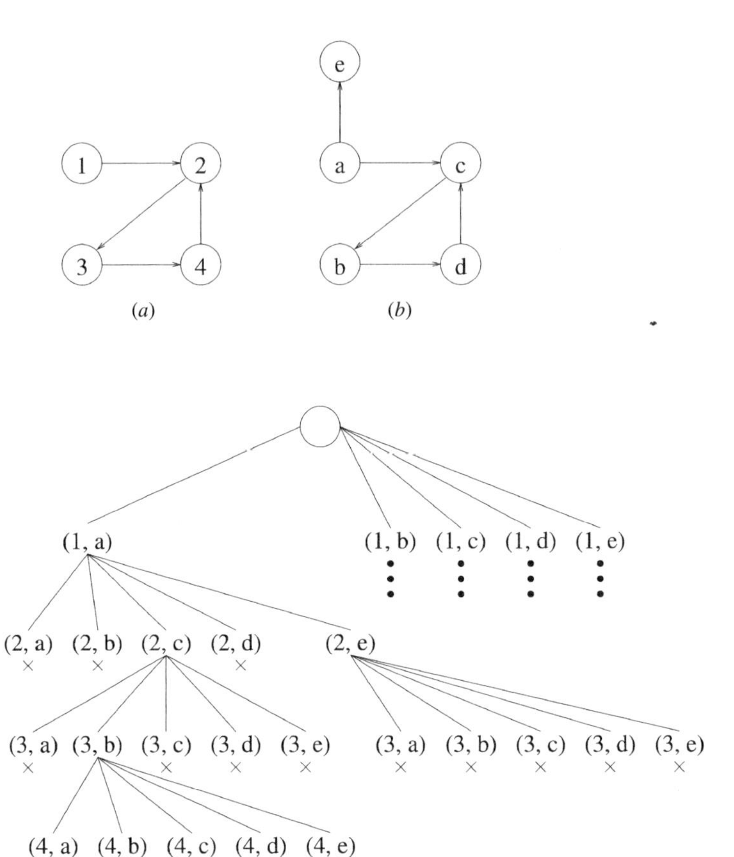

FIGURE 12.5 Backtracking tree search to find a relational homomorphism from $R = \{(1, 2), (2, 3), (3, 4), (4, 2)\}$ to $S = \{(a, c), (c, b), (b, d), (d, c), (a, e)\}$. An "×" under a node indicates failure. The only homomorphism found is $f = \{(1, a), (2, c), (3, b), (4, d)\}$.

```
  for each N-tuple r in R containing a component
    and whose other components are all in domain(f)
    if r ∘ f is not in S
    then { OK := false; break; } endif
  if OK then
    {
    A′ := A - {a}
    if isempty(A);
    then output(A′)
    else treesearch(A′, B, f′, R, S)
    }
  endif
  }
end treesearch;
```

12.3.2 Backtracking with Forward Checking

The backtracking tree search has exponential time complexity. Although there are no known polynomial algorithms in the general case, there are a number of discrete relaxation algorithms that can cut down search time by reducing the size of the tree that is searched. Forward checking is one such method. It is based on the idea that once a unit-label pair (a, b) is instantiated at a node in the tree, the constraints imposed by the relation cause instantiation of some future unit-label pairs (a', b') to become impossible. Suppose that (a, b) is instantiated high in the tree and that the subtree beneath the node contains nodes with first components $a_1, a_2, \ldots, a_n, a'$. Although (a', b') is impossible for any instantiations of $\{a_1, a_2, \ldots, a_n\}$, it will be tried in every path that reaches its level in the tree. The principle of forward checking is to rule out (a', b') at the time that (a, b) is instantiated and to keep a record of that information.

The data structure used to store the information is called a *future error table* (FTAB). There is one future error table for each level of recursion in the tree search. Each table is a matrix having one row for each element of A and one column for each element of B. For any uninstantiated or *future unit* $a' \in A$ and potential label $b' \in B$, $FTAB(a', b') = 1$ if it is still possible to instantiate (a', b') given the history of instantiations already made. $FTAB(a', b') = 0$ is (a', b') has already been ruled out due to some previous assignment. When a pair (a, b) is instantiated by the backtracking tree search, an updating procedure is called to examine all pairs (a', b') of future units and their possible labels. For each pair (a', b') that is incompatible with the assignment of (a, b) and the previous instantiations, $FTAB(a', b')$ has become 0. If for any future unit a', $FTAB(a', b')$ becomes 0 for all labels $b' \in B$, then instantiation of (a, b) fails immediately. The backtracking tree search with forward checking is as follows.

```
procedure forward_checking_treesearch(a, b, f, FTAB, R, S)
a := first(A);
for each b ∈ B
  if (FTAB(a, b) == 1)
  then
    {
    f' := f ∪ {(a, b)};
    A' := A - {a}
    if isempty(A');
    then output(f')
    else
      {
      NEWFTAB := copy(FTAB);
      OK := update(NEWFTAB, a, b, A', B, R, S, f');
      if (OK) forward_checking_treesearch(A', B, f',
        NEWFTAB, R, S);
      }
    endif
    }
endif
end forward_checking_treesearch
```

```
procedure update(FTAB, a, b, future_units, B, R, S, f')
update := false;
for each a' ∈ future_units
  for each b' ∈ B with FTAB(a', b') == 1
  if compatible(a, b, a', b', R, S, f');
  then update := true;
  else FTAB(a', b') := 0;
  endif
end update
```

For binary relation R and S, the utility function *compatible,* which determines whether an instantiation of (a', b') is possible for a given instantiation (a, b), is very simple. Units a and a' only contain one another when either (a, a') or (a', a) is in R. Thus the algorithm for function *compatible* for binary relations R and S is as follows:

```
procedure b_compatible(a, b, a', b', R, S, f')
if ((a, a') ∈ R and not ((b, b') ∈ S)) or
   ((a', a) ∈ R and not ((b', b) ∈ S))
then b_compatible := false;
else b_compatible := true; endif
end b_compatible
```

Note that for binary functions the last argument f' to functions b_compatible is not used but is included here for consistency.

For N-ary relations R and S, $N > 2$, those N-tuples of R where a and a' are among the components and all other components that are already instantiated must be examined. The code for N-ary relations R and S is as follows:

```
procedure compatible(a, b, a', b', R, S, f')
f'' := f' ∪ {(a, b)};
for each r ∈ R containing a and a' whose other components
 are in domain(f')
  if r ∘ f'' is not in S
  then { compatible := false; break; } endif
end compatible
```

The binary procedure is very fast because its time complexity is constant. The general procedure, if implemented as stated here, would have to examine each N-tuple of R. For a software implementation it would be desirable to design the data structures for R, S, and f so that only the appropriate N-tuples of R are tested.

12.4 PROBABILISTIC RELAXATION

The major flaw in relational matching, especially for low-level image analysis, is the irrevocable commitment they make by deleting hypotheses about the assignments of labels to units. Once a label is deleted from some units, there is no way to reinstantiate it. Instead of deleting hypotheses, we will design a relaxation process to enhance some initial estimates of the probabilities that the various units should have specific labels. Then these enhanced probabilities can be used to choose some small set of labels at each unit.

Let U be the set of M units (variables) and L be the set of labels. Let $P_i(l_j)$ be the probability that the unit i has the label l_j, and let $\{P_i(l_1), \ldots, P_i(l_r)\}$ be a vector of probabilities at each unit $i = 1, \ldots, M$, such that

1. $P_i(l_j) \geq 0$ for $j = 1, \ldots, r$;
2. $\sum_{j=1}^{r} P_i(l_j) = 1$.

These initial probabilities come from measurements made on the units and models for the class conditional densities of the different labels conditioned on those measurements.

The probabilistic relaxation process enhances the probability of a unit-label pair through consideration of the distribution of probabilities on labels at neighboring units, and the *compatibilities* between labels at those units and the given unit-label pair. These compatibilities specify the degree of compatibility between two unit-label pairs.

Let $r_{ij}(k, l)$ denote the compatibility of label k at unit i with label l at unit j. We restrict $r_{ij}(k, l)$ to the range $[-1, 1]$ and interpret its values as follows:

1. $r_{ij}(k, l) = -1$. This means that the presence of label k at unit i is extremely incompatible with the presence of label l at unit j.
2. $r_{ij}(k, l) = 0$. The presence of label k at unit i will have no effect on our confidence that unit j should have label l.
3. $r_{ij}(k, l) = 1$. Label k at unit i strongly supports the presence of label l at unit j.

Intermediate values have intermediate interpretations. It should be noted that these compatibilities need not be symmetric, namely $r_{ij}(k, l) \neq r_{ji}(l, k)$. It still remains to devise a method for combining the compatibilities. The formula proposed by Rosenfeld et al. [12] is as follows.

Given the initial set of probabilities $\{P_i^{(0)}(l_j) \mid i = 1, \ldots, M; j = 1, \ldots, r\}$, the set of probabilities is updated to get a better set by combining the compatibilities and probabilities:

$$P_i^{(K+1)}(l) = \frac{P_i^{(K)}(l)\{1 + q_i^{(K)}(l)\}}{\sum_{l'} P_i^{(K)}(l')\{1 + q_i^{(K)}(l')\}}, \tag{12.10}$$

where

$$q_i^{(K)}(l) = \sum_j \sum_{l''} P_j^{(K)}(l'') \cdot r_{ij}(l, l'') \tag{12.11}$$

and j is a neighbor of i.

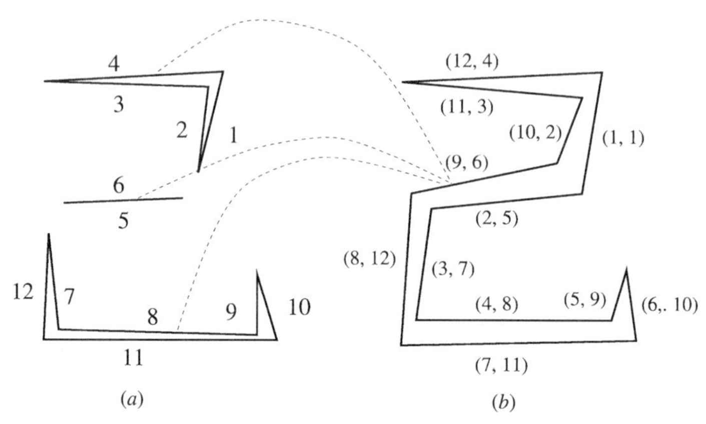

FIGURE 12.6 Application of probabilistic relaxation to handprinted Kanji recognition: (a) input; (b) prototype and correspondence of line segments between the prototype and the input.

The overall support for label l is measured by $q_i^{(K)}(l)$ at node i. It is based on the distribution of probabilities at other unit-label pairs and on the compatibilities. To compute q, we look at each node j and at each label l', multiplying $p_j^{(K)}(l')$ and $r_{ij}(l, l')$. Notice that if $p_{j'}^{(K)}(l') = 1$ and $r_{ij'}(l, l') = 1$ at some node j', then the contribution to q from node j' is 1, the highest possible. Thus we see that we can interpret values of q in the same way as we interpreted the r_{ij}'s: $q_i^{(K)}(l) = 0$ indicates that the neighborhood is indifferent to label l, and $q_i^{(K)}(l) < 0$ indicates that the neighborhood is opposed to label l.

These $q_i^{(K)}$'s are then used to get a weighted measure of the neighborhood support for label l relative to all other labels l'. The normalization also ensures that $\sum_l p_i^{(K)}(l) = 1$ after any iteration.

Here is an example of application of probabilistic relaxation to Kanji recognition conducted by Yamamoto [3]. Figure 12.6 (a) gives an object (input) that has been approximated to a polygon with an index assigned to each line segment of the polygon. Figure 12.6 (b) gives the prototype, and the mapping f is shown by pairs of line segment indexes for corresponding points on the prototype and the object. Table 12.1 presents part of the interactive process in the revision of values $p_i^{(K)}(l)$. In terms of length and direction, line segments 4, 6, and 8 in the object correspond to line segment 9 in the prototype, so $P_4^{(0)}(9)$, $P_6^{(0)}(9)$, and $P_8^{(0)}(9)$ are set to ⅓. There are many ambigu-

TABLE 12.1 Example of $P_i(k')$

i	k	$P_i^{(0)}(k')$	$P_i^{(1)}(k')$	$P_i^{(2)}(k')$	$P_i^{(3)}(k')$	$P_i^{(4)}(k')$	$P_i^{(5)}(k')$
4	9	0.33	0.28	0.18	0.07	0.02	0.00
6	9	0.33	0.49	0.71	0.89	0.97	1.00
8	9	0.33	0.22	0.09	0.02	0.00	0.00

ities at the start, because the initial probabilities must be derived from the local information. The probabilities are updated with compatibilities and the combining rule. After the fifth iteration, $P_6^{(0)}(9)$ becomes 1, whereas $P_4^{(0)}(9)$ and $P_8^{(0)}(9)$ are 0.

12.5 BIBLIOGRAPHICAL REMARKS

For this chapter there are a number of good books. *Computer Vision* [2] by Ballard and Broun is readable and useful text for understanding the basic idea/scheme of the flexible matching based on graphic representation. *Structural Pattern Recognition* [3] by Pavlidis introduces the fundamental graph theory and the basic algorithms for graphs. The graph matching problem is discussed using scene analysis as an example. Relation methods are also treated by concrete examples. In the last chapter, graph language and graph grammar are introduced but not developed in the book. On the relaxation method, Rosenfeld and Kak's book *Digital Picture Processing* [11] is recommended, and in it other relaxation methods such as fuzzy relaxation and discrete relaxation are explained. Readers will find many related topics from this book there. An extensive survey was done by Kittler and Illingworth [13], in which 127 papers are introduced being classified appropriately.

Ullmann treats subgraph isomorphoism both theoretically and experimentally [4]. Based on his result, Haralick derived the characterization theorem for binary relation homomorphisms [5]. By this theorem he showed that all the homomorphisms could be found by a depth-first search. Mackworth gave an effective backtracking way to solve constraint satisfaction problems [6], and he noted three problematic consistencies in backtracking, which are node consistency, arc consistency, and path consistency. To eliminate the problem, he devised a looking-ahead operator. This operator was further improved by Haralick and Shapiro [7]. They formulated a general consistent labeling problem based on a unit constraint relation T containing N-tuple of units that constrain one another, and a compatibility relation R containing a N-tuple of unit-label pairs specifying which N-tuple of units are compatible with which N-tuple of labels. They discussed the various methods that researchers had used to speed up the tree search required to find consistent labelings that use look-ahead operators. He also gave the so-called Φ_{kp} two-parameter class of looking-ahead operators. This is a generalization of the looking-ahead operators that had been proposed by other researchers and Haralick himself, Davis, and Rosenfeld [8]. For example, Mackworth's operator is just Φ_{23}. Furthermore, they elaborated their method in their companion paper [9]. After deriving the mathematical properties, they introduced another related look-ahead operator Φ_{kp}, which is a generalization of the Waltz filtering operator [10]. They showed that the fixed-point power of Φ_{kp} is the same as that of Φ_{kp}, that operators with greater look-ahead are more powerful. Nevertheless, the consistent labeling problem is still N P-complete. These look-ahead methods cannot improve the worst-case problems. However, in practice, the look-ahead operators work quite effectively. In this same work, Haralick, Davis, and Rosenfeld also explored characterizing the difficulty of a given consistent labeling problem. They defined for the consistent labeling problem a minimal compatibility relation by which the look-ahead methods are speeded up in finding all consistent labelings.

Among the relaxation methods applied to OCR, Yamamoto's and Rosenfeld's are the better known ones as mentioned in the chapter. A comparison study was conducted by the ETL OCR research group [14]. The results are shown below:

Matching method	Correct recognition rates (%) for 34th data set
Pixel gray level	44.2
Orientation	73.6
Cell feature	83.9
DP matching	84.0
Background	88.0
Relaxation	93.0

The test data set was the 34th data set which had very sloppily written samples among the ETL 8 (handwritten Kanji database). As can be seen above, the relaxation method applied to handwritten Kanji yields best results at least among the other mentioned methods. The big problem is that the method is time-consuming due to the iterations. Lam and Suen [15] applied the relaxation method to reading real zip codes in a hierachical recognition scheme. That is, well-shaped characters were recognized by structural classification, while sloppily shaped characters were fed into the relaxation matching system in order to save computing time. It was shown that only one iteration is enough. Therefore the necessary computing time of relaxation matching took about twice the time of the thinning process. The results were quite good; overall recognition, rejection, and substitution rates were 89.55%, 7.01%, and 3.45% respectively. Sekita, Toraichi, R. Mori, Yamamoto, and Yamada conducted a Kanji and Hirakana recognition experiment [16] based on relaxation matching in which they used corner features. It had proved difficult to apply relaxation matching to Hirakana, which is more cursive than Kanji; earlier Yamamoto and Rosenfeld had succeeded in applying the relaxation matching method in Kanji recognition. Sekita and his colleagues showed that in using corner features, relation matching can be applied to Hiragana also.

Later Yamamoto, Yamada, Saito, and Oka [16] conducted systematic experiments on Kanji and Hirakana handwritten characters. First an input character is classified using the cellular feature (Chapter 10) and then it is identified based on a relaxation matching scheme, in which the extreme point list method is used as a polygonal approximation of a character contour. The recognition rate is the best so far among the several Kanji and Hirakana recognition systems reported so far. They achieved a 93% correct recognition rate for the sloppy 34th data set, as shown in the table. For the good 2nd data set, the correct recognition rate is 99.0%. The tenth preclassification rates are 98.4% and 99.8%, respectively. However, Xie and Suk [17] pointed out a weak point of Yamamoto and his colleagues' approach. They named it polygonal relaxation matching. That is, first they demonstrated the weak point in some similar classes as 負 and 員. From these observations they claimed that in the polygonal relaxation matching the formation rules and distinct characteristics of the Chinese characters are not fully utilized. Therefore they proposed a new approach to select a small number of critical features and use them in distinguishing characters within a group. Xie and Suk conducted very small scale experiments and showed some cases that obtain better results than the polygonal relaxation matching. However, we cannot conclude that their method is the best, because of the very small scale of their experiment and the data set used is not the same. Further experiment is expected. On the other hand, Hayes [18] demonstrated a kind of hierarchical relation matching scheme that works well in cursive character recognition. He showed that the word "book" with initially 3960 possible letter paths can be reduced to 2 corresponding to the words "book" and "hook" with merits of 0.87 and 0.13, respectively.

BIBLIOGRAPHY

[1] K. Yamamoto, "Handwritten Kanji character recognition using relaxation method," *IE Trans. IECE Japan,* vol. J65-D, no. 9, pp. 1167–1174, 1982.

[2] D. H. Ballard and C. M. Broun, *Computer Vision,* Englewood Cliffs, NJ: Prentice-Hall, 1982.

[3] T. Pavlidis, *Structural Pattern Recognition,* New York: Springer-Verlag, 1977.

[4] J. R. Ullmann, "An algorithm for subgraph isomorphism," *ACM,* vol. 23, no. 1, pp. 31–42, January 1976.

[5] R. M. Haralick, "The characterization of binary relation homomorphisms," *Int. J. General System,* vol. 4, pp. 113–121, 1978.

[6] A. Mackworth, "Consistency in networks of relations," *Art. Intell.,* vol. 8, pp. 99–118, 1977.

[7] R. M. Haralick and L. G. Shapiro, "The consistent labeling problem: Part I," *IEEE Trans. Pattern Anal. Machine Intell.,* vol. PAMI-1, no. 2, pp. 173–184, April 1979.

[8] R. M. Haralick, L. S. Davis, and A. Rosenfeld, "Reduction operations for constraint satisfaction," *Info. Sci.,* vol. 14, pp. 199–219, 1978.

[9] R. M. Haralick and L. G. Shapiro, "The consistent labeling problem: Part II," *IEEE Trans. Pattern Anal. Machine Intell.,* vol. PAMI-2, no. 2, pp. 193–203, May 1980.

[10] D. L. Waltz, "Generating semantic descriptions from drawings of scenes with shadows," MIT Technical Report A1271, November 1972.

[11] A. Rosenfeld and C. Kak, *Digital Picture Processing,* vol. 2, New York: Academic Press, Inc, 1982.

[12] A. Rosenfeld, R. Hummel, and S. Zucker, "Scene labeling by relaxation operations," *IEEE Trans. Syst., Man, Cybern.,* vol. 6, pp. 420–433, 1976.

[13] J. Kittler and J. Illingworth, "Relaxation labeling algorithm—a review," *Image and Vision Computing,* Buttersworth and Co., Ltd., no. 4, pp. 206–216, 1985.

[14] K. Yamamoto, H. Yamada, and T. Saito, "Current state of recognition method for Japanese characters and database for research of handprinted character recognition," *From*

Pixels to Features III, S. Impedovo and J. C. Simon eds., pp. 105–116, Amsterdam Elsevier Science Publishers B. V., 1992. In International Workshop of Frontiers in Handwriting Recognition, Chateau de Bonas, France, pp. 81–92, September 23–27, 1991.

[15] L. Lam and C. Y. Suen, "Structural classification and relaxation matching of totally unconstrained handwritten zip-code nembers," *Pattern Recogn.,* vol. 21, no. 1, pp. 19–31, 1988.

[16] K. Yamamoto, H. Yamada, T. Saito, and R. Oka, "Recognition of handprinted Chinese characters and Japanese cursive syllabary," *Proc. of the 7th Inter. Conf. on Pattern Recognition,* Montreal, pp. 385–388, July 30–August 2, 1984.

[17] S. L. Xie and M. Suk, "On machine recognition of hand-printed Chinese characters by feature relaxation," *Pattern Recogn.,* vol. 21, no. 1, pp. 1–7, 1988.

[18] K. C. Hayes Jr., "Reading handwritten words using hierarchical relaxation," PhD dissertation, University of Maryland, 1979.

CHAPTER THIRTEEN

Elastic Matching

13.1 CONTINUITY AND NONLINEARITY

When we use the term *elastic matching,* it implies some *continuous* and *nonlinear* treatment of shape information of an object. Recognition methods have been classified into two approaches: pattern matching and structural analysis. Elastic matching includes both of these approaches. It offers the continuity of pattern matching and the nonlinearity of the structural method. In this sense elastic matching aims for a unification of the two approaches. Note here that continuity in pattern matching is related to that of intensity change. However, the continuity we primarily want to have is that of shape deformation. At any rate continuity contributes to the stability of the system.

Within the framework of elastic matching, there are two further approaches. One is elastic matching without correspondence. This regards the matching as a direct optimization of an objective function by spatially warping one image to match another image. In this case a potential field is created from the other image, and the former image is warped along the force of the field, which is where iteration and coarse-to-fine techniques are often used. The other is elastic matching with correspondence. This regards the matching as a combinatorial problem seeking the optimum correspondence between feature elements of two images.

Just as the entire field of pattern recognition is divided into pattern matching and structural analysis, so also are our methods for elastic matching. Elastic matching without correspondence is primarily a form of pattern matching, while elastic matching with correspondence is primarily a structural method. However, there is some overlap between the pattern recognition and structural approaches in our formulation of elastic matching. Elastic matching without correspondence takes the two-dimensional structure (spatial nonlinearity) into account, and the elastic matching with correspondence takes the continuity of each feature element into account.

To realize elastic matching, especially elastic matching with correspondence, dynamic programming and relaxation methods have been used. The relaxation method has as its strong point a full two-dimensional evaluation of the object, while dynamic programming gives a globally optimal solution. In this chapter elastic matching using dynamic programming is explained. Basic concepts of dynamic programming are explained first, and then several methods are introduced.

13.2 CORRESPONDENCE PROBLEM

We now consider the problem of finding the optimum correspondence between two sets of features: one comes from our reference model of shape, and the other comes from an input image. Each shape is expressed by a polygonal approximation of the contour as in Fig. 13.1. A *contour line segment* is directed so that it proceeds at the inside of the polygon on the left-hand side. In this section we assume that the pattern consists of only one contour, namely the original pattern is simply connected.

The *reference model* is a set of contour line segments r_m expressed by an arrow from the *starting point* (*sp*) to the *end point* (*ep*):

$$R = \{r_m = (\lambda_m^{(r)}, \tau_m^{(r)}) \mid m = 1, 2, \ldots, M\}. \tag{13.1}$$

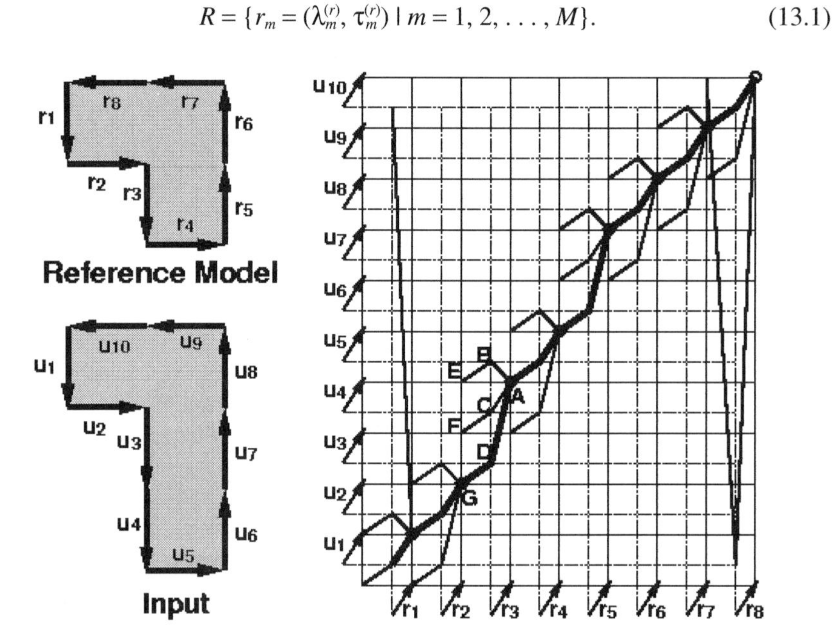

FIGURE 13.1 Correspondence of the end segments (front solid grid) and the starting segments (back dotted grid). *BA*, *CA*, and *DA* express 1:0, 1:1, and 1:2 correspondence, and *EB*, *FC*, *GD* express C_{m-1} to which C_m must be connected by Eq. (13.5). This kind of calculation is performed on every grid point of a graph at the stage of forward calculation. The bold curve shows the optimal correspondence decided by back-tracking the segments.

Each line segment r_m has two attributes, *length* $\lambda_m^{(r)}$ and *direction* $\tau_m^{(r)}$ $(-\pi < \tau_m^{(r)} \le \pi)$. Axes x and y are defined rightward and downward, respectively, and the direction is measured clockwise from rightward (x-axis). We introduce operators $\oplus$ and $\ominus$ for addition and subtraction where the cyclicity is taken into account. These are used for two cases, contour and direction. In the case of contour, usually $m + 1 = m \oplus 1$. However, $1 \ominus 1 = M$ and $M \oplus 1 = 1$ because r_M and r_1 are connected. In the case of direction, $\tau_m^{(r)} \oplus 2\pi = \tau_m^{(r)} \ominus 2\pi = \tau_m^{(r)}$, and the results of the operations $\oplus$ and $\ominus$ are always between $-\pi$ and π, that is, $\tau_m \in [-\pi, \pi]$. In addition to these two attributes, sometimes information of position of the starting point $(x_m^{(r)(sp)}, y_m^{(r)(sp)})$ and that of the end point $(x_m^{(r)(ep)}, y_m^{(r)(ep)})$ are used. When two segments are concatenated, $x_m^{(r)(sp)} = x_{m-1}^{(r)(ep)}$ and $y_m^{(r)(sp)} = y_{m-1}^{(r)(ep)}$.

The *input* is also an ordered sequence that has the same attributes as the reference model:

$$U = \{u_i = (\lambda_i^{(u)}, \tau_i^{(u)}) \mid i = 1, 2, \ldots, I\}. \tag{13.2}$$

Sometimes positional information $(x_i^{(u)(sp)}, y_i^{(u)(sp)})$ and $(x_i^{(u)(ep)}, y_i^{(u)(ep)})$ are used as an attribute of u_i.

Now we think about a *correspondence* between a model segment r_m and input segments which is represented by C_m as

$$C_m = \{i'_m, i'_m \oplus 1, \ldots, i_m\}. \tag{13.3}$$

For the moment we assume that one model segment r_m can correspond to zero (ϕ), one (u_i), or two ($\{u_{i\ominus 1}, u_i\}$) input segments. This constraint is specified by a set of *allowable* segments $A(i_m)$:

$$i'_m \in A(i_m) = \{i_m \oplus 1, i_m, i_m \ominus 1\}. \tag{13.4}$$

This means that the following three kinds of correspondences are allowed: (1) $i'_m = i_m \oplus 1$, one-to-nothing; (2) $i'_m = i_m$, one-to-one; (3) $i'_m = i_m \ominus 1$, one-to-two. The segments i' and i are called the *starting segment* (ss) and the *end segment* (es) of the correspondence. Therefore, more specifically, the correspondence $i'_m = i_m \oplus 1$ means that the starting segment i'_m is preceded by one segment to its end segment i'_m. This is a contradiction, and it is interpreted as one-to-nothing correspondence. From now on, m or i is sometimes used for the same meaning as r_m or u_i, respectively.

When m corresponds to $C_m = \{i'_m, i'_m \oplus 1, \ldots, i_m\}$ and $m - 1$ corresponds to $C_{m-1} = \{i'_{m-1}, i'_{m-1} \oplus 1, \ldots, i_{m-1}\}$, there must exist a relation between i'_m and i_{m-1}. This is called the precedent segment(s) $P(i'_m)$. For the moment

$$P(i') \equiv i' \ominus 1, \tag{13.5}$$

namely $i_{m-1} = i'_m \ominus 1$. This means that the model segments are connected to each other, so the corresponding input segments must also be connected.

The *local distance* $d(m; C_m)$ of the correspondence m versus C_m is a weighted sum of the difference of line lengths and their directions, as will be defined later. The *global distance* between two images is a sum of these values. We want to have the minimum of this value over all possible combinations of sequences of correspondences:

$$D(R, U) = \min_{\substack{\forall(C_1, \ldots, C_M) \\ C_m = (i'_m, \ldots, i_m) \\ i'_m \in A(i_m),\, i_{m-1} = i'_m \ominus 1}} \sum_{m=1}^{M} d(m; C_m). \tag{13.6}$$

At this moment the first correspondence and the last correspondence are neither restricted to be $i'_1 = 1$ nor $i_M = M$.

When we calculate this by brute force, we have to calculate the distance $(3 \times I)^M$ times. This number is huge. To reduce the calculation while preserving the optimality of the solution, we introduce dynamic programming.

13.3 INTRODUCTION OF DYNAMIC PROGRAMMING

Let us consider the following function:

$$f(m, i) = \min_{\substack{\forall(C_1, \ldots, C_m) \\ C_{m'} = (i'_{m'}, \ldots, i_{m'}) \\ i'_{m'} \in A(i_{m'}),\, i_{m'-1} = i'_{m'} \ominus 1 \\ i_m = i}} \sum_{m'=1}^{m} d(m'; C_{m'}). \tag{13.7}$$

The difference from the global distance $D(R, U)$ of equation (13.6) is that the summation is taken only up to m not M, and the correspondence of the last segment m is restricted to be i. In other words, the function f holds for an optimum distance up to m with respect to every i.

When we consider the property of optimality up to m, we can rewrite it by a recurrent form:

$$f(m, i) = \min_{i' \in A(i)} \{f(m-1, i' \ominus 1) + d(m; i', \ldots, i)\}. \tag{13.8}$$

This reformulation is based on the *principle of optimality* of dynamic programming [5]; that is, "an optimal policy has the property that whatever the initial state and initial decision are, the remaining decisions must constitute an optimal policy with regard to the state resulting from the first decision." This equation is called a *recurrence relation.* The reason that this reformalization is possible comes from the Markov property that the original object (constraint) has; that is, the state at $m + 1$ is decided only by the mth state and the history up to mth stage does not affect any decision beyond that point. Thus the problem is solved by the following recurrent calculations.

Initial values at $m = 0$ are set,

$$f(0, i) = 0 \qquad \text{for all } i = 0, 1, \ldots, I. \tag{13.9}$$

Then the recurrence relation at $m = 1, 2, \ldots, M$ stages can be rewritten as

$$f(m, i) = \min \begin{cases} f(m-1, i) + d_0 & \text{(1 vs. 0)}, \\ f(m-1, i \ominus 1) + d_1(m; i) & \text{(1 vs. 1)}, \\ f(m-1, i \ominus 2) + d_2(m; i \ominus 1, i) & \text{(1 vs. 2)}. \end{cases} \tag{13.10}$$

where d_0 is a constant for the penalty of 1 to 0 correspondence, and d_1 and d_2 are defined as follows:

$$d_1(m; i) = |\lambda_i^{(u)} - \lambda_m^{(r)}| + \alpha \cdot |\tau_i^{(u)} \ominus \tau_m^{(r)}|,$$

$$d_2(m; i \ominus 1, i) = |\lambda_{i \ominus 1}^{(u)} + \lambda_i^{(u)} - \lambda_m^{(r)}| + \alpha \cdot \left| \frac{\tau_{i \ominus 1}^{(u)} + \tau_i^{(u)}}{2} \ominus \tau_m^{(r)} \right|. \tag{13.11}$$

Note that $\oplus$ and $\ominus$ are used in two different ways, one for directional cyclicity and the other for contour cyclicity.

At the same time in this forward recurrent calculation, information as to which case among the three in (13.10) is taken is preserved. That is,

$$c(m, i) = \underset{i'}{\operatorname{argmin}}\, f(m, i), \tag{13.12}$$

where argmin specifies the value of the argument i' for which the function $f(m, i)$ attains its minimum value.

Finally the optimum value of (13.6) is obtained by

$$D(R, U) = \min_i f(M, i). \tag{13.13}$$

The correspondence of each model segment is obtained by back-tracking $c(m, i)$:

$$i_M^{opt} = \underset{i}{\operatorname{argmin}}\, f(M, i), \tag{13.14}$$

and

$$i_{m-1}^{opt} = c(m, i_m^{opt}) \qquad \text{for } m = M, M-1, \ldots, 2. \tag{13.15}$$

On introducing this decomposition, the number of calculations is decreased to $3 \times I \times M$, while preserving the optimality of the solution of 13.6. Note that we need a full matrix of $M \times I$ for c, but we only need two columns of $m - 1$ and m, each of which has length I for f.

At this moment the reader may have questions. What do we do when the number of contours is greater than one, for example? Some such matters will be explained in the following sections. At any rate, in some cases it may be easy to solve the problem, and in some other cases it may be difficult. The important thing is that we must formulate a solution that addresses the critical problem and not just pick a simple mathematically convenient answer.

13.4 CORRESPONDENCE WITH END POINT CONSTRAINTS

When the initial and the last correspondences are given (i.e., $m = 1$ vs. $i' = 1$ and $m = M$ vs. $i = I$), the problem can be solved by adding the following constraint into the calculations:

At the initial stage

$$f(0, i) = \begin{cases} 0 & \text{for } i = 0, \\ \infty & \text{for } i = 1, \ldots, I. \end{cases} \tag{13.16}$$

At the final stage we obtain the final value by

$$D(R, U) = f(M, I). \tag{13.17}$$

In this way a subproblem is solved by embedding it into a larger framework of problem solving. This is one of the characteristics of dynamic programming.

13.5 SYMMETRIC CORRESPONDENCE

The one- to -zero, -one, and -two correspondences used in Section 13.3 are not symmetric with respect to the reference model and input. Symmetric correspondences are natural and sometimes advantageous. In this section we think about a symmetric correspondence using one-to-one, one-to-two, and two-to-one correspondences. In order to express each correspondence, we have to use the pair $(m; i)$ and another index k rather than m like $C_k \in \{(m; i), (m; i \ominus 1, i), (m - 1, m; i)\}$.

$$D(R, U) = \min_{\substack{\forall (C_1, \ldots, C_K) \\ C_k = (m'_k, \ldots, m_k; i'_k, \ldots i_k) \\ (m'_k, i'_k) \in A(m_k, i_k) \\ m_{k-1} = m'_k - 1,\, m'_1 = 1,\, m_K = M \\ i_{k-1} = i'_k \ominus 1}} \sum_{k=1}^{K} d(C_k). \tag{13.18}$$

The recurrence relation of (13.10) is rewritten by the following equation:

$$f(m, i) = \min \begin{cases} f(m-1, i \ominus 1) + d_1(m; i) & (1 \text{ vs. } 1), \\ f(m-1, i \ominus 2) + d_2(m; i \ominus 1, i) & (1 \text{ vs. } 2), \\ f(m-2, i \ominus 1) + d_3(m-1, m; i) & (2 \text{ vs. } 1), \end{cases} \tag{13.19}$$

where

$$d_3(m-1, m; i) = |\lambda_i^{(u)} - (\lambda_{m-1}^{(r)} + \lambda_m^{(r)})| + \alpha \cdot \left| \tau_i^{(u)} \ominus \frac{\tau_{m-1}^{(r)} \oplus \tau_m^{(r)}}{2} \right|. \tag{13.20}$$

Now (13.8) is rewritten as

$$f(m, i) = \min_{(m', i') \in A(m, i)} \{f(m' - 1, i' \ominus 1) + d(m', \ldots, m; i', \ldots, i)\}. \quad (13.21)$$

Note that although the correspondence is counted by k as C_k, the stages of the recurrent calculation are divided into M.

13.6 PRECEDENT SEGMENTS FOR BREAK IN INPUT

A question here is: Is it possible to order the segments? This is a good question especially for the dynamic programming because the ordering is effectively used to reduce the calculation in dynamic programming. We are using contour line segments, so the direction of each line segment can be set uniquely. In our case the contour is directed so that the contour proceeds with the background on the right side and the normal direction is from black to white. In using this system, we can order both model and input segments. However, when we consider the correspondence between model and input segments, there is no guarantee that these orders are consistent with each other. When there is a break in an input line as shown in Fig. 13.2, the order will not be preserved. Here we maintain the constraint that the model still has only one contour. This constraint will be dropped later.

One method to allow this disorder in the correspondence is to take all segments as precedent segments. That is, the recurrence relation is rewritten from (13.21) by

$$f(m, i) = \min_{(m', i') \in A(m, i)} \min_{\forall i''} \{f(m' - 1, i'') + e(i'', i') + d(m', \ldots, m; i', \ldots, i)\}, \quad (13.22)$$

where e is a penalty for a gap between i'' and i',

$$e(i'', i') = |x_{i'}^{(u)(sp)} - x_{i''}^{(u)(ep)}| + |y_{i'}^{(u)(sp)} - y_{i''}^{(u)(ep)}|. \quad (13.23)$$

This time the number of calculations increases to $3 \times I^2 \times M$. Another disadvantage of this form is that the constraint is too loose.

To decrease the calculations and to make the constraint suitable, we constrain the precedent segments by local ordering [26]. Suppose that $(m'_{k-1}, \ldots, m_{k-1})$ correspond to $(i'_{k-1}, \ldots, i_{k-1})$ and that $(m'_k, \ldots, m_k)$ correspond to $(i'_k, \ldots, i_k)$, where $m_{k-1} \oplus 1 = m'_k$. Then there must be relation between i_{k-1} and i'_k. For example, in Fig. 13.2, if the starting point of r_3 corresponds to the starting point of u_7 (B in the figure), then the assumption that r_2 can correspond to u_1 is unnatural because u_2 is more plausible than u_1 as a partner of r_2 under the assumption that r_3 corresponds to u_7.

These are *precedent* segments P. Precedent segments i'' for segment i' are selected at a local minimum distance with respect to the preceding segment and the succeeding segment:

$$P(i') = \{i'' \mid e(i'', i') \leq e(i'' \ominus 1, i') \text{ and } e(i'', i') \leq e(i'' \oplus 1, i')\}. \quad (13.24)$$

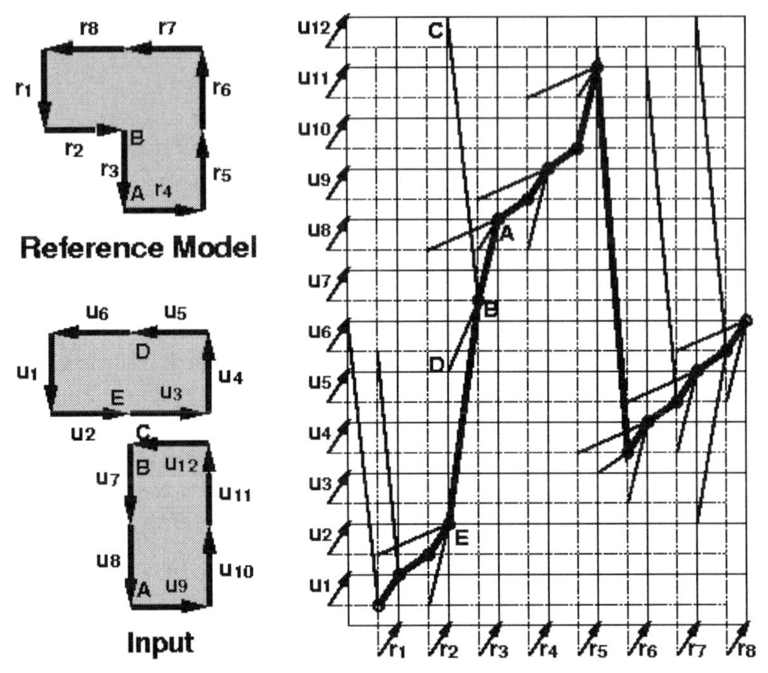

FIGURE 13.2 Break in input line showing order of correspondence not preserved. Note the symmetric correspondence. *BA* shows 1:2 correspondence. When the starting point of r_3 corresponds to that of u_7 like *B*, the end point of r_2 can correspond to that of u_{12}(C), u_5(D), and u_2(E). These are the *precedent segments* of u_7. They are selected by the locally minimum distance from their end points to the starting point of u_7.

This time the global distance is rewritten as

$$D(R, U) = \min_{\substack{\forall(C_1,\ldots,C_K) \\ C_k = (m'_k,\ldots,m_k;\, i'_k,\ldots,i_k) \\ (m'_k, i'_k) \in A(m_k, i_k) \\ m_{k-1} = m'_k - 1,\, m'_1 = 1,\, m_K = M \\ i_{k-1} \in P(i'_k)}} \sum_{k=1}^{K} d(C_k). \tag{13.25}$$

The recurrence relation is rewritten as

$$f(m, i) = \min_{(m', i') \in A(m, i)} \min_{i'' \in P(i')} \{f(m' - 1, i'') + e(i'', i') + d(m', \ldots, m; i', \ldots, i)\}, \tag{13.26}$$

or

$$f(m, i) = \min \begin{cases} \min_{i'' \in P(i)} \{f(m-1, i'') + e(i'', i) + d_1(m; i)\}, \\ \min_{i'' \in P(i \ominus 1)} \{f(m-1, i'') + e(i'', i \ominus 1) + d_2(m; i \ominus 1, i)\}, \\ \min_{i'' \in P(i)} \{f(m-2, i'') + e(i'', i) + d_3(m-1, m; i)\}. \end{cases} \tag{13.27}$$

The number of calculations this time is $3 \times \text{const} \times I \times M$. The constant is the average number of precedent segments, and this number is at most 6 or 7 in the case of Kanji characters, so the total number of calculations becomes the same order as the ordered case of (13.8), (13.10), (13.19), or (13.21). Figure 13.3 shows an example of application to Kanji characters [26]. We can see many disorderings because of the touch/untouch phenomena between lines.

13.7 CYCLIC MODEL

The methods up to the preceding section did not take into account whether the initial segment and the last segment of a model contour are connected in a closed loop. It is not so easy to evaluate this property in a dynamic programming calculation because the Markov property is not preserved between the initial segment and the last segment of the contour if cyclicity has to be taken into account.

One fundamental approach is a brute force method. If we calculate all the recurrence relations for all the cases of initial segment u_i instead of u_0 in equation (13.16), it can be accomplished [24]. However the calculation becomes quite large.

In the case of a symbolic sequence, by constraining the upper bounds and the lower bounds of each correspondence of the initial points, the calculation can be reduced to the same order as the noncyclic case [8]. But this cannot be applied to our

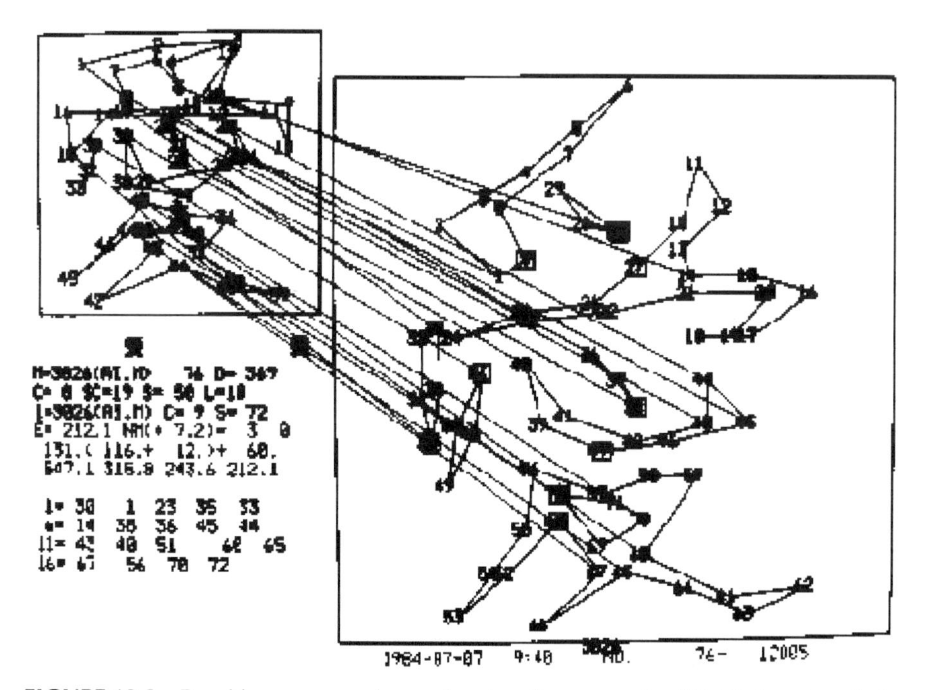

FIGURE 13.3 Resulting correspondence of contour line segments of Kanji characters. Left-side pattern is a reference model R, and the right-side pattern is an input U.

case, since we need to allow a break not only in the model sequence but also in the input sequence. Clearly it is not possible to calculate the upper bounds and the lower bounds.

In [28] a method is presented where the calculation of recurrence relation is performed more than once and then the overcalculated part is subtracted at the final evaluation stage. For simplicity, we use the form of (13.8). The recurrence relation is rewritten as follows. For $m = 0$,

$$f^{(1)}(0, i) = 0, \tag{13.28}$$

and for $m = 1, 2, \ldots, M$,

$$f^{(1)}(m, i) = \min_{i' \in A(i)} \{f^{(1)}(m-1, i' \ominus 1) + d(m; i', \ldots, i)\}. \tag{13.29}$$

Then, for $m = 1$ again,

$$f^{(2)}(1, i) = \min_{i' \in A(i)} \{f^{(1)}(M, i' \ominus 1) + d(1; i', \ldots, i)\}, \tag{13.30}$$

and again for $m = 1, 2, \ldots, M$,

$$f^{(2)}(m, i) = \min_{i' \in A(i)} \{f^{(2)}(m-1, i' \ominus 1) + d(m; i', \ldots, i)\}. \tag{13.31}$$

Finally

$$\begin{aligned} D(R, U) = \min_{i} \{ & f^{(2)}(M, i) - f^{(1)}(M, i^{(tb)}) \\ & - (|x_{i^{(tb)}}^{(u)(ep)} - x_i^{(u)(ep)}| + |y_{i^{(tb)}}^{(u)(ep)} - y_i^{(u)(ep)}|)\}, \end{aligned} \tag{13.32}$$

where $i^{(tb)}$ is the corresponding segment of M at the first cycle when M at the second cycle corresponds to i, and this $i^{(tb)}$ is obtained by back-tracking the second forward calculation. Figure 13.4 shows the effects of introducing cyclicity. We can see the effects of the gap evaluation term and the precycling. However, in this case we cannot obtain the ideal correspondence (r_1, r_5): u_1. Note the case Fig. 13.4 (*a*) is not (r_1, r_5):u_1, but r_1:u_1 and r_5:u_1 independently. In this method the number of forward calculations becomes twice as much as for the noncyclic case, although the retracing must be done at all locations of (i, j) in the final stage. Alternatively, this procedure is avoided if we transfer the information of $c(M, i)$ at the first cycle during the second forward calculation.

This technique can be easily extended to the symmetric case and the case of precedent segments. They can also be applied to the RS matching explained in succeeding sections. Actually this technique was first developed in comparing line segment with pixel matching [28].

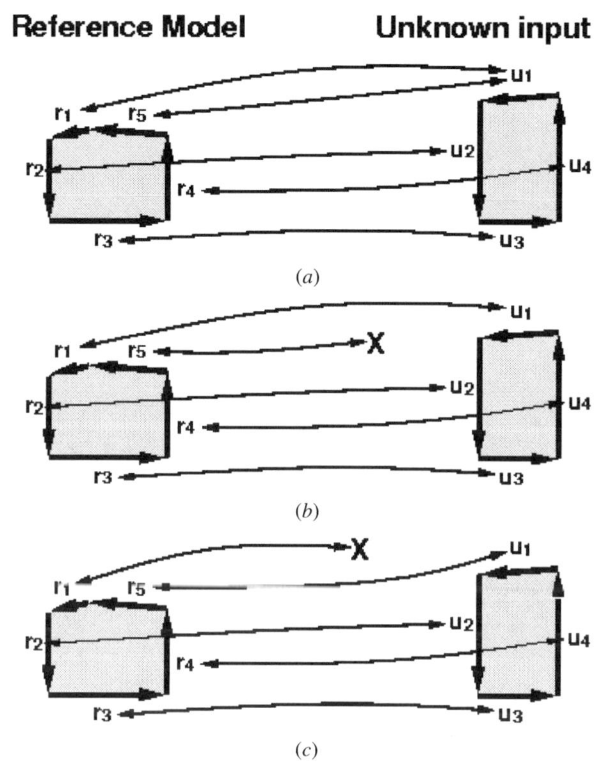

FIGURE 13.4 Effects of introducing cyclicity: (*a*) Without evaluation of cyclicity, r_1 and r_5 search local minimal evaluation independently. (*b*) With a gap between r_1 and r_5, which should be small by $(|x_{i(tb)}^{(u)(ep)} - x_i^{(u)(ep)}| + |y_{i(tb)}^{(u)(ep)} - y_i^{(u)(ep)}|)$, this criterion works at the final stage of the dynamic programming. (*c*) Precycling added to case (*b*) sets up consistency at the initial stage.

Other than this, Maes [13] proposed a method whose computation order is $M \times I \times \log I$ by partitioning a matrix "channel" established with initial width I by optimal alignment of r_m and u_i at each recursive calculation stage.

13.8 HIERARCHY FOR BREAKS IN THE MODEL

In this section the case where a reference model has more than one contour is treated. This technique can be applied not only to contours but also to subcontours. Obviously, since a subcontour is a part of a contour, the case of subcontours is explained without the loss of generality.

Let us assume that there are L model subcontours, and the last segment of each subcontour is denoted by M_l. We make a spanning tree whose node expresses the last segment. In a spanning tree all nodes are connected, and there is no loop. We decide the root node suitably. Usually, each node has a parent and child(ren). The node that has no offspring is called the terminal node. The node that has no parent is the root node. The evaluation within each subcontour is executed by the same procedure as described until now, and the results $f(M_l, i)$ can be obtained. Then the evaluation between subcontours is performed as follows:

$$F(l, i) = \begin{cases} f(M_l, i) & \text{if } M_l \text{ is terminal node,} \\ f(M_l, i) + \sum\limits_{l' \in Child.of(l)} \min\limits_{i'} \{F(l', i') + E(M_{l'}, M_l; i', i)\} & \text{if } M_l \text{ is nonterminal node,} \end{cases} \tag{13.33}$$

where

$$E(m', m; i', i) = |(x_i^{(u)(ep)} - x_{i'}^{(u)(ep)}) - (x_m^{(r)(ep)} - x_{m'}^{(r)(ep)})| + |(y_i^{(u)(ep)} - y_{i'}^{(u)(ep)}) - (y_m^{(r)(ep)} - y_{m'}^{(r)(ep)})|. \tag{13.34}$$

This calculation is performed from the bottom (terminals) to the top (root) of the tree. Finally the result is obtained as

$$D(R, U) = \min_i F(M_{k^{root}}, i). \tag{13.35}$$

Thus the total evaluation is obtained as a hierarchical combination of low-level calculation of each subcontour and high-level calculation between subcontours. Figure 13.5 shows an example of the spanning tree of a Kanji character. This is similar to the multi-level processing in speech recognition where each level corresponds to a phoneme, word, and so on.

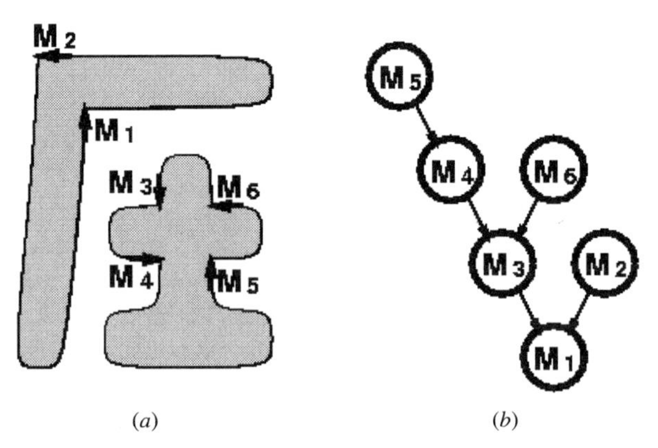

FIGURE 13.5 (*a*) Reference model divided into multiple contours or subcontours. (*b*) Spanning tree constructed whose node is the last segment of each subcontour. The minimum spanning tree may be reasonable where the weighting of each edge is defined as the distance between end points of subcontours.

13.9 LINE SEGMENT VERSUS PIXEL MATCHING

In this section we turn to matching input images expressed by an array of pixels. In this case input to be matched is expressed by

$$U = \{u_{ij} = (v^{(u)}(i, j), \mu^{(u)}(i, j)) \mid i = 1, \ldots, I; j = 1, \ldots, J\}, \tag{13.36}$$

where $\mu^{(u)}(i, j)$ is the magnitude of the inverted gray-level gradient, and $v^{(u)}(i, j)$ is its direction. The inversion is to be from the inside-out direction. The model is a polygon treated in the same manner as in equation (13.1). The length of each side can change within a certain range, but its direction will not change. The neighboring sides are connected to each other except between the initial side and the final side.

The global distance is expressed as

$$D(R, U) = \min_{\substack{\forall(C_1, \ldots, C_M) \\ (i_{m-1}, j_{m-1}) \in A(m, i_m, j_m)}} \sum_{m=1}^{M} d(i_{m-1}, j_{m-1}, i_m, j_m), \tag{13.37}$$

where C_m is a set of pixels on the line segment from (i_{m-1}, j_{m-1}) to (i_m, j_m) excluding (i_{m-1}, j_{m-1}):

$$C_m = \{(i, j) = \lambda'' \cdot (i_{m-1}, j_{m-1}) + (1 - \lambda'') \cdot (i_m, j_m) \mid 0 \le \lambda'' < 1\}. \tag{13.38}$$

When $|i_m - i_{m-1}| \ge |j_m - j_{m-1}|$, one pixel per x-axis is taken. When $|i_m - i_{m-1}| < |j_m - j_{m-1}|$, one pixel per y-axis is taken, and the pixel (i_{m-1}, j_{m-1}) is excluded. The set $A(m, i_m, j_m)$ is an *allowable* location of the end point of segment r_{m-1} when the end point of segment r_m is located at (i_m, j_m):

$$A(m, i; j) = \left\{(i', j') = (i, j) - \lambda' \cdot (\cos \tau_m, \sin \tau_m) \mid \frac{1}{\lambda} \le \lambda' \le \lambda\right\}. \tag{13.39}$$

Here A fulfills the role as if A and P in Sections 13.2 to 13.6 are put together. Note that the definition is somewhat inversed. While $(i_m, j_m) \in A(m, i_{m-1}, j_{m-1})$ may be natural, the definition itself comes from a way of thinking in dynamic programming (see Fig. 13.7). The local distance function is calculated as a weighted sum of angular difference of the pixels of C_m:

$$d(i', j', i, j) = \sum_{(i'', j'') \in C_m} \mu^{(u)}(i'', j'') \cdot \left| v^{(u)}(i'', j'') \ominus \left(\tau_m^{(r)} + \frac{\pi}{2}\right) \right|. \tag{13.40}$$

The criterion of (13.37) is explained using Fig. 13.6. First, we generate a shape by changing the lengths of each side. The generated shape is not necessarily closed. Second, we place this shape over all of the input image. Third, we sum up the difference in directions between the model segment and the input gradient and obtain a total summation with respect to the model shape and the input location. Finally, the minimum of the summations with respect to all the shapes and locations is the required result.

If we were to calculate by the brute force method, the number of combinations would become too large. The number of generated shapes at the first stage would be $|A|^M$, where $|A|$ is an average number of pixels in A. The number of locations at the second stage would be $I \times J$. So we would have to evaluate $|A|^M \times I \times J$ times.

To decrease the calculation, while preserving the optimality of the solution, we have to introduce dynamic programming again. The recurrence relation then can be written as

$$f(m, i, j) = \min_{(i', j') \in A(m, i, j)} \{f(m-1, i', j') + d(i', j', i, j)\}. \tag{13.41}$$

The final value is obtained by,

$$D(R, U) = \min_{(i, j)} f(M, i, j). \tag{13.42}$$

Cyclicity can be introduced in the same manner as in the discussion of segment correspondence in Section 13.7. We can also add a penalty for the gap between the initial point of the contour and the last point of the contour by

$$D(R, U) = \min_{(i_M, j_M)} \{f(M, i_M, j_M) + |(i_1' - i_M)| + |(j_1' - j_M)|\}, \tag{13.43}$$

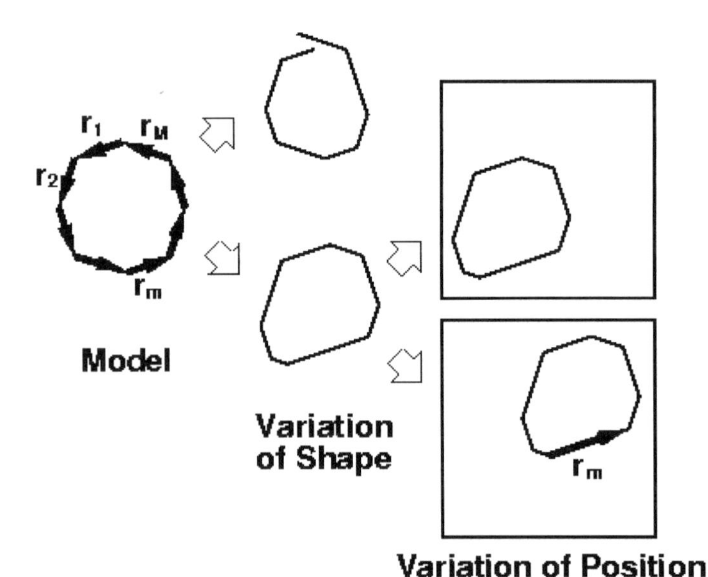

FIGURE 13.6 Length of model segment changes independently while preserving its direction. This yields many shapes. Each shape is placed over an input image, and we evaluate the distance at each location. Dynamic programming is introduced to decrease the calculation time while preserving the optimality of the solution.

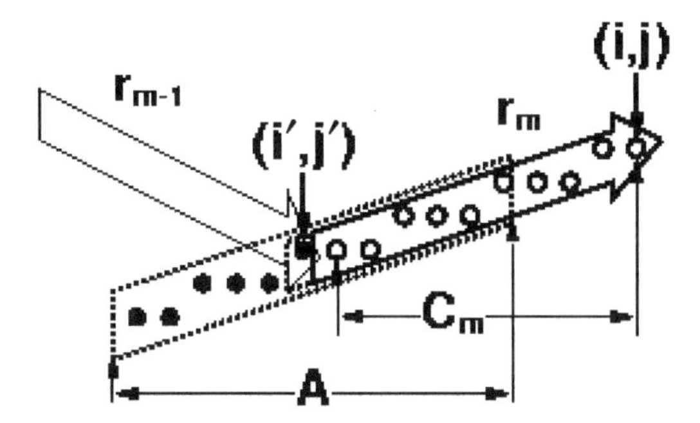

FIGURE 13.7 First the end point of a model segment m is located at (i, j) of the input image. Then the end point of segment $m - 1$ is restricted by $(i', j') \in A$, because the variation of the length is limited. For each (i', j') the local distance d is calculated as a sum of C_m, the pixels along the line segment (i', j') to (i, j).

where (i_1', j_1') is a retraced initial point from (i_M, j_M). Figure 13.8 shows an application of this method to object detection from complex background.

13.10 ORIGINAL RS MATCHING

The original version of this method was independently proposed by Kovalevsky and named the reference sequence method [11, 12] and by Sakoe who named it rubber string matching [21]. In both RS methods the specific conditions are that the input image is binary, directional features are not introduced, direction of model segments are restricted to every 45 degrees, and the cyclic property is not considered. Yamada improved several of the points above and applied the method to gray-level images [28, 29]. Otherwise, the intrinsic aspects of this method conform with the introduction of dynamic programming explained in the preceding section.

13.11 PARALLEL COMPUTATION OF RS MATCHING

In this section the parallel computation method of line segment versus pixel matching is explained [30, 31]. First we take a reference model that can be expressed not by line segments but by a sequence of points along a contour (see Fig. 13.9). That is,

$$R = \{r_m = (v_m^{(r)}, \delta_{Xm}, \delta_{Ym}) \mid m = 1, 2, \ldots, M\}, \tag{13.44}$$

$$(\delta_{Xm}, \delta_{Ym}) = (x_m - x_{m-1}, y_m - y_{m-1}), \tag{13.45}$$

$$(x_0, y_0) \equiv (x_M, y_M) = (0, 0),$$

where $v_m^{(r)}$ $(= -3\pi/4, -2\pi/4, \ldots, 3\pi/4, 4\pi/4)$ is a quantized normal direction at contour point (x_m, y_m). Each r_m is called an *evaluation point,* and especially $r_0 = r_M$ is

(*a*)

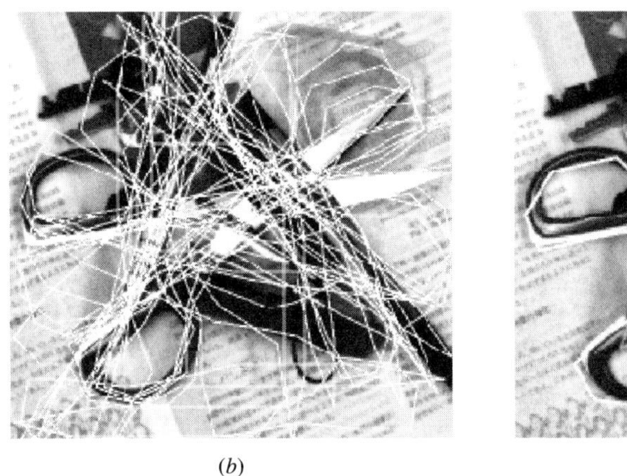

(*b*)

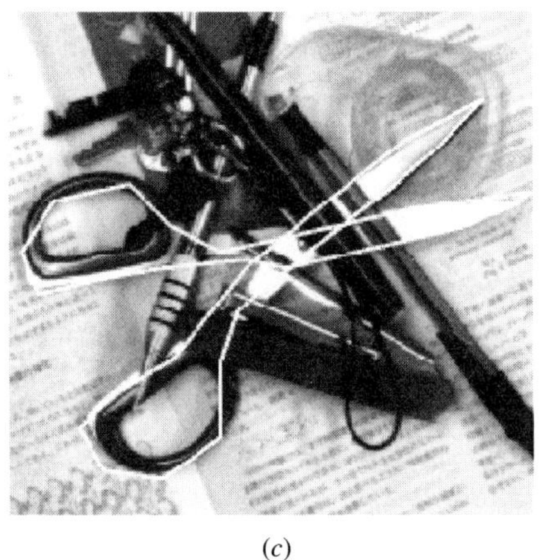

(*c*)

FIGURE 13.8 (*a*) Overlapping nonunique background and irregular specular reflection make recognition difficult in this image. (*b*) This matching model is not rotation invariant, so from a given model of each side of the scissors, whose number of sides is 16, 36 models (one for every ten degrees) are generated, and the matching is performed with each generated model. (*c*) From every pair of both sides, the best pair is chosen that satisfies the condition of crossing point. Although the object is rigid, a flexible model is effective to decrease the number of generated models (i.e., to increase the angle step).

called a *reference point.* (We can take a different point than r_M as the reference point, e.g., the center of the model.) Here $(\delta_{Xm}, \delta_{Ym})$ is the *displacement* from the preceding point r_{m-1} to the current point r_m.

Directional information of input is expressed using multiple *directional planes,* which is called the multi-angled field:

$$U = \left\{ \mu_v^{(u)}(i, j) \mid i = 1, \dots, I;\ j = 1, \dots, J;\ v = \frac{-3\pi}{4}, \frac{-2\pi}{4}, \dots, \frac{3\pi}{4}, \frac{4\pi}{4} \right\}, \tag{13.46}$$

where $\mu_v^{(u)}$ is a component of direction v calculated as a partial derivative of the inverted gray level along the direction v. The reason for the inversion is to be in the same direction as the model. These can be extracted by a 3×3 edge operator in each direction.

The global evaluation is expressed as follows.

$$D(R, U) = \max_{\substack{\forall(C_1, \ldots, C_M) \\ C_m = (i_m, j_m) \\ (i_{m-1}, j_{m-1}) \in A(m, i_m, j_m)}} \sum_{m=1}^{M} \mu_{v(r_m)}^{(u)}(i_m, j_m), \tag{13.47}$$

where the set $A(m, i_m, j_m)$ is an *allowable* location of the evaluation point r_{m-1} when the evaluation point r_m is located at (i_m, j_m):

$$A(m, i, j) = \{(i', j') \mid i' = i - \delta_{Xm} - \varepsilon_x, j' = j - \delta_{Ym} - \varepsilon_y\}. \tag{13.48}$$

Note here that D is not a "distance" but a "similarity" in this case, and it is "maximized" during the computations.

This is calculated by the following procedures by introducing an intermediate function (image) $f_m(i, j)$ called the *counterplane*. First, for $m = 0$,

$$f_0(i, j) = 0, \qquad i = 1, \ldots, I; j = 1, \ldots, J. \tag{13.49}$$

For $m = 1, 2, \ldots, M$,

$$f_m(i, j) = \max_{\substack{-1 \le \varepsilon_x \le 1 \\ -1 \le \varepsilon_y \le 1}} f_{m-1}(i - \varepsilon_x, j - \varepsilon_y) + \mu_{v(r_m)}^{(u)}(i + x_m, j + y_m). \tag{13.50}$$

Note that the origin of each (i, j) of f_m is different according to m. From the relation

$$f_m(i - x_m, j - y_m) = \max_{\substack{-1 \le \varepsilon_x \le 1 \\ -1 \le \varepsilon_y \le 1}} \{f_{m-1}((i - x_{m-1}, j - y_{m-1}) - (x_m - x_{m-1}, y_m - y_{m-1}) - (\varepsilon_x, \varepsilon_y)) + \mu_{v(r_m)}^{(u)}(i, j)\}, \tag{13.51}$$

the coordinates (i, j) of f_m are shifted by (x_m, y_m) from those of $\mu_v^{(u)}(i, j)$ (while $f_{m-1}(i, j)$ is displaced by (x_{m-1}, y_{m-1})).

So, when $f_m(i, j)$ is treated as an image F_m as a whole, the computation above is rewritten by

$$F_m = \max_{\substack{-1 \le \varepsilon_x \le 1 \\ -1 \le \varepsilon_y \le 1}} F_{m-1}[\delta_{Xm} + \varepsilon_x, \delta_{Ym} + \varepsilon_y] + \mu_{v(r_m)}^{(u)}, \tag{13.52}$$

where the notation $F[\delta_x, \delta_y]$ means a translation (shift) of an image by (δ_x, δ_y).

During this iteration, only translation, addition, and local maximum are processed independently for all the pixels, and these operations are executed in parallel with

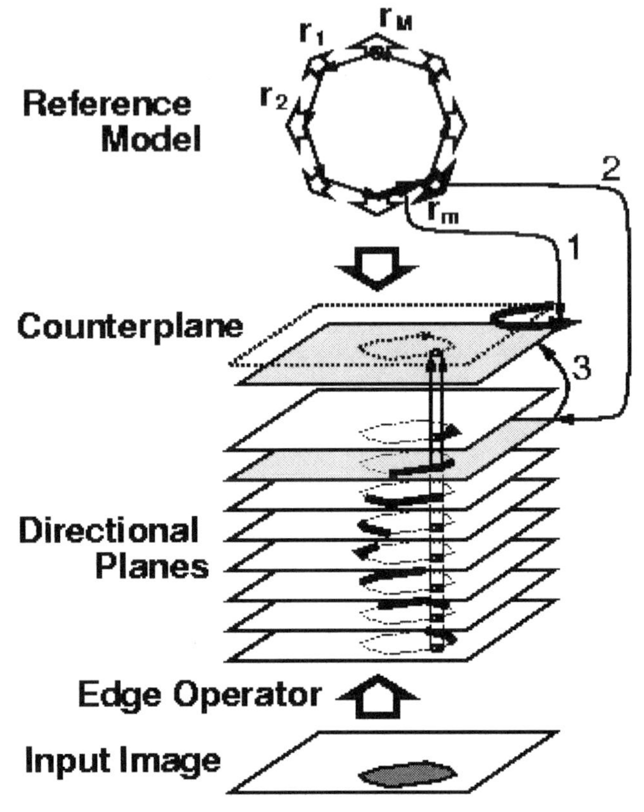

FIGURE 13.9 Directional magnitude of gradient (directional derivative) from input image is set into each directional plane. Counterplane is cleared initially. Operations of translation and renewal are iterated from $m = 1$ to $m = M$: (1) Translation of the counterplane by the displacement of r_m; (2) renewal of the counterplane by the addition of the maximum in neighbors of old counterplane and the directional plane specified by the direction of r_m.

respect to pixels. For dedicated image processors, these operations are very fundamental and executed at high speed.

At last we have the final result by

$$D(R, U) = \max_{(i,j)} f_M(i, j). \tag{13.53}$$

This method is called the MAP (multi-angled parallelism) matching method. Here we take a conversion

$$f(m, i, j) = f_m(i - x_m, j - y_m). \tag{13.54}$$

Then, from (13.51), this method is proved to be equivalent with the RS matching in Sections 13.9 and 13.10. Now the global evaluation is not a distance but a similarity, and the allowable region is not along line but in a $\pm\varepsilon_x$ by $\pm\varepsilon_y$ region. This is a parallel implementation of dynamic programming matching.

Moreover, when $(\varepsilon_x, \varepsilon_y) = (0, 0)$, that is, when only translation and addition are iterated, the iteration is expressed by

$$f_m(i, j) = f_{m-1}(i, j) + \mu^{(u)}_{\nu(r)_m}(i + x_m, j + y_m); \tag{13.55}$$

in other words, the model becomes a rigid body. This is a realization of template matching using directional information, namely an inner product of directions:

$$D(R, U) = \max_{(i, j)} \sum_{m=1}^{M} \mu^{(u)}_{\nu(r)_m}(i + x_m, j + y_m). \tag{13.56}$$

It is also a reformalized realization of the generalized Hough transformation proposed by Ballard [3]. Compared with the generalized Hough transformation, this method is superior in lack of voting error and ease of parallel implementation.

This method was applied to extract symbols in a topographic map. Conventionally, symbols in a map are extracted by a process that separates a region (including its segmentation) first and follows with a classification. But this kind of segment-then-recognize process cannot be applied when the background is complex and there is touching between symbols and ground as in Fig. 13.10. The other advantage of this method is the elasticity of the model. Figure 13.11 shows a result for a medical image where the object is elastic and the quality is poor.

13.12 LOCAL DISTANCE BY MULTIPLE SEGMENTS

In the optimization problem of global distance in RS matching (13.37) between a model shape expressed by a sequence of line segments (13.1) and the input image expressed by pixels (13.36), the generated sequence $C_1, C_2, \ldots, C_M$ can be considered as a transformed shape of the original model shape (13.1).

Sometimes we may want to use more complex features in an evaluation, for example, the curvature of the shape. Curvature is measured by three succeeding vertices like $(i_{m-1} - 2 \cdot i_m + i_{m+1})^2 + (j_{m-1} - 2 \cdot j_m + j_{m+1})^2$. Thus we have to use two succeeding segments in a recurrent calculation (see Fig. 13.12), and the recurrence relation becomes

$$f(m, i_m, j_m) = \min_{\substack{(i_{m-1}, j_{m-1}) \in A(i_m, j_m) \\ (i_{m-2}, j_{m-2}) \in A(i_{m-1}, j_{m-1})}} \{f(m-2, i_{m-2}, j_{m-2})$$

$$+ d(m; i_{m-2}, j_{m-2}, i_{m-1}, j_{m-1}, i_m, j_m)\}, \tag{13.57}$$

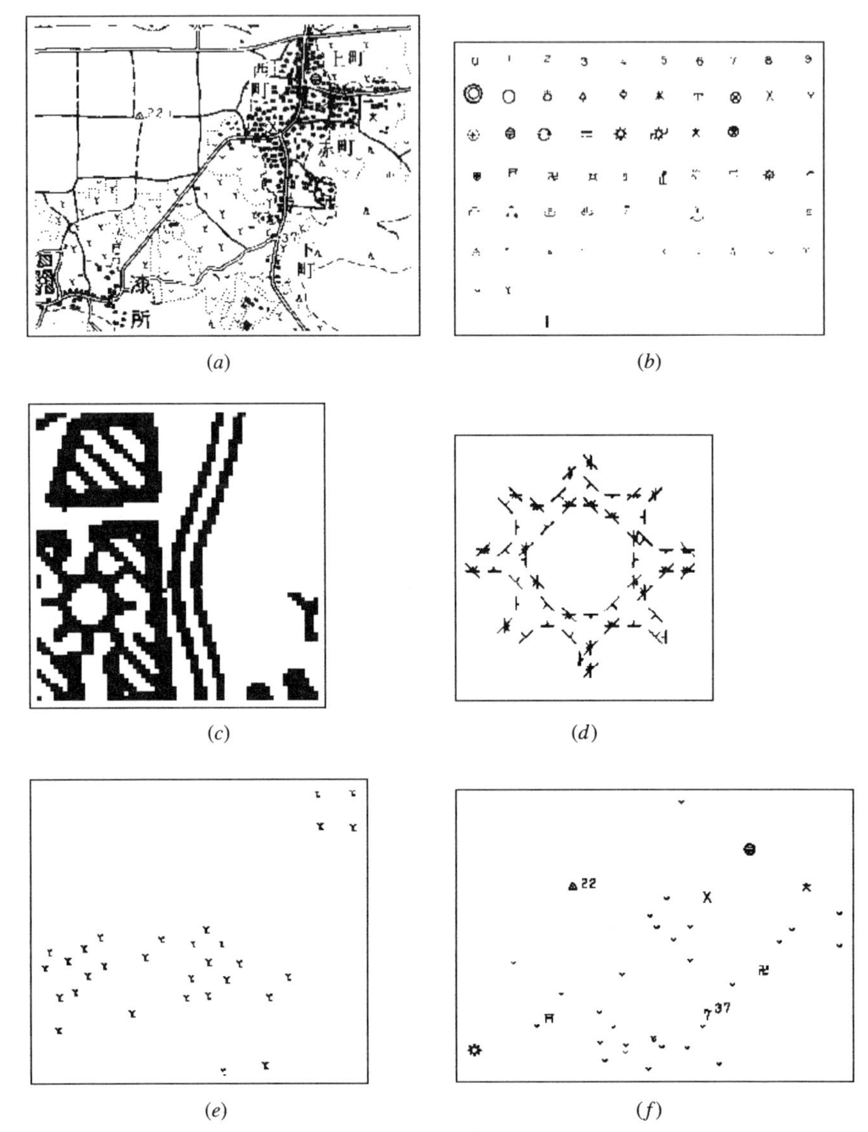

FIGURE 13.10 Symbol recognition from a topographic map (*a*). The reference models are made from the legend (*b*). The reference model of a factory (*d*) is expressed by a sequence of evaluation points along a contour. Each point has information of edge direction and displacement from the preceding point. The MAP matching method extracts successfully even when there are many touching elements with the background (*c*) which is enlarged from lower-left corner of the original image (*a*). Many instances of the same symbol are extracted simultaneously by parallel computation, as in the case of the symbol of a mulberry field (*e*). (*f*) shows results of other symbols and numerals. In this case the model is rigid.

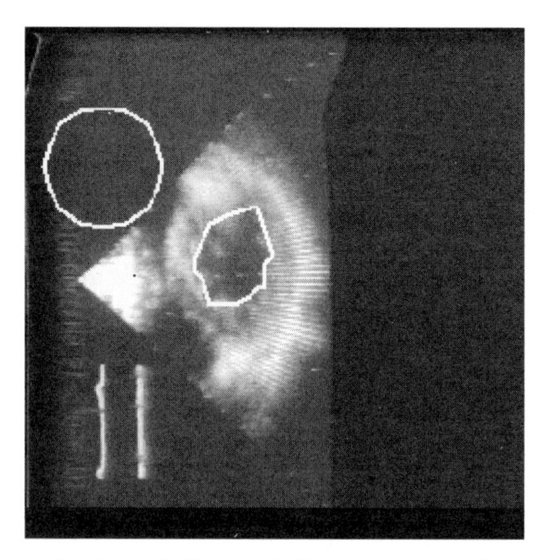

FIGURE 13.11 Example of an elastic model. The input is an echocardiogram showing an ultrasonic image of the left ventricle of the heart. The model at the upper left is a 16-sided regular polygon. At each iteration of matching, a local maximum of the counterplane is taken. Then it becomes a parallel implementation of dynamic programming matching with the elastic model.

or

$$f(m, i_{m-1}, j_{m-1}, i_m, j_m) = \min_{\substack{(i_{m-1}, j_{m-1}) \in A(i_m, j_m) \\ (i_{m-2}, j_{m-2}) \in A(i_{m-1}, j_{m-1})}} \{f(m-1, i_{m-2}, j_{m-2}, i_{m-1}, j_{m-1})$$

$$+ d(m; i_{m-2}, j_{m-2}, i_{m-1}, j_{m-1}, i_m, j_m)\} \quad (13.58)$$

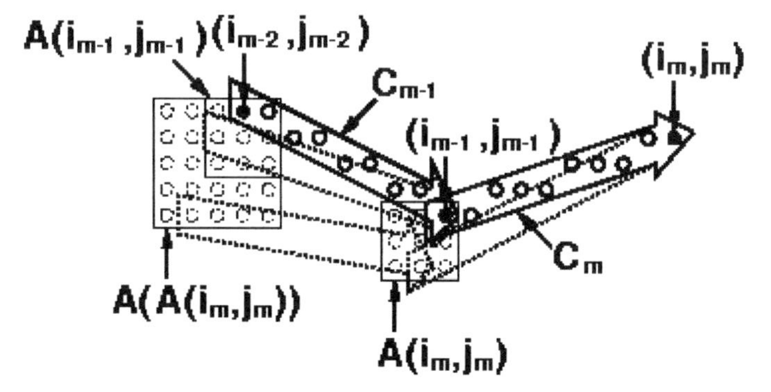

FIGURE 13.12 When succeeding multiple elements are needed for the evaluation of local distance d, the number of arguments of the function f increase. The number of calculations and storage increase by their combination. The allowable starting points are specified by a 3 by 3 region as in MAP matching, not by a linear region as in RS matching.

Such a recurrence relation was used for comparing segment and segment matching of the edge of polyhedral objects taken as a range image, where the angular difference between the two succeeding segments was taken as the invariant feature with respect to the object's rotation [27].

Such recurrent calculation was also used in an energy minimization framework of the Snake model [10]. The original work used a numerical calculation method to minimize the energy; however, Amini et al. introduced dynamic programming to solve the problem [1]. There the total energy is expressed as a summation of internal energy and external energy. The internal energy evaluates the smoothness of the shape, which is calculated by the first derivative and the second derivative of the contour shape. The external energy evaluates the information that the contour has a tendency to rest on the stronger edge magnitude. This corresponds to the distance d in Section 13.9 which includes curvature information. As stated, in order to evaluate the curvature, the order of calculation increases M times. In the case of the application of dynamic programming to the Snake model, this increase is compensated by limiting the range of x- and y-positions of each vertex.

However, this means that the method has no explicit shape model. In the case of shape matching in Section 13.9, the system has an explicit shape model. In this case, initial information is used to constrain the position, while in the case of shape matching intrinsically there is no constraint of the position, namely a globally optimal solution is searched.

13.13 BIBLIOGRAPHICAL REMARKS

In this chapter dynamic programming was introduced in terms of shape matching with an elastic model. It is not easy to make the target problem solvable by dynamic programming. However, once it can be done, the problem is guaranteed to be solved perfectly; that is, a global solution can be obtained. In the process of breaking down into the dynamic programming process, some two-dimensional properties are lost. Introduction of hierarchy and iteration sometimes can help eliminate these weak points.

Before ending this chapter, we need to look back at some other examples of dynamic programming introduced into pattern recognition problems. We know that dynamic programming works well for one-dimensional objects, as in speech recognition where features form a line along the time axis [20, 25]. Even in two-dimensional patterns like characters the same framework can be used when the features are arranged in line, such as on-line character recognition [6, 23] and matching of stroke sequences where the order can be assumed to be given. In the case of stereo matching, if we use an epipolar constraint, the total problem can primarily be broken down into one-dimensional problems along epipolar lines, although in the actual case consistency between epipolar lines has to be taken into account [9, 19].

In two-dimensional images dynamic programming has been used for feature extraction. Montanari [16] used it to extract a smooth curve from a noisy picture.

Compared with the methods explained in this chapter, there is no explicit model. Similar usage can be seen in several other works [4, 15, 18].

The methods of this chapter have explicit shape information, namely application to the problem of matching. We can classify applications to matching problems into three groups according to how the image is represented.

The first type uses representation by feature elements like line segments for both the model and the input as in Sections 13.3 to 13.8. The most important improvement for this application is to allow a break in a line—in other words, nonmonotonic and noncontinuous correspondence.

The second type uses feature elements for the model while input is expressed pixel by pixel as in RS matching of Sections 13.9 to 13.11. In this case, ordering in the input image is completely ignored. The order is decided as a result of correspondence; that is, the order is "guided" by the model (sequence) information.

Maitre et al. [14] introduced another way of allowing a break in a curve with the example of registering satellite images on a map. They introduced the idea of freezing to allow a temporal break of correspondence along a recurrent calculation. This idea assumes that the break in an input image can be detected in a bottom-up process. In this sense such a method can be positioned between segment/segment matching and segment/pixel matching.

The third type uses pixel-based expression for both model and input. Fischler proposed a DP-like method along this line, where the input image is pixel based, and the model description is a subtemplate graphical representation connected by springs [7]. The algorithm was developed by Moore [17], and Tanaka [22] proposed a more two-dimensional method along this line, but the smoothness (continuity) of the warping has not been satisfactory. Because of the wide range of application this framework is attractive. Further development is expected.

BIBLIOGRAPHY

[1] A. A. Amini, T. E. Weymouth, and R. Jain, "Using dynamic programming for solving variational problems in vision," *IEEE Trans. PAMI,* vol. 12, no. 9, pp. 855–867, 1990.

[2] H. H. Baker and T. O. Binford, "Depth from edge intensity based stereo," *Proc. 7th Int. Joint Conf. Artificial Intell.,* pp. 631–636, 1981.

[3] D. H. Ballard, "Generalizing the Hough transform to detect arbitrary shapes," *Pattern Recogn.,* vol. 13, no. 2, pp. 111–122, 1981.

[4] D. H. Ballard and C. M. Brown, *Computer Vision,* Englewood Cliff, NJ: Prentice-Hall, 1982.

[5] R. Bellman, *Dynamic Programming,* Princeton: Princeton University Press, 1957.

[6] D. J. Burr, "A technique for comparing curves," *Proc. IEEE Comp. Soc. Conf. Pattern Recogn. Image Processing,* pp. 271–277, 1979.

[7] M. A. Fischler and R. A. Elschlager, "The representation and matching of pictorial structures," *IEEE Trans. Comp.,* vol. C-22, no. 1, pp. 67–92, January 1973.

[8] J. Gregor and M. G. Thomason, "Dynamic programming alignment of sequences representing cyclic patterns," *IEEE Trans. PAMI,* vol. 15, no. 2, pp. 129–135, February 1993.

[9] Y. Isomichi and T. Takemasa, "Generation of height map from aerial photograph," *Proc. 7th Joint Conf. Imaging Tech.,* no. 3–1, pp. 31–34, 1976. (in Japanese)

[10] A. Kass, A. Witkin, and D. Terzopoulos, "Snakes: Active contour models," *Int. J. Comp. Vision,* vol. 1, no. 3, pp. 321–331, 1988.

[11] V. A. Kovalevsky, "Sequential optimization in pattern recognition," *Proc. IFIP Congress 68,* pp. 1603–1607, 1968.

[12] V. A. Kovalevsky, *Image Pattern Recognition,* New York: Springer-Verlag, 1980.

[13] M. Maes, "On a cyclic string-to-string correction problem," *Infor. Processing Lett.,* vol. 35, pp. 73–78, 1990.

[14] H. Maitre and Y. Wu, "Improving dynamic programming to solve image registration," *Pattern Recogn.,* vol. 20, no. 4, pp. 443–462, 1987.

[15] A. Martelli and U. Montanari, "Optimal smoothing in picture processing: an application to fingerprints," *Info. Processing 71,* pp. 173–178, Amsterdam: North-Holland, 1972.

[16] U. Montanari, "On the optimal detection of curves in noisy pictures," *Comm. ACM,* vol. 14, no. 5, pp. 335–345, 1971.

[17] R. K. Moore, "A dynamic programming algorithm for the distance between two finite areas," *IEEE Trans. Pattern Anal. Machine Intell.,* vol. PAMI-1, no. 1, pp. 86–88, 1979.

[18] H. Ney, "Dynamic programming as a technique for pattern recognition," *Proc. 6th Int. Conf. Pattern Recogn.,* pp. 1119–1125, 1982.

[19] Y. Ohta and T. Kanade, "Stereo by intra- and inter-scanline search using dynamic programming," *IEEE Trans. Pattern Anal. Machine Intell.,* vol. PAMI-7, no. 2, pp. 139–154, 1985.

[20] H. Sakoe and S. Chiba, "A dynamic programming approach to continuous speech recognition," *Proc. 7th Int. Congress Acoust.,* Budapest, Paper 20C-13, 1971.

[21] H. Sakoe, "Handwritten character recognition by rubber string matching method," *Technical Report of IECE Japan,* vol. PRL74-20, 1974. (in Japanese)

[22] E. Tanaka, and Y. Kikuchi, "Distance between pictures," *Trans. IEICE Japan,* vol. J63D, no. 12. (in Japanese)

[23] C. C. Tappert, "Cursive script recognition by elastic matching," *IBM J. Res. Dev.,* vol. 26, no. 6, pp. 765–771, 1982.

[24] N. Ueda, K. Mase, and Y. Suenaga, "A contour tracking method using elastic contour model and energy minimization approach," *Trans. IEICE Japan,* vol. J75-D-II, no. 12, pp. 111–120, 1992. (in Japanese)

[25] V. Velichko and N. Zagoruyko, "Automatic recognition of 200 words," *Int. J. Man-Machine Studies,* vol. 2, pp. 223–234, 1970.

[26] H. Yamada, "Contour DP matching method and its application to handprinted Chinese character recognition," *Proc. 7th Int. Conf. Pattern Recogn.,* pp. 389–392, July 31, 1984.

[27] H. Yamada, M. Hospital, and T. Kasvand, "Rotation-invariant contour DP matching method for 3D object recognition," *Proc. 1986 IEEE Int. Conf. Syst. Man, and Cybern.,* Atlanta, GA, pp. 997–1001, October 14–17, 1986.

[28] H. Yamada, C. Merritt, and T. Kasvand, "Recognition of kidney glomerulus by two-dimensional dynamic programming matching method," *IEEE Trans. Pattern Anal. Machine Intell.,* vol. 10, no. 5, pp. 731–737, September 1988.

[29] H. Yamada and K. Yamamoto, "Recognition of echocardiograms by dynamic programming matching method," *Pattern Recogn.,* vol. 24, no. 2, pp. 147–155, February 1991.

[30] H. Yamada, "MAP matching—Elastic shape matching by multi-angled parallelism," *Syst. Comp. in Japan,* vol. 22, no. 6, pp. 55–66, 1991 (trans. from *TIEICEJ,* vol. 72-DII, no. 5, pp. 678–685, May 1989).

[31] H. Yamada, K. Yamamoto, and K. Hosokawa, "Directional mathematical morphology and reformalized Hough transformation for the analysis of topographic maps," *IEEE Trans. PAMI,* vol. 15, no. 4, pp. 380–387, April 1993.

APPENDIX A

Determination of the Functional Φ

Let us assume that Φ_0 satisfies the following equation:

$$\Phi_0[g(x'); x, \sigma] = \int_{-\infty}^{\infty} \phi\{g(x'); x, x', \sigma\} dx'. \tag{A.1}$$

Auxiliary Theorem A.1 *If the functional Φ_0 given in the form of (A.1) satisfies condition 1, then*

$$\Phi_0[g(x'); x, \sigma] = \int_{-\infty}^{\infty} g(x')\phi(x, x', \sigma) dx' \tag{A.2}$$

holds.

Proof Formula (A.1) is substituted into condition 1 (see Section 1.2.6):

$$\int_{-\infty}^{\infty} \phi\{Ag(x'); x, x', \sigma\} dx' \equiv \int_{-\infty}^{\infty} A\phi\{g(x'); x, x', \sigma\} dx'.$$

In order to hold the equation above regardless of $g(x')$,

$$\phi[Ag(x'); x, x', \sigma] = A\phi\{g(x'); x, x', \sigma\}$$

must hold. For simplicity let $y(z)$ be set as

$$y(Az) = Ay(z).$$

This is derived by A once, and then A is set to 1:

$$zy'(z) = y(z).$$

We solve the expression above as

$$y(z) = Cz,$$

where C can be any function of x, x', and σ. As a result

$$\phi\{g(x'); x, x', \sigma\} = g(x')\phi(x, x', \sigma)$$

is obtained.

If this is substituted into (A.1), then (A.2) is given. ■

Auxiliary Theorem A.2 *If the functional Φ_0 given in the form of (A.2) satisfies condition 2, then*

$$\Phi_0[g(x'); x, \sigma] = \int_{-\infty}^{\infty} g(x')\phi(x - x', \sigma)dx' \tag{A.3}$$

holds.

Proof Formula (A.2) is substituted into condition 2 (see Section 1.2.6):

$$\int_{-\infty}^{\infty} g(x' - a)\phi(x, x', \sigma)dx' = \int_{-\infty}^{\infty} g(x')\phi(x, x' + a, \sigma)dx'$$

$$\equiv \int_{-\infty}^{\infty} g(x')\phi(x - a, x', \sigma)dx'.$$

In order to hold the equation above regardless of $g(x')$,

$$\phi(x, x' + a, \sigma) = \phi(x - a, x', \sigma)$$

must be held. For simplicity we set it as

$$y(x, x') \equiv \phi(x, x', \sigma).$$

Then we have

$$y(x, x' + a) = y(x - a, x').$$

This is derived by a, and then a is set to 0:

$$\frac{\partial}{\partial x'} y(x, x') + \frac{\partial}{\partial x} y(x, x') = 0.$$

It is solved as

$$y(x, x') = y(x - x').$$

Therefore, $\phi(x, x', \sigma)$ must have the following form:

$$\phi(x, x', \sigma) = \phi(x - x', \sigma).$$

If this is substituted into (A.2), then (A.3) is given. ■

Auxiliary Theorem A.3 *If the functional Φ_0 given in the form 2 satisfies condition 3, then*

$$\Phi_0[g(x'); x, \sigma] = \int_{-\infty}^{\infty} g(x')\phi(\nu(\sigma)(x - x'))\nu(\sigma)dx' \tag{A.4}$$

holds, where $\nu(\sigma)$ is any function.

Proof Formula (A.3) is substituted into condition 3 (see Section 1.2.6):

$$\int_{-\infty}^{\infty} g\left(\frac{x'}{\lambda}\right) \phi(x - x', \sigma)dx' = \int_{-\infty}^{\infty} \lambda g(x')\phi(x - \lambda x', \sigma)dx'$$

$$\equiv \int_{-\infty}^{\infty} g(x')\phi\left(\frac{x}{\lambda} - x', \sigma'\right) dx'.$$

In order to hold the equation above regardless of $g(x')$,

$$\lambda\phi(x - \lambda x', \sigma) = \phi\left(\frac{x}{\lambda} - x', \sigma'\right)$$

must be held. For simplicity we set u as

$$u \equiv \frac{x}{\lambda} - x'.$$

Then the equation above becomes

$$\lambda\phi(\lambda u, \sigma) = \phi(u, \sigma').$$

Both sides are derived by λ, and then we set $\lambda = 1$:

$$\phi(u, \sigma) + u\frac{\partial}{\partial u}\phi(u, \sigma) = \mu(\sigma)\frac{\partial}{\partial \sigma}\phi(u, \sigma),$$

where $\mu(\sigma) \equiv (\partial\sigma'/\partial\lambda)_{\lambda=1}$. Note here that in the condition 3, when $\lambda = 1$, then $\sigma' \equiv \sigma$ holds. This partial differential equation is solved as

$$\phi(u, \sigma) = \nu(\sigma)\phi(\nu(\sigma)u),$$

where $\nu(\sigma) \equiv k \exp\{\int d\sigma/\mu(\sigma)\}$ and k is a constant. If this expression is substituted into (A.3), formula (A.4) is obtained. Notice here that we did not specify the form of σ' which is a function of σ and λ. Therefore the function $\nu(\sigma)$ can be any form. ■

Auxiliary Theorem A.4 *If the functional Φ_0 given in the form of (A.4) satisfies condition 4, then*

$$\Phi_0[g(x'); x, \sigma] = \int_{-\infty}^{\infty} g(x')\phi(\nu(\sigma)(x - x'))\nu(\sigma)dx', \tag{A.5}$$

where

$$\phi(u) = \frac{1}{2\pi}\int_{-\infty}^{\infty} e^{-k^{2m}\zeta^{2m} + i\zeta u}d\zeta \qquad (m = 1, 2, \ldots)$$

or

$$\Phi_0[g(x'); x, \sigma] \equiv 0 \tag{A.6}$$

holds, where k is a constant.

Proof For simplicity, let τ be

$$\tau \equiv \frac{1}{\nu(\sigma)}.$$

The formula (A.5) becomes

$$\Phi[g(x'); x, \sigma] = \int_{-\infty}^{\infty} g(x')\phi\left(\frac{x - x'}{\tau}\right)\frac{dx'}{\tau}.$$

This is substituted into condition 4 (see Section 1.2.6):

$$\int_{-\infty}^{\infty} \phi\left(\frac{x-x'}{\tau_2}\right)\frac{dx'}{\tau_2}\int_{-\infty}^{\infty} g(x'')\phi\left(\frac{x'-x''}{\tau_1}\right)\frac{dx''}{\tau_1}$$

$$= \int_{-\infty}^{\infty} g(x'')dx''\int_{-\infty}^{\infty} \phi\left(\frac{x-x'}{\tau_2}\right)\phi\left(\frac{x'-x''}{\tau_1}\right)\frac{dx'}{\tau_2\tau_1}$$

$$\equiv \int g(x'')\phi\left(\frac{x-x''}{\tau_3}\right)\frac{dx''}{\tau_3},$$

where

$$\tau_1 \equiv \frac{1}{\nu(\sigma_1)}, \quad \tau_2 \equiv \frac{1}{\nu(\sigma_2)}, \quad \text{and} \quad \tau_3 \equiv \frac{1}{\nu(\sigma_3)}.$$

In order to hold this equation regardless of $g(x'')$,

$$\phi\left(\frac{x-x''}{\tau_3}\right)\frac{1}{\tau_3} = \int_{-\infty}^{\infty} \phi\left(\frac{x-x'}{\tau_3}\right)\phi\left(\frac{x'-x''}{\tau_1}\right)\frac{dx'}{\tau_2\tau_1} \tag{A.7}$$

must be held. Now by the Fourier transformation of $\phi(u)$, we can set $\Psi(\zeta)$ as

$$\Psi(\zeta) = \int_{-\infty}^{\infty} \phi(u)e^{-iu\zeta}du. \tag{A.8}$$

Then

$$\int_{-\infty}^{\infty} \phi\left(\frac{x-x''}{\tau_3}\right)\frac{1}{\tau_3}\, e^{-i(x-x'')\zeta}d(x-x'') = \int_{-\infty}^{\infty} \phi(u)e^{-i(\tau_3\zeta)u}du$$

$$= \Psi(\tau_3\zeta)$$

and

$$\int_{-\infty}^{\infty} e^{-i(x-x'')\zeta}d(x-x'')\int_{-\infty}^{\infty} \phi\left(\frac{x--x'}{\tau_2}\right)\phi\left(\frac{x'-x''}{\tau_1}\right)\frac{dx'}{\tau_2\tau_1}$$

$$= \int_{-\infty}^{\infty} e^{-i(x-x'')\zeta}d(x-x'')\int_{-\infty}^{\infty} \phi\left(\frac{x-x''-x'}{\tau_2}\right)\phi\left(\frac{x'}{\tau_1}\right)\frac{dx'}{\tau_2\tau_1}$$

$$= \int_{-\infty}^{\infty} \phi\left(\frac{x'}{\tau_1}\right)\frac{1}{\tau_1}\, e^{-ix'\zeta}dx'\int_{-\infty}^{\infty} \phi\left(\frac{x'''}{\tau_2}\right)\left(\frac{1}{\tau_2}\right)e^{-ix'''\zeta}dx'''$$

$$= \Psi(\tau_1\zeta)\,\Psi(\tau_2\zeta)$$

hold.

After the Fourier transformation of both sides of the relation (A.7),

$$\Psi(\tau_3\zeta) = \Psi(\tau_1\zeta)\Psi(\tau_2\zeta) \tag{A.9}$$

is obtained. In order to solve the above function equation, first we set $\zeta = 0$. Then

$$\Psi(0) = 0 \quad \text{or} \quad 1$$

is obtained. If we take "0," then $\Psi(\zeta) \equiv 0$ by (A.9). Therefore $\phi(u) \equiv 0$, and this is substituted into (A.4). Then (A.6) is obtained. On the other hand, if we take

$$\Psi(0) = 1, \tag{A.10}$$

denoting the rth derivative of $\Psi(\zeta)$ as $\Psi^{(r)}(\zeta)$, we can assume that

$$\begin{cases} \Psi^{(r)}(0) = 0 \quad (r = 1, 2, \ldots, n-1), \\ \Psi^{(n)}(0) \neq 0. \end{cases} \tag{A.11}$$

Then, if both sides of the equation (A.9) are differentiated by ζ n times and then ζ is set to 0, we obtain the following equation from (A.1) and (A.11):

$$(\tau_3^n - \tau_1^n - \tau_2^n)\Psi^{(n)}(0) = 0.$$

That is, the relation

$$\tau_3^n = \tau_1^n + \tau_2^n \tag{A.12}$$

is obtained. This is substituted into (A.9), differentiated by τ_1 n times, and $\tau_1 = 0$ and $\tau_2 = 1$ are set to the result. Then the differential equation

$$(n-1)!\zeta\Psi(\zeta) = \zeta^n\Psi^{(n)}(0)\Psi(\zeta)$$

is obtained. If this is solved under the condition (A.11),

$$\Psi(\zeta) = e^{(\Psi^{(n)}(0)/n!)\,\zeta^n} \tag{A.13}$$

is obtained. Now, taking an inverse Fourier transformation, we have

$$\phi(u) = \frac{1}{2\pi}\int_{-\infty}^{\infty} e^{(\Psi^{(n)}(0)/n!)\zeta^n} e^{i\zeta u}\, d\zeta. \tag{A.14}$$

Let us consider the condition that the integral in the above form exists. It is shown as

$$\begin{cases} n = 2m \qquad (m = 1, 2, \ldots), \\ \Psi^{(n)}(0) < 0. \end{cases} \tag{A.15}$$

Therefore, if we take $\Psi^{(n)}(0)$ as

$$\Psi^{(n)}(0) \equiv (-1)n!k^n, \tag{A.16}$$

the general solution of $\phi(u)$ is

$$\phi(u) = \frac{1}{2\pi}\int_{-\infty}^{\infty} e^{-k^{2m}\zeta^{2m}} e^{i\zeta u} d\zeta.$$

From the above result (A.5) is obtained. Now from (A.5) the following relation holds:

$$\phi(\nu(\sigma)(x - x'))\nu(\sigma) = \frac{1}{2\pi}\int_{-\infty}^{\infty} e^{-k^{2m}\zeta^{2m} + i\zeta\nu(\sigma)(x - x')}\nu(\sigma)d\zeta$$

$$= \frac{1}{2\pi}\int_{-\infty}^{\infty} e^{\{(-k^{2m}/\nu(\sigma)^{2m})\zeta^{2m} + i\zeta(x - x')\}}d\zeta.$$

So far we considered σ as a parameter and did not impose any limitation on it. Therefore the generality is not lost by taking $\nu(\sigma)$ as

$$\nu(\sigma) \equiv \frac{k}{\sigma}. \tag{A.17}$$

Then the above formula becomes

$$\phi\left(k\frac{x - x'}{\sigma}\right)\frac{k}{\sigma} = \frac{1}{2\pi}\int_{-\infty}^{\infty} e^{-\sigma^{2m}\zeta^{2m} + i\zeta(x - x')}d\zeta.$$

Furthermore, setting $\phi(ku)k$ newly as $\phi(u)$,

$$\phi\left(\frac{x - x'}{\sigma}\right)\frac{1}{\sigma} = \frac{1}{2\pi}\int_{-\infty}^{\infty} e^{-\sigma^{2m}\zeta^{2m} + i\zeta(x - x')}d\zeta$$

is obtained. ■

Obviously (A.8) is useless physically. Therefore, by the auxiliary theorems 1–4, and from the above result, we haved reached the following theorem:

Theorem 1 *If a functional Φ_0 given in the form of (A.1) satisfies the conditions* 1–4, *in general, the Φ_0 has the form*

$$\Phi_0[g(x'); x, \sigma] = \int_{-\infty}^{\infty} g(x')\phi_m\left(\frac{x - x'}{\sigma}\right)\frac{dx'}{\sigma}, \tag{A.18}$$

where

$$\phi_m(u) = \frac{1}{2\pi}\int_{-\infty}^{\infty} e^{-xi^{2m} + i\zeta u}d\zeta \qquad (m = 1, 2, \ldots).$$

Thus the transformation of Φ_0 denotes that a given pattern is blurred as observed under conditions 1–4. However, the transformation form is not unique, and there are a countable number of forms allowed.

Under conditions 3 and 4, the relations between σ' and σ, λ and between σ_3, σ_1, and σ_2 were not given explicitly. Now they can be explicitly stated according to the above theorem as

$$\sigma' = \frac{\sigma}{\lambda} \tag{A.19}$$

and

$$\sigma_3^{2m} = \sigma_1^{2m} + \sigma_2^{2m}. \tag{A.20}$$

These can be derived from the auxiliary theorems 3 and 4 and (A.17).

APPENDIX B

ETL Character Database

The ETL[1] character databases have been made by ETL character recognition researchers cooperating with the character standardization committee sponsored by MITI of the Japanese government since 1971. They consist of approximately 1.2 million handwritten and machine-printed character images which include Japanese, Chinese, Roman, and Arabic numeric characters with gray levels in rectangular arrays. There are nine kinds of databases, ETL1 through ETL9. They are introduced in Table B.1.

Information about ETL character databases can be obtained in the following ways:

1. From WWW (World Wide Web) at the following URLs:
   http://www.etl.go.jp:/etl/gazo/etlcdb/
   ftp://etlport.etl.go.jp:/pub/image/etlcdb/
2. By ftp:
   Anonymous ftp to "etlport.etl.go.jp" (192.31.197.99), and cd (change directory) to /pub/image/etlcdb, and get the file "etlcdb.sha" (about 50KB) there. Where, key-in "anonymous" as login name, your e-mail address as password.
3. By E-mail:
   E-mail to "etlcdb@etl.go.jp" with the subject "request ETLCDB information." Then our mailer will automatically send the information file "etlcdb.sha" (about 50KB).

[1] ETL belongs to the Ministry of International Trade and Industry in Japan; it stands for Electrotechnical Laboratory.

TABLE B.1 List of ETL Character Databases

Name	Class Name	Contents	No. of Categories	Number of Writers	Number of Samples	Date of Creation	½ inch BPI	M N
ETL1	Handwritten	Numeral Roman Special Katakana	10 26 12 *51 99	1,445	141,319	1973-09	1,600	1
ETL2	Machine- printed Kanji	Kanji Hiragana Katakana Alphanumeric Special	*2184	Mincho style Gothic style	52,796	1973-10	1,600	
ETL3	Handprinted	Numeral Roman Special	10 26 *12 48	200	9,600	1974-04	1,600	
ETL4	Handwritten	Hiragana	51	120	6,120	1974-12	1,600	
ETL5	Handprinted	Katakana	51	104	10,608	1975-02	1,600	1
ETL6	Handprinted	Katakana Numeral Roman Special	46 10 26 *32 114	1,383	157,662	1976-12	1,600	1
ETL7L	Handprinted Large Size	Hiragana b-,p-sound	46 *2 48	175	16,800	1977-08	1,600	
ETL7S	Handprinted Small Size	Hiragana b-,p-sound	46 *2 48	175	16,800	1977-08	1,600	
ETL8G	Handprinted Kyoiku-Kanji	Kanji Hiragana	881 *75 956	1,600	152,960	1980-02	1,600	3
ETL8B2	Handprinted Kyoiku-Kanji Binarized	Kanji Hiragana	881 *75 965	1,600	152,960	1981-07	1,600	
ETL9G 5	Handprinted JIS Level-1 Kanji Set	Kanji Hiragana	2965 *71 3036	4,000	607,200	1984-03	6,250	
ETL9B 5	Handprinted JIS Level-1 Binarized	Kanji Hiragana	2965 *71 3036	4,000	607,200	1984-03	6,250	
ETL9B	Handprinted JIS Level-1 Binarized Binarized	Kanji Hiragana	2965 *71 3036 3036	4,000	607,200	1984-03	1,600	1

Note: An asterisk (*) denotes a total.

We recommend reading the corresponding papers when databases are used. The papers describe primarily analyses of databases. For example, in the paper [7] experimental results of recognition using directional matching are given. To see the quality of handprinted characters of the ETL8, all the samples of "田" are shown in Fig. B.1. Some superposed character images written by different subjects are shown in Fig. B.2. Looking at these images, we can recognize the standard shapes, although they are blurred.

FIGURE B.1

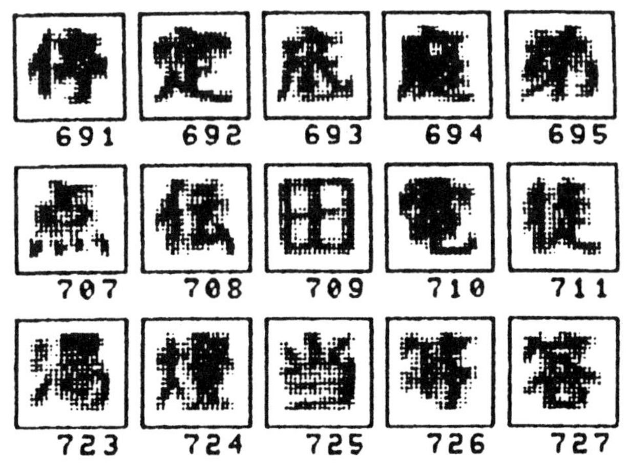

FIGURE B.2

BIBLIOGRAPHY

[1] H. Yamada and S. Mori, "An analysis of the handprinted character data base. I", *Bull. Electrotechn. Lab.,* vol. 39, no. 8, pp. 580–599, August 1975. (in Japanese)

[2] H. Yamama and S. Mori, "An analysis of the handprinted character data base. II," *Bull. Electrotechn. Lab.,* vol. 40, no. 5, pp. 385–434, June 1976. (in Japanese)

[3] T. Saito, H. Yamada and S. Mori, "An analysis of the handprinted character data base. III," *Bulletin Electrotechn. Lab.,* vol. 42, no. 5, pp. 385–434, May 1978. (in Japanese)

[4] S. Mori, K. Yamamoto, H. Yamada, and T. Saito, "On a handprinted Kyoiku-Kanji character data base," *Bull. Electrotechn. Lab.,* vol. 43, no. 11–12, pp. 752–773, 1979. (in Japanese)

[5] S. Mori, K. Yamamoto, H. Yamada, and T. Saito, "An analysis of the handprinted character data base. IV. Statistics of Kyoiku-Kanji characters," *Bull. Electrotechn. Lab.,* vol. 44, no. 4, pp. 219–251, April 1980. (in Japanese)

[6] T. Saito, H. Yamada, K. Yamamoto, and S. Mori, "On a handprinted Kanji data base—Kyoiku-Kanji," *1981 National Convention Record, Institute of Electronics and Communication Engineers,* 1385 pp., April 1981. (in Japanese)

[7] T. Saito, H. Yamada, K. Yamamoto, and S. Mori, "An analysis of the handprinted character data base. V. Evaluation of Kyoiku-Kanji characters by pattern matching approach," *Bull. Electrotechn. Lab.,* vol. 45, no. 1–2, pp. 49–77, 1981. (in Japanese)

[8] T. Saito, H. Yamada, K. Yamamoto, R. Oka, M. Yasuda, and H. Sone, "An intuitive analysis of Kyoiku Kanji characters," *1982 National Convention Record, IECE,* 1342 pp., March 1982. (in Japanese)

[9] T. Saito, H. Yamada, and K. Yamamoto, "An analysis of handprinted Chinese characters by directional pattern matching approach," *Trans. IECE,* vol. J65-D, no. 5, pp. 550–725, May 1982. (in Japanese)

[10] T. Saito, H. Yamada, and K. Yamamoto, "An analysis of handprinted character data base. VI An analysis of Kyoiku Kanji characters by directional pattern matching approach," *Bull. Electrotechn. Lab.,* vol. 46, no. 12, pp. 695–725, December 1982. (in Japanese)

[11] T. Saito, H. Yamada, K. Yamamoto, and M. Yasuda, "An analysis of handprinted character data base. VII. An intuitive analysis of handprinted Kyoiku Kanji characters," *Bull. Electrotechn. Lab.,* vol. 47, no. 4, pp. 261–275, April 1983. (in Japanese)

[12] T. Saito, H. Yamada, and K. Yamamoto, "On the data base ETL9 of handprinted characters in JIS Chinese characters and its analysis," *Trans. IECE.,* vol. J68-D, no. 4, pp. 757–764, April 1985. (in Japanese)

[13] T. Saito, H. Yamada, and K. Yamamoto, "An analysis of handprinted character data base. VIII. An estimation of the data base ETL9 of handprinted characters in JIS Chinese characters by directional pattern matching approach," *Bull. Electrotechn. Lab.,* vol. 49, no. 7 pp. 487–525, July 1985. (in Japanese)

[14] T. Saito, K. Yamamoto, and H. Yamada, "An analysis of handprinted character data base. IX. On the data base ETL9 and its model patterns," *Bull. Electrotechn. Lab.,* vol. 50, no. 4, pp. 259–263, April 1986. (in Japanese)

[15] K. Yamamoto, T. Saito, and H. Yamada, "An analysis of the data base ETL9 of handprinted characters in Kanji characters as observed by humans," *1985 National Convention Record, IECE,* S4-2, pp. 303–304, November 1985. (in Japanese)

APPENDIX C

Data Set of Handprinted Numerals

The handprinted numerals that were used in the experiment on directional feature matching are shown below. The total number of the samples is 2000. Twenty-five subjects were asked to write 40 characters within a specified rectangle 7 mm long and 5 mm wide. The samples printed in Fig. C.1 are about one-quarter the size of the actual characters. The printed data were excerpted from the doctoral thesis of Michio Yasuda "Research on Character Recognition Based on the Correlation Method," August 1981 (in Japanese).

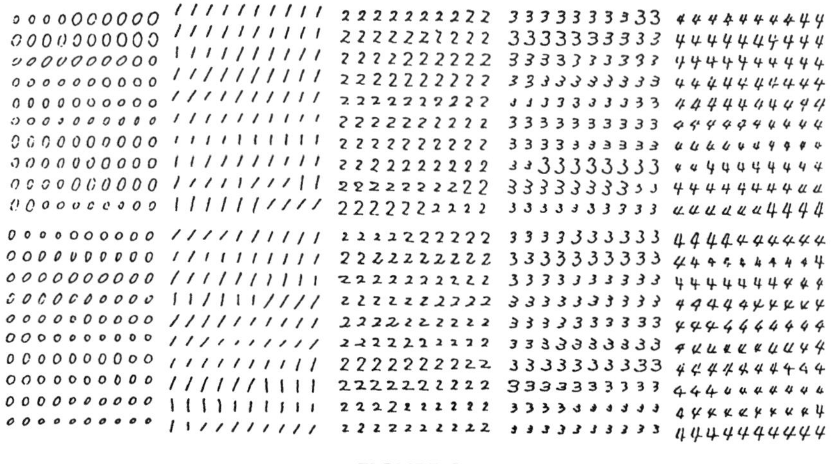

FIGURE C.1

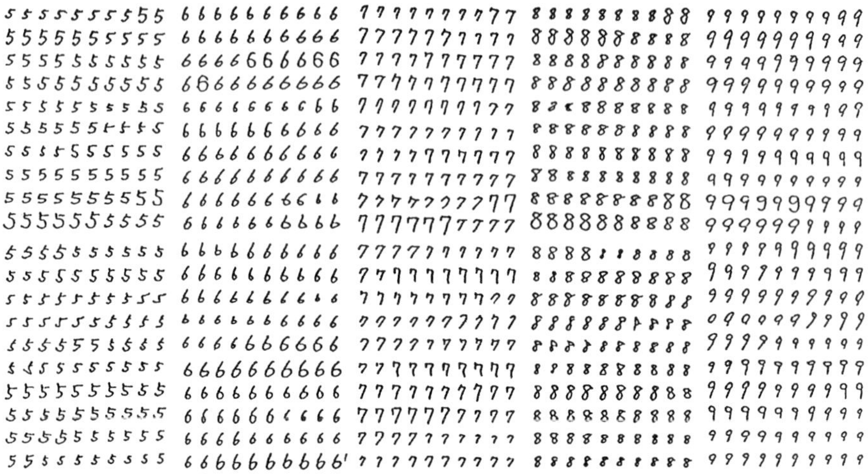

FIGURE C.1 (*continued*)

Index